The Filamentous Fungi

Volume 3   Developmental Mycology

The Filamentous Fungi

Volume 3   Developmental Mycology

Edited by:

JOHN E. SMITH, D.Sc., F.I. Biol. F.R.S.E.

DAVID R. BERRY, Ph.D.

Department of Applied Microbiology, University of Strathclyde, Glasgow

A HALSTED PRESS BOOK

JOHN WILEY & SONS
NEW YORK

© Edward Arnold (Publishers) Ltd. 1978

First Published 1978
by Edward Arnold (Publishers) Limited,
London.

Published in the U.S.A.
by Halsted Press, a Division
of John Wiley & Sons, Inc.
New York.

Library of Congress Cataloging in Publication Data (Revised)
Main entry under title:

The Filamentous fungi.

"A Halsted Press book."
Includes bibliographies and indexes.
CONTENTS: v. 1. Industrial mycology.—v. 2. Biosynthesis and
metabolism.—v. 3. Developmental mycology.
1. Mycology. I. Smith, John E., ed. II. Berry, David Richard, ed.
QK603.F54   1975   589'.2   75-2101
ISBN 0-470-80187-5 (v. 1)
        0-470-15005-X (v. 2)
        0-470-99352-9 (v. 3)

Printed in Great Britain

Preface

The unique morphology of the filamentous fungi inevitably requires that their development involves unique solutions to the problems of growth and differentiation. In this volume we have aimed at bringing together concepts and evidence from the variety of disciplines, *viz.* biochemistry, physiology, genetics and cytology which can contribute to our understanding of fungal development. Furthermore, we have aimed to avoid lengthy and detailed descriptions of morphogenetic events in the fungi which can so easily cloud rather than clarify the principles involved.

We believe that it is self evident that a good understanding of fungal development is essential to a complete appreciation of the behaviour of fungi in an industrial or academic situation. This volume is an attempt to provide research workers in each of these fields with an integrated perspective of the subject of fungal development. As with the previous two volumes in this series we have aimed to make each chapter an entity in itself rather than a survey of recent advances.

1977   J. E. S.
                                                                                 D. R. B.

Contents

List of Contributors

J. G. Anderson,
Department of Applied
  Microbiology,
Strathclyde University,
Glasgow, Scotland.

R. Brambl,
Department of Plant Pathology,
The University of Minnesota,
St. Paul, Minnesota 55108, U.S.A.

Lorna A. Casselton,
Department of Plant Biology and
  Microbiology,
Queen Mary College, Mile End
  Road,
University of London, London,
  England

A. J. Clutterbuck,
Department of Genetics,
University of Glasgow,
Glasgow, Scotland.

A. L. Demain,
Department of Nutrition and Food
  Science,
Massachusetts Institute of
  Technology,
Cambridge, Massachusetts, U.S.A.

L. D. Dunkle,
Department of Plant Pathology,
The University of Nebraska,
Lincoln, Nebraska 68583, U.S.A.

S. K. Dutta,
Department of Botany,
Howard University,
Washington D.C., U.S.A.

Z. Fencl,
Department of Technical
  Microbiology,
Institute of Microbiology,
Czeckoslovak Academy of
  Sciences,
Prague, Czeckoslovakia.

G. W. Gooday,
Department of Microbiology,
Marischal College, University of
  Aberdeen,
Aberdeen, Scotland.

S. N. Grove,
Department of Biology,
Goshen College,
Goshen, Indiana 46526, U.S.A.

K. Gull,
Biological Laboratory,
University of Kent,
Canterbury, England.

G. Lysek,
Lehrstuhl fur Botanik,
Technische Universität Munchen,
D–8080 Freising–Weihenstaphen,
Germany.

J. F. Martin,
Department of Microbiology,
University of Salamanca,
Salamanca, Spain.

M. Ojha,
Laboratory of General
  Microbiology,
University of Geneva,
1211 Geneva 4, Switzerland.

J. F. Peberdy,
Department of Botany,
University of Nottingham,
Nottingham, England.

P. J. Rogers,
School of Science,
Griffith University, Brisbane,
Queensland, Australia.

J. E. Smith,
Department of Applied
 Microbiology,
Strathclyde University,
Glasgow, Scotland.

P. R. Stewart,
Department of Biochemistry,
Faculty of Science,
Australian National University,
Canberra, Australia.

W. J. Sundberg,
Department of Botany,
Southern Illinois University,
Carbondale, Illinois 62901, U.S.A.

K. K. Tan,
Department of Botany,
University College,
Dublin 4, Ireland.

A. P. J. Trinci,
Department of Microbiology,
Queen Elizabeth College,
London.

G. Turian,
Laboratory of General
 Microbiology,
University of Geneva,
1211 Geneva 4, Switzerland.

G. Turner,
Department of Bacteriology,
The Medical School,
University Walk, Bristol, England.

H. Van Den Ende,
Laboratorium voor Algemene
 Plantkunde,
Plantenfysiologie en
 Farmacognosie,
Ijdijk 26, Amsterdam-Oost, The
 Netherlands.

J. L. Van Etten,
Department of Plant Pathology,
The University of Nebraska,
Lincoln, Nebraska 68583, U.S.A.

H. J. Willetts,
Department of Botany,
The University of New South
 Wales,
Kessington, New South Wales,
Australia.

Barbara E. Wright,
Boston Biomedical Research
 Institute,
20 Staniford Street,
Boston, Massachusetts 02114,
 U.S.A.

CHAPTER 1

Concepts of Differentiation
BARBARA E. WRIGHT

1.1 Introduction

The purpose of this introductory chapter is to arrive at a useful definition of 'biochemical differentiation'. The approach will be to examine various types of differentiation in order to see what they have in common, so that we may distil or select those characteristics applicable in most cases. When the word 'differentiation' is in quotation marks, it refers to the word or symbol; when not in quotation marks, it refers to the phenomenon being described. In the latter case, I should be pointing to the phenomenon, rather than speaking or writing the word—to avoid the semantic sin of identifying a symbol with the object it represents. A symbol or definition must of necessity be an abstraction or selection of only a few of the countless characteristics (some unknowable) which together constitute the phenomenon. A symbol *is not* identical with the object. Those who argue about the 'nature' or 'cause' of (') biochemical differentiation (') should realize that such arguments usually stem from a lack of awareness that different definitions are being used, or from a confusion between, and/or identification of, symbol and phenomenon.

Neither phenomena nor definitions can be right or wrong. Phenomena can only be described, and a definition is arbitrarily made, based on criteria such as conventionality, clarity and usefulness. For excellent discussions regarding the nature of definitions see Lotka (1956) and Korzybski (1958). There are different kinds of definition; the one we seek is called a connotative definition, which describes selected characteristics common to similar phenomena.

Phenomena referred to as examples of 'differentiation' can cover a very broad range. Indeed, even enzyme induction in bacteria has been included. For our purposes, however, more complex systems will be used because they are more representative of what most investigators in our field would consider to be cases of 'biochemical differentiation'. As we are interested not in mere correlations, but in biochemical mechanisms underlying 'differentiation', examples have been chosen which offer at least some indication as to the rate-limiting steps (*i.e.*, critical variables) controlling the specific transformation in question. For the purpose of the points to be made, I have taken the liberty of making certain assumptions and simplifications regarding the systems to be described.

1.2 The Accumulation of Fibroin in *Bombyx mori*

The first and in certain respects simplest example is the accumulation of the silk protein, fibroin, in the silk worm *Bombyx mori* (Suzuki, Gage & Brown, 1972; Suzuki & Brown, 1972; Lizardi & Brown, 1975). When the silk gland cells cease cell division, an enormous increase in the amount of DNA occurs; when DNA synthesis stops, only fibroin ($300 \, \mu g/$cell) is made. Elegant analyses have shown that the fibroin gene is not amplified in the silk gland, but is present as a constant fraction of the haploid genome, and represents 0.0022% of the DNA from all parts of the silk worm. The messenger RNA (mRNA) for the fibroin protein has been isolated (80%– 90% pure) and analyzed by partial sequence analysis. The sequence found could actually be predicted, because the protein has a simple primary structure. This mRNA comprises about 1% of the total RNA in the silk gland at the end of the larval life of the animal, when fibroin is the predominant protein synthesized in this tissue. A simplified picture of the sequence of events which culminates in the accumulation of fibroin protein are summarized in Fig. 1.1.

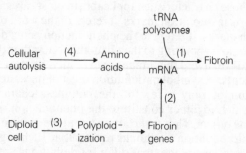

Fig. 1.1 Sequential events essential to the accumulation of silk protein (fibroin) in *Bombyx mori*.

Cellular autolysis, which supplies amino acid precursors for fibroin synthesis (4 in Fig. 1.1) and polyploidization in the silk gland cells, which results in ten times more DNA and hence fibroin gene per cell (3 of Fig. 1.1), occur many days prior to fibroin accumulation (1 of Fig. 1.1). Thus, we shall assume that these events are not rate-limiting when fibroin accumulates, and that the content and activities of cellular tRNA and polysomes are also optimal for the protein synthesis which will occur. Presumably, the critical event occurring during the 3–4 days prior to fibroin accumulation is the production of 10^4 molecules of mRNA by each gene (2 of Fig. 1.1). Each mRNA molecule is then translated into 10^5 silk protein molecules during the last 4 days of larval life (1 of Fig. 1.1). Thus, the critical variable controlling the synthesis and accumulation of this structural protein is gene activation.

1.3 Chitin Synthesis in *Saccharomyces cerevisiae* and *Mucor rouxii*

The next example to be considered is chitin synthesis during septum formation in the budding yeast, *Saccharomyces cerevisiae* (Cabib & Farkas, 1971; Cabib & Ulane, 1973) and during hyphal or cell wall growth in the mould *Mucor rouxii* (Bartnicki-Garcia & Lippmann, 1972; Bartnicki-Garcia, 1973; Ruiz-Herrera & Bartnicki-Garcia, 1974). In both systems, the enzyme is present largely in an inactive state and can be activated by an activating factor (AF) or proteolytic enzyme (2 of Fig. 1.2). The protease, AF, is liberated from vesicles at the site of septum formation and the inactive enzyme or zymogen is then transformed to active chitin synthase.

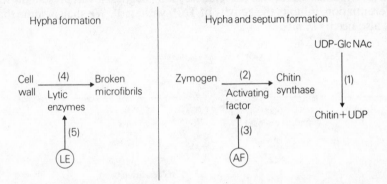

Fig. 1.2 Prior events necessary to the accumulation of chitin in *Mucor rouxii* and *Saccharomyces cerevisiae*.

In the case of hyphal growth, presumptive evidence has been obtained for the critical role of wall lysis coupled with synthesis. Lytic enzymes attack the microfibrillar skeleton by splitting either inter- or intramolecular bonds, thus facilitating the extension of old chains or the production of new ones (4 of Fig. 1.2).

In the case of septum formation, the total amount of chitin synthase, as measured after proteolytic activation, is constant, regardless of growth medium, growth phase or stage of the cell cycle. Therefore, the critical variables for chitin synthesis with respect to this enzyme do not involve gene activation, mRNA synthesis or stabilization. (Zymogen and/or active enzyme may well be turning over, but that would be irrelevant in our present context, as inactive and active enzyme concentration is a constant before and during septum formation). The AF-carrying vesicle fuses with the plasma membrane at the site of septum formation, resulting in the transformation of zymogen to active enzyme. Thus, the production of broken microfibrils (4 of Fig. 1.2) as 'primers' for new synthesis and the liberation of an activating factor (3 of Fig. 1.2) are the critical variables controlling structural polysaccharide formation during differentiation in these two systems.

1.4 Glycogen phosphorylase and trehalose in *Dictyostelium discoideum*

The last two examples are taken from the cellular slime mould, *Dictyostelium discoideum*, and were chosen because they involve two other kinds of products of 'differentiation': an enzyme (glycogen phosphorylase) and a disaccharide (trehalose). Trehalose accumulates rapidly in the terminal stages of differentiation, and represents one of the major end-products formed (Wright & Killick, 1975). It serves as an energy source during germination. Glycogen phosphorylase catalyzes the degradation of glycogen, which is the source of carbon for the synthesis of trehalose and other end-product saccharides. Some of the precursors and enzymes which are essential to trehalose synthesis are shown in the simplified scheme of Fig. 1.3. The accumulation of trehalose-6-phosphate (T6P) *in vivo* cannot be detected; we shall therefore assume that the conversion of T6P to trehalose is not rate-limiting (Sargent & Wright, 1971). It has been observed (K. A. Killick, unpublished results) that glucose, at the highest levels present during differentiation, inhibits the activity of T6P synthetase. We shall assume that this also occurs *in vivo*.

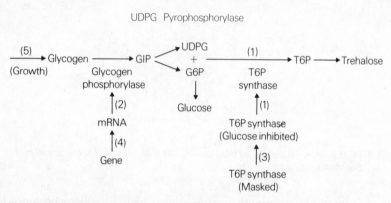

Fig. 1.3 Various events essential to trehalose accumulation in *Dictyostelium discoideum.*

As a function of time during 24 h of differentiation, the two enzymes and five metabolites show the relative changes in concentration per cell as indicated in Fig. 1.4. The numbers for each event in Fig. 1.3 refer to the time of its occurrence, indicated in Fig. 1.4. Trehalose-6-phosphate synthetase is masked (*in vivo*) until the later stages of development, and becomes unmasked prior to trehalose accumulation (Killick & Wright, 1975). However, although the enzyme is unmasked at the time indicated by the arrow, we shall assume in this discussion that it is not active, due to the inhibitory levels of glucose (or some other small molecule) present at that time. As in the case of chitin synthetase, the concentration of masked plus unmasked enzyme is relatively constant (Fig. 1.4). Thus, *with respect to this enzyme protein*, changes at the level of transcription and translation are not critical variables for trehalose accumulation. The same is true for UDP-glucose pyrophosphorylase, which is not rate-limiting and does not in this sense control the rate of UDP-glucose synthesis (Wright & Gustafson, 1972). The latter is determined by the rate of GIP formation, which in turn is determined by an enzyme that is rate-limiting: glycogen phosphorylase (Wright & Park, 1975). Furthermore, the concentration of this enzyme changes significantly during differentiation, due to an increase in the rate of enzyme synthesis (Thomas & Wright, 1976). Let us assume that mRNA

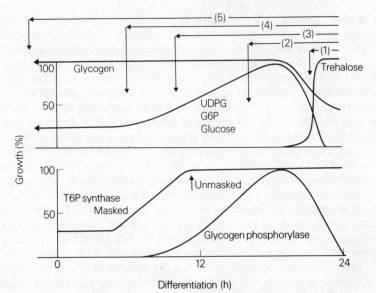

Fig. 1.4 Sequential time frames within which the events described in Fig. 1.3 occur with respect to trehalose accumulation. The numbers correspond to those in Fig. 1.3.

synthesis is the rate-limiting event controlling the synthesis and level of glycogen phosphorylase. Thus, gene activation is the critical variable controlling the accumulation of this enzyme. Is gene activation also the 'cause' of trehalose synthesis? If so, clearly it is not the only 'cause'; the essential role of this enzyme is no more important than, but equally important as, the amount of glycogen and phosphate available to be used as a source of precursors, the presence of UDP-glucose pyroposhorylase at adequate levels, the rate of UDP-glucose and G6P availability when trehalose accumulation is initiated, and the unmasking of T6P synthetase. Whether gene activation is considered to be a critical variable for trehalose synthesis simply depends upon our choice of a time frame. If we include the entire period of multicellular differentiation (time frame 4 of Fig. 1.4), then gene activation is one of the critical variables for trehalose synthesis. If we consider only that period just prior to and during trehalose accumulation (time frame 1 of Fig. 1.4), gene activation is not rate-limiting. In fact, glycogen phosphorylase has played its role, and is actually decreasing in concentration (Fig. 1.4). In this time frame there are two obvious candidates as critical variables just prior to trehalose accumulation (See Fig. 1.3): the flux of UDPG and G6P and the fall of glucose levels to relieve inhibition of T6P synthetase. Clearly, the flux of the two precursors at this time in part determines the rate of trehalose synthesis, especially in view of the simultaneous fall in the concentrations of UDPG and G6P. Assuming the enzyme is fully unmasked but completely inhibited by glucose at 18 h, then the fall in glucose level is the critical variable closest in time controlling the rate of trehalose synthesis.

Time frame 5 of Fig. 1.4 includes the growth phase of differentiation, during which glycogen accumulates. The amount of glycogen made prior to, and present during differentiation is as critical as any other event in determining the extent of trehalose accumulation. We may also consider time frames back into the evolution of *Dictyostelium*, when its ancestors were selected because they could accumulate glycogen and survive starvation, and then selected because they could form a primitive cellulose cell wall and accumulate trehalose as an energy source for future generations.

The above example clearly illustrates that we must include the element of time in our definition of 'biochemical differentiation' in order to make it meaningful. Referring again to the fibroin system, under the conditions of our assumptions, gene activation is the only critical variable if the time frame is restricted to 1 and 2 in Fig. 1.1. As amino acid precursors are essential to fibroin synthesis, the cellular autolysis which occurred prior to gene activation is also rate limiting, if the time frame chosen is expanded to 4 in Fig. 1.1. Indeed, in any system, the series of essential prior events finally resulting in 'differentiation' at some time will be found to include rate-limiting changes at the substrate, enzyme, coenzyme, organelle, RNA and DNA levels. One may logically (but not usefully) include the evolution of these systems in the time frame of events essential to any kind of differentiation. Thus, although an enzyme or precursor may not be rate-limiting during product accumulation, its formation or concentration was at one time rate-limiting during the history of the system.

The specialized products formed in the differentiating systems examined have included a structural protein (fibroin), a structural polysaccharide (chitin), a disaccharide (trehalose) and an enzyme (glycogen phosphorylase). The mechanisms operative included gene activation, enzyme unmasking, substrate and primer availability, inhibitor disappearance organelle fusion and the liberation of lytic enzymes. The more we learn about the details of the mechanisms operative in differentiating systems, the more complex will our thoughts become as to the nature of 'differentiation'. The concept of a 'trigger' for 'differentiation' will appear naive; a hierarchical scheme of control with the gene at the top or in the centre will become meaningless. The problem is to describe the multitudinous events essential to any differentiation process and to understand the interaction of these events by defining those that are rate-limiting at particular points in time in the living organism. We cannot accomplish this goal simply by breaking open the cells at various times during development and measuring an enzyme or a messenger or a metabolite. Such correlations have marginal relevance to the dynamics of metabolism in the living cell, and yet there are very few ways to examine a living cell without destroying it. Under these circumstances, one recourse is to simulate the living system with kinetic models, as we have done in our analysis of *Dictyostelium* (Wright & Gustafson, 1972; Wright & Park, 1975; Wright, Tai & Killick, 1977).

Through an examination of a number of differentiating systems, it is clear that a variety of products can accumulate under the control of many kinds of rate-limiting events. It would be convenient to formulate a definition to encompass all of these systems. One criterion of a good definition of 'biochemical differentiation' is that it be useful. We must therefore not assume a knowledge of the mechanisms involved, as in definitions such as 'differential gene activation' or 'sequential patterns of protein synthesis'. These definitions are prejudiced; moreover, they restrict and inhibit our thinking and our approaches to understanding very complex phenomena. Thus, a broad, descriptive definition has been chosen: 'biochemical differentiation' is a process of cellular specialization in which a substance accumulates to a unique extent over a specified period of time.

This work was supported by the Public Health Research Grants Number AG00433 and AG00260 from the National Institute on Aging and by the General Research Grant 5S01RR05711.

1.5 References

BARTNICKI-GARCIA, S. (1973). Fundamental aspects of hyphal morphogenesis. *Symposium of the Society of General Microbiology* 23, 245–67.

BARTNICKI-GARCIA, S. & LIPPMAN, E. (1972). The bursting tendency of hyphal tips of fungi: presumptive evidence for a delicate balance between wall synthesis and wall lysis in apical growth. *Journal of General Microbiology* 73, 487–500.

CABIB, E. & FARKAS, V. (1971). The control of morphogenesis: an enzymatic mechanism for the initiation of septum formation in yeast. *Proceedings of the National Academy of Science* 68, 2052–6.

CABIB, E. & ULANE, R. (1973). Yeast chitin synthetase. Separation of the zymogen from its activating factor and recovery of the latter in the vacuole fraction. *Journal of Biological Chemistry* 248, 1451–8.

KILLICK, K. A. & WRIGHT, B. E. (1975). Trehalose synthesis during differentiation in *Dictyostelium discoideum. Archives of Biochemistry and Biophysics* 170, 634–43.

KORZYBSKI, A. (1958). *Science and sanity*, 4th Edition. Chapters 2 and 25. Clinton, Mass: The Colonial Press, Inc.

LIZARDI, P. M. & BROWN, D. D. (1975) The length of the fibroin gene in the *Bombyx mori* genome. *Cell* 4, 207–15.

LOTKA, A. J. (1956) *Elements of mathematical biology* New York: Dover Publications, Inc.

RUIZ-HERRERA. J. & BARTNICKI-GARCIA, S. (1974). Synthesis of cell wall microfibrils *in vitro* by 'soluble' chitin synthetase from *Mucor rouxii. Science* 186, 357–9.

SARGENT, D. & WRIGHT, B. E. (1971). Trehalose synthesis during differentiation in *Dictyostelium discoideum.* 11. *In vivo* flux determinations. *Journal of Biological Chemistry* 246, 5340–4.

SUZUKI, Y. & BROWN, D. D. (1972). Isolation and identification of the messenger RNA for silk fibroin from *Bombyx mori. Journal of Molecular Biology* 63, 409–29.

SUZUKI, Y., GAGE, L. P. & BROWN, D. D. (1972). The genes for silk fibroin in *Bombyx mori. Journal of Molecular Biology* 70, 637–49.

THOMAS, D. A. & WRIGHT, B. E. (1976). Glycogen phosphorylase in *Dictyostelium discoideum.* 11. Synthesis and degradation during differentiation. *Journal of Biological Chemistry* 251, 1258–63.

WRIGHT, B. E. & GUSTAFSON, G. L. (1972). Expansion of the kinetic model of differentiation in *Dictyostelium discoideum. Journal of Biological Chemistry* 247, 7875–84.

WRIGHT, B. E. & KILLICK, K. A. (1975). Trehalose metabolism during sporulation in *Dictyostellum. Sixth International Spore Conference*, pp73–84, American Society for Microbiology.

WRIGHT, B. E. & PARK, D. J. M. (1975). An analysis of the kinetic positions held by five enzymes of carbohydrate metabolism in *Dictyostelium discoideum. Journal of Biological Chemistry* 250, 2219–26.

WRIGHT, B. E., TAI, A. & KILLICK, K. A. (1977). Fourth expansion and glucose perturbation of the *Dictyostelium* kinetic model. *European Journal of Biochemistry* 74, 217–25.

CHAPTER 2

Nuclear Control of Differentiation
M. OJHA and S. K. DUTTA

2.1 Introduction

Hybridization experiments of Mendel long ago established the chromosomal location of the genetic determinants. The chromosomes occupy a large part of the nucleus, and are bathed in the nuclear sap. It is therefore logical to consider the influence of the nuclear environment on the genetic activity of the chromosomes. Development of an organism involves growth, intracellular organization, differentiation and morphogenesis. These are most probably interrelated phenomena. It is fascinating to see that a single embryonic cell during development differentiates into many types of cells that differ in morphology, metabolism and function, yet have the same chromosomal genetic constitution. For a cell to develop and differentiate, it must be able to grow, which means it must duplicate its genetic material, synthesize all other cell components and maintain a specific form. In order to maintain genetic continuity during cell differentiation, the chromosomal material must be duplicated and an accurate copy must pass to each daughter cell.

However, cell division involves more than the synthesis of chromosomal genetic material and its equal distribution to two daughter cells. It also means equal or planned distribution of the cytoplasm and the organelles it contains. In the absence of ordered cytoplasmic distribution, chromosomal acquisition would be less meaningful as the cytoplasm, along with its associated components, gives functional meaning to the cell. During development, an organism goes through cycles of gene duplication, organization of the cytoplasm and reorganization of the structure. Besides the primary genetic activity of the chromosomes, the organization of the cytoplasm is probably the most important event in differentiation. The distribution of extrachromosomal material takes place along with the highly specialized process of equal segregation of chromosomal genetic material, but the degree of organization and tendency towards equal distribution may not be the same. In other words, the distribution of cytoplasm during cell division may be unequal.

In some cases there appears to be a special mechanism for distribution and organization of cytoplasmic components. A case in point is the distribution of mitochondria and packing of ribosomes during differentiation of zoospores in the Blastocladiales. In *Blastocladiella* there is a single large mitochondrion at the basal end of the zoospore which rapidly divides and distributes itself randomly into the cytoplasm during germination (Cantino, Lovett, Leek & Lythgoe, 1963; Lovett, 1975). In a related organism *Allomyces*, the mitochondria form a ring around the nuclear cap membrane (Turian & Kellenberger, 1956). During germination, as the nuclear cap disintegrates, the mitochondria are distributed into the cytoplasm. A similar situation has been described by Wilson (1931) in centrurid scorpions where the mitochondria form a closed ring at the onset of spermatogenesis. Organization of the ribosomes into a package around the nucleus has been found in the zoospores of some aquatic fungi. In most phases of development, the cells of higher fungi do not show such an ordered organization of the mitochondria or the ribosomes.

In filamentous fungi, development proceeds along two broad lines, the vegetative and the reproductive. The switch from one type of development to the other, besides being temporal and genetic, is also controlled by external environment. At least in certain organisms the conditions for the two types of development are mutually exclusive. In *Saprolegnia* for example, vegetative development can be maintained for an indefinite period if fresh medium is supplied (Sparrow, 1960). Klebs (see Sparrow, 1960) maintained *Saprolegnia ferax* in the vegetative state for three years by continuous subculturing in fresh medium. In a large number of aquatic fungi, differentiation is induced by starvation. However, some growth appears to be necessary for differentiation to occur. For more details on the induction of sporulation in fungi see Chapter 11.

Waering (1971) has described development as an open system where dynamism towards progression is maintained, whereas morphogenesis is a closed system, *i.e.* resulting in a terminal differentiation and the stabilization of state. During morphogenesis the synthetic pathways are committed towards specific directions until stability is achieved. This commitment is reversible during the initial stages but becomes irreversible later at a point described by Cantino (1966) as 'the point of no return'.

2.2 Nucleocytoplasmic Relationship during Development

If a resting spore with one nucleus contains all the genetic information for development, then what maintains the resting state and what initiates the environmentally mediated response of genetic activity? As the nucleus has a three dimensional relationship with the cytoplasm and the external environment, the first component to detect the change in the environment would be the cytoplasm. The cytoplasm, therefore, will have to be stabilized to maintain the stability of a particular form. Hence, differentiation from one state to another would require the destabilization of the cytoplasm and the formation of a gradient of the inducer. This new environment would then initiate the activity of certain genes which would create another environment that in turn would serve as the activator for other sets of genes ultimately leading to the differentiation of the entire individual.

The influence of the cytoplasm on the nucleus is not only in the induction of transcription but also in its repression. In addition, the cytoplasm translates the genetic information.

Experiments with higher eukaryotes have demonstrated that the nuclei of early embryonic cells are similar and completely interchangeable. Therefore, the fates of individual nuclei during differentiation of tissues, such as nerve, depend on some unknown 'extranuclear' information. Studies on the influence of nuclei on cytoplasm or vice versa, in higher organisms have been elegantly reviewed (Hammerling, 1966; King & Briggs, 1966; Gordon, 1974). The interspecific hybridization experiments in *Acetabularia* have shown the accumulation of specific substances in the cytoplasm, the synthesis of which depends on the nucleus, but which can direct a highly specific type of morphogenesis in the absence of a nucleus or in the presence of a foreign nucleus (Hammerling, 1966). These morphogenetic substances arrange themselves in a gradient, with the region of highest concentration most distant from the nucleus.

Some cytoplasmic components are known to be semi-autonomous, *e.g.* mitochondria, chloroplasts and kinetoplasts. Cytoplasmic DNA other than from mitochondria, has been found in yeasts (Carter, 1975), *Blastocladiella* and *Allomyces* (Myers & Cantino, 1971; Ojha & Turian, 1971.

The influence of cytoplasm on development and differentiation in fungi can be considered under translational control, the metabolic competence of the cytoplasm, and cytoplasmic organelles.

Translational control includes the events where transcription of the genome occurs and mRNA is produced but not translated immediately. One of the most important attributes of these mRNAs is that they are stable and Taylor (1969) has called them masked mRNAs. Stable mRNAs have been thought to be involved in the germination of spores of *Peronospora tabacina* (Holloman, 1969) and *Neurospora crassa* (Bhagwat & Mahadevan, 1970). Preformed mRNA has been indirectly demonstrated in the differentiation of gametes (Fahrnich, 1974*a, b*), female gametangia (Turian & Viswanathan, 1966), zoosporangia and zoospores in *Allomyces arbuscula* (Burke, Seale & McCarthy, 1972), and zoospores of *Blastocladiella emersonii* (Lovett, 1975).

The metabolic competence of the cytoplasm determines the synthetic capacity of the cell. Various cytoplasmic organelles participate in conferring this competence. One of the best known phenomena is the *petite* mutation in yeast which is characterized by the inability of the organisms to grow on non-fermentable sugars (Sherman & Slonimski, 1964). The organisms lack cytochromes a, a_3 and b, but contain cytochrome c. Slow growing (SG), abnormal and respiratory deficient mutants of *Neurospora crassa* also belong to this category of cytoplasmic mutants. The involvement of mitochondria has been demonstrated in the *petite* mutations of yeast and respiratory deficient mutant (*mi*-maternal inheritance) of *N. crassa*, (Sager, 1972). Rowlands & Turner (1974) reported impaired growth ability on drug-free media in an extranuclear mutant (*Str* O^R6) of *Aspergillus nidulans*. They were able to distinguish these types from nuclear mutants. Grimes, Mahler & Pearlman (1974) studied nuclear gene effects on mitochondrial mass and DNA in *Saccharomyces cerevisiae*. They showed a mean number of 22 mitochondria per cell in the diploid and 10 in the haploid strains. The mitochondrial DNA of the diploid and haploid cells accounted for 13% and 14% respectively of the total cellular DNA.

In the aquatic Phycomycete *Blastocladiella*, a cytoplasmic organelle (γ) has been described and shown to be associated with specific developmental stages (Myers & Cantino, 1974). The relevance of the gamma particle in development and differentiation can be summarized as follows:

(i) Spores derived from orange plants are deficient in the gamma particle as compared to OC (ordinary colourless) and LC (late colourless) plants (Cantino & Horenstein, 1956; Matsumae, Myers & Cantino, 1970).
(ii) Cycloheximide induces temporary variants with 8γ particles per zoospore and diphenylamine induces temporary ordinary colourless with 12γ particles per spore (Cantino & Hyatt, 1953).
(iii) Gamma particles decay during encystment but reappear during the final two hours of sporangium and zoospore differentiation (Truesdale & Cantino, 1970; Barstow & Lovett, 1975).

Sexual differentiation in a related organism *Allomyces* is characterized by the swelling of the apex and division of the cytoplasmic territory forming either male or female gametangia. The female cytoplasmic territory is larger and colourless and capable of synthesizing a hormone, sirenin (Machlis, 1958); the male territory is smaller, incapable of sirenin synthesis, but capable of synthesizing β-carotene, like the orange plant of *Blastocladiella emersonii*. That these phenotypes are under genetic control has been demonstrated by Emerson & Wilson (1954). Turian has studied the metabolic differences between the male and female cytoplasmic territories by cytochemical methods and by growing the organism under repressive conditions (Turian, 1969, 1975).

Further, mutations of the *petite* type favour the differentiation of male gametangia (Turian, Ojha, Scheps & Oulevey, 1969). By analogy with *petites* in yeast and based on the selective effects of acridines such as proflavine, Turian (1975) has postulated a hypothetical cytoplasmic factor whose differential distribution, under nuclear control, leads to the differential morphogenesis of the *Allomyces* apex. According to this hypothesis, the differential polar distribution of the factor confers competence on the presumptive male and female cytoplasmic territories.

Genetic analysis of the cytoplasmic control of morphogenesis has been done with mutants of *Neurospora*, *Aspergillus* and yeasts. The most widely studied of these mutants are the respiratory deficient mutants of *Neurospora* and yeast. In yeasts they have been shown to have altered mitochondrial DNA (Carnevali, Morpugo & Tecc, 1969; Goldring *et al.*, 1970). Recently, Faye *et al.* (1973) have shown massive deletions in mitochondrial DNA in the *petite* mutants of yeast. The status of mitochondrial DNA in respiratory deficient mutants of *Neurospora* is not yet clear. Mutants of *N. crassa* (McDougal, 1964; Sager, 1972) and *A. glaucus* (Subak-Sharpe, 1958) have been shown to have cytoplasmic inheritance of protoperithecial differentiation. Barrage formation and senescence in *Podospora* are also phenomena of cytoplasmic inheritance (Rizet, Marco & Acherson, 1958).

Recently, Gunstilleke, Scazzocchio & Arst (1975) have shown that two chloramphenicol resistance mutations out of 123 tested in *Aspergillus nidulans* were inherited extranuclearly as judged by transmissibility in heterokaryons, lack of segregation at meiosis, and independent segregation from all of the eight nuclear linkage groups.

2.3 Nuclear Division, Cell Cycle and Differentiation

Data on the direct kinetic analysis of nuclear division in the different developmental phases are lacking in filamentous fungi. Most studies include random examination of mitotic figures during a particular stage of growth and development, but in terms of synthetic activity of the nucleus, this aspect is perhaps less meaningful than kinetic analysis. Synthetic activity during development means the duplication, transcription, and translation of information and the organization of structure. The relationship of these processes to the rate of division is perhaps more interesting in biochemical terms than the morphological aspect of nuclear division. It is, therefore, essential to correlate the nuclear division cycle during growth and differentiation. The nuclear division cycles have been studied in considerable detail in the bacteria and in unicellular eukaryotic organisms, e.g. yeast, *Tetrahymena*, *Chlorella*, etc. (Mitchison, 1971).

The cell cycle can be divided into distinct phases, G1, S, G2 and M (Howard & Pelc, 1953). The duration of different phases of the cycle varies with the organism. The two phases preceding and following the synthetic period (S) are named G1 and G2. In cells, G1 can vary from a few hours to weeks or months. In a rapidly growing cell like the liver cell, the majority of the cells are in the G1 phase. The lengths of the S and G2 phases are more constant. The centriole begins to reproduce at the end of G1 or at the beginning of S phase. The S phase is characterized by rapid DNA synthesis; however, a small amount of DNA synthesis has been shown to continue during the G2 phase (Kihlman & Hartley, 1967). Studies on the replication of DNA in *Physarum polycephalum* indicate that about 90% of the DNA replicated during the early S phase consists of unique sequences. Further, DNA replicated early in the S phase of the cell cycle contains less than 10% of the repeated sequences (Fouquet, Bierweiler & Saur, 1974). The rate of DNA chain growth in this organism indicates that about 3×10^6 daltons of double-stranded DNA are replicated per minute (Brewer, 1972). In the second half of the S phase a high proportion of repetitive sequences were replicated. Ribosomal DNA is repetitive and late replicating in this organism (Britten & Smith, 1970). Part of this DNA replicates even in the G2 phase (Zellweger, Ryser & Braun, 1972; Newlon, Sonensheim & Holt, 1973). The DNA cycle in *Saccharomyces cerevisiae* has G1, S and G2 periods of approximately equal length in a two hour cycle (Williamson, 1966). Bostock, Donachie, Masters & Mitchison (1966) have studied the cell cycle in *Schizosaccharomyces pombe*. They found absence of G1, a very short S (10 min) and a G2 of 2.5 h. Since there is a gap of about a quarter cycle between nuclear division and cell division, DNA synthesis takes place between these two processes. This DNA cycle can be altered by growing the cells in the presence of phenylethanol (Bostock, 1970) or high glucose concentration (Duffus & Mitchell, 1970). A G1 phase is then inserted and S phase takes a slightly different position in the longer cell cycle. The nucleo-cytoplasmic interaction during cell cycle has been studied by nuclear transplants or fusion. In *Amoeba proteus*, Prescott & Goldstein (1967) have shown that the transplantation of a nucleus engaged in DNA synthesis into a

G2 phase (after DNA synthesis) cell results in inhibition of such synthesis. When the nucleus of a G2 cell is transplanted into an S phase (period of DNA synthesis) cell such a nucleus begins to synthesize DNA. In *Physarum* (Guttes & Guttes, 1968), the transfer of a G2 nucleus to an S phase cytoplasm does not initiate DNA synthesis, whereas S nuclei continue to synthesize DNA in G2 cytoplasm.

In multinucleate hyphae, the nuclear cycle is complicated by the fact that the relationship between cell cytoplasm and nucleus is different. The vegetative hyphae of fungi have incomplete and widely spaced cross walls. Since the nuclei contained in one hypha are suspended in the same cytoplasm, their division in the same cell should be synchronous. Indeed, examination of the living hyphae has given some supporting evidence (Clutterbuck, 1970). Rosenberger & Kessel (1967) have studied the synchrony of nuclear division and have used the total number of nuclei per hyphae, an exponent of 2 as an indication of synchrony. Numbers other than this indicated lack of synchrony. A relationship was also found between the doubling time and dry weight doubling time when the dry weight doubling time was 1.4 to 1.8. The replication of nuclei in the individual hyphae was highly synchronous but nuclear synchrony was lost when increases in doubling time were achieved by alterations in the nitrogen and carbon sources. At the slowest growth rate tested the interval between the division of the fastest and slowest nucleus was 48% of the dry weight doubling time. The active replication rate of some nuclei in a hypha where other nuclei were resting suggested that nuclear duplication in this organism might be controlled by specific inhibitors. King & Alexander (1969) reported that in *Alternaria solani*, the hyphal tip cell was divided by a number of septa after a wave of nuclear divisions had passed from the tip to the rear of the cell. They have also noted that septation in the hyphae occurs from the tip in centripetal succession. Clutterbuck (1970) noted that multinuclear division took about 20 min to pass from one end of the cell to the other: a distance of 440–700 μm. A difference in the timing of division occurs in the tips of branched hyphae as compared to the main hyphal tips and in some cases the nuclear division occurred in two separate waves indicating the sign of breakdown in synchrony that Kessel & Rossenberger (1968) found in poorer medium.

In differentiating uninucleate sterigmata, cross walls are formed in the normal way between pairs of daughter nuclei after nuclear division. The formation of coenocytes can be regarded as due to the suppression of the majority of septa although a relationship between the time of nuclear division and septation is maintained.

The analysis of cellular activity, whether during growth or differentiation, at the level of the individual cell, should implicate the cell cycle since it is the most fundamental response of a cell to a change in the environment. In the preceding paragraph we have discussed nuclear division and cell division; in the following section we shall deal with how the cell cycle of an organism responds to an altered environment, thus committing a cell to a particular type of growth and development. Unfortunately, work on the cell cycle response to development in filamentous fungi is lacking (however see Chapter 8). In the cell cycle, as we have described earlier, an intense metabolic activity takes place in the phases preceding mitosis. The two well characterized events in the cell cycle are the S and M phases. In the S phase most of the DNA is synthesized. The M phase is characterized by the condensation of the chromosome, disappearance of the nucleolus and the nuclear membrane, the formation of the mitotic apparatus, the separation of the chromsomes and the division of the cytoplasm. We would like to emphasize the stage preceding the actual mitosis which occupies 96% of the cell cycle (Mitchison, 1971) because it is the phase where most of the biochemical activity takes place and which prepares the cell for duplication.

Unlike the S and M phases, no unique event characterizes the G1 phase. The G1 could be divided into two parts: early G1 phase and late, or terminal, G1 phase. The temporary or permanent arrest of cell division in eukaryotes is achieved by cells maintaining the G1 phase of the cell cycle (Prescott & Goldstein, 1967). The terminal part of the G1 must contain the events concerned with the intiation of DNA synthesis. The fission yeast, *Schizosaccharomyces pombe* (Bostock *et al.*, 1966), the multinucleate slime mould, *Physarum polycephalum* (Guttes & Guttes, 1968), and *Allomyces neomoniliformia* (Olson & Fuller, 1971) lack a G1 phase. A number of mammalian cells also appear to lack this part of the cycle. The duration of the G1 phase is variable depending on the cultural conditions. When cells growing in culture show an increased generation time, this increase is often accounted for by an expansion of the G1 with little increase in the duration of S, G2 or mitotic phases.

A correlation between cell specialization and differentiation in the inter-mitotic phase exists in the course of evolution. Lark (1963) has pointed out that in eukaryotic cells the intermitotic phase is more distinct than in the prokaryotes. In light of the phenomena just discussed, this can be considered as a prerequisite for elaborate cell specialization. Those organisms which do not possess a G1 phase in the division cycle, *viz.* *Schizosaccharomyces pombe*, *Physarum polycephalum*, *Amoeba proteus*, do not show an elaborate sequence of cell specialization. On the other hand, *Allomyces*, although a primitive eukaryotic organism, shows remarkable cell specialization but the zoospores lack a G1 phase (Olson & Fuller, 1971). However, the lack of G1 may only concern the zoospore and a G1 may be present in the vegetative and differentiating mycelia. The presence of G1 reflects a precise control over the timing of the initiation of nuclear DNA synthesis and suggests that some component of the initiation mechanism may be labile, since the initiation occurs only once in each division cycle. In differentiated tissues it appears that the G1 provides a time in the division cycle where a cell can be held in a non-dividing state, if it be needed, to perform specialized metabolic functions.

The DNA cycle in the multinucleate hyphae from the germinating conidium of *Aspergillus nidulans* has been examined by Kessel & Rosenberger (1968). At fast growth rates (doubling time 87 min), the nuclei have a synchronous S phase lasting about 20 min. At slow growth rate, the S phase remains the same length but is asynchronous between the nuclei in the same hypha. It appears that these nuclei have variable G1 and G2 phases even though they are in the same cytoplasm.

2.4 Nuclear Control of Sex Differentiation

Most fungi are classified into three categories on the basis of sex: (a) hermaphroditic (male and female organs in the same thallus); (b) dioecious (male and female organs in different thalli); and (c) sexually undifferentiated (male and female organs are morphologically similar). On the basis of compatibility, fungi are grouped into: (i) homothallic (sexually self fertile, two mating type genes A and a are in the same thallus); and (ii) heterothallic (when each thallus is sexually self sterile, *i.e.* mating type genes A and a are in different thalli). If two nuclei of opposite mating types A and a are incorporated in the same spore, it is known as a psuedo (or secondary) homothallic fungus. This phenomenon of fungal strain differences based on incompatibility has interested numerous workers (Olive, 1958; Raper, 1960).

Hopper, Kirsch & Hall (1975) have done extensive studies on genetic mechanisms controlling homothallism and heterothallism in *Saccharomyces cerevisiae*. According to them, there are specific genes for mating (aa and AA), and for sporulation. In order to sporulate, diploid cells must be heterozygous at the mating-type locus. They have shown that the gene conferring homothallism is nuclear, recessive and unlinked to mating type. They induced homothallism in a heterothallic strain. Studies made by Harashima, Nogi & Oshima (1974) have described four types of life cycles in *Saccharomyces cerevisiae* based on the genetic system controlling homothallism. These four types include (a) perfect homothallism, (b) two types of semi-homothallism, and (c) heterothallic haploid segregation. Genes controlling mating types in the slime mould *Physarum polycephalum* have been studied by Adler & Holt (1975).

Several species assigned to the genus *Neurospora* are grouped into three broad categories based primarily on their crossing behaviour. These are: (a) heterothallic species like *Neurospora crassa*, *N. intermedia* and *N. sitophila* in which mating types are not in the same strain and in which opposite mating types are crossed; (b) the secondary (or pseudo) homothallic (4-spored asci) species like *N. tetrasperma*, which produce both conidia and viable asco-spores from a single strain; and (c) the third broad group, being true homothallic species like *N. africana*, *N. dodgei* and *N. lineolata*. These are true homothallic species which do not produce conidia but produce perithecia with 8-spored asci and viable ascospores. Most genetic and physiological studies have been concerned with the heterothallic species *N. crassa*. Very little information on genetic background is available for the other heterothallic species *N. intermedia* and *N. sitophila*, nor is much known about homothallic species. Perkins & Turner (1973) have emphasized the importance of crossing behaviour for assignment of unknown natural isolates to a specific species. Problems associated with this approach may be the lack of proper mating type in nature or lack of fertility due to incompatibility. Lack of fertility among these three broad groups of the genus *Neurospora* makes it difficult to study their relatedness.

Attempts have been made to find the relatedness of these organisms using the mean guanine + cytosine (G + C) mole percentage of the DNA, which in practice affords no predictive relationship. For example, it may be noted from the data summarized in Table 2.1 that the G + C mole percentage value of all the *Neurospora* species studied by us are between 55 and 57. Other DNA characteristics, such as thermal denaturation profiles (*i.e.* nature of the change in optical absorption at 260 nm with increasing temperature), and occurrence of repeated DNA sequences (Dutta & Schwartz, 1973) are also unsuitable. These species are, however, clearly distinguishable by DNA: DNA hybridization studies (Table 2.1).

Table 2.1 List of *Neurospora* species studied: their sources, DNA composition and hybridizations

Name of Species	Source and Stock Number		G + C Mole Percentage Major Comp. (mole %)	Percentage DNA : DNA Hybridization (%)
Heterothallic				
Neurospora crassa 74A mycelia	FGSC	987	55	94
N. intermedia A mycelia	DDP	p420	56	86
N. sitophilia 10B-A mycelia	FGSC	580	57	82
Pseudohomothallic				
N. tetrasperma 85-A mycelia	FGSC	1270	56	76
Neurospora sp. (Gianjor) A + a Bali	DDP	p202	55	80
Homothallic				
N. africana mycelia	FGSC	1740	56	60
N. dodgei mycelia	FGSC	1692	57	57
N. lineolata mycelia	FGSC	1919	56	59

S. K. Dutta, I. Sheikh, J. Choppala, A. S. Gurmit and W. H. Nebon (unpublished results).

^{32}P-DNA of *Neurospora crassa* non-repeated DNA, sheared to 400 nucleotides piece size (called fragments), had 90,000 cpm (counts per minute) per microgram DNA. ^{32}P-DNA C_0t used was .06, whereas unlabelled DNA C_0t was at least 1500 in each reaction. FSGC means Fungal Genetics Stock Center. DDP means David D. Perkins of Stanford University, California, and TM means T. Matney of M.D. Anderson Tumor Institute, Houston, Texas. Calf Thymus DNA was obtained commercially from Sigma Chemicals. Values listed here represent one determination. These values were, however, highly reproduceable (with +1.5% deviations) in at least three such independent reactions conducted. The G + C content of the respective DNA fraction was determined from the thermal denaturation curves to the formula: $Tm = 69.3 \times 0.41\, GC$, in 0.12 M phosphate buffer, using Gilford 2400 spectrophotometer.

That DNA controls the so-called phenotypic sex differentiation in *Allomyces* is apparent from the results of interspecific crosses made by Emerson & Wilson (1954) which revealed a dual control of sex ratio and sex positioning by polymeric genes. This DNA control of polar positioning of the sexual organs has been further substantiated by transformation experiments in which extracted DNA from the hypogynous species (*A. arbuscula*) could induce a significant amount of reverse positioning of the gametangia in the epigynous species (*A. macrogynus*) and *vice-versa* (Ojha & Turian, 1971).

2.5 Molecular Organization of Nuclear Genomes

Fungal nuclei are very small and pose serious problems in the evaluation of the DNA content per nucleus by standard colorimetric methods. This is further complicated by the multicellular filamentous organization of the cell. Such problems have been circumvented to a certain extent in organisms that produce single-celled stages in their life cycles (conidia, zoospores, ascospores, basidiospores, *etc.*). Cells containing one or two nuclei have been used to determine the DNA content per nucleus directly by colorimetric methods.

Britten & Kohne (1968) and Wetmur & Davidson (1968) have demonstrasted a relationship between the rate of reassociation of sheared single-stranded DNA and the genome size. We have taken the advantage of this property to analyze certain fungal DNA. Our studies on the rate of reassociation of the DNAs from three organisms representing broad groups of fungi, *Allomyces arbuscula*, *Neurospora crassa* and *Coprinus lagopus* have given a genome-size of 1.4×10^{10}, 2.2×10^{10} and 2.5×10^{10} daltons respectively

Table 2.2 Summary of estimates of fungal genomes by DNA-reassociation kinetics

Organism	G + C Content (Mole %)	Haploid genome size based on reassociation kinetics (daltons 10^{10})		References
		Optical assay	HAP assay	
Chytridiomycetes				
Allomyces arbuscula	64	—	1.4	Ojha (unpublished results)
Zygomycetes				
Phycomyces blakesleeanus	35	—	1.9	Dusenberry (1975)
Mucor bacilliformis	34.5	21.0	—	Dutta & Ojha (1972)
M. racemosus	41	0.7	—	Dusenberry (1975)
Ascomycetes				Christiansen *et al.* (1971)
Saccharomyces carlsbergensis	40.7	0.939	—	"
S. cerevisiae	40–42	0.92	—	"
Debaryomyces hansenii	37.6	0.605	—	"
Kluyveromyces lactis	39.1	0.848	—	"
Hansenula holstii	36.9	0.652	—	"
Torulopsis homii	36.2	1.36	—	"
Candida catenulata	55.2	1.24	—	"
C. macedoniensis	41.4	0.831	—	"
Talaromyces vermiculatus		1.4		
Neurospora crassa	52		2.2	Dutta & Ojha (1972)
Basidiomycetes				
Coprinus lagopus	52		2.5	Dutta & Ojha (1972)
Cellular slime moulds				
Dictyostelium discoideum	23	3.0	3.0	Firtel & Bonner (1972)
Slime moulds				
Physarum polycephalum	40	25–70	—	Fouquet *et al.* (1974)

The rate of reassociation depends upon a fragment size, concentration of the reacting single-stranded DNAs, monovalent cation concentration, pH and temperature. Given optimum conditions of reassociation and in the absence of repetitive sequences, it depends upon the genome-size of the organism. The smaller the genome-size, the faster the rate of reassociation.

(Dutta & Ojha, 1972, Chattopadyha, Kohne & Dutta, 1972; Dutta, Penn, Knight & Ojha, 1972). Dutta & Ojha (1972) have shown an excellent correlation of the increases of the genome size and evolutionary complexity in fungi. Table 2.2 summarizes the available information on the kinetic analysis of fungal DNAs studied so far.

The reassociation kinetics of sheared single-stranded DNA relates not only to the genomic complexity, but also demonstrates qualitative sequence organization (Britten & Kohne, 1968; Wu, Hurn & Bonner, 1972; Britten & Davidson, 1971; Davidson, Galau, Angerer & Britten, 1975). Experiments with DNAs from higher organisms have revealed that reassociation of single-stranded fragments under ideal conditions of temperature and cation concentration do not show the ideal second order rate. Curves obtained from these reactions are not uniform and show several steps differing in rate. Initially, the rate is rapid followed by medium rate and finally a slow reassociation rate. In kinetics, the reaction rate is governed by the concentration of the reacting substances. Fast reassociation, therefore, implies the increasing concentration of similar sequences, medium reassociation, that of moderate concentration of similar sequences and slow reassociation that of unique sequences only. Thermal denaturation profiles of the reassociated repeated sequences have shown divergence in the sequences indicating that those classes consist of similar but not identical groups (families).

The analysis of fungal DNAs studied so far has indicated the existence of repeated sequences as in other eukaryotes (Table 2.3). Most of the repeated sequences are of nuclear origin and contrary to the repeated sequences of higher eukaryotes do not form a satellite in CsCl gradient centrifugation indicating that either they do not differ in their G–C content from the main nuclear DNA, or that they are present in small amounts. The only exception is the ribosomal DNA which in some fungi appears as satellite and consists of repeated sequences. The existence of repeated sequences in fungi have been studied in our laboratory (Dutta, 1974).

The average size of the repeated DNA in the fungi examined is approximately $3–4 \times 10^7$ daltons (*Neurospora crassa, N. intermedia, N. tetrasperma, Mucor azygospora* and *Coprinus lagopus*) (Dutta, 1974). These are middle repetitive DNAs having approximately 100 copies. In yeast, the average complexity has been reported to be 2×10^7 with a range of 6.7×10^6 to 5.1×10^7 (Christiansen, Lethbak, Stenderup & Christiansen, 1971). Our earlier conclusion (Dutta, 1974) based on the analysis of certain filamentous fungi, that unlike most other eukaryotes fungal repeated sequences are homogenous with respect to kinetics of reassociation, seem to be premature. Dusenberry (1975) has shown heterogeneity of *Phycomyces blakesleeanus* repeated sequences. Similar analysis of the kinetics of reassociation of *Dictyostelium discoideum* has indicated sequence heterogeneity of repeated sequences (Firtel & Bonner, 1972). Although most of the repeated DNAs have nuclear origin, studies with yeast DNAs have shown that mitochondrial DNAs also contain repetitive sequence elements (Faye et al., 1973).

In fungi, most of the cellular DNA is of nuclear origin. The nuclear DNAs, however, consist of DNA sequences, some of which are of higher $G+C$ components and others of high $A+T$ components. In *Neurospora crassa*, high $A+T$ DNA sequences are mostly repeated (Dutta, 1974). The high GC fractions are not separable from AT rich fractions by bio-gel column chromatography or CsCl density gradient centrifugation. These observa-

Table 2.3 Occurrence of repeated DNA sequences in fungi

Organism	% Repeated sequences Based on ^{32}P CPM	Based on optical absorbancy at 260 nm	$\frac{1}{2}$ Cot values from DNA:DNA reassociation kinetics	Calculated Mol. Wt. in daltons	No. of copies	References
Phycomycetes						
Allomyces						
arbuscula	15–20%	—	0.1	9.45×10^6	250	Ojha, unpublished results
Mucor						
azygospora		15–18	0.058	3.4×10^7	103	Dutta (1974)
Phycomyces						
blakesleeanus	10, (I)	15–18	—	—	—	"
	20 (II)	—	—	6.4×10^7	—	Dusenberry (1975)
Ascomycetes						
Neurospora						
crassa	10–12	12–15	0.055	3.2×10^7	100	Dutta (1974)
N. intermedia	10–12	12–15	0.055	3.2×10^7	100	"
N. tetrasperma	10–12	12–15	0.058	3.2×10^7	100	"
Saccharomyces						
carlsbergensis	11	—	—	—		Christiansen et al. (1971)
S. exiguus	10.8					"
Debaryomyces						
hansenii	13.8					"
Kluyveromyces						
lactis	13.4					"
Hanseula						
holstii	8.4					"
Torulopsis	16.2					"
T. candida	5.3					"
Candida						
catenulata	14.5					"
C. macedon-iensis	7.1					"
Basidiomycetes						
Coprinus lagopus	12–15	15–18	0.06	3.6×10^7	112	Dutta (1974)
Cellular Slime Moulds						
Dictyostelium						
discoideum	30					Firtel & Bonner (1972)

The discrimination between repeated and unique sequences (sequences occurring only once per nucleus) is based on the rate of reassociation as mentioned earlier. The repeated sequences sould be separated from the unique sequences on HAP columns which discriminate between single double-stranded DNA, depending upon the molarity of phosphate buffer or the temperature of the elution (see Britten & Kohne, 1968). DNA is reassociated at low cot (concentration of nucleotide moles/1 ×t in sec) permitting the reassociation of the fast reassociating fraction only and the reaction mixture is differentiated on the HA columns.

tions suggest that in *N. crassa*, as in other organisms, high AT regions may be interdispersed in high GC rich regions (S. K. Dutta & W. H. Nelson, unpublished results). Detailed studies of interspersion of repetitive sequences in chromosomal DNA in *Dictyostelium discoideum* have been done by Firtel & Kindle (1975), Firtel, Cockburn, Frankel & Hershfield (1976) and Lodish, Firtel & Jacobson, (1973).

The G + C composition of main nuclear DNA varies extensively in fungi, ranging from 23 to 65 mole % (Storck & Alexopoulos, 1970). The G + C mole % of mitochondrial DNA is either the same (Mucorales) or lower (the rest) contrary to the plants and animals where it is either the same or higher (Villa & Storck, 1968). Further, it appears that the G + C mole % in mitochondria increases with increasing G + C ratio of major peak DNA (nuclear) but that the variation is within narrow limits.

Recently whole cell DNA from *Allomyces* has been further characterized by fractionation of DNA on preparative CsCl gradients and subsequent centrifugation of two fractions shows a heavy major and minor light satellites across the profile. Besides the major chromosomal DNA, the two satellites have been identified as mitochondrial and on unidentifiable cytoplasmic DNA in order of their densities (Ojha, Hammerli & Turian, 1976). We have studied the distribution of tDNA and rDNA in *Neurospora erassa* where they appear as heavy satellites (Dutta & Ray (1973)).

2.6 Biochemical Correlates of Gene Expression

Although various factors influence development, it is an undeniable fact that development and differentiation are primarily the results of selective gene activation. It implies that the different cell types contain the same set of genes and their differentiation would depend upon which set of genes are expressed. There is another alternative to this hypothesis, *i.e.* differential gene alteration which proposes that one cell differs from another because of the modification of the genes themselves (Schultz, 1965). This alteration could be due to the base modification of the DNA either due to substitution, deletion, methylation or glucosylation (Brown & David, 1968). The attractiveness of the theory of differential gene expression has come from the experiments of Hoyer, McCarthy & Bolton (1964) who did not find any qualitative difference in the DNA sequences between different cell types by DNA–DNA hybridization. We have analyzed the physical characteristics of the DNA from different cell types of *Neurospora crassa* (Dutta & Ojha, 1972; Dutta & Chaudhuri, 1975a) and found that the GC composition and percentage distribution of major and minor fractions in DNAs from mycelia and conidia are the same. However, the analysis of DNA from ascospores has shown that the minor fractions represent only 7% of the total cellular DNA against 20% in other cell types. If development and differentiation take place due to the selective activation of genes, then what are the factors that confer cell specialization and complexity? One of the most important factors would be the complexity of the genome itself. The kinetic analysis of its sequences in higher organisms has revealed structural organization that may have significance in the control of gene expression (Britten & Davidson, 1969, 1971).

Impressive work has been done on gene expression in prokaryotic organisms like viruses and bacteria. These organisms are easier to work with than eukaryotic organisms because of their small genomic content and primitive cellular organization. The control of gene expression in prokaryotic systems has been reviewed by Gros (1974). Eukaryotic systems present a very complex genome and cellular organization. This requirement for genome complexity is consistent with the evolution in cellular specialization and tissue organization.

Using *Neurospora crassa* developmental cell types, Dutta & Chaudhuri (1975a) have shown that nuclear DNAs of all cell types, conidial sprouts, mycelia and ascospores have identical DNAs but the expression of these DNAs vary remarkably. By DNA–RNA hybridization, they estimated that 34% of the unique sequences are transcribed at mid-log phase of mycelial growth and only 15% in the isolated conidia. Germination of conidia markedly increased the transcription. Conidial sprouts and unbranched mycelial cells showed transcription of 25 and 27% of the unique sequences. Of the 10–12% repeated sequences of *Neurospora crassa*, 7.3% were found to be transcribed in mid-log phase of the mycelial growth. Part of these sequences represent mitochondrial and part nuclear DNA. Most of the nuclear repeated DNAs code for rRNA and tRNA in *Neurospora crassa* (Dutta & Chaudhuri, 1976b). Bhagwat & Mahadevan (1973) have also observed that both amounts and types of RNA synthesized at any growth period are significantly different. Mahadevan & Bhagwat (1974) have provided evidence for the presence of two proteins in *Neurospora crassa* which have regulatory effects on mRNA (pulse labelled RNA) synthesis. According to them, maximum RNA synthesis occurs in 8–10 h mycelial cells. At this stage they find regulatory protein in the cytoplasm. In 16 and 24 h cells they could not find these proteins in the cytoplasm, but instead found them complexed with DNA.

With regard to gene amplification, increase in gene doses like polyploidy, should not have any consequence on differentiation. Selective amplification of genes is still an attractive hypothesis. The important feature of this hypothesis is that commitments on the part of the cell towards certain specific types of metabolism would probably require a sustained supply of the particular genes concerned. Since the structural genes are present only once per genome, it is possible that the rate limiting step in differentiation would be the availability of a particular gene involved. But, if a large scale amplification of single copy sequence is logically not possible, a low degree of amplification of some sequences can occur, as reported in different tissues of calf (Kohne & Byers, 1973). The most striking case of gene amplification has been found for rRNA in oöcytes of amphibia, certain insects and in the puffs of the salivary gland chromosomes. The rDNA in amphibia and insects appears as heavy satellite in CsCl gradient and can easily be separated. Both by neutral CsCl gradient centrifugation and rRNA–DNA hybridization it has been shown that rDNA is amplified in the oöcytes of *Xenopus laevis* (somatic cells contain 1000–2500 cistrons/cell, whereas oöcytes 2×10^6 cistrons/cell) (Birnstiel, Chipchase & Speirs, 1971). In *Acheta* somatic cells have been found to contain 342 cistrons, whereas 1600 rRNA genes occur in ovariant cells (Pero *et al.*, 1973). Recently, Cullis & Davis (1975) have shown that different varieties of *Pisum sativum* (pea) have different proportions of rDNA in their genomes. They have shown an amplification of rRNA genes in the cells having high DNA content of the seed cotyledon and also in the young leaves, but not the old ones.

In *Tetrahymena pyriformis*, preferential replication of the ribosomal RNA genes has been shown to take place when starved cells are shifted to nutrient medium (Engelberg, Mowat & Pearlmann, 1972). By DNA-rRNA hybridization, Engelberg & Pearlmann (1972) have shown that nuclear DNA isolated from starved cells hybridize 30–40% less than that from exponentially growing cells. They concluded that this does not result in amplification, but that it demonstrates that nuclear and rRNA cistron replication are under separate control.

In fungi, rRNA cistrons are either heavier, have the same density or are lighter in CsCl gradient centrifugation, depending upon the main band DNA which has a wide range of density (Storck & Alexopoulos, 1970). The multiplicity and buoyant density of rRNA cistrons in fungi are shown in Table 2.4.

Table 2.4 Ribosomal DNA in fungi

Organization	Mass of DNA/cell	Double-stranded ribosomal DNA complement %	Main band density	Y DNA density	Multiplicity
Phycomycetes					
Allomyces					
arbuscula	1.4×10^{10}	3.2	1.721	1.706	270 (1)
Achlya					
bisexualis	2.8×10^{11}	0.65	1.697	1.697	435 (2)
*Ascomycetes					
Saccharomyces					
carlsbergensis	2.6×10^{10}	3.8	1.698	1.702	240
S. cerevisiae	2.6×10^{10}	2.4	1.699	1.705	140
Neurospora					
crassa	4×10^{10}	2.0	1.715	1.710	200
N. crassa	2×10^{10}	1			100
Cellular slime mould					
Dictyostelium					
discoideum	3×10^{10}	2.2			120 (3)

*Modified from Birnstiel, Chipchase & Speirs (1971)
(1) Ojha *et al.* (1976)
(2) Jaworski & Horgen (1973)
(3) Firtel & Bonner (1972)

Although a strong correlation between rRNA synthesis and differentiation has been noted in fungi (see Lovett, 1975), it is not known whether rRNA cistrons replicate preferentially during growth or differentiation.

The enzyme reverse transcriptase (RNA-dependent DNA polymerase) has been reported to be of significance not only in oncogenic RNA tumour viruses but also in normal cell development and differentiation (Temin, 1971). Beljanski *et al.* (1974) have provided evidence for the presence of this enzyme in *Echerichia coli*. Brown & Tocchini-Valentii (1972) suggested that ribosomal RNA gene amplification in young oöcytes of *Xenopus laevis* involves RNA-directed DNA synthesis. Further, Mishra, Niu & Tatum

(1975) have suggested the implication of this enzyme in their demonstration of RNA induced transformation in inositol requiring mutants of *Neurospora crass*. Dutta, Beljanski & Bougarel (1975) and Dutta & Bhattacharya (1976) have characterized this enzyme in normal *N. crassa* cell types. The molecular size of the ^{3}H DNA synthesized is around 6S.

The studies on the nuclear and nucleolar RNA (site of ribosomal RNA transcription) have revealed that the precursors are large molecules. The mRNA in the nucleus, called hnRNA (heterogenous nuclear RNA), undergoes several maturation steps before functional mRNA molecules are formed (Perry, Bard, Hames & Kelly, 1974). Unfortunately, there are still no data for heterogenous nuclear RNA metabolism in fungal development.

Ribosomal RNA cistrons are transcribed in a similar manner as polycistronic precursor rRNA which follows through intermediate molecules and matures into 28S and 18S rRNA in higher eukaryotes. The process involves: (a) transcription into a high molecular weight 45S precursor RNA; (b) cleavage of 45S into intermediates including a 32S component; and (c) conversion of the intermediate into 28S and 18S RNA of ribosomes.

An understanding of the processing of rRNA has come from short term labelling experiments. In *Blastocladiella emersonii* (Lovett, 1975), the first precursor to appear has a molecular weight of 2.75×10^6 daltons; upon cleavage this yields molecules of 1.47×10^6 and 0.73×10^6 daltons, the latter being 18S RNA. An additional (1.57×10^6 daltons) intermediate has also been found to exist. The conversion of 1.47×10^6 pre-rRNA to 1.32×10^6 daltons 25S rRNA has been supposed to be a rate limiting step. Retel & Planta (1970) have shown that in *Saccharomyces carlsbergensis* precursors with a size corresponding to sedimentation values of 42, 37, 32 and 29S are formed. Under normal conditions of growth, only a rather small amount of ribosomal rRNA precursors are formed as compared to the animal cells. However, an increase in the cellular amount of ribosomal precursor RNA can be brought about by 'shift up', by treatment with cycloheximide and with a methionine requiring mutant. In *Rhizopus stolonifer*, Roheim, Knight & van Etten (1974), have revealed rapid synthesis of 2.4×10^6 and 1.4×10^6 daltons precursor rRNA which are very prominent during the first 30 min of onset of germination. These subsequently disappear and mature rRNAs with high specific activity appear.

Work on part of the unpublished data submitted here was done under contract No. E(40-1)4182 with the U.S. Energy Research and Development Administration. We are grateful to Professor G. Turian of the University of Geneva for his critical reading of the manuscript.

2.7 References

ADLER, P. N. & HOLT, E. (1975). Mating type and the differentiated state in *Physarum polycephalum*. *Developmental Biology* **43**, 240–53.

BARSTOW, W. E. & LOVETT, J. S. (1975). Formation of gamma particles during zoosporogenesis in *Blastocladiella emersonii*. *Mycologia* **67**, 518–29.

BELJANSKI, M., AARON-DA CUNHA, M. I., BELJANSKI, M. S., MANIGAULT, P. & BOURGAREL, P. (1974). Isolation of the tumor-inducing RNA from oncogenic and nononcogenic *Agrobacterium tumefaciens*. *Proceedings of National Academy of Science of the United States of America* **71**, 1585–9.

BHAGWAT, A. S. & MAHADEVAN, P. R. (1970). Conserved mRNA from the conidia of *Neurospora crassa*. *Molecular and General Genetics* **109**, 152–61.

BHAGWAT, A. S. & MAHADEVAN, P. R. (1973). Differential gene action in *N. crassa*. *Journal of Bacteriology* **113**, 572–5.

BIRNSTIEL, M. L., CHIPCHASE, M. & SPEIRS, (1971). The ribosomal RNA cistrons. *Progress in Nucleic Acid Research Molecular Biology* **11**, 351–89.

BOSTOCK, C. J. (1970). The effects of 2-phenyl ethanol on the DNA synthesis cycle of *Schizosaccharomyces pombe*. *Journal of Cell Science* **7**, 523–30.

BOSTOCK, C. J., DONACHIE, W. D., MASTERS, M. & MITCHISON, J. M. (1966). Synthesis of enzymes and DNA in synchronous cultures of *Saccharomyces pombe*. *Nature, London* **210**, 808–10.

BREWER, E. N. (1972). DNA replication in *P. polycephalum*. *Journal of Molecular Biology* **68**, 401–2.

BRITTEN, R. J. & DAVIDSON, E. H. (1969). Gene regulation for higher cells: a theory. *Science* **165**, 349–57.

BRITTEN, R. J. & DAVIDSON, E. H. (1971). Repetitive and non-repetitive DNA sequences and a speculation on the origin of evolutionary novelty. *Quarterly Review of Biology* **46**, 111–38.

BRITTEN, R. J. & KOHNE, D. E. (1968). Repeated DNA sequences. *Science* **165**, 349–57.

BRITTEN, R. J. & SMITH, J. F. (1970). *Carnegie Institution of Washington Year Book*.

BROWN, D. & TOCCHINI-VALENTI, L. (1972). On the role of RNA in gene amplification. *Proceedings of the National Academy of Science of the U.S.A.* **69**, 1746–8.

BROWN, D. D. & DAVID, J. B. (1968). Specific gene amplification in oöcytes. *Science* **160**, 272–80.

BURKE, D. J., SEALE, T. W. & MCCARTHY, B. J. (1972). Protein and ribonucleic acid synthesis during the diploid life cycle of *Allomyces arbuscula*. *Journal of Bacteriology* **110**, 1065–72.

CANTINO, E. C. (1966). Morphogenesis in aquatic fungi. In *The Fungi*. Vol. 2, pp. 283–337. Edited by G. C. Ainsworth and A. S. Sussman. New York: Academic Press.

CANTINO, E. C. & HORENSTEIN, E. A. (1956). Gamma and cytoplasmic control of differentiation in *Blastocladiella*. *Mycologia* **48**, 443–6.

CANTINO, E. C. & HYATT, M. T. (1953). Phenotypic sex determination in the life history of a new species of *Blastocladiella*: *Blastocladiella emersonii*. *Antoni von Leeuwenhoek* **19**, 25–70.

CANTINO, E. C., LOVETT, J. S., LEAK, L. V. & LYTHGOE, J. (1963). The single mitochondrion, fine structure and germination of the spore of *Blastocladiella emersonii*. *Journal of General Microbiology* **31**, 393–404.

CARNEVALI, F., MORPUGO, G. & TECC, G. (1969). Cytoplasmic DNA from *petite* colonies of *Saccharomyces cerevisae*; a hypothesis on the nature of the mutation. *Science* **163**, 1331–3.

CARTER, B. L. A. (1975). The organization and replication of genetic material in yeast. *Cell* **6**, 259–68.

CHATTOPADHYAY, S. K., KOHNE, D. E., DUTTA, S. K. (1972). Ribosomal RNA genes of *Neurospora*. Isolation and characterization. *Proceedings of National Academy of Sciences, United States of America* **69**, 3256–9.

CHRISTIANSEN, C., LETHBAK, A., STENDERUP, A. & CHRISTIANSEN, G. (1971). Repetitive DNA in yeasts. *Nature, New Biology* **231** 176–7.

CLUTTERBUCK, A. J. (1970). Synchronous nuclear division and septation in *Aspergillus nidulans*. *Journal of General Microbiology* **60**, 133–5.

CULLIS, C. A. & DAVIES, D. R. (1975). Ribosomal DNA in a nuclear satellite of tomato. *Genetics* **81**, 485–92.

DAVIDSON, E. H., GALAU, G. A., ANGERER, R. C. & BRITTEN, R. J. (1975). Comparative aspects of DNA organization in Metazoa. *Chromosoma* **51**, 253–9.

DUSENBERRY, R. L. (1975). Characterization of the genome of *Phycomyces blakesleeanus*. *Biochemica et Biophysica Acta* **378**, 363–77.

DUFFUS, J. H. & MITCHELL, C. J. (1970). Effect of high osmotic pressure on DNA synthesis in fission yeast, *Schizosaccharomyces pombe*. *Experimental Cell Research* **61**, 213–6.

DUTTA, S. K. (1974). Repeated DNA sequences in fungi. *Nucleic Acid Research* **1**, 1441–20.

DUTTA, S. K. & BHATTACHARYA, J. (1976). RNA-directed DNA synthesis in *Neurospora crassa* and characterization of the DNA product. *Genetics* **83**, (Abstracts).

DUTTA, S. K. & CHAUDHURI, R. K. (1975a). Differential transcription of non-repeated DNA during development of *Neurospora crassa*. *Developmental Biology* **43**, 35–41.

DUTTA, S. K. & CHAUDHURI, R. K. (1975b). Transcription of repetitive DNA in Neurospora crassa. Molecular and General Genetics 136, 227–323.

DUTTA, S. K. & OJHA, M. (1972). Relatedness between major taxonomic groups of fungi based on the measurement of DNA nucleotide sequence homology. Molecular and General Genetics 114, 232–40.

DUTTA, S. K. & RAY, R. (1973). Partial characterization of tRNA genes isolated from Neurospora crassa. Molecular and General Genetics 125, 295–300.

DUTTA, S. K. & SCHWARTZ, R. E. (1973). Repeated DNA sequences in Neurospora crassa. Neurospora Newsletter 20, 19.

DUTTA, S. K., BELJANSKI, M. & BOURGAREL, P. (1975). RNA-bound reverse transcriptase in Neurospora crassa. Genetics 78 (Abstracts).

DUTTA, S. K., PENN, S. R., KNIGHT, A. R. & OJHA, M. (1972). Characterization of DNAs from Coprinus lagopus and Mucor azygospora. Experientia 28, 582–4.

EMERSON, R. & WILSON, C. A. (1954). Interspecific hybrids and cytogenetics and cytotaxonomy of Euallomyces. Mycologia 46, 393–434.

ENGELBERG, J. & PEARLMAN, R. E. (1972). The amount of ribosomal RNA genes in Tetrahymena pyriformis in different physiological states. European Journal of Biochemistry 26, 393–400.

ENGELBERG, JR., MOWAT, D. & PEARLMAN, R. E. (1972). Preferential replication of ribosomal RNA genes during a nutritional shift-up in Tetrahymena pyriformis. Biochemica et Biophysica Acta 272, 312–20.

FAHNRICH, P. (1974a). Unterschungen zur entwincklung des Phycomyceten Allomyces arbuscula. I. Einfluss von inhibitoren des protein und nucleinsauren synthese auf die differenzierung von gametangien. Archiv für Mikrobiologie 98, 85–9.

FAHNRICH, P. (1974b). Unterschungen zur entwincklung des Phycomyceten Allomyces arbuscula. II. Einfluss von inhibitoren des protein und nucleinsaure synthese auf die gametogenese. Archiv für Mikrobiologie 99, 147–53.

FAYE, G., FUKUHARA, H., GRANDCHAMP, C., LAZOWSKA, J., CASEY, J., GETZ, G. S., COCKER, K., RABINOWITZ, M., BOLOTIN-FUKUHARA, M., COEN, D., DEUTSCH, J., DUJON, B., NETTER, P. & SLONIMSKI, P. P. (1973). Mitochondrial nucleic acids in petite colony mutants: deletion and repetition of genes. Biochimie 55, 779–92.

FIRTEL, R. A. AND BONNER, J. (1972). Characterization of the genome of the cellular slime mold. Journal of Molecular Biology 66, 339–61.

FIRTEL, R. A. & KINDLE, K. (1975). Structural organization of the cellular slime mold Dictyostelium discoideum: interspersion of repetitive and single copy DNA sequences. Cell 5, 401–11.

FIRTEL, R. A., COCKBURN, A., FRANKEL, G. & HERSCHFIELD, V. (1976). Structural organization of the genome of Dictyostelium discoideum: analysis by Eco R 1 restriction endonuclease. Journal of Molecular Biology 102, 831–52.

FOUQUET, H., BIERWEILER, B. & SAUR, H. W. (1974). Reassociation kinetics of nuclear DNA from Physarum polycelphalum. European Journal of Biochemistry 44, 407–10.

GOLDRING, E. S., GROSSMAN, L. T., KONPINCK, D., CRYER, D. R. & MARMUR, J. (1970). The petite mutation in yeast. Loss of mitochondrial DNA during induction of petites with ethidium bromide. Journal of Molecular Biology 52, 323–35.

GORDON, J. B. (1974). The control of gene expression in animal development. Oxford: Clarendon Press.

GRIMES, G. W., MAHLER, A. R. & PEARLMAN, P. S. (1974). Nuclear gene dosage effects on mitochondrial mass and DNA. Journal of Cell Biology 61, 565–74.

GROS, F. (1974). Control of gene expression in prokaryotic systems. FEBS Letters 40, S19–S21.

GUNSTILLEKE, I. A. U. N., SCAZZOCCHIO, C. & ARST, H. N., JR. (1975). Cytoplasmic and nuclear mutations to chloramphenicol resistance in Aspergillus nidulans. Molecular and General Genetics 137, 269–76.

GUTTES, S. & GUTTES, E. (1968). Regulation of the DNA replication in the nuclei of the slime mold Physarum polycephalum: transplantation of nuclei by plasmoidal coalescence. Journal of Cell Biology 37, 761–72.

HAMMERLING, J. (1966). The role of nucleus in differentiation in Acetabularia. In Molecular and cellular aspects of development. pp. 194–201. Edited by E. Bell, New York: Harper & Row.

HARASHIMA, S., NOGI, Y. & OSHIMA, Y. (1974). The genetic system controlling homothallism in Saccharomyces yeasts. Genetics 77, 639–50.

HOLLOMAN, D. W. (1969). Biochemistry of germination in Peronospora tabacina (Adam.) conidia: evidence for the existence of stable messenger RNA.

Journal of General Microbiology **55**, 267–74.

HOPPER, A. K., KIRSCH, JR. & HALL, B. D. (1975). Mating type and sporulation in yeast. II. Meiosis, recombination and radiation sensitivity in an 'aa' diploid with altered sporulation control. *Genetics* **80**, 61–76.

HOWARD, A. & PELC, S. R. (1953). Synthesis of DNA in normal and irradiated cells and its relation to chromosome breakage. *Heredity* **6**, 261.

HOYER, B. H., MCCARTHY, B. J. & BOLTON, E. T. (1964). A molecular approach to the systematics of higher organisms. *Science* **144**, 959–67.

JAWORSKI, A. J. & HORGEN, P. A. (1973). The ribosomal cistrons of the water mold *Achlya bisexualis*. *Archives of Biochemistry and Biophysics* **157**, 260–7.

KESSEL, J. & ROSENBERGER, R. F. (1968). Regulation and timing of DNA synthesis in the hyphae of *Aspergillus nidulans*. *Journal of Bacteriology* **95**, 2275–81.

KIHLMAN, B. A. & HARTLEY, B. (1967). Sub-chromatoid exchanges and the 'folded fiber' model of chromosome structure. *Hereditas* **57**, 289–94.

KING, S. B. & ALEXANDER, L. J. (1969). Nuclear behaviour, septation and hyphal growth of *Alternaria soloni*. *American Journal of Botany* **56**, 249–53.

KING, T. J. & BRIGGS, R. (1966). Serial transplantation in embryonic nuclei. In *Molecular and cellular aspects of development. pp.* 171–93. Edited by E. Bell. New York: Harper & Row.

KOHNE, D. E. & BYERS, M. J. (1973). Amplification and evolution of DNA sequences expressed as RNA. *Biochemistry* **21**, 2373–8.

LARK, K. G. (1963). Cellular control of DNA biosynthesis. In *Molecular genetics*, Part 1, pp. 153–206. Edited by J. H. Taylor, New York: Academic Press.

LODISH, H. F., FIRTEL, R. A. & JACOBSON, A. (1973). Transcription and structure of the genome of the cellular slime mold *Dictyostelium discoideum*. *Symposia on Quantitative Biology Cold Spring Harbor* **38**, 899–914.

LOVETT, J. (1975). Growth and differentiation of *Blastocladiella emersonii*: cytodifferentiation and the role of ribonucleic acid and protein synthesis. *Bacteriological Reviews* **39**, 345–404.

MCDOUGAL, K. J. (1964). Inheritance of induced and naturally occurring cytoplasmic mutations in *Neurospora crassa*. Ph. D. Thesis, Kansas State University. Ann Arbor, Michigan.

MAHADEVAN, P. R. & BHAGWAT, A. S. (1974). Control of transcription in *Neurospora crassa*. *Basic Life Science* **3**, 223–39.

MATSUMAE, A., MYERS, R. B. & CANTINO, E. C. (1970). Comparative number of particles in the flagellate cells of various species and mutants of *Blastocladiella*. *Journal of General and Applied Microbiology* **16**, 443–53.

MISHRA, N., NIU, M. C. & TATUM, E. L. (1975). Induction by RNA of inositol independence in *Neurospora crassa*. *Proceedings of National Academy of Science of the United States of America* **72**, 642–5.

MITCHISON, J. M. (1971). *The biology of the cell cycle*. London: Cambridge University Press.

MYERS, R. B. & CANTINO, E. C. (1971). DNA profile of the spore of *Blastocladiella emersonii*: evidence for γ-particle DNA. *Archiv für Mikrobiologie* 252–67.

MYERS, R. B. & CANTINO, E. C. (1974). The gamma particle: a study of cell organelle interactions in the development of the water mold *Blastocladiella emersonii*. *Monographs in Developmental Biology*. **8**, 1–117.

NEWLON, C. S., SONENSHEIM, G. E. & HOLT, C. E. (1973). Time of synthesis of genes for ribosomal ribonucleie acid in *Physarum*. *Biochemistry* **12**, 2338–45.

OLIVE, L. S. (1958). On the evolution of heterothallism in fungi. *American Naturalist* **92**, 233–51.

OLSON, L. W. & FULLER, MS. S. (1971). Leucine-lysine synchronization of *Allomyces* germlings. *Archiv für Mikrobiologie* **78**, 76–91.

OJHA, M. & TURIAN, G. (1971). Interspecific transformation and DNA characteristics in *Allomyces*. *Molecular and General Genetics* **78**, 252–67.

OJHA, M., HAMMERLI, J. P. & TURIAN, G. (1976). Ribosomal DNA in *Allomyces*. *Experientia* (Abstract).

PERKINS, D. C. & TURNER, B. C. (1973). Reference strains of *Neurospora intermedia*. *Neurospora Newsletters* **20**, 41.

PERRY, R. P., BARD, E., HAMES, B. D. & KELLEY, D. E. (1974). Messenger RNA: transcription, processing and turnover. In *Biochemistry of cell-regulation of gene expression*. pp. 17–34. Edited by Hidvegi, E. J., Sumegi, J. & Venetianer, P. *Federation of European Biochemical Societies Proceedings of Ninth Meeting*.

PERO, R., LIMA-DE-FARIA, A., STAHLE, U., GRANSTROM, H. & GHATNEKAR, R.

(1973). Amplification of ribosomal DNA in *Acheta*. IV. The number of cistrons for 28S and 18S ribosomal RNA. *Hereditas* **73**, 195–210.

PRESCOTT, D. M. & GOLDSTEIN, L. (1967). Nuclear cytoplasmic interaction in DNA synthesis. *Science* (U.S.) **155**, 469–70.

RAPER, J. R. (1960). The control of sex in fungi. *American Journal of Botany* **47**, 794–808.

RETEL, J. & PLANTA, R. J. (1970). On the mechanism of the biosynthesis of ribosomal RNA in yeast. *Biochimica et Biophysica Acta* **224**, 458–69.

RIZET, G., MARCOUR, D. & ACHERSON, J. (1958). Deux phenomenes d'heredite cytoplasmique chez l'ascomycète *Podospora anserina*. *Bulletin Societe Francaise—Physiologie Vegetale* **4**, 136.

ROHEIM, J. R., KNIGHT, R. H. & VAN ETTEN, J. (1974). Synthesis of ribonucleic acid during the germination of *Rhizopus stolonifer* sporangiospores. *Developmental Biology* **41**, 137–45.

ROSENBERGER, R. F. & KESSEL, M. (1967). Synchrony of nuclear replication in individual hyphae of *Aspergillus nidulans*. *Journal of Bacteriology* **94**, 1464–9.

ROWLANDS, R. T. & TURNER, G. (1974). Physiological and biochemical studies of nuclear and extranuclear oligomycin resistant mutants of *Aspergillus nidulans*. *Molecular and General Genetics* **132**, 73–88.

SAGER, R. (1972). *Cytoplasmic genes and organelles*. New York: Academic Press.

SCHULTZ, J. (1965). In *Genetic control of differentiation*. pp. 116–47. Brookhaven National Laboratory, Upton, New Jersey.

SHERMAN, F. & SLONIMSKY, P. P. (1964). Respiration deficient mutants of yeast. *Biochimica et Biophysica Acta* **90**, 1–15.

STORCK, R. & ALEXOPOULOS, C. J. (1970). DNA of fungi. *Bacteriological Reviews* **37**, 126–54.

SPARROW, F. K. (1960). *Aquatic Phycomycetes*. Ann Arbor, Michigan: University of Michigan Press.

SUBAK-SHARPE, J. H. (1958). A closed system of cytoplasmic variations in *Aspergillus glaucus*. *Proceeding of Royal Society Series B* **148**, 355–9.

TAYLOR, A. (1969). 'Masked' messenger RNA and the determination process in embryonic development. In *Gentics and developmental biology*. pp. 107–127. Edited by H. J. Teas. Kentucky: University of Kentucky Press.

TEMIN, H. M. (1971). Endogenous RNA-directed DNA polymerase activity in uninfected chicken embryos. *Proceedings National Academy of Science, U.S.A.* **69**, 1550–3.

THOMAS, D. S. & MULLINS, J. T. (1967). Role of enzymatic wall-softening in plant morphogenesis. *Science* **156**, 84–5.

TRUESDALE, L. C. & CANTINO, E. C. (1970). Decay of gamma particles in germinating spores of *Blastocladiella emersonii*. *Mycologia* **67**, 518–29.

TURIAN, G. (1969). *Differenciation fongique*. Paris: Masson et Cie.

TURIAN, G. (1975). Differentiation in *Allomyces* and *Neurospora*. *Transactions of the British Mycological Society* **64**, 367–80.

TURIAN, G. & KELLENBERGER, E. (1956). Ultrastructure du corps paranucleaires des mitochondries et de la membrane nucleare des gametes d' *Allomyces macrogynus*. *Experimental Cell Research* **11**, 417–22.

TURIAN, G. & VISHWANATHAN, M. A. (1966). Facteurs cytoplasmiques et differenciation sexuelle des *Allomyces*. Inhibition selectives par les acridines et l'Actenomycine D. *Pathologia Microbiologia* **29**, 705–29.

TURIAN, G., OJHA, M., SCHEPS, R. & OULEVEY, N. (1969). Oxidative deficiencies and some ultrastructural features of acridine-induced male strains of *Allomyces arbuscula*. *Archiv für Mikrobiologie* **69**, 92–100.

VILLA, V. D. & STORCK, R. (1968). Nucleotide composition of nuclear and mitochondrial DNA of fungi. *Journal of Bacteriology* **96**, 184–90.

WAERING, (1971). Some aspects of differentiation in plants. *Symposium Society of Experimental Biology* **25**, 323–44.

WETMUR, J. G. & DAVIDSON, N. (1968). Kinetics of renaturation of DNA. *Journal of Molecular Biology* **31**, 349–70.

WILLIAMSON, D. H. (1966). Nuclear events in synchronously dividing yeast cultures. In *Cell synchrony*. pp. 81–101. Edited by I. L. Cameron and G. M. Padilla. New York: Academic Press.

WILSON, E. B. (1931). Sperm forming materials in scorpions. *Journal of Morphology* **52**, 429–83.

WU, J. R., HURN, J. & BONNER, J. (1972). Size and distribution of the repetitive segments of the Drosophilia genone. *Journal of Molecular Biology* **64**, 211–9.

ZELLWEGER, A. RYSER, U. & BRAUN, R. (1972). Ribosomal genes of *Physarum*: their isolation and replication in mitotic cycle. *Journal of Molecular Biology* **64**, 681–91.

CHAPTER 3

The Cytology of Hyphal Tip Growth

S. N. GROVE

3.1 Introduction

Hyphal tip growth is a highly polarized type of cell growth which is responsible for morphogenesis in the filamentous fungi. The fundamental relationship between this apical growth and hyphal morphogenesis leads to the conclusion that 'the key to the hypha lies in the apex' (Robertson, 1965). Thus, whether one is ultimately interested in inhibition or stimulation of growth in medically or economically important species, an understanding of hyphal tip growth is necessary. Because of the polarized organization of growing hyphae, many fundamental questions about cell growth in general can be most easily answered using growing hyphae as experimental systems.

The phenomenon of hyphal tip growth has long been observed by light microscopy and repeatedly established by a varieity of techniques (Ward, 1888; Reinhardt, 1892; Burgeff, 1915; Smith, 1923; Castle, 1958; Robertson, 1965, 1968; Bracker, 1967; Marchant & Smith, 1968; Grove, Bracker & Morré, 1970; Katz & Rosenberger, 1970, 1971; Gooday, 1971; Bartnicki-Garcia, 1973), but the mechanisms of apical extension are incompletely understood. Hyphal tips have been searched by a variety of techniques for clues to their ability to support apical growth. Information from the light microscope with selective staining methods, phase-contrast microscopy, electron microscopy, autoradiography, biochemical methods, and physiological studies have all added to our current understanding of the cytology of hyphal tip growth.

An early clue that the protoplasm of growing hyphal tips is structurally and functionally different from that in the rest of the mycelium was the discovery of an apical iron-haemotoxylin positive area (Brunswick, 1924) designated the Spitzenkörper. Girbardt (1955, 1957), using phase contrast optics, found a dark Spitzenkörper in growing hyphae of *Aspergillus*, *Penicillium* and several Basidiomycetes. He found that growth was closely associated with the presence of the Spitzenkörper. When growth is arrested it vanishes and it reappears again just before growth resumes. The position of this area in the apical dome is related to the subsequent direction of growth. A central position precedes straight growth while eccentric location is followed by a change in direction of growth. These observations have been confirmed for many Basidiomycetes, Ascomycetes, and Deuteromycetes. In Zygomycetes a dark cap-like (cresentic) region is seen but in Oömycetes no similar dark region has been found by phase contrast observation (Fig. 3.1; McClure, Park & Robinson, 1968; Girbardt, 1969; Grove & Bracker, 1970). Thus, evidence from light microscopy suggests that in the septate fungi the Spitzenkörper is closely associated with apical growth.

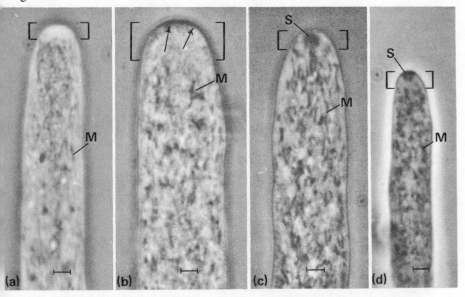

Fig. 3.1 Phase contrast micrographs of growing hyphae representing (a) Oömycetes (*Pythium aphanidermatum*); (b) Zygomycetes (*Gilbertella persicaria*); (c) Ascomycetes (*Neurospora crassa*); and (d) Basidiomycetes (*Rhizoctonia solani*). Compare the organization in the apical zones (brackets). The apex of *G. persicaria* contains numerous small particles which are easily resolved (arrows). Smaller particles, not resolved here, can often be observed in the apices of hyphae from the other groups also. The growing apices of ascomycete and basidiomycete hyphae have a Spitzenkörper (S) while in oömycete hyphae no similar phenomenon can be demonstrated. Contributed by Grove & Bracker, 1970. Scale lines = 1 μm.

Key to labelling of Figures: ER, *endoplasmic reticulum*; G, *Golgi cisternae*; M, *mitochondria*; MT, *microtubules*; MV, *microvesicles*; N, *nucleus*; PM, *plasma membrane*; R, *ribosomes*; S, *Spitzenkörper*; SW, *spore wall*; V, *vesicles*; W, *hyphal wall*.

An additional feature of growing hyphal tips is the presence of small, just resolvable, particles which can be seen in favourable preparations of growing hyphae observed with phase contrast optics (Fig. 3.2; McClure *et al.*, 1968; Grove & Bracker, 1970). These particles seem to originate in the subapical region and then migrate to the apex where they accumulate in great numbers in the apical cytoplasm. Such particles are most easily seen in large hyphae, particularly in some zygomycete species, but also in Oömycetes and the septate fungi. In the latter group, the vesicles are joined in the apex by the Spitzenkörper (Fig. 3.1c & d). In all fungi investigated, an apical zone of cytoplasm is nearly devoid of mitochondria, nuclei and most

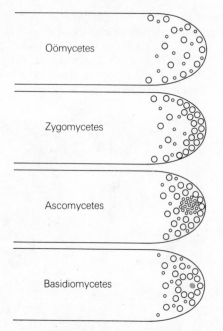

Fig. 3.2 Summary diagram comparing the principal forms of apical organization in hyphae based on evidence from light and electron microscopy.

other inclusions except the small particles, which are in constant rapid motion. Among the fungi examined, a common feature of growing hyphae is an accumulation of these very small particles in the apex where extension is occurring. The much greater resolution of the transmission electron microscope allows one to visualize these particles as well as determine their origin and fate in the growing hyphal tip.

3.2 Fine Structure of Growing Hyphal Tips

The protoplasm in tips of growing hyphae is characteristically divided into an apical zone containing numerous cytoplasmic vesicles and a subapical zone containing other organelles and inclusions in addition to vesicles (McClure *et al.*, 1968; Girbardt, 1969; Grove *et al.*, 1970; Grove & Bracker, 1970). The apical accumulation of cytoplasmic vesicles in hyphae can be easily observed by electron microscopy. This is true also for other stages of the fungal life cycle where tip growth occurs. Often a cluster of vesicles is seen prior to the initiation of growth in spores (Bracker, 1971; Grove, 1972, 1976; Hess *et al.*, 1975), cysts (Grove, 1971), direct sporangial germination (Grove *et al.*, 1970), and at sites where clamp connections develop (Girbardt, 1969; Raudaskoski, 1970). These vesicles are present in the apices to the exclusion of most other types of cytoplasmic organelles and inclusions (Girbardt, 1969; Grove *et al.*, 1970; Grove & Bracker, 1970). The fine structural equivalent of the small apical particles must then be these vesicles, because of their location in hyphal tips that were growing when prepared for electron microscopy and because no other inclusions of

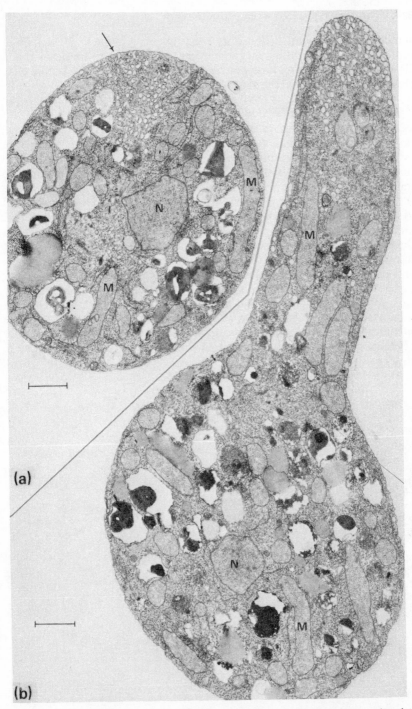

Fig. 3.3(a) A transmission electron micrograph of *Pythium aphanidermatum* showing an encysted zoospore just prior to germ tube emergence. Note the large cluster of cytoplasmic vesicles near the wall (arrow). Scale line = 1 μm. (b) A germling of *Pythium aphanidermatum* showing vesicles in apex of the germ tube. Scale line = 1 μm. Contributed by Grove & Bracker, 1977 (*For key to lettering see Fig. 3.1*)

similar size are found in these growing apices. Although the details of protoplasmic organization vary somewhat from species to species within a class of fungi, several major types can be recognized (Fig. 3.2). Within each class, most species have the same type of organization.

Types of protoplasmic organization

OÖMYCETOUS ORGANIZATION   Prior to the initiation of the mycelial stage in Oömycetes by germination of encysted zoospores a large number of cytoplasmic vesicles accumulate near the site of subsequent germ-tube initiation at the periphery of the cyst (Fig. 3.3a). The cyst wall expands to initiate a germ-tube (Fig. 3.3b). As the germ-tube develops into a new hypha the apical cluster of vesicles is continuously maintained by a supply of vesicles from the subapical cytoplasm (Fig. 3.4; Hoch & Mitchell, 1972; Gotelli, 1974). In the subapical region of a hypha numerous dictyosomes of the Golgi apparatus are observed in association with cytoplasmic vesicles of the type found in hyphal apices. Between these dictyosomes and the apex and often near the periphery of the cytoplasm rows or small groups of vesicles are commonly seen (Fig. 3.4). These cytoplasmic vesicles probably correspond to the particles that are seen migrating to the apex from the subapical region in hyphae viewed by phase contrast microscopy. The subapical region of growing oömycete hyphae also contains numerous other organelles and inclusions typical of all cells (Grove et al., 1970; Grove & Bracker, 1970).

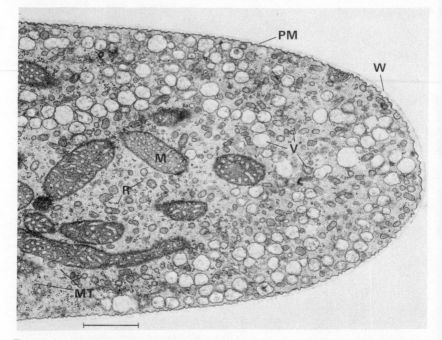

Fig. 3.4 A near median thin section of a hyphal tip of *Pythium aphanidermatum*. The apical region contains many vesicles distributed more or less randomly. Ribosomes, mitochondria and other organelles are scarce or excluded from the apical cytoplasm. Contributed by Grove & Bracker, 1970. Scale line = 1 μm. (*For key to lettering see Fig. 3.1*)

The oömycete type of tip organization can be distinguished from other types by the loose accumulation of a large number of apical vesicles usually of two kinds (Fig. 3.3b, 3.4). One is larger and contains a moderately electron-opaque material of granular appearance while the other is much smaller (microvesicle) and its contents are more compact. The vesicles have no preferred location in the apex but seem to be distributed randomly in the apical cytoplasm. This is in contrast to a more definite arrangement in other groups of fungi.

ZYGOMYCETOUS ORGANIZATION In Zygomycetes, the mycelial stage is initiated by the germination of sporangiospores which occurs by a two-step process, an initial period of spherical growth followed by the initiation of a germ-tube (germination stage I and II Bartnicki-Garcia, Nelson & Cota-Robles, 1968b). During the initiation of germ-tubes in the sporangiospores

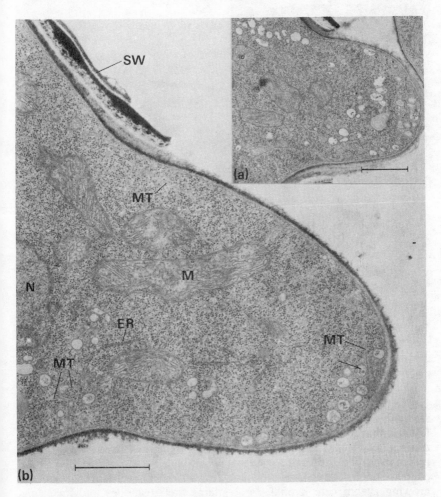

Fig. 3.5 The germ tubes of two germlings of *Choanephora curcurbitans* showing apical vesicles. An occasional multivesicular body (arrow) occurs among the vesicles. Contributed by Grove, 1976. Scale lines = 1 μm. (*For key to lettering see Fig. 3.1*)

of *Gilbertella persicaria* (Bracker, 1971), *Phycomyces blakesleanus*, *Choanephora cucurbitans*, and *Rhizopus stolonifer* (Grove, 1976) a few cytoplasmic vesicles are found in a layer next to the plasma membrane. As the germ-tube elongates and becomes a hypha, progressively larger numbers of vesicles are found in the apical zone (Fig. 3.5). The sporangiospore germ-tubes (Jeffries & Young, 1976), appresoria, and infection pegs (P. Jeffries and T. W. K. Young, personal communication) of *Piptocephalis unispore* also contain numerous apical vesicles during their infection of *Cokeromyces recurvatus*.

After the mycelium is established, the growing hyphae at the periphery of the colony extend rapidly and contain large numbers of apical vesicles (Fig. 3.6; Thornton, 1968; Girbardt, 1969; Grove & Bracker, 1970; Syrop, 1973;

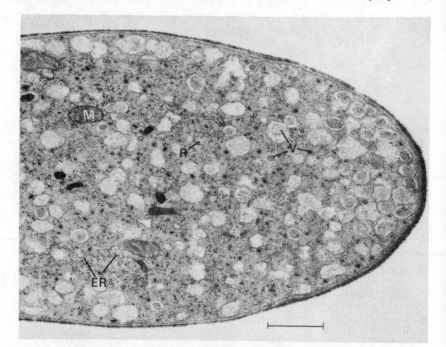

Fig. 3.6 Hyphal tip of *Gilbertella persicaria*. The apical cytoplasm contains many vesicles. Note the band of vesicles near the apical plasma membrane. Ribosomes and other organelles are scarce in apex. Contributed by Grove & Bracker, 1970. Scale line = 1 μm. (*For key to lettering see Fig. 3.1*)

Grove, 1976). These vesicles are of two sizes, a large 0.2–0.4 μm diameter size with granular contents of medium electron opacity and a small 0.1 μm or less size that contains an electron opaque granule. The large vesicles form a distinct layer near the plasma membrane at the apex corresponding to the crescent-shaped cluster of particles seen in living hyphae observed by phase contrast microscopy (Grove & Bracker, 1970). The subapical region of growing zygomycete hyphal tips contains the organelles and inclusions typically found in protoplasm. Thus, the hyphal tips of this group also are organized into an apical zone containing numerous vesicles, and a subapical region which contains the synthetic components for generation of materials for tip growth.

ASCOMYCETOUS ORGANIZATION   Hyphal tip growth in the ascomycete life cycle is initiated by spore germination. In conidia of *Aspergillus parasiticus* the emerging germ-tube contains a few cytoplasmic vesicles near the plasma membrane at the apex (Grove, 1972). As the new hypha extends, the group of apical vesicles increases in number and resides in a distinct ribosome-free region of cytoplasm. Whether the size of the cluster of apical vesicles is related to the speed of growth or some other factor has not yet been determined, but after a mycelium is established the growing hyphae contain many more apical vesicles (Fig. 3.7).

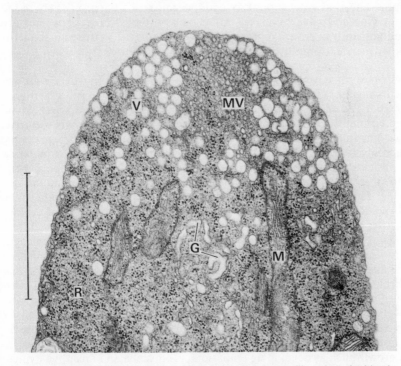

Fig. 3.7 Median longitudinal section of a hyphal tip of *Aspergillus niger*. Amidst the cluster of apical vesicles (V) is an aggregation of microvesicles (MV) and a few ribosomes. Note the abrupt decline in the concentration of ribosomes (R) in the apical zone as compared to the subapical cytoplasm. Smooth-surfaced Golgi cisternae (G) and mitochondria (M) also occur in the subapical region. Contributed by Grove & Bracker, 1970. Scale line = 1 μm.

The growing hyphae from the mycelia of many Ascomycetes have been examined with sufficient care to determine the protoplasmic organization in detail. In *Aspergillus niger* (Fig. 3.7; McClure *et al.*, 1968; Grove & Bracker, 1970), *Sclerotinia sclerotiorum* (Maxwell, Williams & Maxwell, 1972), *Neurospora crassa* (Fig. 3.9; Grove & Bracker, 1970; Collinge & Trinci, 1974), *Ascodesmis nigricans* (Grove & Bracker, 1970), and *Fusarium oxysporium* (Grove & Bracker, 1970) a relatively ribosome-free region in the apex contains a cluster of vesicles about 0.1 μm in diameter. Within this cluster of vesicles is a distinct area where microvesicles (<0.05 μm) and small membranous tubules are found along with a few ribosomes. The apical vesicles are generally excluded from this area so that, within the apex, this central region containing microvesicles is enclosed by a layer of apical vesicles.

The subapical zone, as noted in other groups, contains the protoplasmic components typical of living cells. Numerous groups of vesicles, like those in the apex, are found in ribosome-free regions of cytoplasm which contain single cisternae of the Golgi apparatus (Fig. 3.8). The long narrow mitochondrial profiles are usually oriented parallel to the long axis of the hypha. The moderate amounts of endoplasmic reticulum profiles are of the sheet type and have numerous attached ribosomes. Free ribosomes are predominantly in groups representing polyribosomes. An occasional microbody, vacuole, or multivesicular body also may be observed.

BASIDIOMYCETOUS ORGANIZATION   The initiation of hyphal tip growth occurs regularly in the basidiomycete life cycle not only in spore germination, but also each time a clamp connection forms between adjacent hyphal compartments. Germination of uredospores of *Puccinia graminis* var. *tritici* is accompanied by an accumulation of cytoplasmic vesicles in the apex of the emerging germ-tube (Fig. 3.11; Hess *et al.*, 1975). This initial accumulation of vesicles presumably develops into the highly structured apical zone similar to those found in growing hyphae of other Basidiomycetes and similar fungi such as *Armillaria mellea* (Fig. 3.12) and *Rhizoctonia solani* (Grove & Bracker, 1970). In these fungi a large group of apical vesicles encloses a vesicle-free region in which a dense core can be observed (Fig. 3.12). A large cluster of vesicles can also be observed in the hyphal apices of *Polystictus versicolor* (Fig. 3.13; Girbardt, 1969) and *Volvariella volvacea* (Tanaka & Chang, 1972). In dikaryotic hyphae of *Polystictus versicolor* (Girbardt, 1969) and *Schizophyllum commune* (Raudaskoski, 1970) cytoplasmic vesicles aggregate at the site where a clamp connection subsequently develops. The elongating hook retains a cluster of vesicles during its development (Fig. 3.11). No doubt a cluster of apical vesicles is characteristic of many other stages, yet to be examined, where morphogenesis involves tip growth such as developing epibasidia, promycelia and receptive hyphae in the Basidiomycetes and trichogynes in the Ascomycetes. In all species studied, the apical vesicles are apparently derived from special structures in the subapical cytoplasm.

The subapical zones of *Armillaria mellea* hyphae contain ribosome-free regions in which groups of vesicles are associated with tubular membranous structures that correspond to single Golgi cisternae (Grove & Bracker, 1970). Girbardt (1969) provides serial section views through such a Golgi cisterna in *Polystictus versicolor*. The similarity of these membranes to dictyosome cisternae in other organisms can best be seen in face views where anastomosing tubules of smooth membrane and their associated vesicular profiles are evident. The subapical region in *A. mellea* contains a full complement of other organelles characteristic of hyphal cytoplasm. The mitochondrial profiles are often long and narrow, being most numerous in

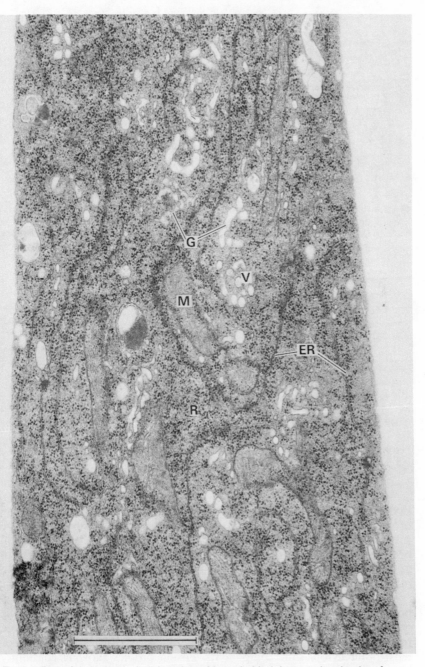

Fig. 3.8 Part of the subapical region about 20 μm behind the apex in a hypha of *Aspergillus niger*. Several Golgi cisternae (G) and their associated vesicles (V) occur in zones of exclusion that contain few ribosomes (R). Endoplasmic reticulum (ER) and mitochondria (M) are common in the subapical region. Contributed by Grove & Bracker, 1970. Scale line = 1 μm.

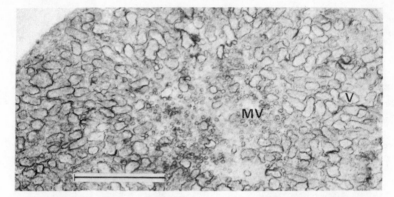

Fig. 3.9 Part of the apical cytoplasm in a hypha of *Neurospora crassa* showing an aggregation of microvesicles surrounded by numerous apical vesicles. Contributed by Collinge & Trinci, 1974. Scale line = 1 µm. (*For key to lettering see Fig. 3.1*)

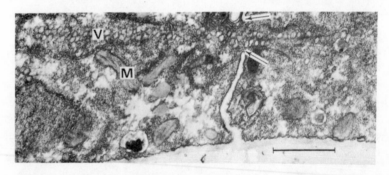

Fig. 3.10 Median longitudinal section of subapical region of a hypha of *N. crassa* showing rows of vesicles that traverse the septal pore (arrows). Contributed by Trinci & Collinge, 1973. Scale line = 1 µm. (*For key to lettering see Fig. 3.1*)

that part of the subapical zone which is adjacent to the apical zone. Cisternae of rough endoplasmic reticulum become prominent farther back in the subapical zone. Both free and bound ribosomes are predominantly organized into polyribosomes of up to 10 or 12 subunits. Ribosomes are not as abundant in the Basidiomycetes as they are in the cytoplasm of some other groups.

From the discussion above it is clear that growing hyphal tips are highly polarized cells which are distinctly organized, on the basis of cytoplasmic components, into an apical zone and a subapical zone. The characteristic feature of the apical cytoplasm is an accumulation of vesicles and the lack of most other cytoplasmic components. Fusion profiles showing continuity between vesicle membranes and plasma membranes are often found on the curved portion of the hyphal apex (Fig. 3.4, 3.12; Grove & Bracker, 1970). This type of organization is typical of tip growing systems in non-fungal organisms as well and may be a general adaptation for this phenomenon.

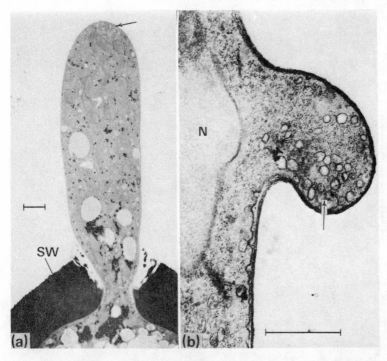

Fig. 3.11(a) Germling from uredospore of *Puccinia graminis* var. *tritici* showing a group of vesicles (arrow) in apex of germ-tube. Contributed by Hess *et al.*, 1975. Scale line = 1 μm. (b) Hook formation in dikaryotic hypha of *Schizophyllum commune*. Numerous vesicles (arrow) occur in apex of hook. Contributed by Raudaskoski, 1970. Scale line = 1 μm. (*For key to lettering see Fig. 3.1*)

The subapical zone contains representatives of all protoplasmic components including nuclei, mitochondria, ribosomes, and an endomembrane system (Girbardt, 1969; Grove & Bracker, 1970; Grove *et al.*, 1970; Bracker, 1974) composed of rough endoplasmic reticulum, smooth-surfaced cisternae, smooth-surfaced tubules, and cytoplasmic vesicles. Stacks of cisternae (dictyosomes) in the Oömycetes, and single cisternae in the other fungi examined, produce the cytoplasmic vesicles. Each site of vesicle production resides in a zone of exclusion from which ribosomes and other cytoplasmic components are usually excluded. Vesicles of the apical type are found throughout the subapical and apical cytoplasm unassociated with either the Golgi apparatus or the apical cluster. Often these are arranged in longitudinal rows near the lateral plasmalemma (Fig. 3.4, 3.5a; Grove *et al.*, 1970; Grove & Bracker, 1970) or traversing septal pores (Fig. 3.10; Trinci & Collinge, 1973).

These observations support a model for hyphal tip growth which depends on the production of vesicles and their migration to the apex, where they are involved in making new hyphal wall and plasma membrane. Additional support for this model comes from a number of experimental studies which provide evidence for the function of many hyphal components.

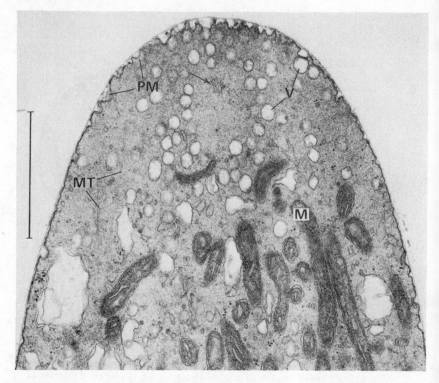

Fig. 3.12 Median longitudinal section through the apical zone of a hypha of *Armillaria mellea*. A vesicle-free region among the apical vesicles contains a small electron-opaque core (arrow). Numerous invaginations of plasma membrane (PM) at apex may represent sites where vesicles (V) have fused. Contributed by Grove & Bracker, 1970. Scale line = 1 μm. (*For key to lettering see Fig. 3.1*)

3.3 Experimental Examination of Growing Hyphae

The nature of growing hyphal tips has been studied by a variety of methods including cytochemistry, autoradiography, physiology and biochemistry. Information from these investigations complements our knowledge of form and structure by providing information about the function of complete hyphae or their components. For this discussion, we are primarily interested in the intrahyphal components which are most closely associated with hyphal tip growth, the apical vesicles and the endomembranes from which they originate, as well as the apical portion of the intact hypha.

Cytochemical methods for the localization of polysaccharides indicate that apical vesicles, the maturing face of dictyosomes and the hyphal wall contain these as constituents. Species for which this has been demonstrated include *Saprolegnia ferax* (Heath, Gay & Greenwood, 1971), *Pythium aphanidermatum* (Grove, 1971), *Achlya bisexualis* (Fig. 3.14; Dargent, 1975), and *Sclerotium* spp. (Fig. 3.15; R. Dargent, personal cummunication). Similar vesicles in the postmeiotic basidium of *Coprinus lagopus* (McLaughlin, 1972) are also polysaccharide positive. Verification of the

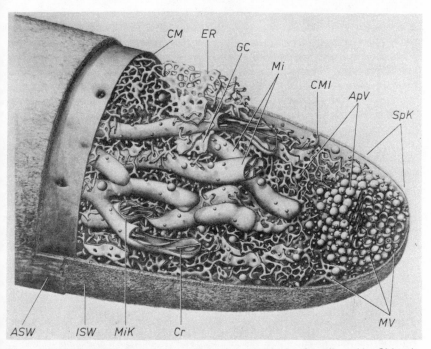

Fig. 3.13 Model of the apical region of *Polystictus versicolor*. Contributed by Girbardt, 1969. Scale line = 1 μm. (*For key to lettering see Fig. 3.1*)

exact nature of the chemical constituents which react in these studies must await biochemical characterization of isolated and purified vesicles.

Enzyme cytochemistry has provided evidence for the activities of several enzymes in the vesicles and dictyosomes of fungal hyphae. Dictyosomes and vesicles similar to apical vesicles give a positive reaction for cellulase activity in *Achlya ambisexualis* (Fig. 3.16; Nolan & Bal, 1974). In *A. bisexualis*, Dargent (1975) found alkaline phosphatase activity in the maturing face of dictyosomes and in cytoplasmic vesicles (Fig. 3.17). The contents of the small vesicles (90 nm) and the dictyosome cisternae have a uniform reaction, while in the larger vesicles (240 nm) an uneven reaction is observed. In germlings of *Phytophthora palmivora*, part of the apical population gives a positive reaction for acid phosphatase activity (Fig. 3.18; M. Powell, personal communication). The maturing face of the dictyosomes in *A. bisexualis* also have acid phosphatase activity (Fig. 3.17b; Dargent & Denise, 1976). Clearly, this technique has great potential for locating the sites of various enzyme activities, but it has not yet been fully utilized for many enzymes because of technical difficulties.

Recently an integrated ultrastructural–biochemical investigation of microfibril biogenesis has revealed the fine structure of isolated chitin synthase particles (chitosomes) from *Mucor rouxii* (Bracker, Ruiz-Herrera & Bartnicki-Garcia, 1977). These organelles are spheroidal (Fig. 3.19a) and usually 45–65 nm in diameter. In thin sections, isolated chitosomes appear as microvesicular structures with a membrane-like shell 6.5–7.0 nm thick (Fig. 3.19b). Morphologically similar structures are found in intact yeast cells

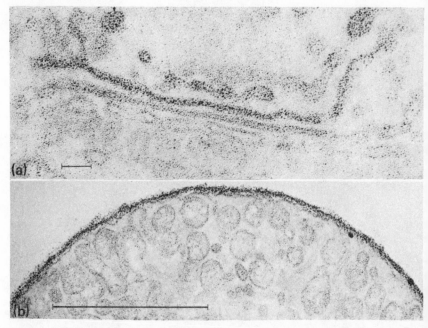

Fig. 3.14 Dictyosome and apical vesicles, respectively, of *Achlya bisexualis* treated by a method which localized polysaccharides by depositing silver at reactive sites. The maturing face of the dictyosome, the vesicles, the plasma membrane, and the hyphal wall show deposits indicating polysaccharides. Contributed by Dargent, 1975. Scale line = 0.1 μm for (a) and 1 μm for (b).

of *Mucor rouxii*. These may be either in multivesicular bodies (Fig. 3.20b) or free in the cytoplasm (Fig. 3.20a). Such microvesicles are also common in hyphae of other fungi (Fig. 3.5b). In the presence of substrate and activators, isolated chitosomes undergo a series of transformations and give rise to microfibrils similar to those found in hyphal walls (Bartnicki-Garcia, 1973). Presumably the chitosome is the cytoplasmic container and conveyor of chitin synthase *en route* to its destination at the cell surface (Bracker *et al.*, 1976).

Another technique with great potential for aiding our understanding of the interaction of structure and function in hyphal growth is autoradiography. At the light microscope level such studies have demonstrated that the structural materials for hyphal wall synthesis are deposited at the apex in *Mucor rouxii* (Bartnicki-Garcia & Lippman, 1969; Bartnicki-Garcia, 1973); *Aspergillus nidulans* (Fig. 3.21; Katz & Rosenberger, 1970, 1971); and *Phytophthora parasitica*, *Neurospora crassa*, and *Schizophyllum commune* (Gooday, 1971). When the walls of *Mucor rouxii* germlings are freed of cytoplasm and incubated with uridine diphosphate-N-acetyl-[^{3}H]-D-glucosamine, a precursor for chitin, the pattern of labelling shows that chitin synthase operates at the apex (Bartnicki-Garcia, 1973). Information available from autoradiographic experiments, then, points to the apex as the site where new material for hyphal tip growth is deposited.

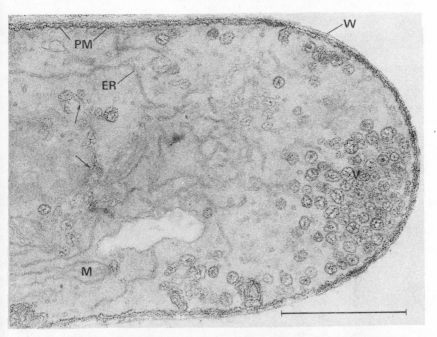

Fig. 3.15 Evidence for polysaccharide in hyphal tip of *Sclerotium* spp. Note that the hyphal wall (W), plasma membrane, apical vesicles (V), and some tubules or cisternae (arrows) in the subapical cytoplasm show the deposits. Endoplasmic reticulum and mitochondrial membranes are relatively unreactive. Micrograph contributed by R. Dargent. Scale line = 1 μm. (*For key to lettering see Fig. 3.1*)

Hyphal tip growth may be inhibited or interrupted by a variety of environmental changes including osmotic shock with distilled water or dilute solutions of various chemicals (Katz & Rosenberger, 1971; Bartnicki-Garcia & Lippman, 1972; Bartnicki-Garcia, 1973), the presence of cycloheximide, an inhibitor of protein synthesis (Katz & Rosenberger 1971; Sternlicht, Katz & Rosenberger, 1973) and temperature changes (Hawker & Syrop, 1973). Even though hyphae stop elongating when cycloheximide is added to growing cultures of *Aspergillus nidulans*, polysaccharide synthesis continues and the walls become thicker. Similar results are observed when tip growth is interrupted by osmotic shock. Since production of essential materials is not inhibited along with tip growth, one might suspect that the transport of materials from the sources within the hypha to the site of deposition at the apex is the key process which is interrupted.

Based on a study of the bursting tendencies of growing hyphal tips when subjected to environmental changes, Bartnicki-Garcia & Lippman (1972) concluded that the activity of enzymes involved in weakening the hyphal wall was somehow increased. These lytic enzymes are likely to be carried in vesicles along with synthetic enzymes and wall precursors (Girbardt, 1969; Grove *et al.*, 1970; Cortat, Matile & Wiemken, 1972). Lytic enzymes are presumably required to weaken the structural microfibrils in the wall so that the hyphal surface can be enlarged. It is interesting to note that an 'apical

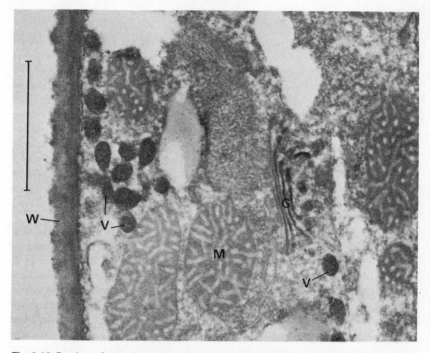

Fig. 3.16 Portion of cytoplasm near the hyphal tip of *Achlya ambisexualis* which was treated to localize cellulase activity. Electron-opaque reaction product occurs on the dictyosome cisternae and secretory vesicles similar to apical vesicles. Contributed by Nolan & Bal, 1974. Scale line = 1 μm. (*For key to lettering see Fig. 3.1*)

pore' can be demonstrated in the wall of enzymatically treated hyphal tips of *Polystictus versicolor* (Strunk, 1968). During normal tip growth a delicate balance must exist between deposition of new material, synthetic activity, lytic activity, and turgor pressure which provides the force for elongation.

3.4 Conclusions

Model for hyphal tip growth

Based upon the preceding information and the literature cited, a model for the mechanism of hyphal tip growth can be outlined (see Bartnicki-Garcia, 1973 for a detailed version). An early step in this mechanism is the incorporation of wall subunits, lytic enzymes, and synthetic enzymes into a vesicle or vesicles at specialized regions of the endomembrane system (Golgi cisternae) in the subapical region of a hyphal tip. The vesicles are then transported through the cytoplasm until they reach the apex where they join many other vesicles each carrying its load of enzymes and/or wall precursors. When a vesicle comes in contact with the plasma membrane at the apex, the two membranes fuse, thus depositing the vesicle contents into the wall region at the apex. The lytic enzymes attack the structural microfibrils of the wall making it unable to withstand the high turgor pressure. The microfibrils become stretched out or separated which increases the surface

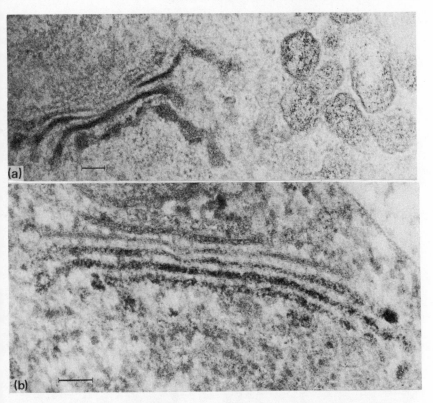

Fig. 3.17(a) Part of a hypha of *Achlya bisexualis* treated to localize alkaline phosphatase activity. Reaction product is deposited on the cisternae at the maturing face of the dictyosome and on the secretory vesicles. Contributed by Dargent, 1975. (b) Part of the cytoplasm in a hypha of *Achlya bisexualis* treated to localize acid phosphatase activity. Reaction product is deposited most heavily on the cisternae at the maturing face of the dictyosome. Contributed by Dargent & Denisse, 1976. Scale lines = 0.1 μm.

area of the wall. Synthetic enzymes then use the wall precursors to produce new microfibrils and extend the old ones. This process is repeated each time a vesicle or the several vesicles deposit the necessary components in the wall region.

Pitfalls in interpretation of cytological information

Erroneous interpretations about hyphal tip structure and the functions of the cytoplasmic components may arise in various ways including faulty methods, incorrect analysis of results, and/or failure to correlate information derived from more than one technique.

No doubt one of the most common pitfalls in cytological research is the production of artifacts in preparing fungal material for observation. A primary consideration should be to ensure that the first change in conditions for the growing hyphae is fixation itself. Many investigators have not found apical vesicles in germlings or hyphae simply because fixation followed a

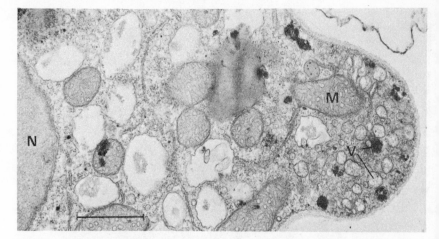

Fig. 3.18 The emerging germ tube on a cyst of *Phytophthora palmivora* treated to localize acid phosphatase activity. Reaction product is deposited on some of the population of vesicles accumulated in the apex. Unpublished micrograph contributed by M. Powell and C. E. Bracker. Scale line = 1 μm. (*For key to lettering see Fig. 3.1*)

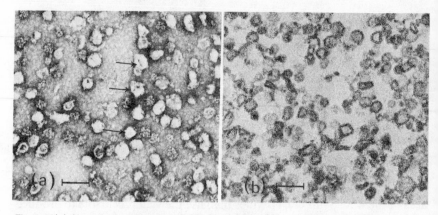

Fig. 3.19(a) Negatively stained sample of isolated chitin synthase preparation showing chitosome (arrows). (b) Thin section of fixed and embedded pellet of isolated chitin synthase preparation. Contributed by Bracker, Ruiz-Herrera & Bartnicki-Garcia, 1976. Scale lines = 1 μm.

previous interruption of growth due to washing the cells or some other change in the environment. An illustration of this problem is the so-called apical corpuscle, which is alleged to be responsible for apical growth in germlings of *Mucor rouxii* (Bartnicki-Garcia, Nelson & Cota-Robles, 1968*a*). The cells were washed three times prior to fixation thereby ensuring that apical growth had ceased some minutes before the time of fixation. In *Rhizopus stolonifer* structures similar to apical corpuscles can be found at random locations on the spore wall but not in growing germ-tube apices

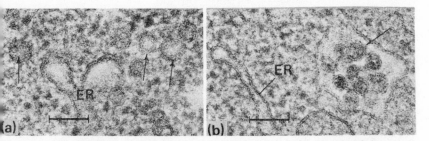

Fig. 3.20. Thin sections of intact yeast cells of *Mucor rouxii* showing microvesicles morphologically similar to isolated chitosomes inside a multivesicular body (arrowed in (a)) or free in the cytoplasm (arrowed in (b)). Contributed by Bracker, Ruiz-Herrera & Bartnicki-Garcia, 1976. Scale lines = 1 μm.

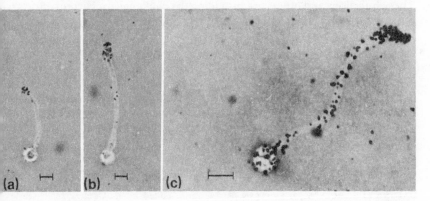

Fig. 3.21 Radioautograms of hyphal walls of *Aspergillus nidulans* incubated with D [1-^{3}H] galactose for 10 min, 20 min, and 40 min respectively. Contributed by Katz & Rosenberger, 1970. Scale line = 1 μm.

(Grove, 1976). Cell growth is apparently maintained by a delicate balance between numerous cellular mechanisms. This balance is so easily upset that one should exercise the utmost care in doing any kind of prefixation manipulation of the hyphae.

Even if the proper preparation techniques are used, one may collect incorrect or at least biased information about the specimen. The unusual feature will often draw more attention than a consistent characteristic. For example, lomasomes which have not been demonstrated in living hyphae and are subject to variation in size, shape, or distribution in hyphae depending on the method of preparation have received credit for numerous possible functions including storage, secretion, and synthesis of various materials such as hyphal wall. Many errors of interpretation could be eliminated if investigations included correlations between several methods of observing the specimens. Where possible, light microscopic observation of living materials greatly facilitates electron microscopy. More than one type of fixation is also recommended as well as a combination of scanning electron microscopy, transmission electron microscopy, and freeze-etching when available and applicable. One should never be satisfied to collect information from the vantage point of only one method.

Major unanswered questions

Several major questions have emerged which have at best only partial answers. Although a great deal of circumstantial evidence points to enzymes and wall precursors as components of the apical vesicles in growing hyphae, the exact function of these structures in hyphal tip growth must await biochemical characterization. A related problem is the nature of the Spitzenkörper which is so closely associated with tip growth in septate fungi. Additional questions include the mechanism of vesicle migration from the subapical zone to the apical zone in growing hyphae, the nature of the specificity of the vesicle membranes for the apical plasma membrane, and the molecular events in the construction of hyphal wall at the growing apex. These questions and others are providing the challenge for current efforts to more completely understand hyphal tip growth in the filamentous fungi.

3.5 References

BARTNICKI-GARCIA, S. (1973). Fundamental aspects of hyphal morphogenesis. *Symposium Society General Microbiology* **23**, 245–67.

BARTNICKI-GARCIA, S. &. LIPPMAN, E. (1969). Fungal morphogenesis: cell wall construction in *Mucor rouxii*. *Science, New York* **165**, 302–8.

BARTNICKI-GARCIA, S. &. LIPPMAN, E. (1972). The bursting tendency of hyphal tips of fungi: presumptive evidence for a delicate balance between wall synthesis and wall lysis in apical growth. *Journal of General Microbiology* **13**, 487–500.

BARTNICKI-GARCIA, S., NELSON, N. &. COTA-ROBLES, E. (1968a). A novel apical corpuscle in hyphae of *Mucor rouxii*. *Journal of Bacteriology* **95**, 2399–402.

BARTNICKI-GARCIA, S., NELSON N. & COTA-ROBLES, E. (1968b). Electron microscopy of spore germination and cell wall formation in *Mucor rouxii*. *Archiv für Mikrobiologie* **63**, 242–55.

BRACKER, C. E. (1967). Ultrastructure of fungi. *Annual Review of Phytopathology* **5**, 343–74.

BRACKER, C. E. (1971). Cytoplasmic vesicles in germinating spores of *Gilbertella persicaria*. *Protoplasma* **72**, 381–97.

BRACKER, C. E. (1974). The endomembrane system of fungi. *Proceedings of the Eighth International Congress on Electron Microscopy, Canberra*, Vol II, 558–9.

BRACKER, C. E., RUIZ-HERRERA, J. & BARTNICKI-GARCIA, S. (1976). Structure and transformation of chitin

synthetase particles (chitosomes) during microfibril synthesis. *In vitro. Proceedings of the National Academy of Science, USA* **73**, 4570–4.

BRUNSWICK, H. (1924). Untersuchungen über Geschlechts und Kernverhaltnisse bei der Hymenomyzetengattung *Coprinus*. In *Botanische Abhandlungen*, Vol. 5, edited by K. Goebel. Jena: Gustav Fisher.

BURGEFF, H. (1915) Untersuchungen über Variabilität, Sexualität und Erblichkeit bei *Phycomyces nitens* Kuntze. *Flora*, N.F. **108**, 353–488.

CASTLE, E. S. (1958). The topography of tip growth in a plant cell. *Journal of General Physiology* **41**, 913–26.

COLLINGE, A. J. & TRINCI, A. P. J. (1974). Hyphal tips of wild-type and spreading colonial mutants of *Neurospora crassa*. *Archiv für Mikrobiologie* **99**, 353–68.

CORTAT, M., MATILE, P. & WIEMKEN, A. (1972). Isolation of glucanase-containing vesicles from budding yeast. *Archiv für Mikrobiologie* **82**, 189–205.

DARGENT, R. (1975). Sur l' ultrastructure des hyphes en croissance de l' *Achlya bisexualis* Coker. Mise en évidence d'une sécrétion polysaccharidique et d'une activité, phosphatasique alkaline dans l'appareil de Golgi et au niveau des vésicules cytoplasmiques apicales. *Comptes Rendus* **280**, 1445–8.

DARGENT, R. &. DENISSE, J. (1976). Sur l'ultrastructure apicale de l'*Achlya bisexualis* Coker. Mise en evidence d'une activité phosphatasique acide. *Comptes Rendus Hepdomadaires des Séances de L'Academie des Sciences, Paris* **282**, 1601–4.

GIRBARDT, M. (1955). Lebend-beobachdungen an *Polystictus versicolor* (L.). *Flora* **142**, 540–63.

GIRBARDT, M. (1957). Der Spitzenkörper von *Polystictus versicolor* (L.). *Planta, Berlin* **50**, 47–59.

GIRBARDT, M. (1969). Die Ultrastruktur der Apikalregion von Pilzhyphen. *Protoplasma* **67**, 413–41.

GOODAY, G. W. (1971). An autoradiographic study of hyphal growth of some fungi. *Journal of General Microbiology* **67**, 125–33.

GOTELLI, D. (1974). The morphology of *Lagenidium callinectes*. I. Vegetative development. *Mycologia* **64**, 639–47.

GROVE, S. N. (1971). *Ph.D. Thesis, Purdue University.*

GROVE, S. N. (1972). Apical vesicles in germinating conidia of *Aspergillus parasiticus*. *Mycologie* **64**, 638–41.

GROVE, S. N. (1976). Form and function in zygomycete spores. In. *The fungal spore: form and function.* pp. 559–592. Edited by D. J. Weber and W. M. Hess. New York: John Wiley & Sons, Inc.

GROVE, S. N. & BRACKER, C. E. (1970). Protoplasmic organization of hyphal tips among fungi: vesicles and Spitzenkörper. *Journal of Bacteriology* **104**, 989–1009.

GROVE, S. N. & BRACKER, C. E. (1977). Protoplasmic changes during zoospore encystment and cyst germination in *Pythium aphanidermatum*. *Exp. Mycol.* In press.

GROVE, S. N., BRACKER, C. E. & MORRÉ, D. J. (1970). An ultrastructural basis for hyphal tip growth in *Pythium ultimum*. *American Journal of Botany* **59**, 245–66.

HAWKER, L. E. & SYROP, M., (1973). The effect of low temperatures on ultrastructure of hyphae and young zygospores of *Rhizopus sexualis* (Smith) Callen. *Protoplasma* **78**, 57–68.

HEATH, I. B., GAY, J. L. & GREENWOOD, A. D. (1971). Cell wall formation in the Saprolegniales: cytoplasmic vesicles underlying developing walls. *Journal of General Microbiology* **65**, 225–32.

HESS, S. L., ALLEN, P. J., NELSON, D. & LESTER, H. (1975). Mode of action of methyl cis-ferulate, the self-inhibitor of stem rust uredospore germination. *Physiological Plant Pathology* **5**, 107–12.

HOCH, H. C. & MITCHELL, J. E. (1972). The

ultrastructure of zoospores of *Aphanomyces euteiches* and of their encystment and subsequent germination. *Protoplasma* **75**, 113–38.

JEFFRIES, P. & YOUNG, T. W. K. (1976). Physiology and fine structure of sporangiospore germination in *Piptocephalis unispora* prior to infections. *Archives of Microbiology* **107**, 99–107.

KATZ, D. &. ROSENBERGER, R. F. (1970). The utilization of galactose by an *Aspergillus nidulans* mutant lacking galactose phosphate-UDP glucose transferase and its relation to cell wall synthesis. *Archiv für Mikrobiolgie* **74**, 41–51.

KATZ, D. & ROSENBERGER, R. F. (1971). Hyphal wall synthesis in *Aspergillus nidulans*: effect of protein synthesis inhibition and osmotic shock on chitin insertion and morphogenesis. *Journal of Bacteriology* **108**, 184–90.

MARCHANT, R. & SMITH, D. G. (1968). A serological investigation of hyphal growth in *Fusarium culmorum*. *Archiv für Mikrobiologie* **63**, 85–94.

MAXWELL, D. P., WILLIAMS, P. H. & MAXWELL, M. D. (1972). Studies on the possible relationships of microbodies and multivesicular bodies to oxalate, endopolygalacturonase, and cellulose (C_x) production by *Sclerotinia sclerotiorum*. *Canadian Journal of Botany* **50**, 1743–8.

MCCLURE, W. K., PARK, D. & ROBINSON, P. M. (1968). Apical organization in the somatic hyphae of fungi. *Journal of General Microbiology* **50**, 177–82.

MCLAUGHLIN, O. J. (1972). Golgi apparatus in the postmeiotic basidium of *Coprinus lagopus*. *Journal of Bacteriology*. **110**, 739–42.

NOLAN, R. A. & BAL, A. K. (1974). Cellulase localization in hyphae of *Achlya ambisexualis*. *Journal of Bacteriology* **117**, 840–3.

RAUDASKOSKI, M. (1970). Occurrence of microtubules and microfilaments, and origin of septa in dikaryotic hyphae of *Schizophyllum commune*. *Protoplasma* **70**, 415–22.

REINHARDT, M. O. (1892). Das Wachsthum der Pilzhyphen. *Jahrbucher für Wessenschaftliche Botanik* **23**, 479–566.

ROBERTSON, N. F. (1965). The fungal hypha. *Transactions of the British Mycological Society* **48**, 1–8.

ROBERTSON, N. F. (1968). The growth process in fungi. *Annual Review of Phytopathology* **6**, 115–36.

SMITH, J. H. (1923). On the apical growth of

fungal hyphae. *Annals of Botany* **37**, 341–3.

STERNLICHT, E., KATZ, D. & ROSEN-BERGER, R. F. (1973). Subapical wall synthesis and wall thickening induced by cycloheximide in hyphae of *Aspergillus nidulans. Journal of Bacteriology* **114**, 819–23.

STRUNK, C. (1968). Zur Darstellung des Apicalporus bei *Polystictus versicolor. Archiv für Mikrobiologie* **60**, 255–61.

SYROP, M. (1973). The ultrastructure of the growing regions of aerial hyphae of *Rhizopus sexualis* (Smith) Callen. *Protoplasma*, **76**, 309–14.

TANAKA, K. & CHANG, S.-T. (1972). Cytoplasmic vesicles in the growing hyphae of the Basidiomycete, *Volvariella volvacea. Journal of General and Applied Microbiology* **18**, 165–79.

THORNTON, R. M. (1968). The fine structure of *Phycomyces*. II. Organization of the stage I sporangiophore apex. *Protoplasma* **66**, 269–85.

TRINCI, A. P. J. & COLLINGE, A. J. (1973). Structure and plugging of septa of wild type and spreading colonial mutants of *Neurospora crassa. Archiv für Mikrobiologie* **91**, 355–64.

WARD, H. M. (1888). A lily-disease. *Annals of Botany* **2**, 317–82.

CHAPTER 4

The Enzymology of Hyphal Growth

G. W. GOODAY

4.1 Introduction

The physical manifestation of a fungus—its form of growth—is the result of the activities of the enzymes that produce its hyphal surfaces. The fungus, saprophyte or pathogen, continually invades new substrates as it grows, adding new material to the continuum of the mycelium, and leaving older parts as vacuolated hyphal shells. Clearly, a multitude of metabolic activities contribute to the synthesis of components of cytoplasm, nuclei, and mitochondria, but these are common to all eukaryotic cells, and so are not dealt with here. This chapter is concerned with those enzymic activities that are solely involved in the formation of the structure of the hypha. This means the biosynthesis of the wall and membrane, which in effect means a consideration of cell wall synthesis, as very little is known yet of the biochemistry of fungal cell membranes.

4.2 Localization of Synthesis

Vegetative hyphae grow by deposition of new material at their extreme apices. This observation has been confirmed for a wide range of fungi, using several techniques. Reinhardt (1892) gives a detailed account of his experiments showing this apical growth. Particles of red lead placed on the tip of the apex moved apart laterally as new material was incorporated, while those placed further back showed progressively less movement, until there was no movement where the hyphae were cylindrical. That this apical growth does indeed represent deposition of newly synthesised material can be shown by autoradiography. Radioactive substrates such as *N*-acetylglucosamine, glucose, galactose, and acetate, are very quickly incorporated into chitin, glucans and other components in the extreme apical walls of growing hyphae, and a 'pulse-chase' experiment results in a sub-apical annulus of radioactivity (Fig. 4.1; Bartnicki-Garcia & Lippman, 1969; Katz & Rosenberger, 1970*a*, 1971*a*, Gooday, 1971; Galun, 1972).

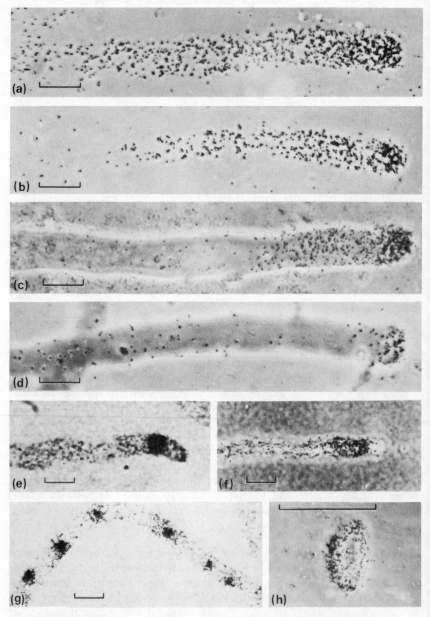

Fig. 4.1 Light microscopic autoradiographs of deposition of wall material. (a)–(f)
Growing hyphal tips showing apical incorporation: *Schizophyllum commune*
incubated 10 min with (a) (^{3}H)-glucose, (b) N-(^{3}H)-acetylglucosamine, (f) (^{3}H)-
glucose followed by 10 min 'chase' with glucose; *Neurospora crassa*
incubated 1 min with (c) (^{3}H)-glucose, (d) N-(^{3}H)-acetylglucosamine;
Phytophthora parasitica incubated 10 min with (e) (^{3}H)-glucose followed by
10 min 'chase' with glucose (Gooday, 1971). (g) *Phytophthora parasitica*
incubated 1 min with (^{3}H)-glucose, showing sub-apical formation of cellulose
wall ingrowths (Gooday & Hunsley, 1971). (h) *Neurospora crassa* incubated
10 min with N-(^{3}H)-acetylglucosamine showing septal formation (Hunsley &
Gooday, 1974). Scale bar, 10 μm.

As shown by Reinhardt (1892) and further calculated by Burnett (1976), the transformation from an apical dome to a cylinder implies a steep fall-off of rate of incorporation of material from the apex backwards. Quantitative autoradiography allows the kinetics of apical incorporation to be estimated. For example the silver grain density over the apical 1 μm of hyphae of *Neurospora crassa* was more than 50 times higher than that over the hyphae 50 to 75 μm behind the apex following a one minute incubation in N-(^{3}H)-acetylglucosamine (Gooday, 1971). Bartnicki-Garcia (1973a) has shown that the relative incorporation of N-acetylglucosamine at a point on the nearly hemispherical apex of hyphae of *Mucor rouxii* closely follows the relative value of cos α, where α is the angle between the radius at that point and the axis of the hypha. Collinge & Trinci (1974) point out that the relative incorporation of N-acetylglucosamine at the apices of hyphae of *Neurospora crassa* parallels the density of the vesicles seen by electron microscopy, which they can predict using the equation $Vp = R^2/r^2$, where Vp is the relative vesicle concentration, R is the final radius of the hypha and r is the radius at the point in question. The relative rates of deposition of wall material are clearly under very precise control, as evidenced by the precision with which the polarity of growth and the resultant hyphal diameter are maintained in any particular cell, even though different hyphae in any one culture may have very different diameters. It would appear that once a particular diameter of hyphal cylinder is produced this then physically establishes the diameter of the new synthesized wall produced by the apex, perhaps partly by acting as a template, and partly by controlling the translocation forward of precursor materials.

Electron microscopy of hyphal apices shows the characteristic aggregation of vesicles (McLure, Park & Robinson, 1968; Girbardt, 1969; Grove & Bracker, 1970), which must have major roles in the apical biosynthesis of the hyphal wall, but the details of these roles remain elusive.

Sub-apical deposition of wall components without hyphal extension does occur in vegetative hyphae. Electron micrographs show a progressive thickening of the walls of some fungi (Hunsley, 1973; Trinci & Collinge, 1975), and increase in the size of chitin and cellulose microfibrils (Hunsley & Burnett, 1968). Autoradiographs often show a uniform low incorporation of radioactive precursors into sub-apical hyphal walls (*e.g.* Hunsley & Gooday, 1974). In *Neurospora crassa*, at least one component, the glycoprotein reticulum, is exclusively formed sub-apically in actively growing hyphae (Hunsley & Burnett, 1970). Katz & Rosenberger (1971a) and Sternlicht, Katz & Rosenberger (1973) have shown that treatments that completely inhibit apical extension of *Aspergillus nidulans*, namely osmotic shock or application of cycloheximide, result in the replacement of apical deposition by a uniform deposition of polymers along the hyphae. For at least an hour after treatment with cycloheximide of concentration sufficient to inhibit protein synthesis almost completely, there was little change in the rate of synthesis of wall polysaccharides. The result after five hours was a doubling in thickness of the hyphal walls.

Hyphae in specialized structures can show intercalary growth. The best example is the elongation of cells of the stipes of Agaric fruit-bodies. The stipe cells of *Coprinus cinereus* have a high activity of chitin synthase, show intercalary elongation, and incorporate radioactive N-acetylglucosamine uniformly along their length (Gooday, 1973, 1975).

Autoradiographs of vegetative cultures do reveal localized wall deposition other than at the leading hyphal apices, most obviously in the sites of developing branches, where an apex must form *de novo* in the lateral hyphal walls. Although the proportions of components are probably different in septum and wall it is to be expected that there are some polymers common to both. Thus the septum of *Neurospora crassa* is very rich in chitin, but lacks or has little of the glucan or glycoprotein reticulum of the lateral walls (Fig. 4.1; Hunsley & Gooday, 1974), and the septum of *Schizophyllum commune* has chitin and β-1, 3-, and β-1,6-glucan in common with the lateral walls but lacks or has little of the α-1,3-glucan (Wessels & Marchant, 1974). *Phytophthora parasitica* hyphae have localized wall ingrowths leading to 'false septa' that are rich in cellulose (Fig. 4.1; Gooday & Hunsley, 1971). Therefore septa must be formed by the activity of enzymes that are also involved in the apical deposition of hyphal wall components.

In the growing hypha, no other metabolic activities have yet been shown to have the extreme apical localization shown by the incorporation of wall precursors. Autoradiographs suggest that both RNA and protein synthesis have higher activities in sub-apical regions (Zalokar, 1959, 1965; Nishi, Yanagita & Maruyama, 1968). Zalokar used a range of cytochemical techniques and concluded that apical growth was supported by a supply of precursors from the hyphae at a considerable distance behind the tips. This distance corresponds to the peripheral growth zone, the minimum width of hyphae that can be excised with no diminution of apical extension rate (Trinci, 1971). Turian (1975), using cytochemical methods, describes the hyphal apex, lacking mitochondria, as being a site of strong reducing power and high fermentative ability, while Chang & Trevithick (1974) suggest it as the chief site of secretion of exoenzymes such as invertase; and so it must not be thought of solely in terms of being the site of wall deposition.

4.3 Synthesis of Wall Components

Chitin synthesis

Chitin synthase is the best characterized enzyme involved in the biosynthesis of hyphal walls. First described from *Neurospora crassa* by Glaser and Brown (1957), it has now been obtained from a wide taxonomic spectrum of chitin-containing fungi (Table 4.1). Its properties are remarkably similar from all of these sources, bearing in mind that conditions of preparation and assay and expression of results vary widely between different authors. It would seem that the constraints put on the enzyme so that it makes a crystalline homopolymer under tight spatial and temporal control have not allowed any significant variations in its properties.

The equation of the enzyme activity is:

$$\text{UDP-GlcNAc} + (\text{GlcNAc})_n \xrightarrow{\text{Mg}^{2+}} (\text{GlcNAc})_{n+1} + \text{UDP}$$

This reaction will be discussed with particular reference to the enzyme from *Coprinus cinereus* (Gooday & de Rousset-Hall, 1975; de Rousset-Hall & Gooday, 1975) but the features seem common to the enzyme from all sources. The sugar donor, UDP-*N*-acetylglucosamine, is a major component of the nucleotide pools of chitin-producing fungi. It allosterically activates the enzyme, and so its availability may be a major factor in

Table 4.1 Properties of some chitin synthase preparations from fungi

Organism	Cell wall category*	'Km' or $[S]_{0.5}$ for UDP-GlcNAc mmol l^{-1}	'Km' for GlcNAc activation mmol l^{-1}	Optimum pH	Optimum temperature °C	Optimum $[Mg^{2+}]$ mmol l^{-1}	% Inhibition by 0.5 mmol l^{-1} UDP	Reference
Chytridiomycetes								
Allomyces macrogynus	V. Chitin–glucan	1.2	—	7.8	30	—	—	Porter & Jaworski (1966)
Blastocladiella emersonii	V. Chitin–glucan	1.8–4.1	3–4	8	20–25	6–20	—	Camargo et al. (1967)
Zygomycetes								
Mucor rouxii	IV. Chitin–chitosan	0.5–1.5	12.5	6.5	24–27	30	53	McMurrough & Bartnicki-Garcia (1971)
Phycomyces								
blakesleeanus	IV. Chitin–chitosan	0.6	—	6.5	28	20	75	Jan (1974)
Mortierella vinacea	IV. Chitin–chitosan	1.8	—	6	32	20	—	Peberdy & Moore (1975)
Ascomycetes								
Saccharomyces cerevisiae	VI. Mannan–glucan	0.6–0.9	4.7	6.2	37	10	62	Keller & Cabib (1971)
Neurospora crassa	V. Chitin–glucan	1–2	4.5	7.5	27	—	—	Glaser & Brown (1957)
Basidiomycetes								
Coprinus cinereus	V. Chitin–glucan	0.9	0.6	8	30	30	44	Gooday & de Rousset-Hall (1975)

* From Bartnicki-Garcia (1970).

controlling enzyme activity. The co-substrate, the acceptor or 'primer', is more problematic. The purified enzyme from *C. cinerous* is active without added acceptor and has no obvious endogenous primer. Once an oligomeric glycan is produced this must act as acceptor, but how does it start? A likely explanation is that the enzyme preparation carries tightly-bound primer molecules. *In vivo* there is presumably usually no shortage of part-formed or recently lysed chitin chains to act as acceptors. However, the zoospore of *Blastocladiella emersonii* is an apparently naked protoplast which rapidly produces a chitin wall on settlement, but Myers & Cantino (1974) conclude that there is as yet insufficient evidence to decide whether or not it does have small amounts of pre-formed chitin.

The glycan product has been identified as macromolecular chitin by a variety of chemical and physical techniques. It is insoluble in water, alkali and weak acids, and is hydrolyzed by strong acids to glucosamine and by chitinase to diacetylchitobiose. The most convincing evidence for its identity with the chitin of the cell wall comes from the demonstration that the enzyme product can be visualized as crystalline microfibrils (Ruiz-Herrera & Bartnicki-Garcia, 1974; Ruiz-Herrera, Sing, van der Woude & Bartnicki-Garcia, 1975) (Fig. 4.2). These authors solubilized the enzyme from a

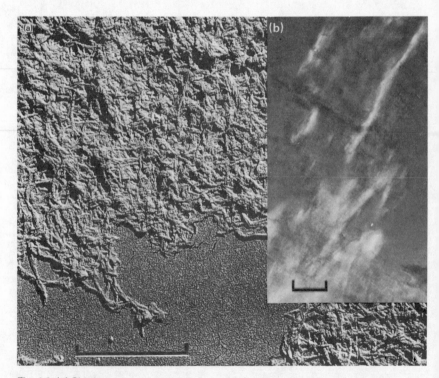

Fig. 4.2 (a) Shadow cast electron micrograph of chitin microfibrils synthesized by a solubilized preparation of chitin synthase from *Mucor rouxii* (Ruiz-Herrera & Bartnicki-Garcia, 1974; copyright 1974 by the American Association for the Advancement of Science). Scale bar, 1 μm. (b) Polarized light micrograph showing birefringence of chitin synthesized by a solubilized preparation of chitin synthase from *Coprinus cinereus*. Scale bar, 10 μm.

membrane fraction of *Mucor rouxii* by treatment at 0°C with the substrate, UDP-*N*-acetylglucosamine, and activator, *N*-acetylglucosamine. Subsequent incubation gave a product giving a powder X-ray diagram characteristic of crystalline α-chitin. Electron microscopy showed single microfibrils emanating from granules, strongly suggesting a granular enzyme feeding out the macromolecular product. The purified preparation of chitin synthase from *Coprinus cinereus* gives a product which is birefringent in polarized light, as is crystalline chitin *in vivo*, and has an identical infra-red spectrum to pure chitin (Gooday, 1977) (Fig. 4.2).

Diacetylchitobiose has also been reported as a minor product of the enzyme preparations from *Blastocladiella emersonii* and *Mucor rouxii* (Camargo, Dietrich, Sonneborn & Strominger, 1967; McMurrough & Bartnicki-Garcia, 1971). In the former case is was formed particularly at low concentrations of UDP-*N*-acetylglucosamine, and in the latter case only in the presence of high *N*-acetylglucosamine concentrations.

The other enzyme product is uridine diphosphate. This is an inhibitor of chitin synthase activity, having an apparent value of K_i of 0.6 mmol 1^{-1} with the enzyme from *Coprinus cinereus*. The enzyme preparations commonly contain a nucleoside diphosphatase that acts on the UDP to give UMP, which is not nearly so inhibitory. This enzyme from *C. cinereus* has a K_m of 0.35 mmol 1^{-1} UDP, and clearly its activity must be considered in any overall view of the kinetics of chitin deposition.

In common with other enzymes with a nucleotide sugar as substrate, chitin synthase preparations have an absolute requirement for Mg^{2+} as a co-substrate. Mn^{2+} can often replace the Mg^{2+}, but as maximal activity is usually obtained between 10 and 30 mmol l^{-1}, magnesium ions must be the active cations *in vivo*. The reported pH optima vary between 6 and 8, but the values should not be compared too closely, as they have been obtained in a wide variety of conditions.

The monomer of chitin, *N*-acetylglucosamine, is an allosteric activator of chitin synthase. This might be considered an observation of purely laboratory significance, as this sugar apparently does not occur in measurable quantities in fungal cells. However, the one place that it will occur is at the site of enzymic lysis of chitin, by a concerted action of chitinase and diacetylchitobiase. Thus it could activate chitin synthase at those sites where lysis and synthesis must go hand in hand, namely at hyphal branch formation and almost certainly at the growing hyphal apex. The dimer, diacetylchitobiose, and higher chitin oligomers, both the products of chitinase activity, also activate the enzyme, although usually not as powerfully as the monomer. It has been suggested that these molecules could act directly as primers for the enzyme activity, but the evidence has not been conclusive, and in most enzyme preparations this has been shown not to occur to any significant extent.

When chitin synthase has been prepared from hyphae or tissues actively synthesizing chitin, the major enzyme activity has appeared in cell fractions containing membranes. In some cases considerable activity appears in the fraction rich in cell wall debris, but this is bound to contain membrane fragments, and in the case of *Coprinus cinereus* the enzyme activity associated with the walls can be solubilized by treatment with digitonin solutions, which disrupt eukaryotic membranes. Jan (1974) further

separated the membrane fractions from *Phycomyces blakesleeanus* by centrifugation in a sucrose density gradient, and found that the chitin synthase
activity correlated well with the activity of 5'-nucleotidase, which is considered a marker enzyme for the cell membrane of eukaryotes. Durán,
Bowers and Cabib (1975) prepared purified cell membranes from cells of
Saccharomyces cerevisiae, and these contained nearly all of the recoverable
chitin synthase activity. Their technique involved a membrane-reinforcing
pre-treatment of protoplasts with the lectin, concanavalin A, prior to
disruption. Microscopy showed that this resulted in cell membrane ghosts.
These could be labelled with (^{3}H)-concanavalin A or ^{125}I, and in both cases
subsequent recovery of the radioactivity coincided with chitin synthase
activity. The active site of the enzyme was located on the inner face of the
cell membrane, as treatment with the fixative, glutaraldehyde, before lysis
did not affect enzyme activity, but treatment after lysis resulted in a great
reduction of activity. In the light of these experiments, nearly all previous
reports are consistent with the chitin synthase activity being located chiefly
or perhaps exclusively in cell membrane. One exception is the zoospore of
Blastocladiella, discussed later.

Electron microscopic autoradiography, using (^{3}H)-*N*-acetylglucosamine
as substrate, has never revealed any evidence for cytoplasmic pre-
polymerization of chitin, as the silver grains are always almost totally
localized over the cell membrane and newly synthesized wall (Gooday,
1977). Thus it is reasonable to picture chitin synthase *in vivo* as an integral
protein in the cell membrane, incorporating substrates from the cytosol into
polymer in the wall. The solubilized enzyme preparation from *Coprinus
cinereus* can be purified to give an aggregate with molecular exclusion
chromatographic behaviour consistent with a molecular weight of several
million, but which can be dissociated by high ionic strength to sub-units
behaving with a molecular weight of 150 000. One might imagine an
assemblage of these sub-units grouped together in the membrane to give rise
directly to the ordered arrangement of chitin chains in a single crystalline
microfibril.

Chitosan, a homopolymer of β-1, 4-linked gluosamine, is a major characteristic component of the walls of the Zygomycetes but there is no obvious
substrate for its polymerization *de novo*. Araki & Ito (1974) have shown
that it can be biosynthesized by deacetylation of pre-formed chitin by an
enzyme extract from *Mucor rouxii*. The major source of the enzyme was the
cell supernatant fraction, but there was detectable activity in the culture
filtrate. The enzyme specifically deacetylated chitin, and had no activity on
N-acetylglucosamine, but showed increasing activity with increasing chain
length of chitin oligomers. Acetate inhibited its activity (50% inhibition by
20 mmol l^{-1}). If, as seems likely from these results, the conversion from
chitin to chitosan takes place outside the cell membrane, then there are
interesting problems in the control of this process. Which chitin microfibrils
are deacetylated? Can the external medium specifically affect the rate of
deacetylation and so lead to altered ratios of chitin: chitosan?

Synthesis of UDP-N-acetylglucosamine

Chitin formation is dependent on a supply of UDP-*N*-acetylglucosamine.
Four enzyme activities make up the biosynthetic pathway to this nucleotide

sugar *de novo* from glutamine and fructose-6-phosphate:

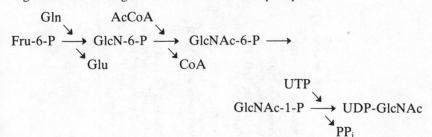

An accessory pathway occurs when pre-existing chitin is hydrolyzed or when exogenous *N*-acetylglucosamine is supplied:

$$\text{Chitin} \longrightarrow \text{GlcNAc-GlcNAc} \longrightarrow \text{GlcNAc} \overset{\text{ATP}}{\underset{\text{ADP}}{\longrightarrow}} \text{GlcNAc-6-P}$$

The enzymes involved in these pathways have not been studied in as much detail as has chitin synthase itself. However, although the resultant UDP-*N*-acetylglucosamine is not uniquely the substrate of chitin synthase (also being a substrate for glycoprotein and polygalactosamine biosyntheses), nevertheless this must be its major fate in most chitin-containing fungi. So a knowledge of the regulation of these pathways is likely to be of importance, in understanding chitin biosynthesis.

The glutamine:fructose-6-phosphate aminotransferase has been described from several fungi (Table 4.2). The enzyme occurs in the soluble cytoplasmic fraction. In common with other eukaryotic systems, UDP-*N*-acetylglucosamine, the end-product of the pathway, is a powerful specific inhibitor of the enzyme from *Blastocladiella emersonii* and *Neurospora crassa*. Norrman, Giddings & Cantino (1975) report it as competitive with respect to fructose-6-phosphate (with a K_i of 18.7 μmol l^{-1}), but noncompetitive with respect to glutamine (K_i of 190 μmol l^{-1}); while Endo, Kakiki & Misato (1970a) report it as noncompetitive with respect to both substrates, for example giving 40% inhibition at 50 μmol l^{-1} in the presence of 6 mmol l^{-1} fructose-6-phosphate and 9 mmol l^{-1} glutamine. This feedback inhibition explains the observation made by Endo *et al.* (1970a) that inhibition of chitin synthesis by addition of polyoxin D to *N. crassa* did not lead to any increase in endogenous level of UDP-*N*-acetylglucosamine, whereas feeding with exogenous glucosamine did give an increase, which was accentuated in the presence of polyoxin D. The exogenous glucosamine must bypass the feedback control step. Of the subsequent enzymes in the pathway, only glucosaminephosphate acetyl transferase is at all well characterized from fungi (Table 4.2).

The other source of UDP-*N*-acetylglucosamine is via phosphorylation of *N*-acetylglucosamine. The enzyme concerned, *N*-acetylglucosamine kinase, has a role *in vivo* in recycling the monomer formed by lysis of pre-existing chitin, but also can be induced by exogenous *N*-acetylglucosamine (Table 4.2).

Table 4.2 Properties of enzymes involved in the biosynthesis of UDP-acetylglucosamine in fungi

Enzyme	Source	Properties*	References
Glutamine fructose-6-phosphate aminotransferase (EC 5.3.1.19)	Blastocladiella emersonii, zoospores	Km: Fru-6-P 0.7; Gln 0.7	Norrman et al. (1975)
	Neurospora crassa, mycelium	Km: Fru-6-P 2.9; Gln 0.6	Endo et al. (1970a)
	Phycomyces blakesleeanus, spores	Produced de novo on germination	Van Laere, Carlier & Van Assche (1976)
	Candida albicans, yeast and mycelium	Activity greater in mycelium	Chattaway et al. (1973).
Glucosaminephosphate acetyltransferase (EC 2.3.1.4)	B. emersonii, zoospores	Km: GlcN-6-P 0.2; AcCoA 0.04	Giddings & Cantino (1974)
	N. crassa, mycelium	Km: GlcN-6-P 0.8; AcCoA 0.8	Davidson (1966)
	P. blakesleeanus, spores	Activity increases on germination	Van Laere et al. (1976)
Acetylglucosamine phosphomutase (EC 2.7.5.2)	N. crassa, mycelium	Regenerated donor: GlcNAc-1,6-diP	Reissig & Leloir (1966)
UDP Acetylglucosamine pyrophos-phorylase (EC 2.7.7.23)	Saccharomyces cerevisiae		Glaser & Brown (1955)
	Mucor rouxii		McMurrough et al. (1971)
N-Acetylglucosamine kinase (EC 2.7.1.59)	Aspergillus parasiticus, mycelium	Induced by GlcNAc	McGarrahan & Maley (1965)
	C. albicans, yeast cells	Induced by GlcNAc	Bhattacharya, Puri & Datta (1974)

* Km values, mmol l^{-1}, values rounded off.

Glucan synthesis

Glucans are major components of many hyphal walls, but little is known of their biosynthesis. One system is the synthesis of glucan in the Oömycete, *Phytophthora cinnamomi*. Wang & Bartnicki-Garcia (1966) found that a preparation of washed walls from this fungus efficiently incorporated glucose from UDP-glucose into β-1,3 and β-1,6-glucans. Mg^{2+} ions were required, and other nucleotide sugars were not substrates. By growing the fungus in conditions to maximize the glucan synthase yield (in a high-glucose medium to reduce endogenous protease activity) a membrane fraction was obtained that gave fibrillar β-1,3-glucan as the sole product (Wang & Bartnicki-Garcia, 1976). Electron microscopy showed its microfibrillar nature, powder X-ray diffraction showed a poorly crystalline material, and enzymic digestion showed the absence of linkages other than β-1,3.

The Oömycetes also have cellulose as a wall component, occurring as a microfibrillar network homologous with that of chitin in other fungi (Hunsley & Burnett, 1970). However, the synthesis of cellulose by a hyphal cell-free system has not yet been reported.

Mishra & Tatum (1972) report the preparation of glucan synthases from *Neurospora crassa*. β-1,3-Glucan synthase activity was associated with their cell wall preparation, and a β-1,3-glucan is a major component of the hyphal wall. These authors also characterized a membrane-bound α-1,4-glucan synthase, and suggest that it may participate in apical cell wall biosynthesis, but the status of the resultant polymer in the wall is unclear. Both enzyme activites utilized UDP-glucose as substrate, and the products were characterized by enzymic hydrolysis.

Cells of *Saccharomyces cerevisiae* that had been made permeable by treatment with toluene and ethanol were able to utilise UDP-glucose to synthesize wall glucans, chiefly β-1,3-linked (Sentandreu, Elorza & Villanueva, 1975). There was no direct evidence for the involvement of a lipid carrier.

Mannan synthesis

Membrane preparations from hyphae of *Aspergillus niger* (Barr & Hemming, 1972), and from yeast cells of *Saccharomyces cerevisiae* (Sentandreu & Lampen, 1972; Jung & Tanner, 1973) and *Hansenula holstii* (Bretthauer & Tsay, 1974) will utilize GDP-mannose to synthesize polyprenol monophosphate mannose, the 'lipid intermediate' which will then act as a glycosyl donor for the synthesis of wall mannans and mannan-glycoproteins. Lehle & Tanner (1975) also demonstrate the biosynthesis of polyprenol pyrophosphate *N*-acetylglucosamine and polyprenol pyrophosphate diacetylchitobiose by yeast membrane preparations as intermediates in the formation of cell wall mannan–glycoproteins. In *Saccharomyces*, mannan synthase activity is associated with intracellular membranes (Cortat, Matile & Kopp, 1973), and autoradiography suggests that these are its site of action (Farkaš, Kovařík, Košinová & Bauer, 1974). Thus mannan biosynthesis is likely to occur within the endoplasmic reticulum, and the products are then transported to the cell surface, in contrast to the chitin and glucan biosyntheses discussed above, which probably occur 'through' the cell membrane, and which do not involve lipid glycosyl carriers.

4.4 Role of Lytic Enzymes

Marshall-Ward (1888) watched the growth and branching of *Botrytis* hyphae and wrote: 'I have tried to figure the process to my own mind somewhat as follows. The protoplasm, confined in a segment, goes on forming the ferment, until, there being no substance for the ferment to employ its energies on, the quantity of the latter becomes so great that it can no longer be retained, and the cellulose-wall undergoes softening at some point and is pressed forwards as a protuberance, a young branch. I imagine, moreover, that the continuous forward growth of the apex of any hypha takes place in a similar way, that is to say, the ferment-substance at the apex keeps the cellulose of the hypha at that place in a soft, extensible condition, and the pressure from behind stretches it and drives the tip forwards'. What experimental evidence do we have with which to assess these ideas from nearly 100 years ago?

Electron microscopy of chemically or enzymically treated hyphae shows that the major recognizable wall components, particularly the structurally important chitin or cellulose, occur as mechanically intact layers right over the growing hyphal apex (Hunsley & Burnett, 1970), as had been surmised by Park & Robinson (1966) who emphasized the resistance of the apex to internal hydrostatic pressure. Thus the insertion of new material must involve some loosening of these layers. The extreme apex is the most fragile part of the hypha, bursting following a variety of treatments, such as flooding with water or solutions of chemicals (Robertson, 1968; Bartnicki-Garcia, 1973a). Robertson describes how hyphal apices cease elongating and start to swell when carefully flooded with water. This cessation of elongation is accompanied by a 'rigidification' of the wall, in that regrowth requires the formation of new apices, often formed sub-apically. He interprets these observations as showing that the growing hyphal apex is 'plastic', and that the sub-apical hypha or an arrested apex becomes rigidified either by the addition of new material, or by chemical cross-linking, but that branch formation or regrowth can become initiated by a resumption of plasticity.

What is the biochemical nature of the plasticity of the apex? It is very tempting to suppose, as did Marshall-Ward, that it is the result of a delicate balance between wall synthesis and the action of wall lytic enzymes. Bartnicki-Garcia (1973a) and Bartnicki-Garcia & Lippman (1973) argue strongly in favour of this thesis, especially from experiments on the bursting of hyphal tips. They observed that the range of chemical solutions that caused bursting included several otherwise innocuous chemicals, and supposed that the action of these was to cause a relative activation of the wall lytic enzymes. This interpretation has been questioned by Dow & Rubery (1975), who observed that pretreatment with $CaCl_2$ solution lengthened the time taken for treatment at pH 4.0 to elicit bursting of *Mucor rouxii* hyphae. They concluded that it is not necessary to invoke a direct effect of such factors as acidic pH or chelating agents on the lytic enzymes to explain these specific burstings, rather that the state of matrix components such as acidic polysaccharides is important for the plasticity of the apical wall, and this state is controlled by interactions of Ca^{2+} and H^+ ions. However, Park & Robinson (1966) had earlier suggested a rationalization of the action of acidic pH, pointing out that it could prevent the formation of new cross-linked material at the apex, while not affecting lysis, and so lead to bursting.

What is lacking is direct evidence of the effect of lytic enzymes in normal hyphal growth. There is plenty of circumstantial evidence, in addition to that given above. Lytic enzymes do occur associated with cell walls. Mahadevan & Mahadkar (1970) described their occurrence in *Neurospora crassa*, and found higher activities in the spreading colonial mutant *spco-1*, which has a higher frequency of branching than the wild type. Increased activity of lytic enzymes has been observed during morphogenesis involving increased branching of hyphae. For example, Mullins (1973) found that cellulase activity and branching in cultures of *Saprolegniales* showed a positive correlation following different treatments. The genetic and biochemical control of enzymic lysis of septa of basidiomycete hyphae to allow nuclear migration during dikaryotization is now well understood (Niederpruem & Jerslid, 1972). Elongating stipe cells of *Coprinus* spp. have an increased chitinase activity (F. I. Eilers; G. W. Gooday; unpublished observations).

It should be possible to visualize the site and action of these lytic enzymes by appropriate cytochemical techniques, but to date results have been disappointing with growing vegetative hyphae. However, Nolan and Bal (1974) report the localization of cellulase activity in vesicles in hyphae of *Achlya ambisexualis* responding to the sex hormone antheridiol by antheridial branching. Mullins & Ellis (1974) found aggregates of vesicles at these sites of antheridial branching, and attributed to these the release of cellulase to produce a localized thinning of the wall. Matile, Cortat, Wiemken & Frey-Wyssling (1971) describe the occurrence of the wall lytic β-1,3-glucanase in lysomal vesicles concentrated at the site of bud formation in *Saccharomyces cerevisiae*. Protease activity is released from lysosomes to act on the spore wall during germination of conidia of *Microsporium gypseum* (Page & Stock, 1972), and β-1,3-glucanase is associated with germination of conidia of *Neurospora crassa* (Mahadevan & Rao, 1970). The probability that the apical vesicles of a growing hypha release lytic enzymes at the apex is high, but there is not yet direct evidence of an identifiable enzyme acting on an identifiable wall component at a recognizable site.

Polacheck & Rosenberger (1975) have shown measurable autolytic activity in freshly prepared walls of *Aspergillus nidulans*. Fairly rapid release of glucose, mannose, galactose, N-acetylglucosamine and oligosaccharides occurred until about 3% of the wall polymers had been hydrolyzed, when the rate of release markedly slowed. Radioactive labelling experiments showed that the rapid release was preferentially of recently synthesized wall, both in the case of the normal apical deposition of wall material, and of the characteristic sub-apical deposition observed following treatment with cycloheximide. Thus, walls from hyphae grown for one hour with (^{14}C)-glucose and then allowed to autolyze for one hour gave glucose, galactose, mannose, and glucosamine with respective specific activities 4.1, 3.1, 2.1, and 2.1 higher than those of the acid hydrolysis products of the same walls. Monomers produced by autolysis represented all of the major wall polymers. These authors concluded that 'in hyphae, there appears to be no localization of autolysin, but the wall acquires resistance shortly after its deposition'. However, their results may be interpreted to suggest that the autolysin is localized at the site of wall synthesis, so endorsing the postulated intimate relationship between lysis and synthesis.

4.5 Mutants with Altered Hyphal Growth

Experimentation with mutants has led to great achievements in unravelling cellular controls of metabolism. Unfortunately, hopes that the study of developmental mutants would lead to the elucidation of molecular mechanisms of growth and differentation have rarely been fulfilled. Indeed, Wright (1975) argues that developmental mutants can play little role in helping to understand the critical variables in differentation, but instead they can be used only to confirm or modify ideas primarily obtained from studies of intact functioning systems. More hope is given by Scott, Mishra & Tatum (1973): '. . . a study of pleiotropic effects quickly becomes complex and convoluted. However, by examining these effects in a step-wise manner one can ascertain the relationships between various critical pathways and cellular ultrastructures, the controls involved, and the degree to which each of these variables influences morphology . . . .'

Despite these strictures, some interesting systems are available for study. The largest number is from *Neurospora crassa*, where about 120 of the 500 known loci can produce morphological abnormalities (Scott *et al.*, 1973). These occur in all seven linkage groups, appear as point mutations from recombination analysis, and behave as recessives in heterokaryons. The most studied mutants are those giving colonial growth, characterized by compact colonies made of highly branched hyphae. These can be divided into true colonial mutants (*e.g. col-2, balloon*) and spreading colonial mutants (*e.g. frost*). Their hyphal walls show changes in composition compared to wild type strains. In the colonial mutants, the level of chitin is often increased, while that of peptide is decreased, and that of glucan is either decreased or increased (Mahadevan & Tatum, 1965; Wrathall & Tatum, 1974). The primary enzymatic lesions in these mutants have proved difficult to locate. The observation that they had altered chitin to glucan ratios in their walls, and the assumption that this was to some extent responsible for their growth form, led to the strategy of screening them for defects in carbohydrate metabolism. Defects in one of three enzymes were found to correlate with morphological defects: glucose-6-phosphate dehydrogenase (*col-2, balloon, frost*); 6-phosphogluconate dehydrogenase (*col-3, col-10*); phosphoglucomutase (*ragged-1, ragged-2*). In the case of glucose-6-phosphate dehydrogenase the three mutants have point mutations in three linkage groups (VII, II, I), suggesting the participation of at least three different sub-units in the enzyme (Scott, 1971). In all cases, the mutation results in changes of properties such as K_m values, heat stability, and electrophoretic behaviour, as compared to the wild-type enzyme.

Clearly, the morphological phenotypes are the summations of pleiotropic effects resulting from the properties of these altered enzymes. Attempts to elucidate these pleiotropic effects have yielded interesting but inconclusive results. As yet, the resultant changes in levels of metabolites (Table 4.3) show no 'cause and effect' correlation with the observed changes in morphology. There is even no clear reason why the particular enzyme deficiency gives rise to the observed changes to levels of some of these metabolites.

The metabolism of the cell membrane is implicated in the colonial growth observed when auxotrophs for inositol or choline are grown on suboptimal levels (Brody, 1973; Scott *et al.*, 1973). In these cases there is perhaps a more direct relationship between enzyme defect and morphological effect,

Table 4.3 Some biochemical properties of morphological mutants of *Neurospora crassa*

Mutants	Enzyme defective	Level raised	Level lowered	Level unchanged
Col-2, Balloon, Frost	Glucose 6-phosphate dehydrogenase	Glucose-6-phosphate Fructose-6-phosphate Neutral lipids	Pentose phosphate pathway Pyridine nucleotides Reduced glutathione Linolenic acid (*Fr.* cyclic AMP)	
Col-3	6-Phosphogluconate dehydrogenase	Neutral lipids	Pyridine nucleotides Linolenic acid	
Ragged	Phosphoglucomutase	Glucose-6-phosphate Glucose-1-phosphate	β-1,3-Glucan Glycogen	Pyridine nucleotides Linolenic acid UDP-Glucose

References: Brody & Tatum (1967); Brody (1970); Brody & Nyc (1970); Scott *et al.* (1973); Wrathall & Tatum (1974); Scott & Mahoney (1976).

as these compounds are precursors of phospholipids in the membrane, and so an altered membrane composition may be implicated in altered control of wall deposition.

The most dramatic morphological variant of *Neurospora crassa* is *slime*. This strain produces no recognizable cell walls, but grows chiefly as protoplasts (Emerson, 1963; Bigger, White & Braymer, 1972). The enzymic lesions of *slime* are not known, but at least three independently inherited characteristics, *fz*, *sg*, and *os*, are essential for the production of its phenotype.

Katz & Rosenberger (1970*b*, 1971*b*) describe a temperature sensitive mutant of *Aspergillus nidulans*, Ts 6, that produces hyphal walls lacking chitin when grown at the nonpermissive temperature (41°C). This phenotype is due to a single recessive mutation, and can be reversed by the addition of glucosamine or *N*-acetylglucosamine, suggesting that the deficient enzyme is involved in synthesis of amino sugars, for example glutamine-fructose-6-phosphate aminotransferase. At 41°C, the mutant only forms hyphae in the presence of an osmotic stabilizer, and hydrolysis of its walls gives only about 10% glucosamine and 20% galactosamine compared to its parent strain. Such hyphae do not burst if transferred to dilute buffer, but do so anywhere along their length if transferred to growth medium lacking osmotic stabilizer. Katz and Rosenberger suggest that this is consistent with the ideas of (i) the intrinsic involvement of wall autolytic enzymes in hyphal growth and (ii) the capability of wall metabolism not being confined to the hyphal apex.

4.6 Inhibition

A large number of factors can inhibit hyphal growth as might be expected of a complex summation of many interdependent metabolic events. It is appropriate here to discuss those inhibitory substances whose mechanism can be directly ascribed to effects on the rates of reactions specifically concerned with hyphal extension.

Polyoxins and related antibiotics

The polyoxins are a group of peptidyl pyrimidine nucleoside antibiotics, produced by *Streptomyces cacoi* var. *asoenis*. They were discovered and developed in Japan, where they are now formulated into agricultural fungicides. They are used to control rice sheath blast, *Piricularia oryzae*, and black spot disease of pears, *Alternaria kikuchiana*. The fungicidal activity of polyoxin D on *Neurospora crassa* was investigated by Endo, Kakiki & Misato (1970*b*). They found that treated hyphae had a greatly reduced incorporation of (^{14}C)-glucosamine into chitin, and further, that the polyoxin was a competitive inhibitor of chitin synthase. This specific action has been confirmed for a wide range of fungi (Table 4.4) and the specificitity of polyoxin has been endorsed by its lack of inhibition of a wide range of other cellular activities, in fungi and in other organisms.

All experiments confirm that the active polyoxins are competitive inhibitors, binding very efficiently at the same site(s) as the substrate UDP-*N*-acetylglucosamine. Their structures, and that of nikkomycin, a closely related antibiotic (Dähn *et al.*, 1976) are shown in Fig. 4.3. Hori, Kakiki & Misato (1974*a*) suggest a plausible mechanism of interaction between

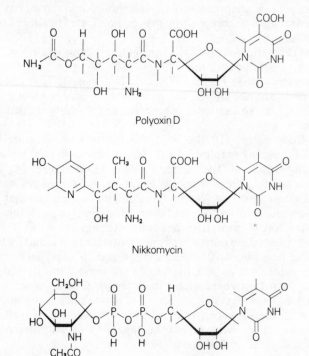

Polyoxin D

Nikkomycin

UDP—*N*—acetylglucosamine

Fig. 4.3 Structure of polyoxin D, nikkomycin and UDP-N-acetylglucosamine.

Table 4.4 Effects of polyoxins and related antibiotics on chitin synthase and fungal growth

Organism	Agent	Apparent K_i μmol l^{-1}	Inhibitory concentration in vivo μmol l^{-1}	Morphological effects	Reference
Chytridiomycetes					
Blastocladiella emersonii	Polyoxin B	—	50	—	Matsumae & Cantino (1971)
Zygomycetes					
Mucor rouxii	Polyoxin D	0.6	19	Apices burst	Bartnicki-Garcia & Lippman (1972)
M. hiemalis	Nikkomycin	—	10	Growth inhibition	Dähn et al. (1976)
Ascomycetes					
Saccharomyces cerevisiae	Polyoxins A, D	0.5	2000	Bursting at buds	Keller & Cabib (1971); Bowers et al. (1974)
Neurospora crassa	Polyoxin D	1.4	190	Germ tubes distorted	Endo et al. (1970b)
Cochliobolus miyabeanus	Polyoxin D	—	10	'Protoplast-like' structures	Endo et al. (1970b)
Alternaria kikuchiana	Polyoxin B	0.9–1.9	25	Bulbous hyphae,	Hori, Eguchi, Kakiki & Misato (1974); Ishizaki, Mitsuoka & Kunoh (1974)
Piricularia oryzae	Polyoxins A, B, D–H, J–M	3.4–33	—	Apices swell	Hori, Kakiki, Suzuki & Misato (1971)
Basidiomycetes					
Coprinus cinereus	Polyoxin D	3	0.5	No stipe elongation, autolysis	Gooday, de Rousset-Hall & Hunsley (1976)
Schizophyllum commune	Polyoxin D	—	10	No protoplast reversion	Vries & Wessels (1975)
Agaricus bisporus	Polyoxin D	—	0.1	Growth inhibition	D. A. Wood & J. B. W. Hammond (Unpublished results)

polyoxin or UDP-GlcNAc and a binding site on the enzyme. The active polyoxins and nikkomycin are remarkably efficient inhibitórs of enzyme activity *in vitro*, with estimates of K_i values being of the order of several hundred-fold less than the apparent K_m values.

Not all chitin-continng fungi are susceptible to polyoxins when tested *in vivo*. In the case of *Alternaria kikuchiana*, this has been due to an acquired immunity in some isolates from orchards that had been sprayed with polyoxin for several years. However, the chitin synthase of these resistant isolates is still sensitive to polyoxin (Hori, Eguchi, Kakiki & Misato, 1974), and so the resistance may be caused by the acquisition of a barrier to entry of the antibiotic. Originally, *Saccharomyces cerevisiae* was reported to be resistant to polyoxin, but Bowers, Levin & Cabib (1974) showed that the yeast cells were fully sensitive when tested in a defined medium. This phenomenon was ascribed to antagonism by peptides in the complex media that had been used previously. Hori, Kakiki & Misato (1974b) showed that these peptides, for example glycyl-valine, did not affect the inhibition of chitin synthase by polyoxin B *in vitro*, but did so *in vivo*. They suggest that the peptides act competitively at the site of penetration of polyoxin through the cell membrane.

The morphological effects of polyoxin on susceptible cells can be satisfy-ingly rationalized with its potent inhibition of chitin synthase (Table 4.4). Growing hyphal apices burst or at least balloon. Dividing cells of *Saccharo-myces cerevisiae* burst at the site of bud scar formation. Elongating stipe cells of *Coprinus cinereus* (normally with intercalary growth and deposition of chitin) show almost total autolysis.

Polyoxins have a very similar effect on the chitin synthase from a wide range of fungi (Table 4.4). For polyoxins A, B, D, in all cases the values for apparent K_i (although not strictly comparable as they have been obtained with different preparations under different conditions) are about one thousand fold less than the values of apparent K_m for UDP-GlcNAc. Thus all these enzymes have a very strong affinity for the antibiotics. This also endorses the remark made earlier that the properties of chitin synthase seem very similar throughout the taxonomic spectrum of the chitin-containing fungi.

Other agents that affect the incorporation of amino-sugars into fungal walls include kitazin, S-benzyl O,O'-diisopropyl phosphorothioate, which may inhibit chitin synthase (Maeda, Abe, Kakiki & Misato, 1970), and tunicamycin, an antibiotic that inhibits the biosynthesis of glycoproteins via inhibition of formation of the lipid intermediate, polyisoprenyl-*N*-acetylglucosaminyl pyrophosphate (Kuo & Lampen, 1974).

Sorbose

Sorbose (a ketohexose) has a specific paramorphogenetic effect on *Neurospora crassa*, causing colonial growth (Mahadevan & Tatum, 1965) and disintegration of hyphal apices (Rizvi & Robertson, 1965). The walls of hyphae grown in the presence of sorbose have a much reduced content of β-1,3-glucan, and this may be attributed in part to a direct effect of the sorbose on the activity of glucan synthesizing enzymes (Mishra & Tatum, 1972). When grown in the presence of sorbose the activities of α-1,4-glucan synthetase and β-1,3-glucan synthetase were reduced by factors of 3.6 and

5.3 respectively compared to control cultures grown with sucrose. Sorbose inhibited both enzymes in assays *in vitro*. A mutant resistant to sorbose, *patch*, did not show these responses *in vivo* or *in vitro*.

Autoinhibition

There are suggestions that specific metabolites produced by older hyphae may have a role *in vivo* in regulating the growth of hyphal apices. Park & Robinson (1967) describe widespread 'vaculation factors' that cause young hyphae to stop growing, apparently by resulting in loss of the hyphal polarity of the control of water movement.

Reissig (1974) proposes that a galactosaminoglycan produced by *Neurospora crassa* may have a natural regulatory role in controlling hyphal growth. This mucopolysaccharide stops hyphal extension when added to growing cultures, and Reissig identifies it with the fibrous reticulum seen by Hunsley & Burnett (1970) embedded in the sub-apical walls of *Neurospora crassa*. Reissig correlates the restricted growth of a *cot* mutant (temperature-sensitive colonial) with over-production of this material

Lectins

The plant lectins are a group of proteins that have strong specific binding affinities for carbohydrate-containing ligands. Their role to the plants producing them is unclear, but they may be protective against pathogens (Callow, 1975). Evidence for this comes from Mirelman, Galun, Sharon & Lotan (1975), who show that wheat germ agglutinin inhibits hyphal growth and chitin deposition in *Trichoderma viride*. This lectin (WGA) specifically binds to β-linked N-acetylglucosamine units, and fluorescein-labelled WGA was shown to bind to the hyphal apices and also to septa. Surprisingly, WGA has no significant effect on chitin synthase from *Coprinus cinereus*, the earlier report of some inhibition (Gooday & de Rousset-Hall, 1975) being due to impurities in the preparation. If this is also true for chitin synthase from *T. viride* the inhibition of growth is presumably mediated by binding of WGA to nascent chitin chains at the apex, perhaps in turn inhibiting loosening of these by chitinase?

4.7 Control of Synthesis

The form of growth of a fungal cell is controlled by the triumvirate:

$$\text{Cytoplasm} \longleftrightarrow \text{Membrane} \longleftrightarrow \text{Wall}$$

There are numerous examples of such interrelationships. For example, hyphal growth kinetics, branching, septum formation, and clamp connection formation can all be correlated with behaviour of nuclei in particular cases. However, it is not possible as yet to distinguish cause and effect.

Nevertheless, some of the finer points of control are becoming more clear. A good example is provided by studies on chitin synthase. In the yeast cells of *Saccharomyces* spp. a ring of chitin is synthesized at the time of budding, specifically to form part of the bud scar (Cabib & Bowers, 1975). Cabib (1975) and Cabib, Ulane & Bowers (1974) describe how chitin synthase occurs in the cell membrane as latent zymogen. It can be activated by the action of a protease that is present in vesicles in the cytoplasm. The protease is specifically inhibited by a protein present in the cytosol. Thus the site and timing of initiation of chitin deposition may be controlled by the selective activation of zymogen molecules in the membrane by release of the protease from the vesicles at that point. The inhibitor will then ensure that the activation remains localized. Hasilik (1974) further shows that the protease (which proves to be identical with 'proteinase B', already known as an inactivator of tryptophan synthase) as well as very rapidly activating the zymogen, will also slowly (at about one seventieth of the rate) inactivate the resultant chitin synthase, thus providing a mechanism for termination of chitin synthesis during the budding cycle.

Chitin synthase in *Mucor rouxii* and *Phycomyces blakesleeanus* is probably controlled in a very similar manner (McMurrough & Bartnicki-Garcia, 1973; Ruiz-Herrera & Bartnicki-Garcia, 1976*a, b*; P. Fischer, personal communication); some of the chitin synthase in cell homogenates is in an inactive latent form and can be activated by proteolysis; this protease is inactivated by an inhibitor and the active chitin synthase is inactivated by further incubation with the protease. *M. rouxii* is a dimorphic fungus, and this chitin synthase system has markedly different properties in yeast cells and mycelium. The yeast cells have a synthase that is predominantly latent, a low protease activity, and a high inhibitor level. The mycelial cells are just the opposite. Ruiz-Herrera & Bartnicki-Garcia (1976*b*) rationalize these differences with the two different growth forms: the apical extension of hypha is maintained by a sharp gradient of active chitin synthase that is rapidly inactivated during transition from tip to lateral wall, while the isodiametric expansion of the yeast cell does not require critical control of inactivation of the enzyme.

It is probable that other chitin synthases and fungal wall synthesizing enzymes will prove to have the irreversible controlling steps of:

$$\text{Latency} \longrightarrow \text{Activity} \longrightarrow \text{Inactivity}$$

The β-1,3-glucan synthase of *Phytophthora cinnamomi* is obtained with greater activity by treatment of cell extracts with proteases, so may be partially latent (Wang & Bartnicki-Garcia, 1976). The formation of a septum in a hypha is likely to be controlled by a very similar mechanism to that controlling the formation of the homologous bud scar in *Saccharomyces* spp. The formation of side branches may also involve activation of latent enzymes. The sub-apical chitin deposition observed in hyphae of *Aspergillus nidulans* following treatment with cycloheximide (Katz & Rosenberger, 1971*a*) may be explained by postulating the activation of latent chitin synthase occurring uniformly along the hyphae following a loss of control of polarity and inhibition of protein synthesis. In the presence of cycloheximide, protoplasts of *Schizophyllum commune* produced chitin for about 5 h at a rate similar to that of control protoplasts (Vries & Wessels, 1975). This synthesis must have been by pre-existing chitin synthase as protein synthesis was almost totally inhibited.

Autoradiography of chitin synthase activity in fragments of young hyphae of *Mucor rouxii* showed an apical localization, mirroring that of chitin deposition *in vivo* (McMurrough, Flores-Carreon & Bartnicki-Garcia, 1971); but there was considerable enzyme activity in the sub-apical non-elongating region of sporangiophore 'ghosts' of *Phycomyces blakesleeanus*

(Jan, 1974), which may represent activity involved in sub-apical intussus-ception or activation of a latent enzyme during preparation.

The hyphal apex receives a continual supply of membrane vesicles, which provide new membrane for insertion, and certainly enzymes and precursors for the apical intussusception. Do the vesicles have latent chitin synthase, which is activated at the moment of its insertion at the apex? An alternative would be to have active enzyme, which meets its substrates at this point. Assuming that the chitin synthase molecule remains 'fixed' in the membrane, its position will change from apical to lateral as the tip advances. Its activity in the lateral membrane falls off very rapidly. This is probably due to irreversible inactivation, but deprivation of a suitable biochemical environment may play a part. If the lateral hyphal membrane does contain zymogen which may be activated for branch or septum formation, has this been inserted at the apex, and if so, how has it escaped an activation/inactivation sequence? In the elongating *Coprinus* stipe there is no evidence for a zymogen activation system, but the cell extracts have endogenous protease activity and the enzyme appears to be distributed throughout the cell membrane in potentially active form.

As discussed earlier, the activity of chitin synthase can be controlled by activators such as N-acetylglucosamine and inhibitors such as UDP. These metabolites may be regulators of enzyme activity *in vivo*, particularly the N-acetylglucosamine and its higher oligomers, which may arise from localized lysis of pre-existing chitin where intussusception is required. Exogenous N-acetylglucosamine induces a yeast to mycelial transition in *Candida albicans*, and Simonetti, St rippoli & Cassone (1974) suggest that this might represent a direct activation of chitin synthase, as the mycelial wall has a highe chitin content. However, this would have to be via uptake of the sugar into the cell, as there is no evidence that cell wall synthetic enzymes are accessible from the outside medium. For example, protoplasts of *Schizophyllum commune* regenerate very well in a solution of magnesium sulphate that would be strongly inhibitory to chitin synthase (Vries & Wessels, 1975).

The settlement of the naked zoospore of *Blastocladiella emersonii* provides a clear example of control of wall synthesis by the comparting of components. The zoospores contain membrane-bound 'gamma particles' that store preformed chitin synthase, which is kept inactive, apparently by being sequestered from its substrates (Myers & Cantino, 1974). Upon settlement the zoospore very quickly synthesizes its new wall, coincident with the apparent fusion of membrane from the gamma particles with the cell membrane, and so presumably with the release of the chitin synthase to meet with UDP-N-acetylglucosamine from the cytosol. The zoospore of *Phytophthora palmivora* is analogous, as it is a swimming protoplast that rapidly forms a wall of cellulose and other glucans by the activation of pre-formed enzymes (Bartnicki-Garcia, 1973b).

Cantino & Myers (1972) report further controls on the chitin synthase activity of the *Blastocladiella* zoospores. Gamma particles of zoospores derived from light- and dark-grown OC plants each have the same chitin synthase activity, but the specific activity per unit protein is half in the former to the latter. Coupled with the observation that the zoospores from light-grown plants have half the number of gamma particles but the same total chitin synthase activity this implies that light lessens the incorporation or retention of chitin synthase in gamma particles. Light can also induce encystment of zoospores, but this effect can also be caused by the addition of 0.45 mmol l^{-1} UDP-N-acetylglucosamine, and so Myers & Cantino (1974) suggest that the light stimulates the supply of this nucleotide sugar, but when added externally it can enter the cell intact to be a precursor of chitin.

The *Phycomyces* sporangiophore provides the best understood response to light by a fungal cell. It reacts to an increase in blue light intensity by an increase in growth rate. Jan (1974) describes a 30% activation of chitin synthase activity in crude cell homogenates that had been illuminated for two hours, compared to dark controls. This phenomenon was not shown by the more purified enzyme preparations, and could represent an activation of a latent enzyme.

Fungal cells show such a variety of responses of growth and differentiation that it is not surprising that we have only scratched the surface in understanding the multifarious mechanisms of these processes. The major task still remains—the biochemical elucidation of apical hyphal growth. As Marshall-Ward speculated in 1888—what precisely are the natures, roles, and controls of the ferments at the apex?

4.8 References

ARAKI, Y. & ITO, E. (1974). A pathway of chitosan formation in *Mucor rouxii*: enzymatic deacetylation of chitin. *Biochemical and Biophysical Research Communications* 56, 669–75.

BARR, R. M. & HEMMING, F. W. (1972). Polyprenol phosphate as an acceptor of mannose from guanosine diphosphate mannose in *Aspergillus niger*. *Biochemical Journal* 126, 1203–8.

BARTNICKI-GARCIA, S. (1970). Cell wall composition and other biochemical markers in fungal phylogeny. In *Phytochemical phylogeny*, pp. 81–103. Edited by J. B. Harborne. London: Academic Press.

BARTNICKI-GARCIA, S. (1973a). Fundamental aspects of hyphal morphogenesis. *Symposium of Society for General Microbiology* 23, 245–67.

BARTNICKI-GARCIA, S. (1973b). Cell wall genesis in a natural protoplast: the zoospore of *Phytophthora palmivora*. In *Yeast, mould and plant protoplasts*. pp. 77–92. Edited by J. R. Villanueva, I. Garćia-Acha, S. Gascón and F. Uruburu, London: Academic Press.

BARTNICKI-GARCIA, S. & LIPPMAN, E. (1969). Fungal morphogenesis: cell wall construction in *Mucor rouxii*. *Science, New York* 165, 302–4.

BARTNICKI-GARCIA, S. & LIPPMAN, E. (1972). Inhibition of *Mucor rouxii* by polyoxin D: effects on chitin synthetase and morphological development. *Journal of General Microbiology* 71, 301–9.

BARTNICKI-GARCIA, S. & LIPPMAN, E. (1973). The bursting tendency of hyphal tips of fungi: presumptive evidence for a delicate balance between wall synthesis and wall lysis in apical growth. *Journal of General Microbiology* 73, 487–500.

BHATTACHARYA, A., PURI, M. & DATTA, A. (1974). Induction of N-acetylglucosamine kinase in yeast. *Biochemical Journal* 141, 593–5.

BIGGER, C. H., WHITE, M. R. & BRAYMER, H. D. (1972). Ultra-structure and invertase secretion of the *slime* mutant of *Neurospora crassa*. Journal of General Microbiology 71, 159–66.

BOWERS, B., LEVIN, G. & CABIB, E. (1974). Effect of polyoxin D on chitin synthesis and septum formation in *Saccharomyces cerevisiae*. *Journal of Bacteriology* 119, 564–75.

BRETTHAUER, R. K. & TSAY, G. C. (1974). Enzymatic transfer of mannose from guanosine disphosphate mannose to yeast mannan-protein complexes. *Archives of Biochemistry and Biophysics* 164, 118–26.

BRODY, S. (1970). Correlation between reduced nicotinamide adenine dinuclectide phosphate levels and morphological changes in *Neurospora crassa*. *Journal of Bacteriology* **101**, 802-7.

BRODY, S. (1973). Metabolism, cell walls, and morphogenesis. In *Developmental regulation: aspects of cell differentiation*, Edited by S. J. Coward, pp. 107-154. New York: Academic Press.

BRODY, S. & NYC, J. F. (1970). Altered fatty acid distribution in mutants of *Neurospora crassa*. *Journal of Bacteriology* **104**, 780-6.

BRODY, S. & TATUM, E. L. (1967). Phosphoglucomutase mutants and morphological changes in *Neurospora crassa*. *Proceedings of National Academy of Sciences U.S.A.* **58**, 923-30.

BURNETT, J. H. (1976). *Fundamentals of mycology*. 2nd Ed. London: Edward Arnold.

CABIB, E. (1975). Molecular aspects of yeast morphogenesis. *Annual Review of Microbiology* **29**, 191-214.

CABIB, E. & BOWERS, B. (1975). Timing and function of chitin synthesis in yeast. *Journal of Bacteriology* **124**, 1586-93.

CABIB, E., ULANE, R., BOWERS, B. (1974). A molecular model for morphogenesis: the primary septum of yeast. *Current Topics in Cellular Regulation* **8**, 1-32.

CALLOW, J. A. (1975). Plant lectins. *Current Advances in Plant Science* **18**, 181-93.

CAMARGO, E. P., DIETRICH, C. P., SONNEBORN, D. & STROMINGER, J. L. (1967). Biosynthesis of chitin in spores and growing cells of *Blastocladiella emersonii*. *Journal of Biological Chemistry* **242**, 3121-8.

CANTINO, E. C. & MYERS, R. B. (1972). Concurrent effect of visible light on γ-particles, chitin synthetase, and encystment capacity in zoospores of *Blastocladiella emersonii*. *Archiv für Mikrobiologie* **83**, 203-15.

CHANG, P. L. Y. & TREVITHICK, J. R. (1974). How important is secretion of exoenzymes through apical cell walls of fungi? *Archives of Microbiology* **101**, 281-93.

CHATTAWAY, F. W., BISHOP, R., HOLMES, M. R. & ODDS, F. C. & BARLOW, A. J. E. (1973). Enzyme activity associated with carbohydrate synthesis and breakdown in yeast and mycelial forms of *Candida albicans*. *Journal of General Microbiology* **75**, 97-109.

COLLINGE, A. J. & TRINCI, A. P. J. (1974). Hyphal tips of wild-type and spreading colonial mutants of *Neurospora crassa*. *Archives of Microbiology* **99**, 353-68.

CORTAT, M., MATILE, P. & KOPP, F. (1973). Intracellular localization of mannan synthetase activity in budding baker's yeast. *Biochemical and Biophysical Research Communications* **53**, 482-9.

DÄHNE, U., HAGENMAIER, H., HÖHNE, H., KÖNIG, W. A., WOLF, G. & ZÄHNER, H. (1976). Stoffwechselprodukte von Mikroorganismen. 154. Mitteilung. Nikkomycin, ein neuer Hemmstoff der Chitinsynthese bei Pilzen. *Archives of Microbiology* **107**, 143-60.

DAVIDSON, E. A. (1966). Glucosamine 6-phosphate N-acetylase. *Methods in Enzymology* **9**, 704-7.

DOW, J. M. & RUBERY, P. H. (1975). Hyphal tip bursting in *Mucor rouxii*: antagonistic effects of calcium ions and acid. *Journal of General Microbiology* **91**, 425-8.

DURÁN, A., BOWERS, B. & CABIB, E. (1975). Chitin synthetase zymogen is attached to the yeast plasma membrane. *Proceedings of the National Academy of Sciences U.S.A.* **72**, 3952-5.

EMERSON, S. (1963). Slime: a plasmodioid variant of *Neurospora crassa*. *Genetica* **34**, 162-82.

ENDO, A., KAKIKI, K. & MISATO, T. (1970a). Feedback inhibition of L-glutamine D-fructose 6-phosphate amidotransferase by uridine diphosphate N-acetylglucosamine in *Neurospora crassa*. *Journal of Bacteriology* **103**, 588-94.

ENDO, A., KAKIKI, K. & MISATO, T. (1970b). Mechanism of action of the antifungal agent polyoxin D. *Journal of Bacteriology* **104**, 189-96.

FARKAŠ, V., KOVAŘÍK, J., KOŠINOVÁ, A. & BAUER, Š. (1974). Autoradiographic study of mannan incorporation into the growing cell wall of *Saccharomyces cerevisiae*. *Journal of Bacteriology* **117**, 265-9.

GALUN, E. (1972). Morphogenesis of *Trichoderma*: autoradiography of intact colonies labelled by (³H) N-acetylglucosamine as a marker for new cell wall biosynthesis. *Archiv für Mikrobiologie* **86**, 305-14.

GIDDINGS, T. H. & CANTINO, E. C. (1974). Partial purification and properties of D-glucosamine 6-phosphate N-acetyltransferase from zoospores of *Blastocladiella emersonii*. *Journal of Bacteriology* **120**, 976-9.

GIRBARDT, M. (1969). Die Ultrastruktur der Apikalregion von Pilzhyphen. *Protoplasma* **67**, 413-41.

GLASER, L. & BROWN, D. H. (1955). The enzymatic synthesis *in vitro* of hyaluronic acid chains. *Proceedings of the National Academy of Sciences U.S.A.* **41**, 253–60.

GLASER, L. & BROWN, D. H. (1957). The synthesis of chitin in cell-free extracts of *Neurospora crassa. Journal of Biological Chemistry* **228**, 729–42.

GOODAY, G. W. (1971). An autoradiographic study of hyphal growth of some fungi. *Journal of General Microbiology* **67**, 125–33.

GOODAY, G. W. (1973). Activity of chitin synthetase during the development of fruit bodies of the toadstool *Coprinus cinereus. Biochemical Society Transactions* **1**, 1105–7.

GOODAY, G. W. (1975). The control of differentiation in fruit bodies of *Coprinus cinereus. Reports of Tottori Mycological Institute (Japan)* **12**, 151–60.

GOODAY, G. W. (1977). The Fleming Lecture. The biosynthesis of the fungal wall mechanisms and implications. *Journal of General Microbiology* **99**, 1–11.

GOODAY, G. W. & HUNSLEY, D. (1971). Cellulose wall ingrowths in *Phytophthora parasitica. Transactions of British Mycological Society* **57**, 178–9.

GOODAY, G. W. & DE ROUSSET-HALL, A. (1975). Properties of chitin synthetase from *Coprinus cinereus. Journal of General Microbiology* **89**, 137–45.

GOODAY, G. W., DE ROUSSET-HALL, A. & HUNSLEY, D. (1976). The effect of polyoxin D on chitin synthesis in *Coprinus cinereus. Transactions of British Mycological Society* **67**, 193–200.

GROVE, S. N. & BRACKER, C. E. (1970). Protoplasmic organization of hyphal tips among fungi: vesicles and Spitzenkörper. *Journal of Bacteriology* **104**, 989–1009.

HASILIK, A. (1974). Inactivation of chitin synthase in *Saccharomyces cerevisiae. Archives of Microbiology* **101**, 295–301.

HORI, M., KAKIKI, K. & MISATO, T. (1974a). Interaction between polyoxin and active center of chitin synthetase. *Agricultural and Biological Chemistry, Japan* **38**, 699–705.

HORI, M., KAKIKI, K. & MISATO, T. (1974b). Further study on the relation of polyoxin structure to chitin synthetase inhibition. *Agricultural and Biological Chemistry, Japan* **38**, 691–8.

HORI, M., EGUCHI, J., KAKIKI, K. & MISATO, T. (1974). Studies on the mode of action of polyoxins. VI. Effect of polyoxin B on chitin synthesis in polyoxin sensitive and resistant strains of *Alternaria kikuchiana, The Journal of Antibiotics* **27**, 260–6.

HORI, M., KAKIKI, K., SUZUKI, S. & MISATO, T. (1971). Studies on the mode of

action of polyoxins. Part III. Relation of polyoxin structure to chitin synthetase inhibition. *Agricultural and Biological Chemistry, Japan* **35**, 1280–91.

HUNSLEY, D. (1973). Apical wall structure in hyphae of *Phytophthora parasitica. New Phytologist* **72**, 985–90.

HUNSLEY, D. & BURNETT, J. H. (1968). Dimensions of microfibrillar elements in fungal walls. *Nature, London* **218**, 462–3.

HUNSLEY, D. & BURNETT, J. H. (1970). The ultrastructural architecture of the walls of some hyphal fungi. *Journal of General Microbiology* **62**, 203–18.

HUNSLEY, D. & GOODAY, G. W. (1974). The structure and development of septa in *Neurospora crassa. Protoplasma* **82**, 125–46.

ISHIZAKI, H., MITSUOKA, K. & KUNOH, H. (1974). Effect of polyoxin on fungi. I. Optical microscopic observations of mycelia of *Alternaria kikuchiana* Tanaka. *Annals of the Phytopathological Society of Japan* **40**, 433–8.

JAN, Y. N. (1974). Properties and cellular localization of chitin synthetase in *Phycomyces blakesleeanus. Journal of Biological Chemistry* **249**, 1973–9.

JUNG, P. & TANNER, W. (1973). Identification of lipid intermediate in yeast mannan biosynthesis. *European Journal of Biochemistry* **37**, 1–7.

KATZ, D. & ROSENBERGER, R. F. (1970a). The utilisation of galactose by an *Aspergillus nodulans* mutant lacking galactose phosphate-UDP transferase and its relation to cell wall synthesis. *Archiv für Mikrobiologie* **74**, 41–51.

KATZ, D. & ROSENBERGER, R. F. (1970b). A mutation in *Aspergillus nidulans* producing hyphal walls which lack chitin. *Biochimica et Biophysica Acta* **208**, 452–60.

KATZ, D. & ROSENBERGER, R. F. (1971a). Hyphal wall synthesis in *Aspergillus nidulans*: effect of protein synthesis inhibition and osmotic shock on chitin insertion and morphogenesis. *Journal of Bacteriology* **108**, 184–90.

KATZ, D. & ROSENBERGER, R. F. (1971b). Lysis of an *Aspergillus nidulans* mutant blocked in chitin synthesis and its relation to wall assembly and wall metabolism. *Archiv für Mikrobiologie* **80**, 284–92.

KELLER, F. A. & CABIB, E. (1971). Chitin and yeast budding: properties of chitin synthetase from *Saccharomyces carlsbergensis. Journal of Biological Chemistry* **246**, 160–6.

KUO, S. C. & LAMPEN, J. O. (1974). Tunicamycin—an inhibitor of yeast glycoprotein synthesis. *Biochemical and Biophysical Research Communications* **58**, 287–95.

LEHLE, L. & TANNER, W. (1975). Formation of lipid-bound oligosaccharides in yeast. *Biochimica et Biophysica Acta* **399**, 364–74.

MAEDA, T., ABE, H., KAKIKI, K. & MISATO, T. (1970). Studies on the mode of action of organophosphorus fungicide, kitizin. *Agricultural and Biological Chemistry, Japan* **34**, 700–9.

MAHADEVAN, P. R. & MAHADKAR, U. R. (1970). Role of enzymes in growth and morphology of *Neurospora crassa*; cell-wall-bound enzymes and their possible role in branching. *Journal of Bacteriology* **101**, 941–7.

MAHADEVAN, P. R. & RAO, S. R. (1970). Enzyme degradation of conidial wall during germination of *Neurospora crassa*. *Indian Journal of Experimental Biology* **8**, 293–7.

MAHADEVAN, P. R. & TATUM, E. L. (1965). Relationship of the major constituents of the *Neurospora crassa* cell wall to wild-type and colonial morphology. *Journal of Bacteriology* **90**, 1073–81.

MARSHALL-WARD, H. (1888). A lily-disease. *Annals of Botany* **2**, 319–82.

MATILE, P., CORTAT, M., WIEMKEN, A. & FREY-WYSSLING, A. (1971). Isolation of glucanase—containing particles from budding *Saccharomyces cerevisiae*. *Proceedings of the National Academy of Sciences U.S.A.* **68**, 636–40.

MATSUMAE, A. & CANTINO, E. C. (1971). Sensitivity of the spores of *Blastocladiella emersonii* and related fungi to antibiotics and some other drugs. *Journal of Antibiotics* **24**, 77–84.

MCGARRAHAN, J. F. & MALEY, F. (1965). Hexose metabolism. IV. Studies on the induction of N-acetylglucosamine kinase in *Aspergillus parasiticus*. *Journal of Biological Chemistry* **240**, 2328–33.

MCCLURE, W. K., PARK, D. & ROBINSON, P. M. (1968). Apical organization in the somatic hyphae of fungi. *Journal of General Microbiology* **50**, 177–82.

MCMURROUGH, I. & BARTNICKI-GARCIA, S. (1971). Properties of a particulate chitin synthetase from *Mucor rouxii*. *Journal of Biological Chemistry* **246**, 4008–16.

MCMURROUGH, I. & BARTNICKI-GARCIA, S. (1973). Inhibition and activation of chitin synthesis by *Mucor rouxii* cell extracts. *Archives of Biochemistry and Biophysics* **158**, 812–6.

MCMURROUGH, I., FLORES-CARREON, A. & BARTNICKI-GARCIA, S. (1971).

Pathway of chitin synthesis and cellular localization of chitin synthetase in *Mucor rouxii*. *Journal of Biological Chemistry* **246**, 3999–4007.

MIRELMAN, D., GALUN, E., SHARON, N. & LOTAN, R. (1975). Inhibition of fungal growth by wheat germ agglutinin. *Nature, London* **256**, 414–6.

MISHRA, N. C. & TATUM, E. L. (1972). Effect of L-sorbose on polysaccharide synthetases of *Neurospora crassa*. *Proceedings of National Academy of Sciences U.S.A.* **69**, 313–7.

MULLINS, J. T. (1973). Lateral branch production and cellulase production in the water molds. *Mycologia*, **65**, 1007–14.

MULLINS, J. T. & ELLIS, E. A. (1974). Sexual morphogenesis in *Achlya*: ultrastructural basis for the hormonal induction of antheridial hyphae. *Proceedings of National Academy of Sciences U.S.A.* **71**, 1347–50.

MYERS, R. B. & CANTINO, E. C. (1974). *The gamma particle*. Monographs in Developmental Biology, Vol. 8. Basel: S. Karger.

NIEDERPRUEM, D. J. & JERSILD, R. A. (1972). Cellular aspects of morphogenesis in the mushroom *Schizophylum commune*. *Critical Reviews in Microbiology* **1**, 545–76.

NISHI, A., YANAGITA, T. & MARUYAMA, Y. (1968). Cellular events occurring in growing hyphae of *Aspergillus oryzae* as studied by autoradiography. *Journal of General and Applied Microbiology* **14**, 171–82.

NOLAN, R. A. & BAL, A. K. (1974). Cellulase localization in hyphae of *Achyla ambisexualis*. *Journal of Bacteriology* **117**, 840–3.

NORRMAN, J., GIDDINGS, T. H. & CANTINO, E. C. (1975). Partial purification and properties of L-glutamine:D-fructose-6-phosphate amino transferase from zoospores of *Blastocladiella emersonii*. *Phytochemistry* **14**, 1271–4.

PAGE, W. J. & STOCK, J. J. (1972). Isolation and characterization of *Microsporium gypseum* lysosomes: role of lysosomes in microconidia germination. *Journal of Bacteriology* **110**, 354–62.

PARK, D. & ROBINSON, P. M. (1966). Internal pressure of hyphal tips of fungi, and its significance in morphogenesis. *Annals of Botany (N.S.)* **30**, 425–39.

PARK, D. & ROBINSON, P. M. (1967). A fungal hormone controlling internal water distribution normally associated with cell ageing in fungi. *Symposium of the Society for Experimental Biology* **21**, 323–36.

PEBERDY, J. F. & MOORE, P. M. (1975). Chitin synthase in *Mortierella vinacea*; properties, cellular location and synthesis in growing cultures. *Journal of General Microbiology* **90**, 228–36.

POLACHEK, Y. & ROSENBERGER, R. F. (1975). Autolytic enzymes in hyphae of *Aspergillus nidulans*: their action on old and newly formed walls. *Journal of Bacteriology* **121**, 332–7.

PORTER, C. A. & JAWORSKI, E. G. (1966). The synthesis of chitin by particulate preparation of *Allomyces macrogynus*. *Biochemistry* **5**, 1149–54.

REINHARDT, M. O. (1892). Das Wachsthum der Pilzhyphen. *Jahrbücher für wissenschaftliche Botanik* **23**, 479–566.

REISSIG, J. L. (1974). Decoding of regulatory signals at the microbial surface. *Current Topics in Microbiology and Immunology* **67**, 43–96.

REISSIG, J. L. & LELOIR, L. F. (1966). Phosphoacetylglucosamine mutase from *Neurospora. Methods in Enzymology* **8**, 175–8.

RIZVI, S. R. H. & ROBERTSON, N. F. (1965). Apical disintegration of hyphae of *Neurospora crassa* as a response to L-sorbose. *Transactions of British Mycological Society* **48**, 468–77.

ROBERTSON, N. F. (1968). The growth process in fungi. *Annual Review of Phytopathology* **6**, 115–36.

DE ROUSSET-HALL, A. & GOODAY, G. W. (1975). A kinetic study of a solubilized chitin synthetase preparation from *Coprinus cinereus. Journal of General Microbiology* **89**, 146–54.

RUIZ-HERRERA, J. & BARTNICKI-GARCIA, S. (1974). Synthesis of cell wall microfibrils in vitro by a 'soluble' chitin synthetase from *Mucor rouxii. Science, New York* **186**, 357–9.

RUIZ-HERRERA, J. & BARTNICKI-GARCIA, S. (1976*a*). Biochemistry of dimorphism. *Abstracts of the American Society for Microbiology*, p. 130.

RUIZ-HERRERA, J. & BARTNICKI-GARCIA, S. (1976*b*). Proteolytic activation and inactivation of chitin synthetase from *Mucor rouxii. Journal of General Microbiology* **97**, 241–249.

RUIZ-HERRERA, J., SING, V. O., VAN DER WOUDE, W. J. & BARTNICKI-GARCIA, S. (1975). Microfibril assembly by granules of chitin synthetase. *Proceedings of the National Academy of Sciences U.S.A.* **72**, 2706–10.

SCOTT, W. A. (1971). Physical properties of glucose 6-phosphate dehydrogenase from *Neurospora crassa. Journal of Biological Chemistry* **246**, 6353–9.

SCOTT, W. A. & MAHONEY, E. (1976). Defects of glucose-6-phosphate and 6-phosphogluconate dehydrogenases in *Neurospora* and their pleiotropic effects. *Current Topics in Cellular Regulation* **10**, 205–36.

SCOTT, W. A., MISHRA, N. C. & TATUM, E. L. (1973). Biochemical genetics of morphogenesis in *Neurospora. Brookhaven Symposium in Biology* **25**, 1–18.

SENTANDREU, R. & LAMPEN, J. O. (1972). Biosynthesis of mannan in *Saccharomyces cerevisiae*. Isolation of a lipid intermediate and its identification as a mannosyl-1-phosphoryl polyprenol. *FEBS Letters* **27**, 331–4.

SENTANDREU, R., ELORZA, M. V. & VILLANUEVA, J. R. (1975). Synthesis of yeast wall glucan. *Journal of General Microbiology* **90**, 13–20.

SIMONETTI, N., STRIPPOLI, V. & CASSONE, A. (1974). Yeast-mycelial conversion induced by N-acetyl-D-glucosamine in *Candida albicans. Nature, London* **250**, 344–6.

STERNLICHT, E., KATZ, D. & ROSENBERGER, R. F. (1973). Subapical wall synthesis and wall thickening induced by cycloheximide in hyphae of *Aspergillus nidulans. Journal of Bacteriology* **114**, 819–23.

TRINCI, A. P. J. (1971). Influence of the width of the peripheral growth zone on the radial growth rate of fungal colonies on solid media. *Journal of General Microbiology* **67**, 325–44.

TRINCI, A. P. J. & COLLINGE, A. J. (1975). Hyphal wall growth in *Neurospora crassa* and *Geotrichum candidum. Journal of General Microbiology* **91**, 355–61.

TURIAN, G. (1975). Differentiation in *Allomyces* and *Neurospora. Transactions of British Mycological Society* **64**, 367–80.

VAN LAERE, A. J., CARLIER, A. R. & VAN ASSCHE, J. A. (1976). Effect of 5-fluorouracil and cycloheximide on the early development of *Phycomyces blakesleeanus* spores and the activity of N-acetylglucosamine synthesizing enzymes. *Archives of Microbiology* **108**, 113–6.

VRIES, O. M. H. DE & WESSELS, J. G. H. (1975). Chemical analysis of cell wall regeneration and reversion of protoplasts from *Schizophyllum commune. Archives of Microbiology* **102**, 209–18.

WANG, M. C. & BARTNICKI-GARCIA, S. (1966). Biosynthesis of β-1,3- and β-1,6-linked glucan by *Phytophthora cin-*

namomi hyphal walls. *Biochemical and Biophysical Research Communications* **24**, 832-7.

WANG, M. C. & BARTNICKI-GARCIA, S. (1976). Synthesis of β-1,3-glucan microfibrils by a cell-free extract from *Phytophthora cinnamomi*. *Archives for* Biochemistry and Biophysics **175**, 351-4.

WESSELS, J. G. H. & MARCHANT, R. (1974). Enzymic degradation of septa in hyphal wall preparations from a monokaryon and a dikaryon of *Schizophyllum commune*. *Journal of General Microbiology* **83**, 359-68.

WRATHALL, C. R. & TATUM, E. L. (1974). Hyphal wall peptides and colonial morphology in *Neurospora crassa*. *Biochemical Genetics* **12**, 59-68.

WRIGHT, B. E. (1975). Usefulness of developmental mutants in the analysis of biochemical differentiation. In *Microbiology—1975*. Edited by D. Schlessinger, pp. 500-507. Washington: American Society for Microbiology.

ZALOKAR, M. (1959). Growth and differentiation of *Neurospora* hyphae. *American Journal of Botany* **46**, 602-10.

ZALOKAR, M. (1965). Integration of cellular metabolism. In *The fungi*, pp. 326-377 Vol. 1. Edited by G. C. Ainsworth and A. S. Sussman, New York: Academic Press.

CHAPTER 5

Form and Function of Septa in Filamentous Fungi

K. GULL

5.1 Introduction

The presence of cross walls in the hyphae of filamentous fungi must have
been detected very early in the study of mycology. In 1884, in his classical
treatise on the fungi, de Bary described pits in the fungal cross walls although
he did not indicate whether he considered these pits to be closed or open.
Wahlich (1893) used fresh mycelium fixed in a watery solution of iodine in
potassium iodide to show that septa of many species of fungi possessed
pores. Subsequently, cytoplasmic flow was seen between hyphal compartments of some fungi, indicating that the septa were perforate. Wahlich also
showed that these perforate septa were formed by growth of an annulus
from a ring-like projection on the inside of the lateral wall. Growth of the
septum ceased before complete closure of the cross wall.

   These early reports were extended by other observers using light microscopy and various septal associated structures were described. Use of the
electron microscope then allowed the fine structural detail of many of these
septal associated organelles to be documented. Studies with the electron
microscope also indicated that a variety of septal types were to be found in
the fungi. Moreover, there appeared to be a reasonable correlation between
the classical taxonomic grouping of the fungus and the organization of its
septum. Much attention has been focused on this relationship between
taxonomic grouping and septal ultrastructure, whilst some of the more basic
areas such as septal development, biochemistry and function have tended to
be neglected.

   The septa of the vegetative mycelium of filamentous fungi do not take part
in cell division and separation like the septa of the yeasts. In budding yeasts a
primary septum develops centripetally at the region connecting the mother
cell to the bud. This primary cross wall is then thickened on both sides by the
addition of further material. When the mother cell and bud separate the

primary septum remains with the mother cell as a component of the bud scar. This process of localized wall growth and lysis is now understood in great detail (Cabib, 1975). Development of such molecular models for the septa of filamentous fungi are not as well advanced, possibly because of the emphasis that has been placed on the *differences* in septal structure encountered among these organisms.

5.2 Ultrastructural Form of Septa in the Fungal Sub-Divisions
Mastigiomycotina

Septa separating the developing zoosporangium from its rhizoid in the chytrids *Rhizophydium spherotheca* and a species of *Entophlyctis* have recently been shown to possess plasmodesmata (Powell, 1974). *Blasto-cladiella emersonii* and the Oömycetes *Phytophthora parasitica* and *capsici* produce imperforate septa at the sporangium base (Lessie & Lovett, 1968; Williams & Webster, 1970).

Zygomycotina

Electron microscope studies of septal types in this sub-division have tended to concentrate on some of the more unusual members. This probably reflects the generally accepted view that the young vegetative mycelium of most species in this sub-division is aseptate and that when septa are formed in older hyphae they are imperforate and complete (Buller, 1933). However, C. Fiddy and A. P. J. Trinci (personal communication) have shown that the mycelium of *Mucor hiemalis* and *M. rammanianus* regularly produces septa even when growing at the maximum specific growth rate under optimum cultural conditions. *Basidiobolus ranarum* is also an interesting exception to the general pattern in that its hyphae are very regularly septate. The septa are complete with no central pore or plasmodesmata (Gull & Trinci, 1975 and Figs. 5.1*a*, *b*). The septa of *Linderina pennispora* have a very different ultrastructural form, in that they possess an organized central pore. During formation of the septum the growing edge bifurcates and forms a central lenticular cavity in which a septal plug is subsequently formed. Septa produced between the pseudophialides and the merosporangia and spores are again modified by the addition of more wall material above and below the septum (Young, 1969; Benny & Aldrich, 1975). Septa with almost identical ultrastructural characteristics to these *Linderina* septa have been well documented in several members of the Trichomycetes (Reichle & Lichtwardt, 1972; Moss, 1975).

The existence of a further septal type in this sub-division of the fungi is indicated by the studies of Hawker & Beckett (1971) on the formation of the two septa which delimit the gametangium in zygospore formation. When still forming, the developing edge of the septum is thin and undulating. Adjacent to this developing edge and in the same plane is a line of flattened or circular vesicles which coalesce, resulting in the centripetal growth of the septum. After closure the completed cross wall is thickened by the addition of new material. Deposition of this material is asymetrical with more thickening occurring on the gametangial side of septum than on the suspensor side. This material on the gametangial face of the septum will eventually form part of the zygospore wall. The mature septum is perforated by numerous plasmodesmata.

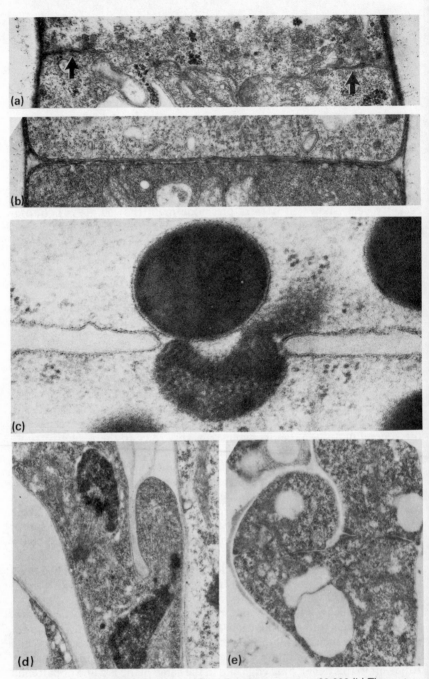

Fig. 5.1 (a) The developing septum of *Basidiobulus ranarum* × 29,000 (b) The mature non-perforate septum of *Basidiobolus ranarum* × 22,700 (Gull & Trinci, 1975) (c) Woronin bodies and the striated septal pore complex of *Aleuria aurantia* × 89,000 (d) Dividing nuclei in a developing clamp connection of the Basidiomycete *Phallus impudicus* × 18,900 (e) Septa developing at the positions previously occupied by dividing nuclei in the clamp connection of *Phallus impudicus* × 16,800.

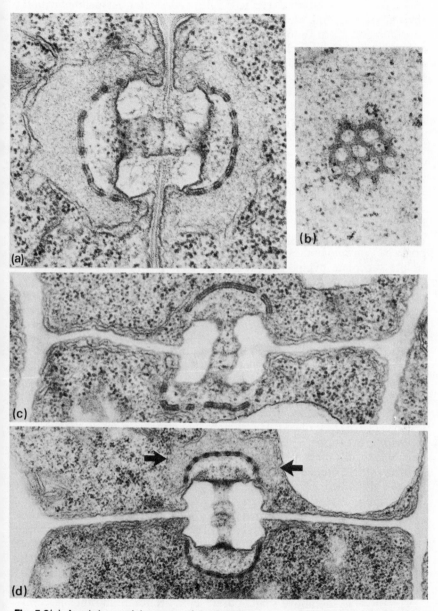

Fig. 5.2(a) A sub-hymenial septum of the Basidiomycete *Agrocybe praecox* × 43,400
(b) Glancing section of a parenthosome × 70,000 (c) Septum of *Agrocybe
praecox* from the hymenial region × 62,500 (d) Septum of *Agrocybe
praecox* from a hymenium/sub-hymenium transition region. Note the
presence of an outer cap (arrows) on the subhymenial side and the
absence of an outer cap on the hymenial side of the septum × 47,500
(Gull, 1976).

Ascomycotina

The 'typical' ascomycete septal type is generally described as a simple plate with a central pore around 0.05 to 0.5 μm in diameter. Associated with the septum are spheroid, electron dense, membrane bound organelles termed Woronin bodies. In 1886 Woronin described these organelles in *Ascobolus* as highly refractile particles on either side of the septum. They remain in the proximity of the pore and are not moved by cytoplasmic streaming. Woronin bodies of some Ascomycetes show an internal lattice structure with a regular repeating unit of between 5 and 40 nm (Bracker, 1967; Brenner & Carroll, 1968; K. Gull, unpublished observations). The presence of this lattice substructure and the enzyme digestion studies of McKeen (1971) and Hoch & Maxwell (1974) suggest that Woronin bodies are proteinaceous and they should not be confused with lysosomes or lipid bodies. Woronin bodies were not observed in various strains of *Neurospora crassa* (Trinci & Collinge, 1973; Hunsley & Gooday, 1974) or in many members of the *Aspergillus glaucus* group (Mason & Crosse, 1975). They are present, however, in a wide range of other species of *Aspergillus*. Woronin bodies generally have a larger diameter than the septal pore and apparently occlude the pore after injury to an adjacent hyphal compartment. It is possible that this closing of pores by Woronin bodies operates in older parts of the mycelium in order to isolate vacuolate hyphal compartments. Both the position and timing of Woronin body formation is unclear. Brenner & Carroll (1968) observed possible stages in the development of Woronin bodies within single membrane bound sacs.

Some fungi which lack characteristic Woronin bodies possess large hexagonal crystals in the cytoplasm which can occlude the septal pores of damaged hyphae (Trinci & Collinge, 1974; Mason & Crosse, 1975). These crystals were originally identified as ergosterol but more recent evidence suggests that they, like Woronin bodies, are composed of protein (Armentrout & Maxwell, 1974; Mason & Crosse, 1975). Elaborations of the septal pore rim have been found in a number of Ascomycetes. Hunsley & Gooday (1974) described tubular or striated structures within the pore rim of *Neurospora crassa*. The septal pore rim of *Ascodesmis sphaerospora* has been shown to contain a striate array of discs spaced 25 to 28 nm apart. Such structures are similar in appearance to the cytoplasmic microfilament bundles described in a *Russula* species by Gull (1975). However, their composition is as yet unknown. The central septal pore may be open providing a clear passage between the two hyphal compartments or it may be closed by an electron dense septal plug (Trinci & Collinge, 1973). Hyphae in the fruit-body of *Aleuria aurantia* possess large Woronin bodies which have a crystalline lattice of periodicity. The septal pore sometimes possesses the striated component with an electron dense plug (K. Gull, unpublished results and Fig. 5.1c).

Variations in the above single pored septa are found in other Ascomycetes. Certain lichenised Ascomycetes such as *Peltigera* possess multiperforate septa (Wetmore, 1973). These pores are much larger than plasmodesmata and have Woronin bodies associated with them. In the ascomycete fruit body septa that delimit asci from ascogenous hyphae often possess elaborate structures not found in vegetative hyphae. The structures formed at these septa vary in morphology from species to species. It is

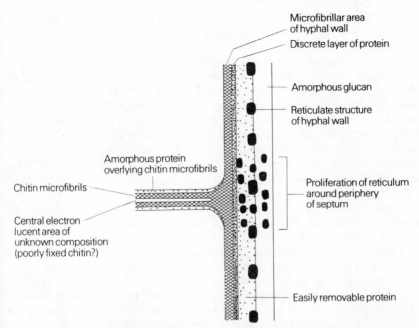

Fig. 5.3 Diagram of the suggested model of septal structure for *Neurospora crassa* (Hunsley & Gooday, 1974).

important to note that in certain species the elaborate structure appears only on the ascus side of the septum. In *Ascodesmis sphaerospora* and *Saccobolus kerverni* (Caroll, 1967) highly differentiated dome shaped arrays of radiating tubules cover the pore inside the ascus.

Basidiomycotina

The most highly structured septal type is found in most Basidiomycetes apart from the Uredinales and Ustilaginales. Using light microscopy Buller (1933) described hemispherical pads on either side of the septal pores of *Rhizoctinia solani* and *Coprinus* sp. Subsequently, the basidiomycete septum has been extensively studied using electron microscopy.

The septa are characterized by a swelling around the central pore (dolipore) and a hemispherical perforate cap (parenthosome) at each side of the pore. The plasmalemma is continuous along both sides of the septum and intercellularly around the dolipore swelling. The membrane of the parenthosome is considered to be a differentiated form of endoplasmic reticulum and is often seen to be continuous with sheets of endoplasmic reticulum that lie parallel to the septal wall. Three types of parenthosome have been described: (i) a cap with large (up to 1 μm) diameter apertures; (ii) a cap with small (40–80 nm) diameter apertures; and (iii) an imperforate parenthosome cap. We have studied the septal structure of many basidiomycete species in this laboratory and although these three major groups probably do exist we have often found variation in parenthosome structure within different cell types in the same fruit body. We are, therefore, very doubtful about the usefulness of such rather arbitrary groupings. Electron dense occlusions in the pore channel have been described in a number of basidiomycete septa (Moore & Marchant, 1972; Setliff, MacDonald & Patton, 1972).

A very precise ultrastructural modification to the basidiomycete septum has been described for septa occurring in certain cell types within the gill of the fruit-body of *Agrocybe praecox* (Gull, 1976). Septa of the subhymenium hyphae in the gill are characterized by an outer hemispherical cap of fibrillar material surrounding the parenthosome. This outer cap is membrane bound. Septa in the tramal and hymenial regions lack this extra outer cap. However, hyphae from the subhymenial/hymenial transition zones possess the outer cap only on the subhymenial side of the septum. This addition of an outer cap only to septa in the subhymenium of *Agrocybe* gills is a clear example of the coordinated control of septal structure within a mycelium. Moreover, the asymetrical septa, having an outer cap only on the subhymenial side emphasises the very precise nature of the control exercised by the adjacent hyphal compartments.

The dolipore/parenthosome septum complex is present in the vegetative hyphae of both monokaryons and dikaryons. During establishment of the dikaryon by fusion of two genetically compatible monokaryons, extensive intercellular migration of nuclei is known to occur. This movement of nuclei is facilitated by the selective removal of the complex septa to form large pored (up to 1.2 μm) simple septa (Giesy & Day, 1965).

The dolipore/parenthosome septum is not found in the phylogenetically primitive rust fungi. Septa of this group of fungi are interesting in that they appear to possess septa with an ultrastructural form intermediate between the central pored Ascomycetes and the dolipore/parenthosome septum of the more advanced Basidiomycetes. The septum of the rust fungi appears to consist of three layers which taper in thickness towards a single central pore. An amorphous electron dense region of cytoplasm surrounding the pores is separated from the main hyphal cytoplasm by a zone of single, membrane-bound vesicles. These vesicles have the ultrastructural appearance of microbodies and contain crystalline material, although histochemical tests indicate that they lack catalase (Coffey, 1975). The septal pore is often blocked by an electron dense plug that has the configuration of a pulley wheel. In addition to this distinct septum Littlefield & Bracker (1971) have described two other septal types in *Melampsora lini*. Partial septa were described which apparently formed by invagination of the existing wall layers and which may be a result of the convoluted growth of the fungus. Hyphae of the uredial primordia of *M. lini* also contained complete septa lacking any perforation.

A detailed study of the ultrastructural changes associated with the haustorial mother cell septum during haustorium formation has been described for the rust *Uromyces phaseoli* var. *vignae* (Heath & Heath, 1975).

Deuteromycotina

The ultrastructure of this group of fungi has received increasing attention recently with the result that a variety of septal types have been documented. Much of the attention, however, has focused on *Geotrichum candidum* whose septa have been shown to possess characteristic plasmodesmata (Steele & Fraser, 1973). The plasmodesmata are cylindrical intercellular connections 8.5 to 9.5 nm in diameter and the number per septum varies up to a maximum of around 50. They are most clearly seen in negatively stained or shadowed preparations of isolated septa (Hashimoto, Morgan & Conti, 1973). It should be noted that certain micrographs by Hashimoto *et al.* (1973) of the *Geotrichum* septum are most probably of contaminating valva from a centric diatom (Schnepf, 1974).

A number of fungi within this group possess a septum with a single central pore (Cooper, Grove, Mims & Szaniszlo, 1973; Hammill, 1974*b*). Reichle & Alexander, (1965) have, however, published an interesting account of certain *Fusarium* species which have multiperforate septa. These septa do not contain plasmodesmata but are multiperforate in that they possess a number (around 5 to 10) of pores of the same size as that in the more usual single pored septa. The pores appear to be arranged in a circle in the septum near the hyphal wall, possibly with one central pore. All the pores appear to have associated Woronin bodies. The Woronin bodies of another species— *Fusarium oxysporum* f.sp. *lycopersici* (Wergin, 1973) appear to be formed from organelles resembling microbodies and their development shows certain similarities to the outline of Woronin body development proposed by Brenner & Carroll (1968) for the Ascomycete *Ascodesmis*. Woronin bodies, pore plugs and elaborations of the septal pore rim have all been described for the septa of *Trichoderma saturnisporum*. A more complex septal type has been shown in the Deuteromycete *Wallemia sebi* (Terracina, 1974). Although the septa of this organism possess a single central pore the pore rim is expanded and resembles the pore rim of the homobasidiomycete septa. The septum of *Wallemia*, however, does not possess the parenthosome cap characteristic of the higher basidiomycete septum.

5.3 Biochemical Nature of Septa

Most ultrastructural reports of septa are concerned with the gross structure and few have attempted to define in any detail the biochemical nature of the septal components. However, the septa of two organisms—one an Ascomycete, *Neurospora crassa*, and the other a Basidiomycete, *Schizophyllum commune*—have been characterized to some extent, ultrastructurally and biochemically. In both organisms the ultrastructure and biochemistry of the lateral hyphal wall are well known and further studies have shown that the septum itself exhibits considerable biochemical specialization.

Neurospora crassa

Hunsley & Gooday (1974) have recently presented a model of septal structure in *Neurospora crassa*. In thin section the septa had a three-layered structure consisting of an electron transparent central layer flanked by an electron dense layer on each side. Ultrastructurally, the isolated septa changed with increasing culture age. The septa from one-day old mycelium often had microfibrils on the surface whilst in five-day old mycelium these microfibrils were obscured by an additional covering of amorphous material. The microfibrils were orientated tangentially and to a lesser extent radially to the single central pore. Various pieces of evidence indicate that these microfibrils are chitin and both light and electron microscope autoradiography showed *N*-acetylglucosamine incorporation into the developing septum. The synthesis of chitin appeared to take place either in the septum or at the plasmalemma and no evidence was found for cytoplasmic prepoly-merization and subsequent transport to the septum. Pronase treatment of the five day old septa removed the amorphous material and revealed the microfibrils thus indicating that this covering layer may be protein in part. The reticulate structure, thought to be a glycoprotein, which is a normal component of the lateral hyphal wall was conspicuously absent from the septum. It did, however, proliferate in the lateral wall around the septum periphery. A diagrammatic model of the *N. crassa* septum based on the work of Hunsley & Gooday (1974) is seen in Fig. 5.3.

Schizophyllum commune

The cell wall biochemistry of this basidiomycete fungus has been extensively studied (Niederpruem & Wessels, 1969; Wessels *et al.*, 1972). An alkali soluble α-1,3 linked glucan (S glucan) is concentrated at the hyphal surface and covers an inner layer of R glucan (an alkali insoluble glucan with β-1,3 and β-1,6 linkages) and chitin. Mechanical breakage of cells followed by extensive washing has been used to produce an isolated wall fraction. Septa in such hyphae retain much of their original form except that the parentho-some cap is removed. A series of chemical and enzymic digestion studies combined with electron microscopy of the treated septa have enabled a biochemical model of the *Schizophyllum commune* septum to be produced. In thin section the main part of the septum, excluding the dolipore swellings, often shows the central electron translucent line typical of many septa and shown by Hunsley & Gooday (1974) for *Neurospora crassa*. In *S. commune*, treatment with alkali or R-glucanase enzyme removes the dolipore swellings, but does not affect the main cross wall. Wessels & Marchant (1974) suggest that, as the R-glucanase preparation does not contain S-glucanase activity, the removal of dolipore swellings by alkali is not indicative of an S-glucan component. Janszen & Wessels (1970) have shown that the main part of the septal cross wall can be removed by an R-glucanase and chitinase, in combination or in succession. It should be noted that the lateral hyphal walls maintain their integrity after this treat-ment, presumably due to the presence of S-glucan. More recently P. Van der Valk and J. G. H. Wessels (personal communication) have completed work on chemical and enzymic treatment of septa which suggests that the dolipore swellings are composed of R-glucan and that the central electron translucent zone of the main septal cross wall is composed of chitin microfibrils in a rather pure form. This central component is overlaid on either side by chitin microfibrils embedded in R-glucan which stains with the Thiery reagent. This septal material appears to be continuous with the inner R-glucan/chitin layer of the lateral walls. One important point from their work is that S-glucan is not found in septa even though it constitutes about 30% of the lateral walls.

A clear understanding of the biochemistry of basidiomycete septa is essential in order to study the processes occurring in septal dissolution and nuclear migration during dikaryotization and fruiting. Differences in suscep-tibility to enzymic degradation between monokaryon and dikaryon septa have been reported (Wessels & Marchant, 1974). If such initial results are substantiated and extended, a rational model for these important develop-mental processes would seem to be in view. The existence of septal dissolution in Common A heterokaryons and B-factor mutants correlates with high R-glucanase activity and suggests that increases in R-glucanase activity may be intimately involved in septal dissolution and nuclear migration in the normal dikaryotization process.

5.4 Septum Formation

Buller (1933) noted that the time required for septum formation in *Rhizopus nigricans* was around 20 min. Septa of *Neurospora crassa* (Hunsley & Gooday, 1974) and *Alternaria solani* (King & Alexander, 1969) are formed somewhat faster—in 4 and 2 min respectively. In all cases where septum formation has been observed, development appears to be by the inward centripetal growth of an initial rim of material. Electron microscopy of developing septa supports this view (Littlefield & Bracker, 1971, Gull & Trinci, 1971). This centripetal growth of the cross wall goes to completion in organisms with complete septa whilst it ceases before complete closure in those septa which have a central pore. It is more difficult to understand the control of septal growth which produces the multiperforate septa of *Fusarium* (Reichle & Alexander, 1965) or the various septa with plasmodesmata. In the latter case it appears that the plasmodesmata are formed whilst the septum is developing, thus partially formed septa do possess plasmodesmata (Steele & Fraser, 1973). An increasing amount is known about the control of wall growth in the filamentous fungi. Although synthesis of new wall is confined to specific regions it does appear that there is potential for growth at almost every point. Katz & Rosenberger (1971) have shown that inhibition of protein synthesis by cycloheximide or by osmotic shock led to a uniform incorporation of N-acetylglucosamine into chitin along the length of a hypha. These treatments also resulted in the formation of increased numbers of both septa and branches.

What is clear about septum formation is that it is not entirely random. In uninucleate fungi such as the yeast *Schizosaccharomyces pombe*, hyphal compartments of basidiomycete monokaryons (Niederpruem & Jersild, 1972) and *Basidiobolus ranarum* (K. Gull, unpublished observations) septa are formed at sites previously occupied by the metaphase plate of the dividing nucleus. In some basidiomycete dikaryons which possess clamp connections this relationship is clearly seen (Fig. 5.1d, e). In the coenocytic fungi this very precise relationship is not seen, although again, septation is not a random event. The septa of *Neurospora crassa* begin to form 100 to 180 μm behind the growing hyphal tip (Hunsley & Gooday, 1974) with one septum forming every 3 to 4 min. In *Alternaria solani*, nuclear division is confined almost exclusively to hyphal tip cells, with mitosis in these cells being almost synchronous (King & Alexander, 1969). 8 to 15 min after nuclear division, septa begin to appear, with the first septum usually forming near the centre of the tip cell. The subsequent septa are then produced in no particular sequence but 3 to 7 septa were formed per septation period. King & Alexander (1969) found good correlation between length of the intercalary hyphal compartments and the numbers of nuclei in those compartments. They suggested that septation may be influenced by nuclear concentration. In *Aspergillus nidulans*, Clutterbuck (1970) could not determine any correlation between sites of septum formation and the positions of nuclei. However, tip cells where nuclei were sparse were divided into longer units than usual. A. P. J. Trinci (personal communication) has shown that in *A. nidulans* the volume of cytoplasm per nucleus in a diploid strain is approximately double the value obtained for a haploid strain. Also, the initial septum formed tended to reduce the apical hyphal compartment to about half its maximum length.

Many of these results suggest that, as has been proposed for other eukaryotes, a factor may accumulate in the hyphae of filamentous fungi during interphase which at a critical concentration triggers mitosis. In some filamentous fungi, septation appears to be closely linked to the division of a nucleus, although in coenocytes there appears to be a diminution of the expected number of septa. However, even in a coenocytic fungus there is a close relationship between the time of mitosis and septum formation (Clutterbuck, 1970; King & Alexander, 1969).

5.5 Function of Septa

The fundamental characteristics of septa have tended to become obscured by the wide variety of septal forms found in the filamentous fungi and this has hindered the development of a reasonable theory for the existence of septa. It has been suggested that septa serve to add rigidity to hyphae. Obviously, the addition of cross walls to a long tube will stabilize it to some extent. So, septa must have some mechanical function but this is probably of secondary importance. Septa protect the mycelium from mechanical damage or possibly from osmotic injury to growing tip areas. Many experiments have shown that septa can serve to localize damage to the tip cell or the few intercalary compartments that sustain damage. The rest of the mycelium is protected and can continue to grow directly or by branching. The large pored septa with associated Woronin bodies have the advantage that intercompartment cytoplasmic movement is possible whilst a mechanism exists for rapid closure of the pore under stress conditions.

I would argue that both of the above two properties of septa are a bonus to their main function which, I believe, is to give the hypha the ability to undergo differentiation. The existence of septa, particularly perforate septa, gives the higher fungi the potential of colony differentiation and tissue production together with the development of specific cell types. It appears that the more complex the septal structure so the more complex is the type and degree of differentiation allowed.

The lack of septa, or the limited capacity to produce the imperforate septa that are found in the lower fungi, means that these organisms tend to rely heavily on cytoplasmic (*i.e. intra*cellular) cleavage to achieve limited differentiation. This is clearly seen in the mechanisms of asexual spore production in the Mucorales. Asexual spore formation in the Ascomycotina and Deuteromycotina, however, is heavily dependent upon *inter*cellular cleavage achieved by means of a septum. Also, in many fungi, spore development is dependent upon the development of a perforate system which can be plugged at some later point in the sporulation process. In *Trichoderma saturnisporum*, Hammill (1974) has clearly shown that spore development depends on growth of a septum which becomes bilayered, the upper layer forms part of the conidial base and the lower layer becomes a portion of the wall of the conidial initial. Thus cell separation is achieved by splitting of the septum down a central line. Recent studies involving careful light and electron microscopy indicate that septal splitting is an important part of asexual spore development in many Ascomycotina and Deuteromycotina. Presumably, this intercellular cleavage involves enzymic lysis of the central zone of these septa. In order to understand fully such localized events we need to know more about the biochemical nature of the layers found in septa, particularly the central electron translucent zone (Hunsley & Gooday, 1974).

Unfortunately, there are very few reports of the spatial and temporal control of septal plugging in the vegetative mycelium of Ascomycotina or Deuteromycotina. Presumably, unplugged septal pores allow intercalary compartments of young hyphae to contribute to tip growth whilst, after septal plugging, these hyphal compartments can redirect their activities into alternative differentiation events. The presence of pores, therefore, maximizes the capacity of the mycelium for vegetative growth, whilst septal pore plugging provides a means of rerouting the activities of an intercalary compartment. Trinci & Collinge (1973) have shown that in various strains of *Neurospora crassa* (both wild type and spreading colonial mutants) there was a good relationship between hyphal density and the plugging of septal pores. Also, in some strains, there was a close relationship between the width of the peripheral growth zone of the colony and the position in the colony at which septa became plugged. The peripheral growth zone of a colony is the region in which hyphae are able to contribute protoplasm to the apical extension of the leading hyphae of the colony. As Trinci & Collinge suggest (1973) it is possible that plugging of septal pores is initiated when adverse conditions such as low oxygen tension, substrate exhaustion or accumulation of inhibitory secondary metabolites reach critical levels close to the hyphal compartment. Much more of this type of research is required to determine the importance of septal plugging in the differentiation of the fungal colony.

A different approach to the problem has been initiated by Morris (1975) with the isolation of mutants blocked in septum formation. Morris has reported five temperature-sensitive mutants of *Aspergillus nidulans* which are defective in septum formation at the restrictive temperature of 42°C. Hyphae of two of the mutants appeared morphologically normal whilst the other three had fat hyphae more than twice the diameter of the wild type. The mutants germinated and grew for a period of 18 h at 42°C. Numerous mitotic nuclear divisions occurred during this time. However, the mutants were unable to form colonies at this temperature. These experiments clearly show that growth and nuclear division in *A. nidulans* are not dependent upon the formation of septa. However, although growth and nuclear division are not limited initially, failure to become septate imposes some eventual restriction such that growth and differentiation of the colony is inhibited.

The most highly differentiated fruit-bodies formed by micro-organisms are those produced by the basidiomycetous fungi. They involve complex arrangements of hyphae to produce tissue-like specializations and cell types. I do not believe that it is coincidental that these organisms also possess the most complex septal form found in the fungi. Much of the fruit-body is designed to support, both mechanically and biochemically, the spore producing surface or hymenium. The hymenial layer of the gills of many basidiomycete fruit-bodies contains two distinct cell types: cystidia and basidia. Basidia are the hyphae in which meiosis occurs and which eventually produce the basidiospores. Cystidia are termed sterile since they do not produce basidiospores; they are usually larger than basidia and can occur in various locations on the gill face and edge. Gull & Newsam (1975) have shown that cystidia of *Agrocybe praecox* show distinct ultrastructural differentiation. They are large coenocytic cells which lack the glycogen storage reserves of the adjacent basidia and contain large amounts of smooth endoplasmic reticulum. Cystidia and basidia both arise from hyphae in the sub-hymenial region of the gill and are separated from these hyphae by a septum. The septa of sub-hymenial hyphae are unique in that they possess an extra outer cap (Gull 1976 and Figs. 5.2a, b, c and d). Presumably, it is the complex dolipore/parenthosome septum and possibly the addition of the outer cap which allows the differentiation of cell types in such fruit-bodies. The elaboration of septa between meiotic cells and supporting hyphae is also found in the Ascomycotina (Carroll, 1967).

Development of differing cell types and coincident development of tissues is a complex problem in higher plant and animal systems. It involves extra difficulties for a fungus in that the filamentous growth form imposes its own peculiar restraints. Septa are, therefore, used to delimit these cell types, and septal structure seems dependent upon the particular region or cell.

Variations in the ultrastructural organization of septa have also been shown to be intimately involved in the process of haustorium development in rust fungi (Heath & Heath, 1975). During infection of a host plant, rust fungi form a haustorium mother cell by laying down a septum near the hyphal apex. Heath & Heath (1975) used the ultrastructural alterations in this haustorial mother cell's development as markers to determine the sequence of changes in this characteristic septum. After its initial formation, the septal wall resembled that found in other parts of the mycelium or in other rusts (Littlefield & Bracker, 1971; Coffey, 1975) in consisting of two electron dense layers separated by a narrow central electron dense zone. Immediately before and during breaching of the host wall by the penetration peg, very refined ultrastructural changes occurred in the organization of the haustorial mother cell septum. The most striking of these changes was the elaboration of the plasmalemma on both sides of the septum. These elaborations in the haustorial mother cell consisted of whorls and tended to disappear prior to the formation of the penetration peg. The more tubular plasmalemma projections on the hyphal side of the septum reached maximum lengths of around 900 nm at the time of the penetration. The original open septal pore became plugged by osmiophilic material during this period, but by the time the haustorium was mature both it and the plasmalemma projections had disappeared. Subsequent to removal of this plug, the septal pore develops in the region of differentiated cytoplasm bounded by microbody-like organelles (Littlefield & Bracker, 1971; Coffey, 1975).

This complex series of ultrastructural changes in a septum which parallels a developmental process again shows the diversity of form which can be expected of septa in different regions of a mycelium. Certain changes described by Heath & Heath (1975) also show that the septal pore complex sometimes varies on opposite sides of a particular septum. These ultrastructural changes considered with the asymmetrical septal pore structures reported by Gull (1976) for a Basidiomycete and Carroll (1967) for Ascomycetes indicate that septal structure can be very profoundly and precisely controlled by the hyphal compartment on either side.

5.6 Conclusions

Large numbers of publications have detailed the wide variety of septal types to be found in the fungi. A few of these reports have shown that the septal form of a particular fungus may be modified according to time, its position in a mycelium or its position in a differentiated structure such as a fruit-body. Awareness of variability means that future work detailing septal ultrastructure must define precisely the temporal and spatial position of the septum in the life cycle of the organism. If this is achieved, studies of septal organisation will retain their taxonomic usefulness whilst making an important contribution to our understanding of fungal physiology and development.

5.7 References

ARMENTROUT, V. N. & MAXWELL, D. P. (1974). Hexagonal crystals in an ergosterol-free mutant of *Neurospora crassa. Canadian Journal of Microbiology* **20**, 1427–8.

BENNY, G. L. & ALDRICH, H. C. (1975). Ultrastructural observations on septal and merosporangial ontogeny in *Linderina pennispora* (Kickxellales; Zygomycetes). *Canadian Journal of Botany* **53**, 2325–35.

BRACKER, C. E. (1967). Ultrastructure of fungi. *Annual Review of Phytopathology* **5**, 343–74.

BRENNER, D. M. & CARROLL, G. C. (1968). Fine structural correlates of growth in hyphae of *Ascodesmis sphaerospora. Journal of Bacteriology* **95**, 658–71.

BULLER, A. H. R. (1933). The translocation of protoplasm through septate mycelium of certain Pyrenomycetes, Discomycetes and Hymenomycetes. In *Researches of fungi.* Vol. 5, pp. 75–167. London: Longmans, Green.

CABIB, E. (1975). Molecular aspects of yeast morphogenesis. *Annual Review of Microbiology* **29**, 191–214.

CARROLL, G. C. (1967). The fine structure of the ascus septum in *Ascodesmis sphaerospora* and *Saccobolus kerverni. Mycologia* **59**, 527–32.

CLUTTERBUCK, A. J. (1970). Synchronous nuclear division and septation in *Aspergillus nidulans. Journal of General Microbiology* **60**, 133–5.

COFFEY, M. D. (1975). Obligate parasites of higher plants, particularly rust fungi. *Symposium of the Society for Experimental Biology* **29**, 297–323.

COOPER, B. H., GROVE, S., MIMS, C. & SZANISZLO, P. J. (1973). Septal ultrastructure in *Phialophora pedrosoi, Phialophora verrucosa* and *Cladosporium carrionii. Sabouraudia* **11**, 127–30.

DE BARY, A. (1884). *Vergleichende Morphologie und Biologie der Pilze.* Leipzig.

GIESY, R. M. & DAY, P. R. (1965). The septal pores of *Coprinus lagopus* (*FR*) sensu Buller in relation to nuclear migration. *American Journal of Botany* **52**, 287–93.

GULL, K. (1975). Cytoplasmic microfilament organisation in two basidiomycetous fungi. *Journal of Ultrastructure Research* **50**, 226–32.

GULL, K. (1976). Differentiation of septal ultrastructure according to cell type in the basidiomycete *Agrocybe praecox. Journal of Ultrastructure Research* **54**, 89–94.

GULL, K. & NEWSAM, R. J. (1975). Ultrastructural organisation of cystidia in the basidiomycete *Agrocybe praecox. Journal of General Microbiology* **91**, 74–8.

GULL, K. & TRINCI, A. P. J. (1971). Fine structure of spore germination in *Botrytis cinerea. Journal of General Microbiology* **68**, 207–20.

GULL, K. & TRINCI, A. P. J. (1975). Septal ultrastructure in *Basidiobolus ranarum. Sabouraudia* **13**, 49–51.

HAMMILL, T. M. (1974a). Electron microscopy of phialides and conidiogenesis in *Trichoderma saturnisporum*. *American Journal of Botany* **61**, 15–24.

HAMMILL, T. M. (1974b). Septal pore structure in *Trichoderma saturnisporum*. *American Journal of Botany* **61**, 767–71.

HASHIMOTO, T., MORGAN, J. & CONTI, S. F. (1973). Morphogenesis and ultrastructure of *Geotrichum candidum* septa. *Journal of Bacteriology* **116**, 447–55.

HAWKER, L. E. & BECKETT, A. (1971). Fine structure and development of the zygospore of *Rhizopus sexualis* (Smith) Callen. *Philosophical Transactions of the Royal Society of London. B. Biological Sciences* **263**, 71–100.

HEATH, M. C. & HEATH, I. B. (1975). Ultrastructural changes associated with the haustorial mother cell septum during haustorium formation in *Uromyces phaseoli* var *vignae*. *Protoplasma* **84**, 297–314.

HOCH, H. C. & MAXWELL, D. P. (1974). Proteinaceous hexagonal inclusions in hyphae of *Whetzelinia sclerotiorum* and *Neurospora crassa*. *Canadian Journal of Microbiology* **20**, 1029–35.

HUNSLEY, D. & GOODAY, G. W. (1974). The structure and development of septa in *Neurospora crassa*. *Protoplasma* **82**, 125–46.

JANSZEN, F. H. A. & WESSELS, J. G. H. (1970). Enzymic dissolution of hyphal septa in a Basidiomycete. *Antonie van Leeuwenhoek* **36**, 255–7.

KATZ, D. & ROSENBERGER, R. F. (1971). Hyphal wall synthesis in *Aspergillus nidulans*: effect of protein synthesis inhibition and osmotic shock on chitin insertion and morphogenesis. *Journal of Bacteriology* **108**, 184–90.

KING, S. B. & ALEXANDER, L. J. (1969). Nuclear behaviour, septation and hyphal growth of *Alternaria solani*. *American Journal of Botany* **56**, 249–53.

LESSIE, P. E. & LOVETT, P. E. (1968). Ultrastructural changes during sporangium formation and zoospore differentiation in *Blastocladiella emersonii*. *American Journal of Botany* **55**, 220–36.

LITTLEFIELD, L. J. & BRACKER, C. E. (1971). Ultrastructure of septa in *Melampsora lini*. *Transactions of the British Mycological Society* **56**, 181–8.

MASON, P. J. & CROSSE, R. (1975). Crystalline inclusions in hyphae of the *glaucus* group of Aspergilli. *Transactions of the British Mycological Society* **65**, 129–34.

MCKEEN, W. E. (1971). Woronin bodies in *Erysiphe graminis* D. C. *Canadian Journal of Microbiology* **17**, 1557–60.

MOORE, R. T. & MARCHANT, R. (1072). Ultrastructural characterisation of the basidiomycete septum of *Polyporus biennis*. *Canadian Journal of Botany* **50**, 2463–9.

MORRIS, N. R. (1975). Mitotic mutants of *Aspergillus nidulans*. *Genetical Research* **26**, 237–54.

MOSS, S. T. (1975). Septal structure in the Trichomycetes with special reference to *Astreptonema gammari* (Eccrinales). *Transactions of the British Mycological Society* **65**, 115–27.

NIEDERPRUEM, D. J. & JERSILD, R. A. (1972). Cellular aspects of morphogenesis in the mushroom *Schizophyllum commune*. *Critical Reviews in Microbiology* **1**, 545–76.

NIEDERPRUEM, D. J. & WESSELS, J. G. H. (1969). Cytodifferentiation and morphogenesis in *Schizophyllum commune*. *Bacteriological Reviews* **33**, 505–35.

POWELL, M. J. (1974). Fine structure of plasmodesmata in a chytrid. *Mycologia* **66**, 606–14.

REICHLE, R. E. & ALEXANDER, J. V. (1965). Multiperforate septa, Woronin bodies and septal plugs in *Fusarium*. *Journal of Cell Biology* **24**, 489–96.

REICHLE, R. E. & LICHTWARDT, R. W. (1972). Fine structure of the Trichomycete, *Harpella melusinae* from black fly guts. *Archiv für Mikrobiologie* **81**, 103–25.

SCHNEPF, E. (1974). Septum of *Geotrichum candidum* or valva of a centric diatom? *Journal of Bacteriology* **119**, 330–1.

SETLIFF, E. C., MACDONALD, W. L. & PATTON, R. F. (1972). Fine structure of the septal pore apparatus in *Polyporus tomentosus*, *Poria latemarginata*, and *Rhizoctonia solani*. *Canadian Journal of Botany* **12**, 2559–63.

STEELE, S. D. & FRASER, T. W. (1973). The ultrastructure of *Geotrichum candidum* hyphae. *The Canadian Journal of Microbiology* **19**, 1507–12.

TERRACINA, F. C. (1974). Fine structure of the septum in *Wallemia sebi*. *Canadian Journal of Botany* **52**, 2587–90.

TRINCI, A. P. J. & COLLINGE, A. J. (1973). Structure and plugging of septa and wild type and spreading colonial mutants of *Neurospora crassa*. *Archiv für Mikrobiologie* **91**, 355–64.

TRINCI, A. P. J. & COLLINGE, A. J. (1974). Occlusion of the septal pores of damaged hyphae of *Neurospora crassa* by hexagonal crystals. *Protoplasma* **80**, 57–67.

WAHRLICH, W. (1893). Zur Anatomie der Zelle bei Pilzen und Fadenalgen. *Script Botanica Horti Universitalis Imperialis*

Petropolitanae **4**, 101–55.

WERGIN, W. P. (1973). Development of Woronin bodies from microbodies in *Fusarium oxysproum* f.sp. *lycopersici*. *Protoplasma* **76**, 249–60.

WESSELS, J. G. H. & MARCHANT, R. (1974). Enzymic degradation of septa in hyphal wall preparations from a monokaryon and a dikaryon of *Schizophyllum commune*. *Journal of General Microbiology* **83**, 359–68.

WESSELS, J. G. H., KREGER, D. R., MARCHANT, R., REGENSBURG, B. A. & DE VRIES, O. M. H. (1972). Chemical and morphological characterisation of the hyphal wall surface of the basidiomycete

Schizophyllum commune. *Biochemica et Biophysica Acta* **273**, 346–58.

WETMORE, C. M. (1973). Multiperforate septa in lichens. *New Phytology* **72**, 535–8.

WILLIAMS, W. T. & WEBSTER, R. K. (1970). Electron microscopy of the sporangium of *Phytophthora capsici*. *Canadian Journal of Botany* **48**, 221–7.

WORONIN, E. (1864). Zur Entwicklungsgeschichte des *Ascobolus pulcheorimus* und einiger Pezizen. *Abhandl Senkenberg Naturfursch Ges* **5**, 333–4.

YOUNG, T. W. K. (1969). Ultrastructure of aerial hyphae in *Linderina pennispora*. *Annals of Botany* (*London*) **33**, 211–16.

CHAPTER 6

Nucleic Acid and Protein Synthesis during Fungal Spore Germination

R. BRAMBL, L. D. DUNKLE AND J. L. VAN ETTEN

6.1 Introduction

The ordered conversion of metabolically quiescent fungal spores to rapidly growing cells is a phenomenon of considerable interest to mycologists; but it also provides developmental biologists with an attractive and convenient system for examining certain processes of eukaryotic cell development and metabolic regulation. There are many problems of common interest to both groups of biologists who want to understand how the germinating fungal spore accomplishes important rearrangements and conversions of molecular and macromolecular materials that are preserved during the dormancy period. However, it is unknown how such metabolic transformations are related temporally or spatially and how modifications of spore materials are related causally or consequentially to the control of spore germination itself. An important subsidiary problem is how these conversions and metabolic changes, as well as control of germination, may be related to the rapid increases in the synthetic rates of proteins and nucleic acids which are characteristic of spore germination. Within recent years a number of investigators have begun to describe some of the molecular events which accompany spore germination, and they have sought to describe how and when proteins and nucleic acids are synthesized during germination. While very little is known about how the transcriptional and translational activities of the spore genetic system are related to other important germination-induced transformations and to control of germination, we now have some information about when certain proteins and nucleic acids are synthesized and the level at which the flow of genetic information is resumed upon germination. In this chapter, we will attempt to summarize the information available about these topics and, where possible, integrate this information with that concerning certain other spore metabolic activities.

We shall use the term dormancy in its most inclusive sense, as any rest period or reversible interruption of the phenotypic development of an organism (Sussman, 1966). Excluded from this discussion is the important research on non-filamentous fungus spore germination of the slime moulds and yeast. A recent review by Lovett (1975) has admirably described spore germination in the water mould *Blastocladiella emersonii* and we will refer to research on this organism and closely related fungi only for supportive or instructive purposes. For supplementary reviews of the topics described here, the reader should see the volume edited by Weber & Hess (1976).

6.2 Composition and Physical Properties of DNA from Dormant and Germinating Spores

The base composition and certain physical properties of fungal DNA have been determined for many species (Storck & Alexopoulos, 1970), but only a few studies have compared directly the properties of DNA in dormant and germinating spores. The following generalities can be made from results of studies on DNA from the vegetative stage of various fungi representing all major taxonomic groups: (i) guanine plus cytosine (GC) contents of fungal DNAs are quite diverse, having a range of 27 to 70 mole % GC (Storck, 1974); (ii) estimates of the size of the haploid genome of several species range from 0.5×10^{10} daltons to 3×10^{10} daltons of DNA distributed over 2 to 18 chromosomes (Storck, 1974; Dusenbery, 1975); (iii) a portion (5 to 20%) of the total cellular DNA consists of repeated sequences, including cistrons encoding tRNAs and rRNAs (Van Etten, Dunkle & Knight, 1976; Dutta, 1974); and (iv) of the total cellular DNA, 80 to 99% is nuclear DNA and 1 to 20% is mitochondrial DNA, which is a closed circular molecule (Clayton & Brambl, 1972; Hollenberg, Borst & Van Bruggen, 1970) containing 4 to 7×10^7 daltons of unique sequences (Schäfer & Küntzel, 1972). In addition, some fungi contain minor DNA species that are compositionally and physically different from nuclear or mitochondrial DNA. Thus, the fungal genome is intermediate in size and complexity between prokaryotes and higher eukaryotes. The mechanisms of DNA replication and regulation of transcription are poorly understood. Results of investigations on the presence of histones and the role of other proteins associated with fungal DNA are equivocal.

A few studies have been concerned directly with comparing the base composition of DNA from dormant and germinated spores. From the results of these studies, we may determine whether a selective amplification or deletion of genetic information occurs during sporulation or spore germination. In a few cases such compositional changes have been reported to occur during the life cycles of prokaryotic and eukaryotic organisms.

Minagawa, Wagner & Strauss (1959) first chemically analyzed the nucleotides in DNA from dormant and germinated conidia and from mycelium of *Neurospora crassa* and found no significant differences in base composition of DNA in the various morphological states. Similar conclusions obtained from DNA:DNA hybridization of *N. crassa* conidial and mycelial DNAs and from thermal elution experiments were reported by Dutta & Chaudhuri (1975). Yanagita and his colleagues (Kogane & Yanagita, 1964; Tanaka, Kogane & Yanagita, 1965) concluded from chemical analyses that DNA is not altered during early stages in germination of *Aspergillus oryzae* conidia. Dunkle & Van Etten (1972) found that the buoyant density in CsCl gradients, thermal denaturation profiles, and spectral properties of dormant spore DNA were indistinguishable from those of germinated spore DNA in *Botryodiplodia theobromae* and *Rhizopus stolonifer*. These results corroborate the observations of Storck & Alexopoulos (1970) and suggest that, in the fungi examined, no loss or amplification of genetic material occurs during vegetative development. A possible exception to this generality may occur in *Blastocladiella emersonii*. Zoospores of this water mould contain 8 to 16 organelles, designated gamma (γ) particles which contain a small proportion of DNA with a distinct buoyant density (Myers & Cantino, 1971). The γ-particles disappear during zoospore germination, but neither the fate of γ-DNA the consequence of γ-particle decay on the composition of the major species of DNA have been determined.

6.3 DNA Synthesis during Spore Germination

Studies on DNA synthesis have been hampered by the failure of fungi to incorporate radioactive thymidine specifically into DNA, apparently because of the absence of thymidine kinase in this group of micro-organisms (Grivell & Jackson, 1968). However, considerable detail about DNA synthesis and its temporal relation to germination has been obtained by other approaches. For example, cytological studies have shown that nuclei divide at some stage of spore germination in most of the fungi examined. The inference from these results is that DNA synthesis occurs prior to nuclear division. However, in spores of *Phycomyces blakesleeanus* an increase in nuclear number occurs 3 to 4 h before significant incorporation of precursor into DNA is detected (Van Assche & Carlier, 1973). Thus, Van Assche & Carlier (1973) concluded that the dormant spore is in the G2 (premitotic) stage of the cell cycle. Similarly, in *Neurospora crassa* conidia, the number of nuclei increases during germination (Bianchi & Turian, 1967). But even when DNA synthesis is inhibited by hydroxyurea, about 30% of the nuclei divide (Loo, 1976). Schmit & Brody (1976) concluded that some of the nuclei in dormant conidia are arrested in the G1 (pre-synthetic) stage and some in the G2 stage of the cell cycle. The spores of most fungi that have been studied are multinucleate, and there is a lack of mitotic synchrony in the nuclear population. On the other hand, conidia of *Aspergillus nidulans* (Bainbridge, 1971) and microconidia of *Fusarium oxysporum* (Kumari, Decallone & Meyer, 1975) are uninucleate, and the nuclei divide in a more synchronous pattern. In these species an increase in spore DNA content and incorporation of radioactive precursors into DNA occur well before the increase in the number of nuclei per spore. Thus, it appears that the nuclei are arrested in the G1 phase and that the G2 phase is relatively short. This is in contrast to the situation in *Blastocladiella emersonii* in which the G1 phase is apparently very short, since DNA is replicated soon after mitosis is completed (Lovett, 1975).

The temporal relation between nuclear-DNA synthesis and germ tube emergence has been studied by pulse-labelling germinating spores with non-specific radioactive precursors and by colorimetric analyses of spore extracts. In general, the results of these studies indicate that DNA synthesis is a late event of the macromolecular biosyntheses accompanying germination. In most fungi, if not all, the onset of DNA synthesis occurs after RNA and protein synthesis and after or only a short time prior to germ tube emergence. For example, in *Fusarium oxysporum*, the first germ tubes were observed after 4 h, and the total DNA of the culture increased sharply after 5 h (Kumari *et al.*, 1975).

The essentiality of nuclear DNA synthesis for spore germination has not been established. Many inhibitors of DNA synthesis have additional effects on cellular biosynthesis or do not penetrate the spore and consequently have no effect. However, because other biosynthetic activities begin before DNA synthesis and because DNA synthesis usually does not occur until about the time of germ tube emergence, this process is probably not required for spore germination.

Uredospores of the obligate parasite bean rust fungus *Uromyces phaseoli* can be induced in the absence of the host to differentiate a discrete series of infection structures, consisting of an appressorium, vesicle, and infection hyphae, which are comparable to the structures formed during parasitic colonization of the host. Nuclear division and nuclear DNA synthesis occur only in germ tubes that are induced to form the infection structures; these events do not occur in germ tubes that are growing linearly. Staples, App & Ricci (1975) studied the essentiality of nuclear division and nuclear DNA synthesis for germ tube differentiation by using a high concentration (1 mg/ml) of the inhibitors hydroxyurea and cordycepin. They found that while neither drug influenced germination, hydroxyurea inhibited DNA synthesis, nuclear division, and infection structure development, whereas cordycepin inhibited nuclear division without affecting DNA synthesis or early stages of infection structure formation. They concluded that infection structure differentiation, but not germination, requires DNA synthesis and that nuclear division is not essential for either process.

Mitochondrial DNA synthesis has been studied in only a few fungi. In conidia of *Botryodiplodia theobromae* mitochondrial DNA synthesis begins at the same time as nuclear DNA synthesis and continues throughout germination and germ tube growth (Dunkle & Van Etten, 1972). In the presence of ethidium bromide, which selectively inhibits mitochondrial DNA synthesis, the conidia germinated at the same rate and to the same extent as conidia incubated in the absence of the drug (Dunkle, Van Etten & Brambl, 1972). Furthermore, ethidium bromide did not have an immediate effect on the synthetic rates of nuclear DNA, RNA, and protein or on aerobic respiration. Thus, nuclear and mitochondrial DNA syntheses are apparently independent processes and the activity of the mitochondrial genetic system is not essential to the spores until the germ tubes begin vegetative growth. Similar conclusions were drawn by Tingle, Küenzi & Halvorson (1974) who found that ascospores of a *petite* strain of yeast which lacks mitochondrial DNA germinated at a rate only slightly less than wild-type spores.

During germination and undifferentiated germ tube growth of *Uromyces phaseoli* uredospores, only mitochondrial DNA is synthesized (Staples, 1974). When the germ tubes stop elongating upon differentiation of appressoria, mitochondrial DNA synthesis ceases and nuclear DNA synthesis begins. Apparently, in the bean rust fungus, mitochondrial DNA synthesis and nuclear DNA synthesis are mutually exclusive processes occurring during germination and germ tube differentiation, respectively.

The enzymes involved in fungal DNA synthesis have not been thoroughly studied, especially as they pertain to spore germination. Gong, Dunkle & Van Etten (1973) isolated DNA polymerase from spores of *Rhizopus stolonifer*. The enzyme from dormant and germinated spores was purified several hundred-fold and found to exhibit identical physical and chromatographic characteristics and to have similar reaction kinetics and requirements for activity. However, during early stages of purification the enzyme from dormant spores was almost completely dependent upon the addition of exogenous DNA for maximum activity, whereas the enzyme fraction from germinated spores did not require exogenous DNA. Furthermore, the germinated spore enzyme sedimented in glycerol density gradients in a heterogeneous manner and more rapidly than the dormant spore enzyme. By treating the enzyme preparation with deoxyribonuclease prior to centrifugation, the sedimentation constant was reduced to about 5S and the DNA independence was completely eliminated. Thus, DNA polymerase from germinated *R. stolonifer* spores, in which DNA synthesis is proceeding, is apparently more tightly bound to endogenous DNA than the enzyme from dormant spores.

Insufficient information is available to determine if fungi contain multiple DNA polymerases with repair and replicative activities. But Jeggo, Unrau, Banks & Holliday (1973), in their studies of DNA polymerase in temperature-sensitive mutants of *Ustilago maydis*, indicated that more than one DNA polymerase exists. If this is the general situation in fungi, it should be determined whether the DNA synthesis observed during spore germination is a repair or a replicative synthesis.

6.4 Composition and Physical Properties of RNA from Dormant and Germinating Spores

Fungi contain all of the major classes of RNA present in other eukaryotic organisms. Large ($1.3–1.4 \times 10^6$ daltons, 25S RNA) and small (*c.* 0.7×10^6, 18S RNA) ribosomal (r)RNAs comprise approximately 75% of the total cellular RNA; ribosome-associated RNA components, 5S RNA (*c.* 3.5×10^4) and 5.8S RNA (*c.* 5.4×10^4), together with transfer (t)RNA usually make up 15–20% and messenger (m)RNA less than 5% of the total RNA. The mitochondria also contain unique species of rRNAs, tRNAs and mRNAs. Additional species of RNA may be associated with DNA in chromatin or with cell membrane and wall fractions, but these species, if they exist, comprise a small percentage of the total RNA. Although all these RNAs are probably present in dormant spores, direct evidence for the existence of any of these RNAs in dormant spores is limited to only a few fungi.

Ribosomal RNAs and ribosomes

Large and small rRNAs isolated from dormant spores appear to be identical to rRNAs in germinated spores. For example, no differences in physical properties have been detected in rRNAs isolated from dormant and germinated spores of *Neurospora crassa* (Henney & Storck, 1963*a*, *b*), *Aspergillus oryzae* (Horikoshi, Ohtaka & Ikeda, 1965) and *Botryodiplodia theobromae* (Knight & Van Etten, 1976*a*). In the last species, the two small ribosomal associated components, 5S RNA and 5.8S RNA, from both spore states also exhibited identical electrophoretic properties in polyacrylamide gels.

Ribosomes which have biological activity when assayed with polyuridylic acid as the mRNA have been isolated from dormant spores. These include spores of *Aspergillus oryzae* (Horikoshi & Ikeda, 1968), *Botryodiplodia theobromae* (Van Etten, 1968), *Fusarium solani* (Rado & Cochrane, 1971) and several rust fungi (Staples & Yaniv, 1973 and references cited therein). With the exception of some of the rust fungi, ribosomes isolated from dormant spores were less active (*c*. 25 to 80% less) than ribosomes isolated from germinated spores. The reason for this apparent lower activity is not known although it could be due to isolation artifacts. For example, Horikoshi and his colleagues (Horikoshi & Ikeda, 1969; Horikoshi, 1971) have shown that ribosomes from *A. oryzae* spores are more unstable, presumably because of more ribonuclease (RNase) activity, than ribosomes from germinated spores. Such increased RNase activity could degrade either the rRNA, mRNA, or both. In the case of some of the rust fungi, ribosomes may actually have lower *in vitro* protein synthetic activity after germination (Staples & Yaniv, 1973).

The question of whether there are physical differences between ribosomes from germinated and ungerminated spores is difficult to answer at the present time. The sedimentation coefficients for ribosomes from dormant and germinated spores are identical for *Neurospora crassa* (Henney & Storck, 1963*b*), *Aspergillus oryzae* (Horikoshi *et al.*, 1965) *Botryodiplodia theobromae* (Brambl & Van Etten, 1970), and several rust fungi (Staples & Yaniv, 1973). Furthermore, ribosomes isolated from ungerminated conidia and hyphae of *N. crassa* are similar serologically and ribosomal proteins isolated from these two fungal states have similar qualitative profiles after acrylamide gel electrophoresis (Rothschild, Itakawa & Suskind, 1967). However, Henney & Strock (1963*b*) found that the ratio of protein to RNA varied in ribosomes isolated from ascospores (33:67), conidia (45:55), and hyphae (42:58). Likewise, Horikoshi & Ikeda (1969) have shown that ribosomes isolated from dormant conidia of *A. oryzae* melt at a lower temperature and are autolyzed more rapidly than vegetative cell ribosomes. Ribosomes isolated from germinated uredospores of *Uromyces phaseoli* responded less to elongation factor 1 and bound less poly U than ribosomes from dormant spores (Yaniv & Staples, 1975). Thus, these investigators suggested that ribosomal proteins in this obligate parasite are more loosely attached to ribosomes during late stages of germination.

Transfer RNA

Transfer RNA has been isolated from dormant and germinated spores of several fungi including *Neurospora crassa* (Henney & Storck, 1963*a*, *b*), *Aspergillus oryzae* (Tanaka *et al.*, 1966*a*; Horikoshi, Ohtaka & Ikeda,

1969), *Botryodiplodia theobromae* (Van Etten, Koski & El-Olemy, 1969), and *Rhizopus stolonifer* (Van Etten *et al.*, 1969; Merlo, Roker & Van Etten, 1972). Base compositions of tRNA isolated from dormant ascospores and conidia of *N. crassa* were indistinguishable from those of tRNA isolated from mycelium (Henney & Storck, 1963*a*, *b*). Likewise, the base composition, thermal denaturation properties, and sedimentation profiles of tRNA isolated from dormant conidia, germinated conidia, and mycelium of *A. oryzae* were nearly identical (Tanaka *et al.*, 1966*a*). Furthermore, the tRNAs from both spore states had amino acid acceptor activity for the 13 amino acids tested (Horikoshi *et al.*, 1969). Amino acid acceptor activity for all 20 amino acids commonly found in protein was observed with tRNA isolated from dormant and germinated spores of *B. theobromae* (Van Etten *et al.*, 1969) and *R. stolonifer* (Van Etten *et al.*, 1969; Merlo *et al.*, 1972).

However, several lines of evidence indicate that at least some differences exist between tRNA species in dormant and germinated spores, although the significance of these observations is unknown. When the 13 aminoacyl-tRNAs synthesized with tRNA and enzyme fractions isolated from conidia and vegetative cells of *Aspergillus oryzae* were chromatographed on methylated albumin-Kieselguhr columns, differences were observed in the isoaccepting species of tRNAs for lysine and methionine. Two isoaccepting tRNA species for lysine were present in *A. oryzae* conidia and three in vegetative cells; a quantitative difference in the two isoaccepting tRNA species for methionine was also observed between the two cell stages (Horikoshi *et al.*, 1969). Ten aminoacyl-tRNAs synthesized with tRNA and aminoacyl-tRNA synthetases isolated from dormant and germinated spores of *Rhizopus stolonifer* were analyzed by co-chromatography on benzoylated-DEAE cellulose columns (Merlo *et al.*, 1972). No significant differences were detected for 7 of the 10 aminoacyl-tRNAs from dormant and germinated spores. The two isoaccepting species of lysyl-tRNA and valyl-tRNA differed quantitatively and isoleucyl-tRNA changed qualitatively during germination. Analysis of the tRNAs isolated from conidia of *Neurospora crassa* revealed a unique species of phenylalanyl-tRNA which was under-methylated and more susceptible to nucleases than the other two phenylalanyl-tRNAs present in both the mycelium and the conidia (Jervis & DeBusk, 1975). These investigators proposed that this unique phenylalanyl-tRNA species might be involved in controlling the initiation of protein synthesis in dormant spores.

A comparison of tRNA methylases and the ability of tRNAs from mycelium and conidia of *Neurospora crassa* to accept methyl groups also indicates that at least some tRNA species may be altered during spore germination (Wong, Scarborough & Borek, 1971). Transfer RNA and tRNA methylases were isolated from conidia and 12 h-old vegetative cells. The enzyme fraction from the hyphae catalyzed significant methylation of conidial tRNA, whereas conidial enzymes were less active with tRNA isolated from vegetative cells. Thus, the tRNA methylase content and the methylated nucleotides in tRNA differ in conidia and vegetative cells. Further evidence for this conclusion is provided by the observation that methylation of tRNAs occurred *in vivo* immediately after *Botryodiplodia theobromae* spores were placed in a germination medium and prior to *de novo* synthesis of tRNA (Knight & Van Etten, 1976*b*).

In summary, both qualitative and quantitative changes in tRNA isoaccepting species occur during fungal spore germination. These changes may reflect increased synthesis of the particular isoaccepting species of tRNA during germination or modification of pre-existing isoaccepting species. However, the significance of these observations can not be assessed until more is known about the metabolic function of isoaccepting species of tRNA.

Messenger RNA

Because the presence of polyribosomes is *a priori* evidence for the presence of mRNA, there has been considerable interest among investigators to establish whether polyribosomes exist in dormant fungal spores. Early studies did not detect polyribosomes in conidia and ascospores of *Neurospora crassa* (Henney & Storck, 1964) and conidia of *Aspergillus oryzae* (Horikoshi *et al.*, 1965), although polyribosomes were detected readily in germinated spores.

In contrast, more recent investigations indicate that at least some fungal spores contain small quantities of polyribosomes and thus contain mRNA. Good evidence exists for polyribosomes in uredospores of *Uromyces phaseoli* (Staples, Bedigian & Williams, 1968), conidia of *Botryodiplodia theobromae* (Brambl & Van Etten, 1970; Brambl, 1975a), and conidia of *Neurospora crassa* (Mirkes, 1974). Data supporting the existence of polyribosomes in basidiospores of *Schizophyllum commune* (Leary, Morris & Ellingboe, 1969), conidia of *Erysiphe graminus* f. sp. *tritici* (Leary & Ellingboe, 1971), and macroconidia of *Fusarium solani* (Cochrane, Rado & Cochrane, 1971) are less convincing.

RNA fractions which exhibited some properties expected for mRNA have been isolated from uredospores of *Uromyces phaseoli* (Ramakrishnan & Staples, 1970a), conidia of *Neurospora crassa* (Bhagwat & Mahadevan, 1970), and conidia of *Botryodiplodia theobromae* (Knight & Van Etten, 1976a). The mRNA fraction from uredospores of *U. phaseoli* sedimented between 4S and 19S and slightly stimulated amino acid incorporation into protein using a cell-free protein synthesizing system from *Escherichia coli*. Base analysis of the RNA fraction revealed that it was rich in AMP. The mRNA fraction from conidia of *N. crassa* hybridized to 2% of the cellular DNA and slightly stimulated amino acid incorporation into protein using a cell-free protein synthesizing system from mycelium of *N. crassa*. Likewise, the polyribosomal fraction isolated from dormant conidia of *B. theobromae* (Brambl & Van Etten, 1970) and conidia of *Peronospora tabacina* (Holloman, 1971) slightly stimulated amino acid incorporation into a trichloracetic acid-insoluble product. Direct evidence for the presence of a poly A(+)RNA (presumed mRNA) was recently obtained from ungerminated conidia of *B. theobromae* (Knight & Van Etten, 1976a); total RNA isolated from the spores was labelled chemically with (^{3}H-methyl) dimethyl sulphate and then fractionated on an oligo (d)T-cellulose column. About 2% of the total RNA consisted of a poly A(+)RNA fraction which sedimented in a heterogeneous fashion upon sucrose density gradient centrifugation.

It appears, therefore, that many, if not all, dormant fungal spores contain a latent or preserved mRNA which, as discussed later, is translated upon the onset of spore germination. It is not as certain, however, that this mRNA is always stored in the form of polyribosomes; it is possible that in some spores this macromolecule is preserved during dormancy in some other form which allows assembly with ribosomes only upon hydration or the initiation of germination.

6.5 RNA Synthesis during Spore Germination

Synthesis of the various RNA species in filamentous fungi probably occurs in a manner analogous to that in yeast and in other eukaryotic organisms. In yeast 5.8S RNA and large and small rRNAs are synthesized in the nucleolus as a single large precursor rRNA molecule (c. 2.5×10^6 daltons, 35S), whereas 5S RNA is synthesized in a separate transcriptional process (Udem & Warner, 1972). The 35S precursor rRNA is cleaved to 27S (c. 1.6×10^6) RNA and 20S (0.8×10^6) RNA; the 27S RNA is then cleaved to 25S rRNA and 5.8S RNA (Helser & McLaughlin, 1975) and the 20S to 18S rRNA. In contrast to 5.8S RNA and 5S RNA, the 25S rRNA and 18S rRNA contain an extensive number of methylated bases; methylation of the 25S and 18S rRNAs occurs post-transcriptionally at the 35S precursor rRNA stage (Taber & Vincent, 1969; Udem & Warner, 1972).

Although information on the biosynthesis of fungal mRNAs is limited, one can make several predictions about their synthesis from studies with animal cells. In animal cells mRNA is derived by post-transcriptional modification of large precursor RNAs (45S to >100S), termed heterogeneous nuclear RNA (HnRNA) (Darnell, 1968; Darnell, Jelinek & Molloy, 1973). After HnRNA is synthesized in the nucleus, adenine residues are added to the 3'-OH end by a stepwise addition resulting in polyadenylate (poly A) segments of 150 to 200 nucleotides. In these cells as much as 90% of the HnRNA stays in the nucleus; the remaining HnRNA, containing most of the polyadenylate segment, moves into the cytoplasm and serves as mRNA. Recently, mRNAs from a range of higher eukaryotic organisms have been shown to contain a 7-methylguanosine triphosphate nucleotide attached to the 5'-OH end of the RNA molecule (Muthukrishnan et al., 1975). The function of this methylated guanosine is not known, although it may be involved in the processing of HnRNA to mRNA and/or required for attachment of mRNA to the ribosomes.

The synthesis of mRNA in fungi probably occurs by a similar process, although the precursor of mRNA may be smaller than it is in animal cells. Firtel & Lodish (1973) reported that the mRNA precursor in the slime mould *Dictyostelium discoideum* is only 20% larger than mature mRNA. The inability to detect large HnRNAs in *Phycomyces blakesleeanus* (Gamow & Prescott, 1972), *Rhizopus stolonifer* (Roheim, Knight & Van Etten, 1974), and *Botryodiplodia theobromae* (Knight & Van Etten, 1976b) also supports the concept that precursor mRNAs in the filamentous fungi are smaller than in higher eukaryotes. Furthermore, the poly A segments attached to fungal mRNAs, unlike the somewhat longer segments in animal cell mRNA, average only about 50–60 residues long (Rosen, Edelman & Galun, 1975; Van Etten, Dunkle & Freer, 1977). As is true for other eukaryotic organisms, recent evidence indicates that not all fungal mRNAs contain poly A segments (Knight & Van Etten, 1976b; Mirkes & McCalley, 1976).

Several investigators have examined the relationship between RNA synthesis and fungal spore germination (Van Etten et al., 1976). Even though RNA synthesis is usually one of the earlier events associated with germination, it may not be essential for germ tube formation. In this discussion only studies which have used radioactive precursors to monitor RNA synthesis during germination will be mentioned since they are much more sensitive than colorimetric analyses. Using these latter techniques, one usually does not detect net increases in RNA until at least the time of germ tube emergence.

The initiation of RNA synthesis during fungal spore germination can be classified into two basic patterns. In some spores there is a simultaneous onset of synthesis of all classes of RNA, whereas in other spores the initial synthesis of the various classes of RNA occurs in a sequential manner. Sporangiospores of *Rhigopus stolonifer* (Roheim *et al.*, 1974) and conidia of *Neurospora crassa* (Mirkes, 1974; Mirkes & McCalley, 1976) initiate *de novo* synthesis of all classes of RNA (rRNA, tRNA, and mRNA) during the first few minutes of germination. However, the percentage of precursor incorporation into each of the 3 classes of RNA may shift with germination time. For example, precursor incorporation into mRNA relative to tRNA and rRNA is highest during the first few minutes of *R. stolonifer* germination and then decreases during later stages of germination (Roheim *et al.*, 1974). In both *R. stolonifer* and *N. crassa* at least some of the newly synthesized mRNA associates with ribosomes within the first 15 min of germination and presumably is translated immediately into protein.

In contrast, at least 60 min of germination elapse before synthesis of all classes of RNA is initiated in conidia of *Aspergillus oryzae* (Ono, Kimura & Yanagita, 1966; Tanaka, Ono & Yanagita, 1966*b*), conidia of *Peronspora tabacina* (Hollomon, 1970), uredospores of *Uromyces phaseoli* (Ramakrishnan & Staples, 1967, 1970*a*, *b*), and conidia of *Botryodiplodia theobromae* (Knight & Van Etten, 1976*b*). However, the order in which the classes of RNA initiates synthesis varies with the organism. For example, in *A. oryzae*, rRNA synthesis begins immediately, tRNA synthesis begins a few minutes later, and the synthesis of a mRNA-like fraction is not detected until approximately 60 min of germination. In *P. tabacina*, tRNA synthesis begins during the first 20 min, mRNA-like material at 20 to 40 min, and rRNA synthesis is first detected after germ tube formation at about 60 min. In *U. phaseoli* uredospores, rRNA synthesis begins 90 min after initiation of germination, whereas tRNA and mRNA syntheses are not detected until after germ tubes appear. In the case of *B. theobromae*, mRNA and tRNA syntheses are first detected at about 30 min, 5S RNA at 45 min, and 5.8S RNA and large and small rRNAs at about 60 min into germination.

At the present time it is not known if any significance can be attached to the variety of patterns exhibited in the initial synthesis of RNA by different species. Some of the results summarized above were obtained before a convenient method for specifically identifying mRNA was available (by the presence of the 3'-OH poly A segment); therefore, the time during germination at which mRNA synthesis was first detected in some of these organisms might be questioned. However, one of the authors (J. L. Van Etten) of this chapter has studied both *Rhizopus stolonifer* (simultaneous onset of RNA synthesis) and *Botryodiplodia theobromae* (sequential onset of RNA synthesis) using identical procedures, and he is reasonably convinced that the differences in patterns of initial RNA synthesis in the two spores are authentic.

The length of the poly A segment in poly A(+)RNA synthesized during various times of germination of *Rhizopus stolonifer* has been examined (Van Etten *et al.*, 1977). When the poly A segment was isolated from total RNA the average length of the poly A segment synthesized during the first 15 min of germination was about 100 nucleotides. The length decreased with germination time until it was about 60 nucleotides for spores pulsed from 225 to 240 min. In contrast, when the poly A segment was isolated from polysomal RNA, the average length of the newly synthesized poly A segment was about 50 adenylic acid residues at all pulse periods. One interpretation of these results is that the majority of the poly A(+)RNA synthesized during the early periods of germination might take longer to be processed to mRNA and, hence, is isolated as HnRNA. After the spores are incubated in a germination medium for longer periods of time, the poly A(+)RNA may be synthesized, processed, and associated with ribosomes faster so that the majority of the poly A segments isolated at later times are associated with mature mRNA. If these interpretations are correct, at least part of the processing of HnRNA to mRNA in *R. stolonifer* may result in a loss of about 40 to 50 adenine residues from the poly A region of the HnRNA.

The nuclei of filamentous fungi, like other eukaryotic organisms, contain two or more DNA-directed RNA polymerases (Van Etten *et al.*, 1976). Evidence obtained from fungi, although not conclusive, is consistent with the concept that RNA polymerase I is responsible for the synthesis of rRNA, RNA polymerase II is responsible for the synthesis of mRNA, and RNA polymerase III may be involved in the synthesis of small (transfer and 5S) RNAs.

Both qualitative and quantitative differences were reported for RNA polymerases isolated from dormant and germinated sporangiospores of *Rhizopus stolonifer* (Gong & Van Etten, 1972). Germinated (6 h) spores contained three RNA polymerases (I, II, and III), whereas dormant and swollen (2 h) spores yielded only RNA polymerases I and III. RNA polymerase III was indistinguishable between the two spore states, whereas RNA polymerase I from the dormant spores responded to divalent cations differently and eluted from a DEAE-cellulose column earlier than the corresponding fraction from germinated spores. RNA polymerase II was first detected 3 h after the spores were placed on a germination medium and were initiating germ tube formation. At the same time the characteristics of RNA polymerase I were altered so that they resembled those of germinated spore enzyme. Consequently, it appeared that the dormant spore might be incapable of synthesizing mRNA until RNA polymerase II was synthesized 2 to 3 h into the germination process. However, subsequent studies revealed that the synthesis of tRNA, rRNA and mRNA all began within the first 15 min of germination (Roheim *et al.*, 1974), and so the significance of these changes in RNA polymerases during germination are presently unknown.

In contrast to the studies on *Rhizopus stolonifer*, only two RNA polymerases were detected in germinated and ungerminated conidia of *Botryodiplodia theobromae*; no differences in the properties of the two enzymes isolated from the two spore states were observed (T. J. Morris and J. L. Van Etten, unpublished data). Likewise, no differences were detected in the three RNA polymerases isolated from vegetative cells and zoospores of the water moulds *Blastocladiella emersonii* (Horgen, 1971) and *Allomyces arbuscula* (Cain & Nester, 1973).

6.6 Components and Regulation of Protein Synthesis in Spores

It is assumed, although very little experimental evidence is available, that protein synthesis is either absent or occurs at a very low rate in dormant spores. It is known, however, that one of the earliest events during germination is a rapid increase in the rate of protein synthesis. Furthermore, protein synthesis is probably essential for germ tube formation in all fungal spores (Lovett, 1976). This conclusion is supported by several lines of evidence: (i) germ tube formation in all fungal spores which have been examined is prevented by low concentrations of cycloheximide (Lovett, 1976); (ii) conidia from a mutant strain of *Neurospora crassa* which is temperature sensitive for protein synthesis do not form germ tubes at the nonpermissive temperature (Loo, 1975); and (iii) conidia from some amino acid auxotrophs of *N. crassa* do not form germ tubes on unsupplemented media (Schmit & Brody, 1976).

Several investigators have been interested in determining how protein synthesis is regulated during germination, *i.e.*, what keeps the system essentially inactive in the dormant spore but still allows the system to become active shortly after the spores are placed in germinating conditions? Protein synthesis in the dormant spore could be repressed because one or more components of the protein synthetic machinery are absent, spatially separated, or present in a non-functional state (Van Etten, 1969). Consequently, ungerminated and germinated spores have been examined to determine if dormant spores are defective in one or more of the components necessary for protein synthesis.

As noted previously, ribosomes which have biological activity when assayed with polyuridylic acid as the mRNA have been isolated from dormant spores of several species. Likewise, tRNA has been isolated from dormant spores of at least three fungi and shown to have amino acid acceptor activity for the amino acids tested. Also, there is good evidence from spores of several fungi that endogenous mRNA is present in the dormant spore. However, the evidence to indicate the biological functionality of this mRNA is very weak at the present time. When the mRNA has been tested with *in vitro* protein synthesizing systems, very low activity has been obtained. With the availability now of improved methods for isolation of mRNA and the refined wheat germ translation system, the capacity of the dormant spore mRNA to support *in vitro* protein synthesis should be re-examined.

Elongation factor activity (polymerizing enzymes) has been detected in dormant spores of several fungi. For example, active enzymes have been isolated from conidia of *Botryodiplodia theobromae* (Van Etten & Brambl, 1968, 1969), uredospores of *Uromyces phaseoli* (Staples & Bedigan, 1967; Yaniv & Staples, 1971), uredospores of *Puccinia graminis* f. sp. *tritici* (Staples, Yaniv & Bushnell, 1972), conidia of *Aspergillus oryzae* (Horikoshi & Ikeda, 1968), and macroconidia of *Fusarium solani* (Rado & Cochrane, 1971).

When aminoacyl-tRNA synthetases were isolated from dormant spores and tested for activity with individual amino acids, they were shown to be present in an active form for all of the amino acids examined. Such experiments have been conducted with dormant sporangiospores of *Rhizopus stolonifer* (Merlo *et al.*, 1972) and conidia of *Aspergillus oryzae* (Horikoshi *et al.*, 1969).

Taken as a group these studies suggest that the lack of protein synthesis in the spore is not the result of defective ribosomes, tRNA, elongation factors, aminoacyl-tRNA synthetases, or mRNA, since all of these components can be isolated from dormant spores. Indirect evidence that all of the protein synthetic components in the ungerminated spore are biologically active is supplied by the observation that the enzyme activities of α-amylase, invertase, and glucose dehydrogenase can be induced in *Aspergillus oryzae* conidia under nongerminating conditions (Sinohara, 1970). However, direct evidence that these enzymes were synthesized *de novo* is still lacking.

Other components of the protein synthetic apparatus, such as initiation factors and release factors, have not been examined in filamentous fungi in relation to spore germination. Suppression of initiation factor activity would provide a convenient way of regulating protein synthesis in the germinating spore.

Other mechanisms for regulating protein synthesis in fungal spores have been proposed. Studies with *Blastocladiella emersonii* revealed that in the zoospore all the ribosomes are stored in a package surrounded by a double membrane which is termed a nuclear cap (Lovett, 1963), suggesting that protein synthesis might be suppressed in these zoospores by the spatial separation of some of the protein synthetic components. While this possibility still exists, more recent studies by Lovett and his colleagues (Lovett, 1975) indicate that the nuclear cap particles also contain tRNA, aminoacyl-tRNA synthases, and elongation factor activity so that at least several components required for protein synthesis may be packaged together. However, packaging of ribosomes has not been observed in spores of filamentous fungi.

Spores of several fungi contain germination inhibitors which have to be removed before the spores can form a germ tube (Macko, Staples, Yaniv & Granados, 1976). Germination inhibitors which function by preventing protein synthesis would provide a convenient method for suppressing protein synthesis in the dormant spore. There is some evidence that a few self-inhibitors may actually function in this manner. For example, zoospores of *Blastocladiella emersonii* apparently contain a self inhibitor since encystment of the spores is prevented by high concentrations of spores (Soll & Sonneborn, 1969; Truesdell & Cantino, 1971). An inhibitor, which has many of the properties of this factor, is associated with the zoospore ribosomes. This inhibitor apparently reduces the activity of ribosomes by preventing translation of mRNA in the zoospores; once zoospore encystment begins the inhibitor is released and protein synthesis commences (Adelman & Lovett, 1974). Likewise, the germination inhibitor of *Glomerella cingulata* may inhibit amino acid incorporation in conidia (Lingappa, Lingappa & Bell, 1973). An inhibitor present in spores of the slime mould *Dictyostelium discoideum* has also been reported to inhibit amino acid incorporation (Bacon, Sussman & Paul, 1973) but its chemical identification as $N'N$-dimethylguanosine has been challenged by other investigators (Tanaka, Yanagisawa, Hashimoto & Yamaguchi, 1974; Katilus & Ceccarini, 1975), and consequently the significance of the germination inhibitor in *D. discoideum* remains to be determined.

With some supporting data Hollomon (1973) has proposed a very interesting model to explain how protein synthesis is activated during the initial stages of germination of *Peronospora tabacina* conidia. He suggests that some of the ribosomes, free of mRNA, are attached to membranes in the dry conidia. When water is added to the spores, the small ribosomal subunit is released and becomes associated with mRNA which is stored elsewhere in the conidia. Attachment of the mRNA to the small ribosomal subunit together with the proper initiation factors results in the formation of an initiation complex. This initiation complex in turn reassociates with the large ribosomal subunit and protein synthesis commences. Furthermore, Hollomon suggests that the spore self inhibitor β-ionone has no effect on the initial release from the membrane of the small ribosomal subunit but that *in vivo* it does prevent the reassociation of the initiation complex with the large ribosomal subunit.

Jervis & DeBusk (1975) have proposed that a novel phenylalanyl-tRNA may be involved in controlling protein synthesis during the germination of *Neurospora crassa* conidia. Analysis of the *N*-terminal amino acids of proteins from conidia and mycelium of *N. crassa* revealed a preponderance of phenylalanine in the proteins of conidia (Rho & DeBusk, 1971). Transfer RNAs isolated from conidia contained a unique species of phenylalanyl-tRNA which was under-methylated and consequently more susceptible to nuclease degradation. Jervis & DeBusk (1975) proposed that this tRNA, which presumably is involved in the synthesis of proteins during conidiation, might occupy a specific site on the ribosomes in dormant spores and thereby prevent initiation of protein synthesis. When the spores are placed under conditions favourable for spore germination, this tRNA is degraded by RNases and protein synthesis commences.

Finally, it is theoretically possible that activation of protein synthesis in dormant spores could result from hydration-induced modifications of pre-existing proteins, possibly proteolytic activity, reduction of disulphide linkages, or other means of protein alteration (see Schmit & Brody, 1976). While several models have been proposed to explain the regulation of protein synthesis in the dormant spore and during spore germination many more experiments will have to be conducted before it is possible to decide if any of these models are valid for even a single spore system

6.7 Nature of Proteins Synthesized during Spore Germination

Cytoplasmic ribosome products

It is well established that the cytoplasmic ribosomes stored in the dormant spore become functional upon the onset of germination, but very little is known about the nature of the earliest proteins synthesized. We are not aware of a demonstration of the *de novo* synthesis of a single enzyme or structural protein during germination. In cases where the specific activity of an enzyme increases during germination, it has not been demonstrated that *de novo* synthesis of the enzyme protein has occurred. This lack of insight into which proteins or enzymes are synthesized early in germination is especially disappointing; if the types of proteins synthesized during germination were known, more sophisticated or incisive hypotheses about the metabolic requirements for germination and about its regulation could be devised and tested.

Only a few polyacrylamide gel electrophoretic analyses have been made of the proteins synthesized early in germinating spores, and we know nothing about the function of these proteins. In *Botryodiplodia theobromae* it was shown that the proteins isolated from the $105\,000\times$ **g** supernatant fraction of spores labelled between 0 and 60 min of germination were relatively stable after synthesis and that there was a difference in the distribution of the soluble proteins synthesized in early and late periods of germination (Van Etten, Roker & Davies, 1972). In this study no attempt was made to examine the labelled proteins in the organelle or cell wall fractions of the spores. In *Blastocladiella emersonii* no obvious differences were observed after gel electrophoretic analyses of proteins labelled in early and late stages of germination; these proteins were solubilized from whole spores with detergent (Silverman, Huh & Sun, 1974). However, if identical extracts were prepared from spores incubated in the presence of actinomycin D (this drug inhibits RNA synthesis *in vivo* in spores of this organism), considerable differences existed between the proteins synthesized early and late in germination. In a study of mitochondrial protein synthesis during conidial germination of *B. theobromae* (Brambl & Handschin, 1976), it was shown that proteins synthesized on cytoplasmic ribosomes were incorporated into or associated with the spore mitochondria during the first 60 min of germination. Again, nothing is known about the functions of these particular proteins, although, considering other types of evidence, it was proposed that these proteins may serve to activate or complement the mitochondrial aerobic respiratory apparatus which is stored in a latent form in these spores.

It is evident, then, that we also have no information about which of the cytoplasmic ribosome products synthesized before germ tube emergence are essential for germination and which are non-essential or are required only for normal vegetative growth. In *Blastocladiella emersonii*, treatment of the spores with cycloheximide within 15 min after the beginning of germination blocked germ tube emergence, but by 30 min nearly all the cells had lost their sensitivity to this drug with respect to germ tube formation which was normally completed by 60 min (Silverman *et al.*, 1974). Similarly, germination of *Botryodiplodia theobromae* spores was prevented completely if cycloheximide was present up to about 135 min of germination; if added at times after this point, at least some of the spores began developing germ tubes at about 150 min and most of them completed germination by the normal time of 240 min (Brambl, 1975*b*). Such experiments indicate times at which proteins essential for germination must be synthesized, and they could provide useful avenues for further study to identify the function of these proteins. However, as pointed out by others (Sussman & Douthit, 1973; Lovett, 1975), detection of the synthesis of one or more essential proteins in the presence of many other proteins whose syntheses perhaps are not essential for germination may prove to be very difficult with conventional techniques.

It seems unlikely that these early essential proteins would be enzymes required for RNA synthesis, since in some fungal spores RNA is synthesized immediately upon the onset of germination with protein synthesis; furthermore, at least in some spores the RNA polymerases appear to be stored in the dormant spore in a form which is active *in vitro* (Gong & Van Etten, 1972). It is also improbable that the early essential proteins are enzymes involved in protein synthesis, because a spore must have a functional protein synthesis system in order to synthesize the essential proteins. If it were not for the facts that protein synthesis invariably is one of the first activities of spore germination and that germ tube formation is blocked by cyclohexicide, there would be no good reason to invoke a role for protein synthesis in germination, with all the proteins required for germination being stored in the dormant spore. How protein synthesis may serve a pivotal role is unknown, of course, but it seems to us that likely essential products of the cytoplasmic ribosomes which may be necessary (but not sufficient) for germination are enzymes of cell wall assembly, enzymes or enzyme subunits required for cellular respiration, enzymes involved in lipid metabolism for membrane assembly, and possibly enzymes which modify macromolecular substrates (proteins or nucleic acids) which are stored in the dormant spore. It might not be necessary to synthesize a complete enzyme *de novo* in any of these examples; rather synthesis of a subunit or regulatory factor might serve for activation of a proenzyme stored in the spore.

Mitochondrial protein synthesis

In most cases mitochondria from dormant spores appear to be morphologically similar to mitochondria from vegetative cells. In a few spores the mitochondria may decrease in size and increase in number during germination; at the same time there may be an increase in cristae with germination (Smith, Gull, Anderson & Deans, 1976). It is not clear, however, whether these dormant spore mitochondria are comparable metabolically to those of growing cells, and it is not understood what is required metabolically for the mitochondria of dormant spores to resume their normal genetic and respiratory activities upon germination. The nature of the cooperation that must exist between the nuclear and mitochondrial genetic systems for transcription and translation of genetic information necessary for mitochondrial assembly during spore germination is incompletely defined, and the contributions of each genetic system toward assembly and function of the mitochondrial aerobic respiratory apparatus have not been identified.

In germinating *Botryodiplodia theobromae* spores both oxygen consumption and germination are blocked by cyanide, antimycin A, and cyclohexicide. This suggests roles for both aerobic respiration and cytoplasmic protein synthesis in the control of germination (Brambl, 1975b). We have shown that function of the mitochondrial genetic system is not required for spore germination and that, if mitochondrial DNA synthesis is inhibited with ethidium bromide, neither the capacity for germination nor the earliest phase of oxygen uptake is inhibited (Dunkle et al., 1972; Brambl, 1975b). Likewise, concentrations of chloramphenicol, which are sufficient to inhibit mitochondrial protein synthesis, do not prevent germ tube formation or early oxygen uptake (Brambl, 1975b). These observations suggested that these spores contain a preserved, potentially functional, aerobic respiratory system which requires an early contribution from the cytoplasmic ribosomes in order to become active in respiration (Brambl, 1975*b*).

In a series of oxygen uptake experiments conducted with spores of *Botryodiplodia theobromae* incubated in several sequential combinations of cycloheximide and chloramphenicol, Brambl (1975*b*) found that if the cytoplasmic ribosomes were inhibited early in germination (through 180 min) mitochondrial ribosome products accumulated. After reversal of ribosome inhibition at 180 min, these accumulated mitochondrial ribosome products could then combine with newly synthesized cytoplasmic ribosome products to permit the development of mitochondrial respiration. These experiments implied that the two translational systems of the spores respond independently to the stimulus or stimuli of germination and demonstrated that function of the mitochondrial genetic system alone is not sufficient for development of respiratory activity in the pre-emergence phase of germination. Further, they predicted that a product of the cytoplasmic ribosomes, possibly translated from the preserved mRNA, was required for initiation of activity of the mitochondrial respiratory system.

In a subsequent study, Brambl & Handschin (1976) showed that mitochondrial ribosomes do not function during the first 60 min of germination; between 60 and 90 min, synthesis of the first products of these ribosomes was detected by gel electrophoresis. During this first 60 min interval, in the absence of mitochondrial ribosome function, labelled products of the cytoplasmic ribosomes were incorporated into or associated with the mitochondria. After 60 min, the first mitochondrial ribosome product synthesized was a low molecular weight peptide which was uniquely soluble in chloroform–methanol; in later stages of germination, higher molecular weight products of the mitochondrial ribosomes appeared so that by 240 min of germination the mitochondria contained the complement of mitochondrial ribosome products characteristic of physiologically mature cells. Respiratory cytochromes appear to be first synthesized and incorporated into the mitochondrial membrane at about 180 min (Brambl & Josephson, 1976).

Hawley & Greenawalt (1975) electrophoresed whole cell extracts of germinating *Neurospora crassa* spores labelled in the presence of cycloheximide; in the 8-h cells they observed synthesis of about 10 proteins presumably synthesized on mitochondrial ribosomes. They observed early synthesis (0–60 min of germination) of two low molecular weight proteins whose relative quantity diminished in comparison to the higher molecular weight proteins as the spores germinated and grew into young mycelial cells. Whether the low molecular weight protein(s) synthesized in the mitochondria of *Botryodiplodia theobromae* and *N. crassa* are identical and have similar functions needs to be explored.

6.8 Patterns and Essentiality of Macromolecular Biosynthesis

The spores of filamentous fungi can be divided into three categories on the basis of the relative times at which RNA and protein syntheses begin during the germination process (Van Etten *et al.*, 1976). In all fungi examined, incorporation of labelled precursors into both types of macromolecules begins prior to any visible germ tube formation. The spores of most fungi studied synthesize RNA and protein concomitantly, usually a short time after the spores are placed in a germination medium. Fungi in this category (category I) include: *Fusarium solani* (Cochrane *et al.*, 1971), *Lenzites saepiaria* (Scheld & Perry, 1970), *Microsporum gypseum* (Barash, Conway & Howard, 1967), *Neurospora crassa* (Bhagwat & Mahadevan, 1970), *Peronospora tabacina* (Holloman, 1969), *Phycomyces blakesleeanus* (Van Assche & Carlier, 1973), and *Rhizopus stolonifer* (Van Etten, Bulla & St. Julian, 1974). In Fig. 6.1 some of the biosynthetic activities in *R. stolonifer* spores are schematically related to oxygen uptake and the morphological changes occurring during germination. Apparently, in this fungus and the others in this category, the metabolic and biosynthetic components are poised in a state which can be activated rapidly by hydration and uptake of essential nutrients.

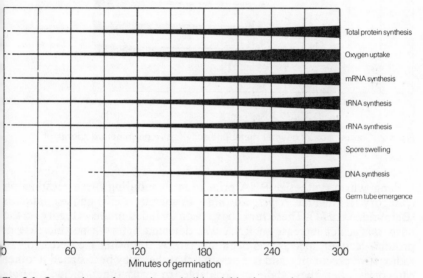

Fig. 6.1. Comparison of several metabolic activities in germinating spores of *Rhizopus stolonifer*.

Fungal spores in which protein synthesis precedes RNA synthesis (category II) exhibit a pattern of macromolecular biosynthesis similar to that observed during development of many higher eukaryotes, *e.g.* activation of sea urchin eggs and *Xenopus* oöcytes (Nemer, 1967; Gross, 1967). *Botryodiplodia theobromae* is the only filamentous fungus included in this category (Brambl & Van Etten, 1970; Knight & Van Etten, 1976b). Fig. 6.2 shows the relation between these biosynthetic events and other selected activities in *B. theobromae* spore germination. It is apparent from the figure and from the previous discussion that dormant spores of this fungus can synthesize protein without simultaneously activating RNA synthesis. Furthermore, the very early increase in the rate of protein synthesis apparently does not require extensive oxygen uptake. Rather, the development of mitochondrial respiratory activity is dependent upon this very early cytoplasmic protein synthesis (Brambl, 1975b).

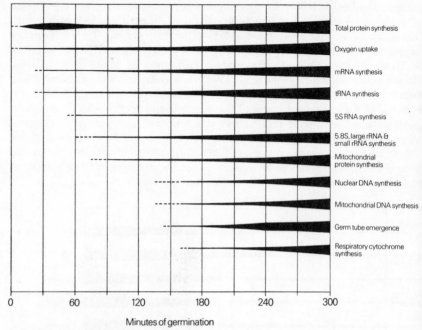

Fig. 6.2. Comparison of several metabolic activities in germinating spores of *Botryodiplodia theobromae.*

Fungi which synthesize RNA prior to protein during spore germination (category III) include *Aspergillus niger* (Yanagita, 1957) and *A. nidulans* (Bainbridge, 1971). These fungi have been included in this category on the basis that a net increase in RNA was detected before a net increase in protein. Neither species has been studied by following incorporation of radioactive precursors, and it is possible that they may be assigned to other categories when these data are available. In many fungi, including *Botryodiplodia theobromae* (unpublished data), net increases in RNA or protein may not be detected by colorimetric analyses until about the time of germ tube emergence. The phenomenon of turnover of proteins and nucleic acids during spore germination has not been adequately studied, although protease activity is associated with germination of *Microsporum gypseum* conidia (Page & Stock, 1971).

Although protein, RNA and DNA are synthesized during spore germination, these biosynthetic processes or their products may not be essential for germ tube emergence in all fungi. The major approach to establish essentiality of macromolecular biosynthesis has been to use antibiotics or analogues that specifically block the synthesis, which is monitored by following incorporation of labelled precursors. It is necessary in this kind of study to establish that the antibiotic penetrates the spore and has a primary effect only on the synthetic process under consideration and that it does not interfere directly with precursor uptake, oxygen uptake, oxidative phosphorylation, or other metabolic activities unrelated directly to protein and nucleic acid synthesis.

Based on the effect of cycloheximide, as well as other evidence discussed above, it is concluded that protein synthesis is required for germ tube formation in all fungi that have been examined. The effect of other inhibitors of eukaryotic protein synthesis, such as puromycin, is not as obvious. The efficacy of cycloheximide may be due to a greater permeability of the spore to this drug compared with other inhibitors. Although cycloheximide is believed to specifically prevent protein synthesis, primary effects on oxidative phosphorylation in plant systems have also been reported (Ellis & MacDonald, 1970).

Evidence that DNA synthesis is not essential for spore germination is less direct, because inhibitors of DNA synthesis are less specific and no precursors are specifically incorporated into fungal DNA.

RNA synthesis is apparently required for spore germination in some species but not in others. Proflavine at concentrations that inhibited RNA synthesis inhibited germination of conidia of *Aspergillus nidulans* and *Neurospora crassa* but had no effect on germination of conidia of *Alternaria solani* and *Peronospora tabacina* (Hollomon, 1970). However, in *N. crassa* proflavine may inhibit precursor uptake rather than incorporation into RNA (Tisdale & DeBusk, 1972). Van Etten *et al.* (1976) summarized experiments on the effect of several inhibitors of RNA synthesis on germination of *Rhizopus stolonifer* and *Botryodiplodia theobromae* spores. Lomofungin and daunorubicin inhibited germination of *R. stolonifer* spores but not that of *B. theobromae* spores. However, since the effects of these drugs on precursor incorporation were not determined, it is not known whether they specifically inhibited RNA synthesis.

6.9 Concluding Remarks

Much information has accumulated in the last few years on the temporal relationships between protein, RNA, and DNA syntheses during fungal spore germination. Some of the properties of the intermediates as well as the enzymes and other components required for the syntheses of these macromolecules have been elucidated for a few fungi. However, it is obvious that the mechanism(s) for regulating these various syntheses in dormant and germinating spores is unknown, although a few potential mechanisms can be eliminated. The conclusions and conceptual generalities available at present have been compiled from research on a very limited representation of the fungi and suggest that during spore germination in different fungi variable patterns exist in macromolecular synthesis as well as in relative activities of the biosynthetic components. The fungi are ecologically, morphologically, and biochemically diverse micro-organisms, and after considerably more species are examined in detail we will be able to derive more definite conclusions about the significance of their different biochemical patterns of spore germination. This valuable descriptive information should be applied simultaneously to the development and testing of hypotheses about the regulation of protein and nucleic acid syntheses during spore germination and to the examination of the specific roles of these processes in this transition from dormancy to rapid growth.

This is paper No. 9543, Scientific Journal Series, Minnesota Agricultural Experiment Station and paper No. 5137, Journal Series, Nebraska Agricultural Experiment Station. Research from the laboratory of RB was supported by NIH Research Grant GM-19398 from the National Institute of General Medical Sciences and from the laboratory of JVE by NIH Research Grant AI-108057 from the National Institute of Allergy and Infectious Diseases.

6.10 References

ADELMAN, T. G. & LOVETT, J. S. (1974). Evidence for a ribosome-associated translation inhibitor during differentiation of *Blastocladiella emersonii*. *Biochimica et Biophysica Acta* **335**, 236–45.

BACON, C. W., SUSSMAN, A. S. & PAUL, A. G. (1973). Identification of a self-inhibitor from spores of *Dictyostelium discoideum*. *Journal of Bacteriology* **113**, 1061–3.

BAINBRIDGE, B. W. (1971). Macromolecular composition and nuclear division during spore germination in *Aspergillus nidulans*. *Journal of General Microbiology* **66**, 319–25.

BARASH, I., CONWAY, M. L. & HOWARD, D. M. (1967). Carbon catabolism and synthesis of macromolecules during spore germination of *Microsporum gypseum*. *Journal of Bacteriology* **93**, 656–62.

BHAGWAT, A. S. & MAHADEVAN, P. R. (1970). Conserved mRNA from the conidia of *Neurospora crassa*. *Molecular and General Genetics* **109**, 142–51.

BIANCHI, D. E. & TURIAN, G. (1967). Nuclear division in *Neurospora crassa* during conidiation and germination. *Experimentia* **23**, 192–7

BRAMBL, R. (1975a). Presence of polyribosomes in conidiospores of *Botryodiplodia theobromae* harvested with nonaqueous solvents. *Journal of Bacteriology* **122**, 1394–5.

BRAMBL, R. (1975b). Characteristics of developing mitochondrial genetic and respiratory functions in germinating fungal.spores. *Biochimica et Biophysica Acta* **396**, 175–86.

BRAMBL, R. & HANDSCHIN, B. (1976). Mitochondrial biogenesis during fungal spore germination: products of mitochondrial protein synthesis *in vivo*. *Archives of Biochemistry and Biophysics* **175**, 606–17.

BRAMBL, R. & JOSEPHSON, M. (1977). Mitochondrial biogenesis during fungal spore germination: respiratory cytochromes of dormant and germinating spores. *Journal of Bacteriology* **129**, 291–7.

BRAMBL, R. M. & VAN ETTEN, J. L. (1970). Protein synthesis during fungal spore germination. V. Evidence that the ungermi-nated conidiospores of *Botryodiplodia theobromae* contain messenger ribonucleic acid. *Archives of Biochemistry and Biophysics* **137**, 442–52.

CAIN, A. K. & NESTER, E. W. (1973). Ribonucleic acid polymerase in *Allomyces arbuscula*. *Journal of Bacteriology* **115**, 769–76.

CLAYTON, D. A. & BRAMBL, R. M. (1972). Detection of circular DNA from mitochondria of *Neurospora crassa*. *Biochemical and Biophysical Research Communications* **46**, 1477–82.

COCHRANE, J. C., RADO, T. A. & COCHRANE, V. W. (1971). Synthesis of macromolecules and polyribosome formation in early stages of spore germination in *Fusarium solani*. *Journal of General Microbiology* **65**, 45–55.

DARNELL, J. E. (1968). Ribonucleic acids from animal cells. *Bacteriological Reviews* **32**, 262–90.

DARNELL, J. E., JELINEK, W. R. & MOLLOY, G. R. (1973). Biogenesis of mRNA: genetic regulation in mammalian cells. *Science* **181** 1215–21.

DUNKLE, L. D. & VAN ETTEN, J. L. (1972). Characteristics and synthesis of deoxy-ribonucleic acid during fungal spore germination. In *Spores V*, pp. 283–9. Edited by H. O. Halvorson, R. Hanson & L. L. Campbell. Washington D.C.: American Society for Microbiology.

DUNKLE, L. D., VAN ETTEN, J. L. & BRAMBL, R. M. (1972). Mitochondrial DNA synthesis during fungal spore germination. *Archiv für Mikrobiologie* **85**, 225–32.

DUSENBERY, R. L. (1975). Characterization of the genome of *Phycomyces blakesleeanus*. *Biochimica et Biophysica Acta* **378**, 363–77.

DUTTA, S. K. (1974). Repeated DNA sequences in fungi. *Nucleic Acids Research* **1**, 1411–9.

DUTTA, S. K. & CHAUDHURI, R. K. (1975). Differential transcription of nonrepeated DNA during development of *Neurospora crassa*. *Developmental Biology* **43**, 35–41,

ELLIS, R. J. & MACDONALD, I. R. (1970). Specificity of cycloheximide in higher plant systems. *Plant Physiology* **46**, 227–32.

FIRTEL, R. A. & LODISH, H. F. (1973). A small nuclear precursor of messenger RNA in the cellular slime mold *Dictyostelium discoideum*. *Journal of Molecular Biology* **79**, 295–314.

GAMOW, E. & PRESCOTT, D. M. (1972). Characterization of the RNA synthesized by *Phycomyces blakesleeanus*. *Biochimica et Biophysica Acta* **259**, 223–7.

GONG, C-S., DUNKLE, L. D. & VAN ETTEN, J. L. (1973). Characteristics of deoxyribonucleic acid polymerase isolated from spores of *Rhizopus stolonifer*. *Journal of Bacteriology* **115**, 762–8.

GONG, C-S. & VAN ETTEN, J. L. (1972). Changes in soluble ribonucleic acid polymerases associated with the germination of *Rhizopus stolonifer* spores. *Biochimica et Biophysica Acta* **272**, 44–52.

GRIVELL, A. R. & JACKSON, J. F. (1968). Thymidine kinase: evidence for its absence from *Neurospora crassa* and some other micro-organisms, and the relevance of this to the specific labeling of deoxyribonucleic acid. *Journal of General Microbiology* **54**, 307–17.

GROSS, P. R. (1967). The control of protein synthesis in embryonic development. In *Current topics in developmental biology*, vol. 2, pp. 1–46. Edited by A. A. Moscona & A. Monroy. New York and London: Academic Press.

HAWLEY, E. S. & GREENAWALT, J. W. (1975). Biogenesis of mitochondrial membranes in *Neurospora crassa*. Mitochondrial protein synthesis during conidial germination. *European Journal of Biochemistry* **54**, 585–601.

HELSER, T. L. & MCLAUGHLIN, C. S. (1975). Small ribonucleic acid molecules produced during ribosome biosynthesis in *Saccharomyces cerevisiae*. *Journal of Biological Chemistry* **250**, 2003–7.

HENNEY, H. & STORCK, R. (1963a). Nucleotide composition of ribonucleic acid from *Neurospora crassa*. *Journal of Bacteriology* **85**, 822–6.

HENNEY, H. R. & STORCK, R. (1963b). Ribosomes and ribonucleic acids in three morphological states of *Neurospora*. *Science* **142**, 1675–6.

HENNEY, H. R. & STORCK, R. (1964). Polyribosomes and morphology in *Neurospora crassa*. *Proceedings of the National Academy of Sciences, U.S.A.* **51**, 1050–5.

HOLLENBERG, C. P., BORST, P. & VAN BRUGGEN, E. F. J. (1970). Mitochondrial DNA. V. A 25-μ closed circular duplex DNA molecule in wild-type yeast mitochondria and genetic complexity. *Biochimica Biophysica Acta* **209**, 1–15.

HOLLOMON, D. W. (1969). Biochemistry of germination in *Peronospora tabacina* (Adam) conidia: evidence for the existence of stable messenger RNA. *Journal of General Microbiology* **55**, 267–74.

HOLLOMON, D. W. (1970). Ribonucleic acid synthesis during fungal spore germination. *Journal of General Microbiology* **62**, 75–87.

HOLLOMON, D. W. (1971). Protein synthesis during germination of *Peronospora tabacina* (Adam) conidia. *Archives of Biochemistry and Biophysics* **145**, 643–9.

HOLLOMON, D. W. (1973). Protein synthesis during germination of *Peronospora tabacina* conidia: an examination of the events involved in the initiation of germination. *Journal of General Microbiology* **78**, 1–13.

HORGEN, P. A. (1971). In vitro ribonucleic acid synthesis in the zoospores of the aquatic fungus *Blastocladiella emersonii*. *Journal of Bacteriology* **106**, 281–2.

HORIKOSHI, K. (1971). Studies on the conidia of *Aspergillus oryzae*. XI. Latent ribonuclease in the conidia of *Aspergillus oryzae*. *Biochimica et Biophysica Acta* **240**, 532–40.

HORIKOSHI, K. & IKEDA, Y. (1968). Studies on the conidia of *Aspergillus oryzae*. VII. Development of protein synthesizing activity during germination. *Biochimica et Biophysica Acta* **166**, 505–11.

HORIKOSHI, K. & IKEDA, Y. (1969). Studies on the conidia of *Aspergillus oryzae*. IX. Protein synthesizing activity of dormant conidia. *Biochimica et Biophysica Acta* **190**, 187–92.

HORIKOSHI, K., OHTAKA, Y. & IKEDA, Y. (1965). Ribosomes in dormant and germinating conidia of *Aspergillus oryzae*. *Agricultural and Biological Chemistry* **29**, 724–7.

HORIKOSHI, K., OHTAKA, Y. & IKEDA, Y. (1969). Properties of ribosomes and transfer ribonucleic acid in dormant conidia of *Aspergillus oryzae*. In *Spores IV*, pp. 175–79. Edited by L. L. Campbell. Washington, D.C.: American Society for Microbiology.

JEGGO, P. A., UNRAU, P., BANKS, G. R. & HOLLIDAY, R. (1973). A temperature sensitive DNA polymerase mutant of *Ustilago maydis*. *Nature New Biology* **242**, 14–6.

JERVIS, H. H. & DEBUSK, A. G. (1975). Rapid loss of a novel phenylalanyl-tRNA on germination of *Neurospora crassa* conidia. *Nature, London* **258**, 160–2.

KATILUS, J. & CECCARINI, C. (1975). Purification and new biological properties of the slime mold germination inhibitor. *Developmental Biology* **42**, 13–8.

KNIGHT, R. H. & VAN ETTEN, J. L. (1976a). Characteristics of ribonucleic acids isolated from *Botryodiplodia theobromae* pycnidiospores. *Archives of Microbiology* **109**, 45–50.

KNIGHT, R. H. & VAN ETTEN, J. L. (1976b). Synthesis of ribonucleic acids during the germination of *Botryodiplodia theobromae* pycnidiospores. *Journal of General Microbiology* **95**, 257–67.

KOGANE, F. & YANAGITA, T. (1964). Isolation and purification of deoxyribonucleic acid from *Aspergillus oryzae* conidia. *Journal of General and Applied Microbiology* **10**, 61–8.

KUMARI, L., DECALLONNE, J. R. & MEYER, J. A. (1975). Deoxyribonucleic acid metabolism and nuclear division during spore germination in *Fusarium oxysporum*. *Journal of General Microbiology* **88**, 245–52.

LEARY, J. V. & ELLINGBOE, A. H. (1971). Isolation and characterization of ribosomes from nongerminated conidia of *Erysiphe graminis* f. sp. *tritici*. *Phytopathology* **61**, 1030–1.

LEARY, J. V., MORRIS, A. J. & ELLINGBOE, A. H. (1969). Isolation of functional ribosomes and polysomes from lyophilized fungi. *Biochemica et Biophysica Acta* **182**, 113–20.

LINGAPPA, B. T., LINGAPPA, Y. & BELL, E. (1973). A self-inhibitor of protein synthesis in the conidia of *Glomerella cingulata*. *Archiv für Mikrobiologie* **94**, 97–107.

LOO, M. (1975). *Neurospora crassa* temperature sensitive mutant apparently defective in protein synthesis. *Journal of Bacteriology* **121**, 286–95.

LOO, M. (1976). Some required events in conidial germination of *Neurospora crassa*. *Developmental Biology* **54**, 201–13.

LOVETT, J. S. (1963). Chemical and physical characterization of 'nuclear caps' isolated from *Blastocladiella* zoospores. *Journal of Bacteriology* **85**, 1235–46.

LOVETT, J. S. (1975). Growth and differentiation of the water mold *Blastocladiella emersonii*: cytodifferentiation and the role of ribonucleic acid and protein synthesis. *Bacteriological Reviews* **39**, 345–404.

LOVETT, J. S. (1976). Regulation of protein metabolism during spore germination. In *The fungal spore: form and function*, pp. 189–242. Edited by D. J. Weber & W. M. Hess. New York and London: John Wiley & Sons.

MACKO, V., STAPLES, R. C., YANIV, Z. GRANADOS, R. R. (1976). Self-inhibitor of fungal spore germination. In *The fungal spore: form and function*, pp. 73–100. Edited by D. J. Weber and W. M. Hess. New York and London: John Wiley & Sons.

MERLO, D. J., ROKER, H. & VAN ETTEN, L. (1972). Protein synthesis during fungal spore germination. VI. Analysis of transfer ribonucleic acid from germinated and ungerminated spores of *Rhizopus stolonifer*. *Canadian Journal of Microbiology* **18**, 949–56.

MINAGAWA, T., WAGNER, B. & STRAUSS, B. (1959). The nucleic acid content of *Neurospora crassa*. *Archives of Biochemistry and Biophysics* **80**, 442–5.

MIRKES, P. E. (1974). Polysomes, ribonucleic acid, and protein synthesis during germination of *Neurospora crassa* conidia. *Journal of Bacteriology* **117**, 196–202.

MIRKES, P. E. & MCCALLEY, B. (1976). Synthesis of polyadenylic acid-containing ribonucleic acid during germination of *Neurospora crassa* conidia. *Journal of Bacteriology* **125**, 175–80.

MUTHUKRISHNAN, S., FILIPOWICZ, W., SIERRA, M., BOTH, G. W., SHATKIN, A. J. & OCHOA, S. (1975). mRNA methylation and protein synthesis in extracts from embryos of brine shrimp, *Artemia salina*. *Journal of Biological Chemistry* **250**, 9336–41.

MYERS, R. B. & CANTINO, E. C. (1971). DNA profile of the spore of *Blastocladiella emersonii*: evidence for γ particle DNA. *Archiv für Mikrobiologie* **78**, 252–67.

NEMER, M. (1967). Transfer of genetic information during embryogenesis. In *Progress in nucleic acid research and molecular biology*, Vol. 7, pp. 243–301. Edited by J. N. Davidson & W. E. Cohn. New York and London: Academic Press.

ONO, T., KIMURA, K. & YANAGITA, T. (1966). Sequential synthesis of various molecular species of ribonucleic acid in the early phase of conidia germination in *Aspergillus oryzae*. *Journal of General and Applied Microbiology* **12**, 13–26.

PAGE, W. J. & STOCK, J. J. (1971). Regulation and self-inhibition of *Microsporum gypseum* macroconidia germination. *Journal of Bacteriology* **108**, 276–81.

RADO, T. A. & COCHRANE, V. W. (1971). Ribosomal competence and spore germination in *Fusarium solani*. *Journal of Bacteriology* **106**, 301–4.

RAMAKRISHNAN, L. & STAPLES, R. C. (1967). Some observations on ribonucleic acids and their synthesis in germinating bean rust uredospores. *Phytopathology* **57**, 826.

AMAKRISHNAN, L. & STAPLES, R. C. (1970*a*). Evidence for a template RNA in resting uredospores of the bean rust fungus. *Contributions from Boyce Thompson Institute* **24**, 197–202.

AMAKRISHNAN, L. & STAPLES, R. C. (1970*b*). Changes in ribonucleic acids during uredospore differentiation. *Phytopathology* **60**, 1087–91.

HO, H. M. & DEBUSK, A. G. (1971). HN$_2$-terminal residues of *Neurospora crassa* proteins. *Journal of Bacteriology* **107**, 840–5.

OHEIM, J. R., KNIGHT, R. H. & VAN ETTEN, J. L. (1974). Synthesis of ribonucleic acids during germination of *Rhizopus stolonifer* sporangiospores. *Developmental Biology* **41**, 137–45.

OSEN, D., EDELMAN, M. & GALUN, E. (1975). Characterization of polyadenylate from the fungus *Trichoderma viride*. *Journal of Bacteriology* **123**, 765–7.

ROTHSCHILD, H., ITIKAWA, H. & SUSKIND, S. R. (1967). Ribosomes and ribosomal proteins from *Neurospora crassa*. II. Ribosomal proteins in different wild-type strains and during various stages of development. *Journal of Bacteriology* **94**, 1800–1.

CHÄFER, K. P. & KÜNTZEL, H. (1972). Mitochondrial genes in *Neurospora*: a single cistron for ribosomal RNA. *Biochemical and Biophysical Research Communications* **46**, 1312–9.

CHELD, H. W. & PERRY, J. J. (1970). Basidiospore germination in the wood-destroying fungus *Lenzites saepiaria*. *Journal of General Microbiology* **60**, 9–21.

CHMIT, J. C. & BRODY, S. (1976). Biochemical genetics of *Neurospora crassa* conidial germination. *Bacteriological Reviews* **40**, 1–41.

SILVERMAN, P. M., HUH, M. M.-O. & SUN, L. (1974). Protein synthesis during zoospore germination in the aquatic phycomycete *Blastocladiella emersonii*. *Developmental Biology* **40**, 59–70.

SINOHARA, H. (1970). Induction of enzymes in dormant spores of *Aspergillus oryzae*. *Journal of Bacteriology* **101**, 1070–2.

SMITH, J. E., GULL, K., ANDERSON, J. G. & DEANS, S. G. (1976). Organelle changes during fungal spore germination. In *The fungal spore: form and function*, pp. 301–51. Edited by D. J. Weber & W. M. Hess. New York and London: John Wiley & Sons.

SOLL, D. R. & SONNEBORN, D. R. (1969). Zoospore germination in the water mold, *Blastocladiella emersonii*. II. Influence of cellular and environmental variables on germination. *Developmental Biology* **20**, 218–35.

STAPLES, R. C. (1974). Synthesis of DNA during differentiation of bean rust uredospores. *Physiological Plant Pathology* **4**, 415–24.

STAPLES, R. C., APP, A. A. & RICCI, P. (1975). DNA synthesis and nuclear division during formation of infection structures by bean rust uredospore germlings. *Archives of Microbiology* **104**, 123–7.

STAPLES, R. C. & BEDIGIAN, D. (1967). Preparation of an amino acid incorporation system from uredospores of the bean rust fungus. *Contributions from Boyce Thompson Institute* **23**, 345–7.

STAPLES, R. C. & YANIV, Z. (1973). Spore germination and ribosomal activity in the rust fungi. II. Variable properties of ribosomes in the Uredinales. *Physiological Plant Pathology* **3**, 137–45.

STAPLES, R. C., BEDIGIAN, D. & WILLIAMS, P. H. (1968). Evidence for polysomes in extracts of bean rust uredospores. *Phytopathology* **58**, 151–4.

STAPLES, R. C., YANIV, Z. & BUSHNELL, W. R. (1972). Spore germination and ribosomal activity in the rust fungi. I. Comparison of a bean rust fungus and a culturable wheat rust fungus. *Physiological Plant Pathology* **2**, 27–35.

STORCK, R. (1974). Molecular mycology. In *Molecular microbiology* pp. 423–77. Edited by J. B. G. Kwapinski. New York, London and Sydney: John Wiley & Sons.

STORCK, R. & ALEXOPOULOS, C. J. (1970). Deoxyribonucleic acid of fungi. *Bacteriological Reviews* **34**, 126–54.

SUSSMAN, A. S. (1966). Dormancy and spore germination. In *The fungi*, Vol. II, pp. 733–64, Edited by G. C. Ainsworth and A. S. Sussman. New York and London: Academic Press.

SUSSMAN, A. S. & DOUTHIT, H. A. (1973). Dormancy in microbial spores. *Annual Review of Plant Physiology* **24**, 311–52.

TABER, R. L. & VINCENT, W. S. (1969). The synthesis and processing of ribosomal RNA precursor molecules in yeast. *Biochimica et Biophysica Acta* **186**, 317–25.

TANAKA, K., KOGANE, F. & YANAGITA, T. (1965). Is deoxyribonucleic acid of *Aspergillus oryzae* conidia modified chemically in the early period of germination? *Journal of General and Applied Microbiology* **11**, 85–90.

TANAKA, K., MOTOHASHI, A., MIURA, K. & YANAGITA, T. (1966a). Isolation and characterization of soluble RNA from dormant and germinated conidia of *Aspergillus oryzae*. *Journal of General and Applied Microbiology* 12, 277–92.

TANAKA, K., ONO, T. & YANAGITA, T. (1966b). Further observations on the carbon dioxide incorporation into RNA in the early phase of conidia germination in *Aspergillus oryzae* with special reference to soluble RNA synthesis. *Journal of General and Applied Microbiology* 12, 329–36.

TANAKA, Y., YANAGISAWA, K., HASHIMOTO, Y. & YAMAGUCHI, M. (1974). True spore germination inhibitor of a cellular slime mold *Dictyostelium discoideum*. *Agricultural and Biological Chemistry* 38, 689–90.

TINGLE, M. A., KÜENZI, M. T. & HALVORSON, H. O. (1974). Germination of yeast spores lacking mitochondrial deoxyribonucleic acid. *Journal of Bacteriology* 117, 89–93.

TISDALE, J. H. & DEBUSK, A. G. (1972). Permeability problems encountered when treating conidia of *Neurospora crassa* with RNA synthesis inhibitors. *Biochemical and Biophysical Research Communications* 48, 816–22.

TRUESDELL, L. C. & CANTINO, E. C. (1971). The induction and early events of germination in the zoospore of *Blastocladiella emersonii*. *Current topics in developmental biology*, Vol. 6, pp. 1–44, Edited by A. Monroy and A. A. Moscona. New York and London: Academic Press.

UDEM, S. A. & WARNER, J. R. (1972). Ribosomal RNA synthesis in *Saccharomyces cerevisiae*. *Journal of Molecular Biology* 65, 227–42.

VAN ASSCHE, J. A. & CARLIER, A. R. (1973). The pattern of protein and nucleic synthesis in germinating spores of *Phycomyces blakesleeanus*. *Archiv für Mikrobiologie* 93, 129–36.

VAN ETTEN, J. L. (1968). Protein synthesis during fungal spore germination. I. Characteristics of an *in vitro* phenylalanine incorporating system prepared from germinated spores of *Botryodiplodia theobromae*. *Archives of Biochemistry and Biophysics* 125, 13–21.

VAN ETTEN, J. L. (1969). Protein synthesis during fungal spore germination. *Phytopathology* 59, 1060–4.

VAN ETTEN, J. L. & BRAMBL, R. M. (1968). Protein synthesis during fungal spore germination. II. Aminoacyl-soluble ribonucleic acid synthetase activities during germination of *Botryodiplodia theobromae* spores. *Journal of Bacteriology* 96, 1042–8.

VAN ETTEN, J. L. & BRAMBL, R. M. (1969). Protein synthesis during fungal spore germination. III. Transfer activity during germination of *Botryodiplodia theobromae* spores. *Phytopathology* 59, 1894–902.

VAN ETTEN, J. L., BULLA, L. A. & ST. JULIAN, G. (1974). Physiological and morphological correlation of *Rhizopus stolonifer* spore germination. *Journal of Bacteriology* 117, 882–7.

VAN ETTEN, J. L., DUNKLE, L. D. & FREER, S. (1977). Germination of *Rhizopus stolonifer* sporangiospores. In *Eucaryotic microbes as model developmental systems*. Edited by D. H. O'Day & P. A. Horgen. New York: Marcel Dekker. pp. 372–401.

VAN ETTEN, J. L., DUNKLE, L. D. & KNIGHT, R. H. (1976). Nucleic acids and fungal spore germination. In *The fungal spore: form and function*, pp. 243–300. Edited by D. J. Weber & W. M. Hess. New York and London: John Wiley & Sons.

VAN ETTEN, J. L., KOSKI, R. K. & EL-OLEMY, M. M. (1969). Protein synthesis during fungal spore germination. IV. Transfer ribonucleic acid from germinated and ungerminated spores. *Journal of Bacteriology* 100, 1182–6.

VAN ETTEN, J. L., ROKER, H. R. & DAVIES, E. (1972). Protein synthesis during fungal spore germination: differential protein synthesis during germination of *Botryodiplodia theobromae* spores. *Journal of Bacteriology* 112, 1029–31.

WEBER, D. J. & HESS, W. M. (1976). *The fungal spore: form and function*. New York and London: John Wiley & Sons.

WONG, R. S. L., SCARBOROUGH, G. A. & BOREK, E. (1971). Transfer ribonucleic acid methylases during the germination of *Neurospora crassa*. *Journal of Bacteriology* 108, 446–50.

YANAGITA, T. (1957). Biochemical aspects on the germination of conidiospores of *Aspergillus niger*. *Archiv für Mikrobiologie* 26, 329–44.

YANIV, Z. & STAPLES, R. C. (1971). The purification and properties of the aminoacyl-tRNA binding enzyme from bean rust uredospores. *Biochemica et Biophysica Acta* 232, 717–25.

YANIV, Z. & STAPLES, R. C. (1975). Changes in ribosomes associated with spore senescence in the bean rust fungus. *Journal of General Microbiology* 87, 189–97.

CHAPTER 7

Protoplasts and their Development

J. F. PEBERDY

7.1 Introduction

During the last decade an important development has been the establishment of a technology for the isolation of protoplasts from wall-bearing cells. The methods used involve the complete removal of the wall, or part of it, leading to the release of the cell contents as discrete spheres of cytoplasm enclosed by plasma membrane. These structures have been variously described as 'protoplasts', 'protoplast-like structures' or 'spheroplasts'. The differences in terminology are a reflection of the uncertainty over the presence or absence of wall residues associated with the plasma membrane. 'Protoplast' is used to describe a truly naked cell completely free of wall material, while 'protoplast-like structure' and 'spheroplast' are used when wall material is present or is thought might be present. Absolute certainty of the status of the material can be obtained only from electron microscopic observations; however, resource to this technique is impracticable for routine purposes. Electron microscopic examinations have been made on material prepared from several different fungi and in every case the structures have proved to be true protoplasts.

The filamentous structure found in fungi is unique and has implications in relation to protoplast isolation. In coenocytic species protoplasts are produced by sub-division of the cytoplasm from whole hyphae, but in septate species, the cytoplasm of individual hyphal segments may be released as a single protoplast or may be broken up into smaller units. This subdivision of cytoplasm is peculiar to filamentous fungi, unlike bacteria, yeasts and plants where the whole cell contents are normally released as a single protoplast.

A major line of research with fungal protoplasts has been related to their development when cultured in osmotically stabilized media. Under these conditions a proprotion of the protoplasts in any preparation exploit what appears to be an innate capacity to rebuild a cell wall and ultimately develop a form identical to the cells from which they were isolated. This process has been of particular interest because it provides a system for the study of cell wall synthesis and of the role of the wall in hyphal morphogenesis. This aspect together with protoplast isolation is dealt with in this chapter.

However, it should be stressed that protoplast systems also provide a spectrum of opportunities in other fields of research. The cell wall is an effective barrier preventing the uptake of large molecules or particles; removing this barrier exposing the plasma membrane thus presents a new situation possibly leading to genetic modification by the uptake of DNA or viral material. Removing the wall also makes cell fusion more feasible. Clearly, success in any such manipulation depends on the regeneration of an identifiable organism, either identical to that used to produce the protoplasts or one that shows some modification due to recombination of genetic material. In either case, the ability to culture the protoplasts is of paramount importance.

7.2 Protoplast Isolation

The most effective procedures for protoplast isolation depend upon the hydrolysis of the fungal cell wall by suitable lytic enzymes. The mycelium is suspended in a buffered solution of a non-electrolyte or an inorganic salt which will act as an osmotic stabilizer for the protoplasts on their release; the osmotic pressure of the stabilizer being equivalent to the internal osmotic pressure of the cytoplasm. The mixture of mycelium, enzyme and stabilizer is incubated under suitable conditions for the particular organism. Each component of the digestion mixture can effect the yield of protoplasts obtained.

Lytic Enzymes

Protoplast release is primarily dependent upon the hydrolysis of polysaccharides which account for approximately 60–80% of the fungal cell wall. Fungi belonging to different taxonomic groups show fundamental differences in the polysaccharide composition of their cell walls (Bartnicki-Garcia, 1968) and this is clearly reflected in the lytic enzyme complexes used in protoplast production.

The isolation of protoplasts from a filamentous fungus was first reported in *Neurospora crassa* using a lytic enzyme preparation from the snail, *Helix pomatia* (Emerson & Emerson, 1958; Bachmann & Bonner, 1959). The enzyme preparation is prepared from the digestive juices of the snail and contains a number of carbohydrases including chitinase and β-glucanases. The use of this enzyme was based on the much earlier report from Giaja (1914) in which the effects of snail digestive juices on the yeast cell were described. This enzyme, which can be obtained from commercial sources, is still used in a few laboratories (Ferenczy, Kevei & Zsolt, 1974) but enzymes of microbial origin are now more commonly used.

During the early 1960s Villanueva and his colleagues carried out extensive studies on the lytic activities of the Actinomycetes (see Villanueva & Garcia Acha, 1971, for review). Arising from these experiments several species of *Streptomyces* and *Micromonospora* were found to produce enzyme complexes which attacked the cell walls of a number of filamentous fungi. Lytic properties are not confined to this group of micro-organisms. Certain species of fungi have also proved to be effective sources of enzymes (Musilkova, Fencl & Seichertova, 1969; de Vries & Wessels, 1972). The lytic enzymes produced by the organisms used so far, are inducible. A variety of inducing substrates have been used in the growth medium including heat-killed mycelium, purified hyphal walls or mixtures of polysaccharides such as chitin, laminarin and pustulan.

In some instances the components of the lytic mixtures playing a key role in protoplast release have been identified. The lytic complex, obtained from a streptomycete, used for the isolation of protoplasts from *Pythium* sp. was shown to contain endo- and exo-laminarinases, cellulase, lipase and protease. After isolation, purification and mixing these individual components in various combinations, it was demonstrated that laminarinase and cellulase were essential for protoplast release. Lipase had a stimulatory effect shortening the incubation period required for protoplast release while protease had no effect at all (Seitsma, Eveleigh & Haskins, 1968). De Vries & Wessels (1973) found that lytic enzymes from *Trichoderma viride* were very effective in releasing protoplasts from *Schizophyllum commune* mycelium. The lytic complex contained chitinase, α-glucanase and β-glucanase, however, only the former two enzymes were essential for protoplast release despite the presence of R-glucan (β-glucan) in the hyphal wall. When the mycelium was treated with chitinase and α-glucanase the R-glucan was hydrolysed by R-glucanase (β-glucanase), present in *S. commune* mycelium.

Osmotic Stabilizers

A wide range of osmotic stabilizers have been used for protoplast isolation. Sugars and sugar alcohols have been used successfully with some fungi, but inorganic salts have generally proved to be the most effective (Lopez-Belmonte, Garcia Acha & Villanueva, 1966; de Vries & Wessels, 1973; Anné, Eyssen & de Somer, 1974; Peberdy, Buckley, Daltrey & Moore, 1976). Magnesium sulphate can be singled out for its interesting properties not found with other stabilizers. During the lytic digestion the mycelium undergoes extensive fragmentation and this results in protoplast release from the open ends of the fragments, including those from older regions of the hyphae. Many of the protoplasts develop large vacuoles and float on centrifugation; the non-vacuolate protoplasts and mycelial debris sediment. This differential property provides a procedure for obtaining clean preparations of protoplasts free of mycelial debris (de Vries & Wessels, 1972).

The Mycelium

By its very nature, the mycelial component of the digestion mixture for protoplast isolation is most difficult to standardize, but it is probably the most important factor in determining protoplast yield. Two aspects of growth of the mycelium are important: the nature of the growth medium used to produce it, and the stage of culture development at which it is harvested. Musilkova & Fencl (1968) compared protoplast production from *Aspergillus niger* mycelium that had been cultured on a variety of media. Mycelium grown on glucose–salts–asparagine medium gave the highest yield of approximately 4.5×10^6 protoplasts/mg dry mycelium during a three-hour digestion. Adding malt extract to this medium, or changing the nitrogen source to NH_4Cl, or culturing on malt extract alone, had a dramatic effect, reducing the yield by 75% or more. Similar effects have been recorded for other fungi, but so far no explanations have been put forward. It is known that environmental conditions have a marked effect on the gross composition of fungi but what changes, if any, occur in the walls are unknown.

The influence of culture age, at harvesting, on protoplast isolation is shown in Fig. 7.1. Protoplast yield is highest from mycelium in the exponential growth phase. The reasons for this are not clear; it is possible that the hyphal wall is more susceptible to lysis at this stage of growth because melanin is absent or present at a low concentration (Carter & Bull, 1969; Bull 1970a, b). An alternative explanation may be that at this time, the activity of wall-bound lytic enzymes, thought to be used by the fungus in wall assembly, may be high and thus enhance the lytic effect of the exogenous enzymes (Bartnicki-Garcia & Lippman, 1972). In many reports, protoplasts have been seen to arise first from hyphal tips, the proposed site for wall-bound lytic enzymes, and later from the distal regions as the hyphae disintegrate.

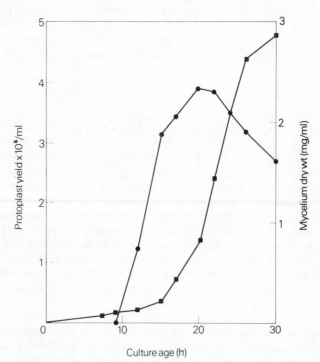

Fig. 7.1 Protoplast formation from mycelium of *Aspergillus flavus* at different stages of growth: (——■——) dry weight of mycelium, (——●——) protoplast yield. From Peberdy *et al.* (1976).

Hyphal organization and protoplast formation

The fungal hypha is characterized by a progressive cytoplasmic differentiation in which three zones can be identified on the basis of organelle distribution (see Chapter 3). During protoplast formation the functional unity of the hypha is broken down, becoming sub-divided into discrete cytoplasmic units. Thus a population of protoplasts produced after extensive mycelium digestion might be expected to show a marked heterogeneity, in

terms of organelle constitution and biochemical function. The first suggestion that such heterogeneity might exist is found in a study on protoplasts of
Phytopthora in which considerable variation in their internal morphology
was reported (Bartnicki-Garcia & Lippman, 1966). Confirmation of such
variation came later in an electron microscopic study of protoplasts of
Aspergillus nidulans, where differences in protoplasts taken from digestion
mixtures at different stages were found (Gibson & Peberdy, 1972). The
protoplasts formed early in the lytic digestion arose from hyphal apices and
only as the digestion proceeded did protoplasts appear from distal regions.
The differences observed in the protoplasts related to the presence of
absence of vacuoles, ribosomal density, and the relative numbers of organelles (Table 7.1). Further evidence for heterogeneity was demonstrated by

Table 7.1 Observations on protoplasts released from *Aspergillus nidulans*
mycelium during a 3 h incubation period (From Gibson & Peberdy,
1972)

Time of release (h)	Light-microscopic observations of hyphae	Protoplast diameter (μm)	Ultrastructural observations of protoplasts
0–1	Swelling of tips	4.1–5.5	Non-vacuolate. Dense granular cytoplasm. Small vesicles underlying the plasmalemma.
1–2	Swelling of distal regions	4.1–8.25	Vacuolate. Cytoplasm diffusely granular. Small vesicles absent.
2–3	Fragmentation	4.1–13.75	Larger vacuole. Other characteristics as in 1–2 h protoplasts.

Buckley (1973) in relation to distribution of nuclei in protoplasts. The
non-vacuolate protoplasts which were formed during the first hour of
digestion had 2–3 nuclei, protoplasts released later had on average 1–2
nuclei. A more recent approach has been to compare biochemical properties
of protoplasts released during a normal digestion period to assess the extent
of fractionation of cytoplasm in relation to the distribution of organelles and
biochemical function in the hypha. Experiments in which the protein
content and acid and alkaline phosphatase levels, in protoplasts prepared
with KCl stabilizer, were estimated, showed a decrease in all these parameters per unit protoplast over the digestion period (Fig. 7.2, S. Isaac unpublished data). This suggested that the protoplasts released in the early stages
of digestion arose from hyphal apices where the protein content and
phosphatase levels, located in mitochondria and nuclei, would be expected
to be the highest (Zalokar, 1965; Reiss, 1973). However, it is clear this
regular fragmentation of cytoplasm into protoplasts during the early stages
of mycelial digestion is affected by the osmotic stabilizer used. In the
presence of $MgSO_4$, which causes extensive fragmentation of the hyphae,
the distribution of protein and enzyme activities in the protoplast fractions
suggests a more irregular pattern of release (Fig. 7.2).

Protoplasts from spores

The asexual spores of fungi are potentially a very useful source of protoplasts since in most cases they are unicells and when mature are more

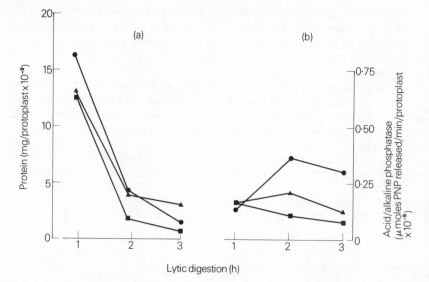

Fig. 7.2 Protein content and acid and alkaline phosphatase activities of protoplasts from *Aspergillus nidulans* at different times during lytic digestion using (a) KCl as stabilizer and (b) $MgSO_4$ as stabilizer (—●—) protein, (—▲—) alkaline phosphatase, (—■—) acid phosphatase (S. Isaac, unpublished data).

uniform in their cytological and biochemical properties. The relatively few reports of protoplasts from spores is probably a reflection of the problems in obtaining them. In some fungi, the spore wall is more resistant to lysis than the mycelium and this limits opportunities for protoplast production (Chu & Alexander, 1972). In general, where protoplast production from spores has been successful, the digestion period necessary is longer than for isolation of mycelial protoplasts (Garcia Acha, Lopez-Belmonte & Villanueva, 1966; Laborda, Garcia Acha & Villanueva, 1974) and the cultural conditions used for spore production may affect protoplast isolation (Moore & Peberdy, 1976*a*).

7.3 Culture and Development of Protoplasts

When suspended in an osmotically stabilized growth medium, part of the protoplast population demonstrates the capacity to return to the normal hyphal form of the organism. This process, which depends on the rebuilding of a new cell wall, has been described as regeneration (Villanueva & Garcia Acha, 1971; Peberdy, 1972) although de Vries & Wessels (1972) identify separate stages described as regeneration and reversion. This difference in terminology is mainly due to fundamental differences in the pattern of events that have been observed in different fungi.

Morphological changes during regeneration

Descriptions of regeneration have been reported for fungal species from most taxonomic groups and although there are some variations in detail it is possible to identify two basic patterns of development when protoplasts are grown in liquid medium. In *Rhizopus nigricans* (Gabriel, 1970) and in *Schizophyllum commune* (de Vries & Wessels, 1975) the protoplasts develop a wall which gradually thickens and one or more hyphae arise directly from the thick-walled spherical cell. The terms regeneration and reversion were used to describe the development in *S. commune* of the thick walled cell and the hyphae respectively. A few protoplasts developed a chain of bud-like cells and hyphae were formed from some of these.

The second pattern of development has been found to occur more widely in different fungal species; the protoplast bulges out to one side developing what has variously been called a budding chain, an aberrant tube and an aberrant hypha (Fig. 7.3a and b). From this structure one or more normal hyphae develop, usually from the growing tip distal to the protoplast, as in *Aspergillus nidulans* (Gibson & Peberdy, 1972) and in *Penicillium chrysogenum* (Anné *et al.*, 1974) or from the original protoplast as in *Trichoderma viride* (Benitez, Ramos & Garcia Acha, 1975). Protoplasts from conidia of *A. flavus* also showed this general pattern of development (Moore & Peberdy, 1976a). As well as this pattern of development, many protoplasts also develop into irregular polymorphic structures without hyphal development. In *A. nidulans* the development of the first formed aberrant structure is influenced by the culture medium. With *N*-acetylglucosamine or yeast extract supplementation to glucose medium, or with *N*-acetylglucosamine alone, the aberrant hypha was much shorter (Fig. 7.4).

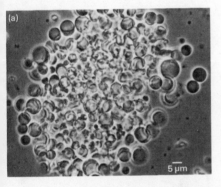
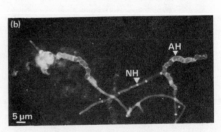

Fig. 7.3 (a) Freshly isolated protoplasts from mycelium of *Aspergillus nidulans*. (b) Protoplasts after culture in an osmotically stabilized medium containing glucose as carbon source. The protoplasts were stained with Tinopal BOPT. The aberrant hypha (AH) which ultimately develops into a normal hypha (NH) are clearly seen.

Quantitative measurements of regeneration have not proved easy. Regeneration frequencies can be determined by plate counting methods. However, in some fungi the frequencies of regeneration of the same protoplast population in agar medium and in liquid medium are often different. Peberdy and Buckley (1973) devised a method to compare regeneration activities of protoplasts cultured under different nutritional conditions. The procedure was based on the property of adsorption of an optical brightener dye from standard solutions; the dyes bind specifically to polysaccharide (Fig. 7.4). Using *Aspergillus nidulans* it was found that the

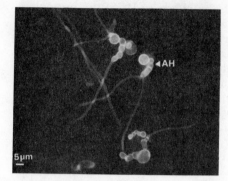

Fig. 7.4 Protoplast from Aspergillus nidulans mycelium cultured in a medium with *N*-acetylglucosamine. The material has been stained with the optical brightener Tinopal BOPT. In the presence of this carbon source the aberrant hypha (AH) is much shorter and the normal hyphae develop more rapidly.

amount of brightener adsorbed increased as regeneration progressed, but the rate of adsorption changed dramatically at a specific time in the development of the culture (Fig. 7.5). This point of change correlated with the timing of the morphological switch from the aberrant to the normal hypha.

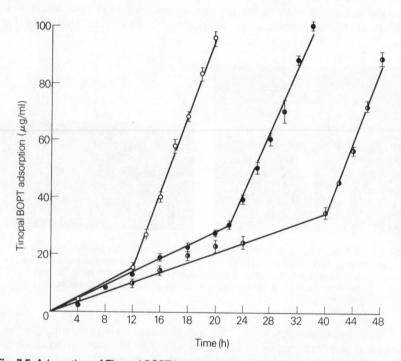

Fig. 7.5 Adsorption of Tinopal BOPT by regenerating protoplasts of *Aspergillus nidulans*. Protoplasts were sampled from three regeneration media (—●—) 10% glucose, (—●—) 10% glucose + 0.25% N-acetylglucosamine; (—○—) 10% glucose + 0.25% yeast extract). After homogenization the debris was suspended in the brightener solution for 5 min. The suspension was centrifuged and residual brightener concentrations in the supernatant determined (From Peberdy & Buckley, 1973).

Wall synthesis during regeneration

The development of protoplasts in culture provides a simple experimental system to examine the relationship between wall composition and the morphogenetic change that occurs. Information on this aspect of regeneration is limited; however, the studies that have been made do indicate that changes in cell morphology might be associated with changes in the pattern or balance of polymer synthesis. In the regeneration stage of *Schizophyllum commune* protoplasts, *i.e.* development of the thick-walled cell, only two of the polymers found in the normal hyphal wall, chitin and S-glucan (α-glucan) are synthesized. The third polymer, R-glucan (β-glucan) was produced as regeneration was nearly completed and hyphal development started. The addition of cycloheximide to the culture at the start of regeneration had no effect on chitin and S-glucan synthesis until 5 h, when a decrease in the rates of polymer deposition was caused. In contrast, a drastic reduction in R-glucan synthesis was observed and the final reversion stage was not reached. These observations suggested that the freshly isolated protoplasts were relatively rich in chitin synthase and S-glucan synthase, but R-glucan synthase was dependent on *de novo* protein synthesis (de Vries & Wessels, 1975).

Analysis of wall material synthesized by protoplasts of *Trichoderma viride* has also been made. The protoplasts were cultured for 11–12 h, at which time the transition to normal hyphal development hyphae had not taken place. A striking feature of the analysis is the absence of chitin in the walls of the aberrant hyphae which were composed solely of β-1, 3- and β-1, 6-glucans. Chitin is present in normal hyphal walls and it must be assumed that this polymer was synthesized when the morphological switch occurs. It was argued that the absence of chitin from walls of the aberrant hyphae was due to the periplasmic location of chitin synthase (Benitez, Villa & Garcia Acha, 1975). This seems unlikely, however, in view of the postulation of Ruiz Herrera, Sing, Van der Woode & Bartnicki-Garcia (1975) that chitin synthase in *Mucor rouxii* exists as granules in association with the plasma membrane. Furthermore, Moore & Peberdy (1976*b*) have shown the presence of chitin synthase in protoplasts of *Aspergillus flavus* and also an increase in its activity during regeneration.

Wall ultrastructure during regeneration

In the fungi so far examined the wall re-building process begins with the deposition of a microfibrillar skeleton composed of chitin or β-glucan depending on the species concerned (Gabriel, 1970; Seitsma, Child, Nesbitt & Haskins, 1975; Wessels, Van der Walk & de Vries, 1976; R. K. Gibson *et al.*, in preparation). In species which show direct development of a hypha from the thick-walled protoplast, the network becomes denser and is slowly packed or covered with loose amorphous material. In the specific case of *Schizophyllum commune* R-glucan deposition was reported in the thickened wall only at the site of inception of the hypha. In contrast to the thick-walled cell, the hypha contains R-glucan and the chitin microfibrils were found to be embedded in this. In both this fungus and *Pythium acanthicum* a sharp demarcation in the surface texture was observed where the thick-walled cell differentiates into the hypha.

A rather different situation exists in *Aspergillus nidulans*. The loose chitin network that develops first appears to become denser as the aberrant hypha develops and amorphous material, thought to be β-glucan, becomes deposited in it (Fig. 7.6). A demarcation zone in the surface of the hypha, marking the morphogenetic switch, was not observed.

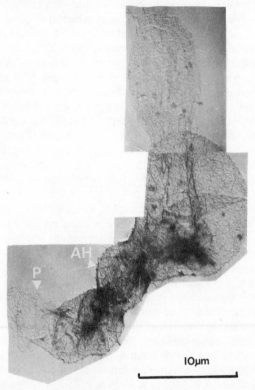

Fig. 7.6 Shadowed preparation of regenerating protoplast of *Aspergillus nidulans*. A loose network of microfibrils has been formed around the protoplast (P) and as the aberrant hypha develops the net gets thicker and amorphous material is also deposited (AH).

Protein and nucleic acid synthesis during regeneration

Synthesis of these macromolecules in *Schizophyllum commune* protoplasts follows a fairly linear pattern during regeneration, following continuous labelling with various ^{14}C precursors. Pulse-labelling showed that the rate of synthesis changed during regeneration. Initially it was low, then rising rapidly after 1–2 h to attain a constant level. Protein synthesis was necessary before the β-glucan component of the wall could be laid down (de Vries & Wessels, 1975).

7.4 Protoplast Fusion and Development

Protoplast fusion can be induced by various chemical agents of which the polyethylene glycols are the most efficient so far known (Anné & Peberdy, 1975, 1976; Ferenczy, Kevei & Szegedi, 1975, 1976). Intraspecific fusions were developed in various *Aspergillus* and *Penicillium* species by treating suspensions of protoplasts derived from nutritionally complementing strains with the fusogen. Heterokaryons were selected on a minimal medium and in the case of *P. chrysogenum* it was possible to isolate diploid colonies directly from fusion products. An important development arising from fusion is the possibility of interspecific hybridization of fungi. So far it has been possible to produce hybrids between closely related species of *Aspergillus* (Fig. 7.7, F. Kevei and J. F. Peberdy, in preparation). Strains produced in this way would provide opportunities for a new approach to the study or morphogenesis by study of the interactions between the different parental controlling mechanisms in the resultant hybrids.

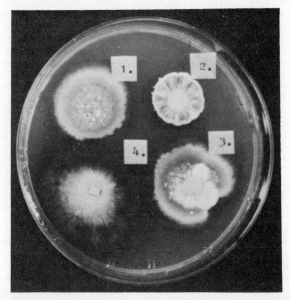

Fig. 7.7 Interspecific 'hybrid' obtained by fusion of protoplasts from auxotrophic strains of *Aspergillus nidulans* and *A. rugulosus*; (1) *A. nidulans* pro⁻, (2) *A. rugulosus* met⁻, (3) Heterokaryon (fusion product), (4) 'Hybrid' type, stable on complete medium.

7.5 Future Considerations

The isolated protoplast is essentially an 'artificial' structure and the evidence presented indicates that its wall-synthetic capacities at the early stages of regeneration are somewhat different from those of the growing hyphal apex. The extent to which the changes are due to isolation procedure or to the fact that protoplasts represent a heterogeneous collection of cytoplasmic units is not clear. The artificial nature of protoplasts does not invalidate their use in the study of fungal morphogenesis; and evidence available so far, allows us to make tentative suggestions as to the relationship between changes in morphology and wall composition and ultrastructure. More detailed studies with other species are required to further substantiate these views, including regeneration experiments involving morphological mutants with impediments in wall synthesis. The possibility of producing interspecific hybrids by protoplast fusion is an exciting challenge. The resulting mixture of genetic information is potentially very valuable in many respects. At the developmental level interspecific hybrids could prove of great interest in the understanding of morphogenesis, and its genetic control, in fungi.

7.6 References

ANNÉ, J., EYSSEN, H. & DE SOMER, P. (1974). Formation and regeneration of *Penicillium chrysogenum* protoplasts. *Archives of Microbiology* **98**, 159–66.

ANNÉ, J. & PEBERDY, J. F. (1975). Conditions for induced fusion of fungal protoplasts in polyethylene glycol solutions. *Archives of Microbiology* **105**, 210–5.

ANNÉ, J. & PEBERDY, J. F. (1976). Induced fusion of fungal protoplasts following treatment with polyethylene glycol. *Journal of General Microbiology* **92**, 413–17.

BACHMANN, B. J. & BONNER, D. M. (1959). Protoplasts from *Neurospora crassa*. *Journal of Bacteriology* **78**, 550–6.

BARTNICKI-GARCIA, S. (1968). Cell wall chemistry, morphogenesis and taxonomy of fungi. *Annual Review of Microbiology* **22**, 87–108.

BARTNICKI-GARCIA, S. & LIPPMAN, E. (1966). Liberation of protoplasts from the mycelium of *Phytopthora*. *Journal of General Microbiology* **42** 411–16.

BARTNICKI-GARCIA, S. & LIPPMAN, E. (1972). The bursting tendency of hyphal tips of fungi: presumptive evidence for a delicate balance between wall synthesis and wall lysis in apical growth. *Journal of General Microbiology* **73**, 487–500.

BENITEZ, T., RAMOS, S. & GARCIA ACHA, I. (1975). Protoplasts of *Trichoderma viride*. Formation and regeneration. *Archives of Microbiology* **103**, 199–203.

BENITEZ, T., VILLA, T. G. & GARCIA ACHA, I. (1975). Chemical and structural differences in mycelial and regeneration walls of *Trichoderma viride*. *Archives of Microbiology* **105**, 277–82.

BUCKLEY, C. E. (1973). *PhD Thesis. University of Nottingham.*

BULL, A. T. (1970a). Inhibition of polysaccharases by melanin: Enzyme inhibition in relation to mycolysis. *Archives of Biochemistry and Biophysics* **137**, 345–56.

BULL, A. T. (1970b). Chemical composition of wild-type and mutant *Aspergillus nidulans* cell walls. The nature of the polysaccharide and melanin constituents. *Journal of General Microbiology* **63**, 75–94.

CARTER, B. L. A. & BULL, A. T. (1969). Studies of fungal growth and intermediary carbon metabolism under steady and non-steady state conditions. *Biotechnology and Bioengineering* **11**, 785–804.

CHU, S. F. & ALEXANDER, M. (1972). Resistance and susceptibility of fungal spores to lysis. *Transactions of the British Mycological Society* **58**, 489–97.

EMERSON, S. & EMERSON, M. R. (1958). Production, reproduction, and reversion of protoplast-like structures in the osmotic strain of *Neurospora crassa*. *Proceedings of the National Academy of Sciences, USA* **44**, 668–71.

FERENCZY, L., KEVEI, F. & SZEGEDI, M. (1975). Increased fusion frequency of *Aspergillus nidulans* protoplasts. *Experientia* **31**, 50–2.

FERENCZY, L., KEVEI, F. & SZEGEDI, M. (1976). Fusion of fungal protoplasts induced by polyethylene glycol. In *Microbial and plant protoplasts*, pp. 178–188. Edited by J. F. Peberdy, A. H. Rose, H. J. Rogers and E. C. Cocking, London and New York: Academic Press.

FERENCZY, L., KEVEI, F. & ZSOLT, J. (1974). Fusion of fungal protoplasts. *Nature, London* **248**, 793–4.

GABRIEL, M. (1970). Cell wall regeneration in *Rhizopus nigricans* protoplasts. In *Yeast protoplasts*, pp. 147–151. Edited by O. Necas and A. Svoboda, University J E Purkyne, Brno, Medical Faculty.

GARCIA ACHA, I., LOPEZ-BELMONTE, F. & VILLANUEVA, J. R. (1966). Preparation of protoplast-like structures from conidia of *Fusarium culmorum*. *Antonie van Leeuwenhoek* **32**, 299–311.

GIAJA, J. (1914). Sur l'action quelques ferments sur les hydrates de carbone de la levure. *Comptes rendues des Seances de la Societe de Biologie* **77**, 2.

GIBSON, R. K. & PEBERDY, J. F. (1972). Fine structure of protoplasts of *Aspergillus nidulans*. *Journal of General Microbiology* **72**, 529–38.

LABORDA, F., GARCIA ACHA, I. & VILLANUEVA, J. R. (1974). Studies on a strepzyme capable of obtaining protoplasts from *Fusarium culmorum* conidia.

Transactions of the British Mycological Society **62**, 509–18.

ـOPEZ-BELMONTE, F., GARCIA ACHA, I. & VILLANUEVA, J. R. (1966). Observations on the protoplasts of *Fusarium culmorum* and on their fusion. *Journal of General Microbiology* **45**, 127–34.

ـMOORE, P. M. & PEBERDY, J. F. (1976a). Release and regeneration of protoplasts from the conidia of *Aspergillus flavus*. *Transactions of the British Mycological Society* **66**, 421–5.

ـMOORE, P. M. & PEBERDY, J. F. (1976b). A particulate chitin synthase from *Aspergillus flavus* Link: the properties, location and levels of activity in mycellium and regenerating protoplast preparations. *Canadian Journal of Microbiology* **22**, 915–21.

ـMUSILKOVA, M. & FENCL, Z. (1968). Some factors affecting the formation of protoplasts in *Aspergillus niger*. *Folia Microbiologica* **13**, 235–9.

ـMUSILKOVA, M., FENCL, Z. & SEICHERTOVA, O. (1969). Release of *Aspergillus niger* protoplasts by *Penicillium purpurogenum*. *Folia Microbiologica* **14**, 47–50.

ـPEBERDY, J. F. (1972). Protoplasts from fungi. *Science Progress, Oxford* **60**, 73–86.

ـPEBERDY, J. F. & BUCKLEY, C. E. (1973). Adsorption of fluorescent brighteners by regenerating protoplasts of *Aspergillus nidulans*. *Journal of General Microbiology* **74**, 281–8.

ـPEBERDY, J. F., BUCKLEY, C. E., DALTREY, D. C. & MOORE, P. M. (1976). Factors affecting protoplast release in some filamentous fungi. *Transactions of the British Mycological Society* **67**, 23–6.

ـREISS, J. (1973). Enzyme cytochemistry of fungi. *Progress in Histochemistry and Cytochemistry* **5**, 1–40.

ـRUIZ HERRERA, J., SING, V. O., VAN DER WOODE, W. J. & BARTNICKI-GARCIA, S. (1975). Microfibril assembly by granules of chitin synthetase. *Proceedings of the National Academy of Sciences, USA* **72**, 2706–10.

SEITSMA, J. H., EVELEIGH, D. E. & HASKINS, R. H. (1968). The purification of cellulase and exo-laminaronase and their role in the formation of *Pythium* sp. 'protoplasts'. *Antonie van Leeuwenhoek* **34**, 331–40.

SEITSMA, J. H., CHILD, J. J., NESBITT, L. R. & HASKINS, R. H. (1975). Ultrastructural aspects of wall regeneration by *Pythium* protoplasts. *Antonie van Leeuwenhoek* **41**, 17–23.

DE VRIES, I. M. H. & WESSELS, J. G. H. (1972). Release of protoplasts from *Schizophyllum commune* by a lytic enzyme preparation from *Trichoderma viride*. *Journal of General Microbiology* **73**, 13–22.

DE VRIES, O. M. H. & WESSELS, J. G. H. (1973). Release of protoplasts from *Schizophyllum commune* by combined action of purified 1,3-glucanase and chitinase derived from *Trichoderma viride*. *Journal of General Microbiology* **76**, 319–30.

DE VRIES, O. M. H. & WESSELS, J. G. H. (1975). Chemical analysis of cell wall regeneration and reversion in protoplasts from *Schizophyllum commune*. *Archives of Microbiology* **102**, 209–18.

VILLANUEVA, J. R. & GARCIA ACHA, I. (1971). Production and use of fungal protoplasts. In *Methods in microbiology*, Vol. 4, pp. 665–718, edited by C. Booth. London and New York: Academic Press.

WESSELS, J. G. H., VAN DER WALK, P. & DE VRIES, O. M. H. (1976). Wall synthesis by fungal protoplasts. In *Microbial and plant protoplasts*, pp. 267–81, edited by J. F. Peberdy, A. H. Rose, H. J. Rogers and E. C. Cocking. London and New York: Academic Press.

ZALOKAR, M. (1965). Integration of cellular metabolism. In *The fungi*, Vol. 1, pp. 377–426, edited by G. C. Ainsworth and A. S. Sussmann. New York and London: Academic Press.

CHAPTER 8

The Duplication Cycle and Vegetative Development in Moulds

A. P. J. TRINCI

8.1 Introduction

Many studies have been made of the morphological, physiological and biochemical events which occur during the cell cycle of prokaryotic (Pritchard, 1974) and eukaryotic (Mitchison, 1971; Hartwell, 1974) microorganisms. Information about the cell cycle may be obtained from microscopic observations of individual cells or studies of populations which have been synchronized by induction or selection methods. Such investigations may eventually lead to an understanding of the mechanisms which regulate and integrate the synthesis of macromolecules, enzymes and organelles during cell growth.

Mechanisms must also be present in moulds to regulate and integrate the synthesis of the various constituents which make up the protoplasm of mycelia. The existence of such mechanisms is suggested by the more or less homogeneous distribution of macromolecules, enzymes and organelles in undifferentiated mycelia or apical compartments of leading hyphae. For example, proteins, ribonucleic acid synthesis, nuclei and mitochondria are distributed fairly evenly throughout actively growing undifferentiated mycelia or leading hyphae (Nishi, Yanagita & Maruyama, 1968: Fencl, Machek & Novak, 1969; Zalokar, 1959b). However, some morphological and biochemical gradients occur in hyphae (Zalokar, 1959b) and, in particular, the tips of hyphae are always morphologically and biochemically differentiated from other parts of the mycelium, *e.g.* they contain vesicles (Grove & Bracker, 1970) and are rich in chitin synthase activity (Gooday, 1971) but lack nuclei and mitochrondria (Zalokar, 1959b). The mechanisms which regulate and integrate the synthesis of cellular components during fungal growth are likely to be similar, if not identical, to those present in unicellular organisms.

Under optimal conditions, growth of an asynchronous population is 'balanced', *i.e.* all extensive properties of the population increase at the same rate (Campbell, 1957). The intermittent synthesis of certain macromolecules, enzymes and organelles observed during growth of synchronized populations (Mitchison, 1971) is thus obscured in asynchronous cultures. Intermittent synthesis may occur in individual compartments or whole mycelia of fungi but remain undetected because it is not possible to synchronize such cultures. Since cell separation does not occur in moulds, the term 'duplication cycle' will be used to describe events during fungal growth which are analogous to those observed during the cell cycle of uninucleate cells. Results obtained with unicellular organisms suggest that the duration of such a duplication cycle will be the same as the doubling time of the mould and during a cycle there will be a doubling of all components of the protoplasm. This kind of concept is implicit in the work of King & Alexander (1969), Clutterbuck (1970) and Valla (1973a, b). As in unicellular organisms, the duration of a duplication cycle can be determined from the interval between a particular discontinuous event (*e.g.* mitosis or septation) in successive duplication cycles. The present review will consider the validity of the duplication cycle concept. This hypothesis may help to integrate some of the observations previously made on fungi and provide a basis for a comprehensive theory of mould growth which can be tested experimentally. In addition, the development of mycelia will be considered from spore germination to colony formation. Part of this article represents speculation rather than a review of established fact. I am sure that the reader will be able to distinguish the one from the other.

8.2 Differentiated and Undifferentiated Mycelia

Steele & Trinci (1975a) compared the morphology and growth kinetics of differentiated and undifferentiated mycelia.

Undifferentiated mycelia

A fungal spore germinates to form an undifferentiated mycelium which initially increases in total length at an exponential rate (Trinci, 1974). The hyphae of such mycelia are narrow, have short compartments and extension zones (Steele & Trinci, 1975a, b) and are usually accelerating towards the linear growth rate characteristic of the strain and environmental conditions (Fig. 8.10). Undifferentiated mycelia cultured on solid media are morphologically and physiologically similar to mycelia cultured in submerged culture (Steele & Trinci, 1975a) since in both cases they are formed under the near constant conditions which prevail during early growth.

Leading hyphae of differentiated mycelia (colonies)

Yanagita & Kogane (1962, 1963) divided mature colonies of *Aspergillus niger* and *Penicillium urticae* into zones (the extending, productive, fruiting and aged zones) which differed in morphology, cytology and metabolic activity. Although it is convenient to recognize such zones, there is actually a continuous differentiation of the colony from the periphery to the centre.

Hyphae in the peripheral growth zone (Trinci, 1971) at the margin of the colony are usually differentiated into wide 'leading' and narrower branch hyphae (Butler, 1961; Trinci, 1973; Steele & Trinci, 1975b). Many studies have been made of the kinetics and cytology of the growth of these leading hyphae; they extend at a constant rate, are orientated radially outwards from the centre of the colony and grow at a more or less fixed distance from each other. They differ in diameter, compartment length, extension zone length and extension rate from the primary and secondary branches which they subtend (Table 8.1).

Table 8.1 Extension rates, extension zone length and diameter of hyphae at the margin of colonies and germ-tube hyphae of *Neurospora crassa* SY7A at 25°C on Vogel's medium (From Steele & Trinci, 1975b)

	Diameter (μm)	Extension zone length (μm)	Extension rate (μm min^{-1})
(a) Hyphae in the peripheral growth zone of colonies			
Leading hyphae	13.1	29.2	38.0
Primary branches	8.9	20.2	27.4
Secondary branches	6.3	11.4	15.2
(b) Germ-tube hyphae	3.7	2.2	2.6

Hyphae at the margin of a colony extend into uncolonized medium. The composition of the medium surrounding a colony which expands at a slow rate may change significantly in advance of colonization due to substances (including nutrients and products of secondary metabolism formed within the colony) diffusing between colonized and uncolonized regions of the substrate (Rieck, Palumbo & Witter, 1973). However, the colonies of most moulds expand at rates which are considerably in excess of the rates at which chemicals diffuse within agar media. Consequently, the leading hyphae of most fungal colonies extend into medium which probably has a very similar, if not identical composition to uninoculated medium. Thus leading hyphae usually extend at a linear rate under the near constant conditions which prevail at the margin of the colony.

The duplication cycle will be described in undifferentiated mycelia and leading hyphae at the margin of colonies, as I consider these to be hyphae which are growing under constant or near constant conditions.

8.3 Duplication Cycle in Leading Hyphae

Monokaryotic apical compartments

Basidiobolus ranarum is a good organism to use to study the relationship between growth, mitosis and septation since its nuclei may exceed 20 μm in length and are easy to observe (Robinow, 1963). Leading hyphae of this mould extend at a linear rate, laying down septa at intervals (about 30 min) at the rear of their uninucleate, apical compartments. Intercalary compartments formed in this way have previously been evacuated of cytoplasm by its contraction into the upper part of the apical compartment prior to septation (Fig. 8.2). Gregory (1966) has defined a fungus as a plasmodium which creeps about in tubes and *B. ranarum* behaves exactly in this manner. The length of the apical compartment immediately after septation increases with time until it has attained an apparently critical length of 300 to 400 μm when the nucleus divides mitotically. During mitosis hyphal extension falls to 30 to 50% of its interphase rate (Fig. 8.1). At the end of telophase, one daughter

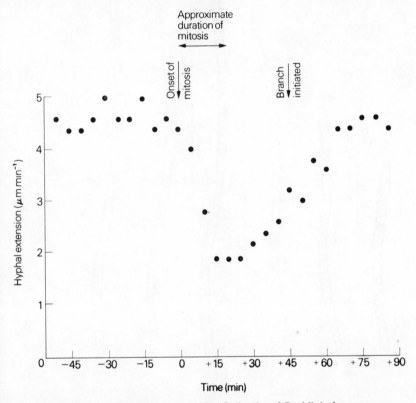

Fig. 8.1 Change in the rate of extension rate of a hypha of *Basidiobolus ranarum* during mitosis. Redrawn from Robinow (1963).

nucleus moves rapidly towards the hyphal tip whilst the other moves a much shorter distance in the opposite direction. A complete septum is formed between the daughter nuclei in a plane which is identical or close to that previously occupied by the metaphase plate of the dividing nucleus. Septation divides the apical compartment into two compartments, each of which contains a nucleus (Fig. 8.2). About 25 min after septation a branch is initiated from the intercalary compartment just behind the newly formed septum. Thus mitosis results in division of the original compartment into a short apical compartment and an intercalary compartment which forms a branch (Fig. 8.2).

Leading hyphae of monokaryotic mycelia of *Schizophyllum commune*, unlike those of *Basidiobolus ranarum*, continue to extend at a linear rate throughout the duplication cycle (Volz & Niederpruem, 1968). It is perhaps significant that *B. ranarum* forms complete septa whilst the dolipore septa of

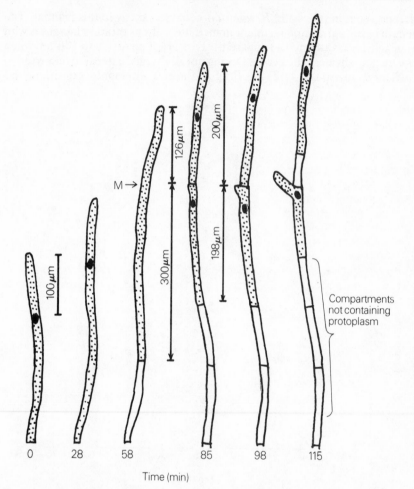

Fig. 8.2 Mitosis, septation and branching in *Basidiobolus ranarum*. Drawings made from photographs taken by Robinow (1963). *M* = position of nucleus during mitosis.

S. commune, at least initially, allow cytoplasmic continuity between adjacent compartments. In *S. commune* the nucleus maintains a more or less central position in the apical compartment during growth by migrating towards the tip at a rate somewhat less than that of hyphal extension. However, nuclear migration decelerates or even stops just prior to mitosis. About 9 to 15 min after mitosis a septum is formed in the compartment on a site previously occupied by the dividing nucleus. Septation divides the apical compartment into apical and intercalary compartments of more or less equal length (Volz & Niederpruem, 1968). However, like *B. ranarum*, *S. commune* does not always form daughter compartments of equal length after cytokinesis. As in *B. ranarum* a branch is initiated from the intercalary compartment just behind the newly formed septum. The duplication cycle in this organism lasts approximately one hour.

Dikaryotic apical compartments

The duplication cycle in leading hyphae of dikaryotic mycelia of *Schizophyllum commune* has been described by Niederpruem, Jersild & Lane (1971). As in monokaryotic hyphae of this fungus, mitosis and septation are not associated with a deceleration in the rate of tip extension. The pair of nuclei in an apical compartment migrate towards the hyphal tip at approximately the same rate as hyphal extension and in so doing they maintain a more or less central position in the compartment; pairs of nuclei do not apparently separate as they migrate towards the tip.

Nuclear migration stops just before or during clamp formation and prior to mitosis. The pair of nuclei divide synchronously and 15 to 20 min later septa are formed on cytoplasmic sites previously occupied by the dividing nuclei (Niederpruem *et al.*, 1971; Butler, 1972). As in the monokaryon, a branch is subsequently formed from the intercalary compartment just behind the newly formed septum.

Data of Niederpuem *et al.* (1971) are presented in Fig. 8.3. They demonstrate the regularity of the duplication cycle in the dikaryon. As in the monokaryon, septation usually divides the apical compartment into two compartments of about equal length, each of which contains a pair of nuclei.

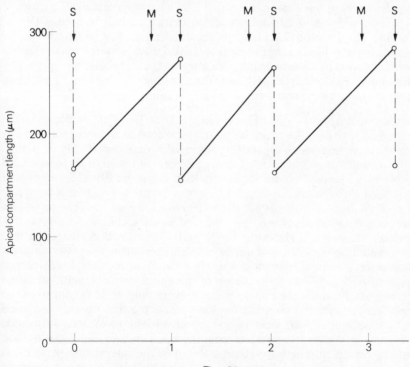

Fig. 8.3 Apical compartment length, mitosis and septation in a single dikaryotic leading hypha of *Schizophyllum commune*. Periods of mitosis (m) and septation (s) indicated by arrows. Graph drawn from data of Niederpruem, Jersild & Lane (1971).

The duplication cycle in the dikaryon, as in the monokaryon, lasts approximately one hour. Apical compartments of *Coprinus disseminatus* are also divided in half by septation Butler (1961).

Coenocytic apical compartments

Duplication cycles have been described in coenocytic apical compartments (containing 25 to 90 nuclei) of leading hyphae of *Alternaria solani* (King & Alexander, 1969), *Aspergillus nidulans* (Clutterbuck, 1970; Fiddy & Trinci, 1976a) and *Polyporus arcularius* (Valla, 1973a, b). As in monokaryotic and dikaryotic hyphae, the duplication cycle in apical compartments of these fungi is analogous to the cell cycle of uninucleate cells. The main morphological events are: (i) septation (usually multiple) reduces the apical compartment to about half its maximum length; (ii) the newly formed apical compartment continues to increase in length at a linear rate; (iii) the volume of cytoplasm per nucleus increases until at a critical ratio the nuclei are induced to divide more or less synchronously; and (iv) mitosis is followed by multiple septation which is completed when the apical compartment is about twice its original length. The duration of such a duplication cycle can be determined by timing the interval between a given discontinuous event (*e.g.* mitosis or septation) in successive duplication cycles. For example, in *P. arcularius* there was a mean interval of 2.09 h between successive periods of septation and one of 2.06 h between successive periods of mitosis (Valla, 1973b). The duration (2.1 ± 0.2 h) of the duplication cycle (as determined by measuring the interval between successive cycles of septation) in apical compartments of *A. nidulans* was identical to the mycelial doubling time of this mould (Fiddy & Trinci, 1976a). Thus, as suggested earlier, the doubling time of a mould is a direct measure of the duration of its duplication cycle.

NUCLEAR DISTRIBUTION   The mean distance between adjacent nuclei increased with distance from the tip in apical compartments of *Ceratocystis fagacearum*, *Polyporus arcularius*, *Aspergillus nidulans* (Fiddy & Trinci, 1976a) and *Alternaria solani*. During interphase, nuclei of *C. fagacearum* constantly changed shape (although they were usually elongated) and migrated throughout the length of compartments and between compartments (Wilson & Aist, 1967; Aist, 1969). However, nuclear migration between compartments was not observed in *A. solani* or *Fusarium oxysporum* (Koenig & Howard, 1962) and Clutterbuck & Roper (1966) observed little circulation of nuclei in apical compartments of *A. nidulans*. Just prior to mitosis the most apically located nucleus in *A. solani* was generally 55 to 70 μm from the tip of the apical compartment, but about 10 min after mitosis the daughter nuclei were only 25 to 30 μm from the tip. The distance between the hyphal tip and the first nucleus increased with time until by the next period of mitosis the most apically located nucleus was again 55 to 70 μm from the tip. However, since the hyphae had increased in length by about 240 μm in the intervening period it is clear that, as in monokaryons and dikaryons, nuclei in coenocytes must migrate towards the tip at a rate similar, if slightly slower than that of hyphal extension. Nuclei in *Thanatephorus cucumeris* certainly migrated forward in the apical compartment at approximately the same rate as hypal extension (Flentje, Stretton & Hawn, 1963). The mean distance between the hyphal tip and the first nucleus was 48 to 75 μm in *P. arcularia* (Valla, 1973*b*) and 17.5 μm in *A. nidulans* (Fiddy & Trinci, 1976*a*).

MITOSIS   Nuclei increase in volume as they approach mitosis (Robinow & Caten, 1969), the first sign of which is the dispersal of the nucleolus (Aist & Williams,. 1972; McCully & Robinow, 1971). Synchronous or near synchronous mitosis has been observed in trinucleate apical compartments of *Coprinus disseminatus* (Butler, 1972), in multinucleate apical compartments of *Penicillium cyclopium* (Rees & Jinks, 1952), *Thanatephorus cucumeris* (Flentje *et al.*, 1963), *Fusarium oxysporum* (Koenig & Howard, 1962; Aist, 1969), *Ceratocystis fagacearum* (Aist & Wilson, 1966; Aist, 1969), *Alternaria solani* (King & Alexander, 1969), *Aspergillus nidulans* (Clutterbuck, 1970), *Ascobolus immersus* (Zickler, 1971) and *Polyporus arcularius* (Valla, 1973*a, b*) and in multinucleate protoplasts of *Schizophyllum commune* (Valk & Wessels, 1973). In *T. cucumeris* (Flentje *et al.*, 1963) and *P. arcularius* (Valla 1973*a*) mitosis in adjacent hyphae was apparently synchronized, but this type of colonial synchrony has not been observed in other fungi. In *P. cyclopium* (Rees & Jinks, 1952), *C. fagacearum* (Aist, 1969), *A. solani* and *A. nidulans* mitosis was usually initiated in an apical compartment by the division of nuclei at its tip and was followed by a wave of nuclear division which passed distally towards the septum. Such mitotic waves took about 20 and 10 to 15 min respectively to traverse apical compartments of *A. nidulans* and *F. oxysporum* (Koenig & Howard, 1962). However, Aist (1969) found that mitosis in his strain of *F. oxysporum* was initiated at a point somewhere between the apical compartment and the fourth or fifth intercalary compartment and travelled as a wave towards or away from the hyphal tip or in both directions. Mitosis in *P. arcularius* (Valla, 1973*a*) was initiated towards the rear of the apical compartment and travelled slowly (taking 20 to 60 min) to the hyphal tip. At the same time a second wave of mitosis was initiated at the same point as the first and travelled for a short distance in the opposite direction *i.e.* towards the septum. However, nuclei at the base of apical compartments of *P. arcularius* did not divide; the number of these non-dividing nuclei was directly related to the total number of nuclei in the compartment.

It took about 10 min or less for a nucleus in *Ceratocystis fagacearum, Fusarium oxysporum, Aspergillus nidulans, Polyporus arcularius* or *Alternarsia solani* to complete mitosis; thus the time required for mitosis in fungi is at least an order of magnitude less than in higher plants or animals (Aist & Williams, 1972). These workers determined the duration of the various stages of mitosis in *C. fagacearum*; the whole process took 5.5 min, being made up of 60 sec for prophase, 121 ± 29 s for metaphase, 13 ± 5 sec for anaphase and 124 s for telophase. After nuclear division in *A. nidulans* one daughter nucleus stayed close to the site of mitosis, whilst the other moved rapidly away, proximally or distally to the tip, over distances of about 10 to 15 μm (Robinow & Caten, 1969). After synchronous mitosis in *Thanatephorus cucumeris* half the daughter nuclei migrated towards the hyphal tip whilst the other half moved in the opposite direction (Flentje *et al.*, 1963). Girbardt (1971) found that kinetochore equivalents in *Polystictus versicolor* were made up of two globular elements which did not change in size during the first 125 min of interphase but doubled in size just before mitosis. He suggested that these elements are made up of proteins and may be involved in the initiation of mitosis.

SEPTATION   Synchronous or near synchronous nuclear division in coenocytic apical compartments of *Fusarium oxysporum* (Koenig & Howard, 1962), *Alternaria solani, Aspergillus nidulans and Polyporus arcularius* was followed after 12 to 18, 8 to 15, 20 to 40 and 11 to 30 min respectively by the formation of groups of 0 to 6, 3 to 7, 2 to 6 and 1 to 11 septa respectively. The intervals between successive septation cycles (*i.e.* duration of the duplication cycles) in *F. oxysporum, A. solani, A. nidulans* and *P. arcularius* were 1, 0.95, 3.7 and 2.1 h respectively. Thus the rather long interval (20 to 40 min) observed between mitosis and septation in *A. nidulans* (Clutterbuck, 1970) is correlated with the prolonged duplication cycle (about 3.7 h) of this organism at 16 to 18°C. At 25°C the interval between successive septation cycles in *A. nidulans* was reduced to 2.1 h (Fiddy & Trinci, 1976a); at this temperature groups of septa in *A. nidulans* were completed in 8.5 ± 6 min.

Septa in *Fusarium oxysporum* (Koenig & Howard, 1962) were apparently only formed on cytoplasmic sites previously occupied by dividing nuclei. However, dividing or daughter nuclei in *F. oxysporum* and *Ceratocystis fagacearum* (Aist, 1969) migrated for distances up to 25 μm before septa were formed (Aist, 1969). In *Thanatephorus cucumeris* eight daughter nuclei migrated towards the germ tube tip after synchronous mitosis, whilst the other eight moved in the opposite direction. A septum was subsequently formed between the two groups (Flentje *et al.*, 1963). Other workers have apparently not observed any spatial relationship in coenocytes between mitotic sites and the subsequent location of septa.

The first septa produced after mitosis in apical compartments of *Aspergillus nidulans, Alternaria solani* and *Polyporus arcularius* were usually formed close to the hyphal tip. The remaining septa in a group were usually formed distally to the first and positioned at random. A newly formed group of septa in *A. solani* or *A. nidulans* (Fig. 8.4) divided the apical compartment of a leading hypha into a distal septated region and a proximal aseptate region (the new apical compartment) of approximately equal length. However, the long (over 1200 μm just prior to septation) apical compartments of leading hyphae of *P. arcularius* (Valla, 1973b) were not usually divided in half by septation (Fig. 8.5). Instead, septation reduced these apical compartments by an amount which was less than half its overall length.

In some hyphae of *Aspergillus nidulans* in which there were relatively few nuclei at the rear of the apical compartment, multiple septation resulted in the formation of intercalary compartments which were longer than those normally formed (Clutterbuck, 1970). King & Alexander (1969) suggest that nuclear concentration may partially determine the location of septa in *Alternaria solani* but failed to detect any significant variation in the length of the intercalary compartments of this mould.

INTERCALARY COMPARTMENTS   Nuclei in intercalary compartments rarely divided (Rees & Jinks, 1952; King & Alexander, 1969; Koenig & Howard, 1962) or divided more slowly (Aist, 1969) than those in apical compartments. Sometimes most, but not all, nuclei in an intercalary compartment of *Ceratocystis fagacearum* divided synchronously (Aist, 1969).

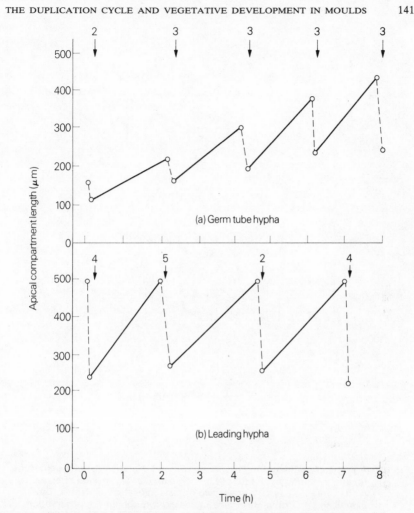

Fig. 8.4 Variation in the length of apical compartments of a germ tube (a) and a leading hypha (b) of *Aspergillus nidulans*. The extension rate of the germ tube increased throughout the observation period, whilst that of the leading hypha was constant. Periods of septation (and numbers of septa formed) indicated by arrows. From Fiddy and Trinci (1976a).

In *Aspergillus oryzae* RNA and protein synthesis occurred at a faster rate in intercalary than apical compartments but the generation time of nuclei was up to two times longer in intercalary than apical compartments (Nishi *et al.*, 1968). However, it may be significant that in these experiments the mould was growing linearly rather than exponentially when the measurements were made.

8.4 Duplication Cycle in Undifferentiated Mycelia

Nuclear distribution

Nuclei are distributed more or less uniformly (Fig. 8.6) throughout the cytoplasm of undifferentiated mycelia (Rees & Jinks, 1952; Zalokar,

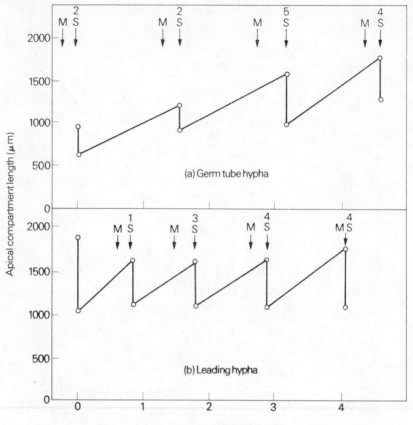

Fig. 8.5 Apical compartment length, mitosis (M) and septation (S) of a germ tube (a) and a leading hypha (b) of *Polyporus arcularius*. Periods of septation (and number of septa formed) and mitosis indicated by arrows. The extension rate of the germ tube increased throughout the observation period, whilst that of the leading hypha was constant. Graphs drawn from data of Valla (1973*b*).

1959*a*; Valla, 1974; Fiddy & Trinci, 1976*a*). In *Aspergillus nidulans* the mean distance between the germ tube tip and the first nucleus was 6.9 ± 1.2 μm; the low standard deviation observed suggests that, as in leading hyphae, nuclei migrate towards the hyphal apex at approximately the same rate as germ tube extension.

Mitosis

In *Thanatephorus cucumeris* the basidial nucleus divides and one daughter nucleus remains in the spore whilst the other migrates into the germ tube. The nuclei continue to divide synchronously until the germ tube contains at least 16 nuclei (Flentje *et al.*, 1963).

The nuclei in germlings formed from uninucleate conidia of *Aspergillus nidulans* also initially divide synchronously (Rosenberger & Kessel, 1967; Robinow & Caten, 1969) but subsequently, mitosis in the mycelium as a

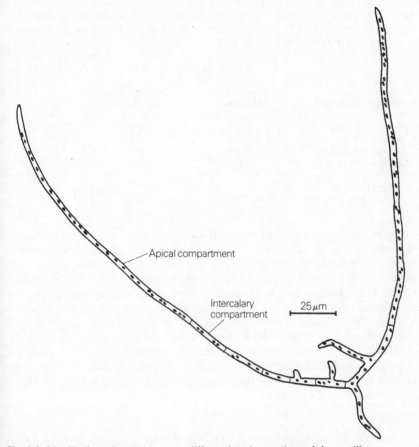

Apical compartment

Intercalary
compartment

25 μm

Fig. 8.6 Distribution of nuclei in an undifferentiated mycelium of *Aspergillus nidulans*. Hypha stained with Giemsa and tracing made from photographs.

whole becomes asynchronous (Fiddy & Trinci, 1976*a*). After nuclear division in *Aspergillus nidulans*, as in *Basidiobolus ranarum* (Robinow, 1963), one daughter nucleus remains close to the site of mitosis whilst the other moves 10 to 15 μm towards or away from the hyphal tip (Robinow & Caten, 1969). The transition from synchronous to asynchronous mitosis in *Aspergillus nidulans* was initiated when the mycelium was 60 to 130 μm long and contained 8 to 16 nuclei and was apparently correlated with septation. However, the asynchronous development of a mycelium masks synchronous mitosis in individual compartments (Robinow & Caten, 1969), *i.e.* there is intra- but not inter-compartmental synchrony. Because compartments develop asynchronously a mycelium eventually assumes the characteristics (Table 8.2) associated with balanced growth of an asynchronous population (Campbell, 1957), *i.e.* all extensive properties of the mycelium increase at the same rate.

The maximum lengths (26, 56, 126 and 230 μm respectively) attained by mycelia of *Aspergillus nidulans* which contained 4, 8, 16 and 32 nuclei were approximately twice the minimum lengths (9, 29, 59 and 120 μm) observed

Table 8.2 Growth kinetics of undifferentiated mycelia of *Aspergillus nidulans* grown on malt extract at 25°C and *Schizophyllum commune* grown on defined medium at 30°C

Parameter measured	Specific rate of increase (h^{-1})	
	*Aspergillus nidulans**	*Schizophyllum commune*†
Mycelial length	0.32	0.26
Number of tips	0.26	0.27
Number of nuclei	0.28	0.16
Number of septa	0.32	0.21

* From Fiddy & Trinci (1976*a*)
† Calculated from the data of Volz & Niederpruem (1968)

for these same numbers of nuclei (Fiddy & Trinci, 1976*a*). These data suggest that, as in apical compartments of leading hyphae, synchronous mitosis is initiated in germlings when the cytoplasmic volume per nucleus ratio obtains a critical value. In 35 of 58 germlings examined by Robinow & Caten (1969) there was a mitotic gradient from the hyphal tip to the conidium (*i.e.* the nuclei in the tip initiated a wave of mitosis which passed towards the conidium) whilst a gradient in the reverse direction was observed in 12 germlings and no gradient was detected in the remainder.

Morris (1975, 1976) has isolated 35 temperature-sensitive mutants of *Aspergillus nidulans* which were unable to complete nuclear division at the restrictive temperature; 26 of these mutants failed to enter mitosis whilst the rest were arrested in mitosis.

Septation

In bacteria there is a constant interval between termination of DNA replication and cell separation (Pritchard, 1974) whilst in eukaryotic organisms mitosis is followed more or less immediately by septation and cell separation. This tight coupling between mitosis and septation was not always observed in germlings of *Aspergillus nidulans* (Fiddy & Trinci, 1976*a*) or *Polyporus arcularius* (Valla, 1973*b*) since septa were only formed after the second round of mitosis *i.e.* when the germ tube contained 4 nuclei. Similarly a septum is not formed after nuclear division during basidiospore maturation in *Schizophyllum commune*, but when the binucleate basidiospore germinates, near synchronous nuclear division is followed by the formation of two septa which divide the germling into three compartments, two uninucleate apical compartments and a binucleate intercalary compartment (Niederpruem & Jersild, 1972). The sites at which septa are formed in the germling coincide with positions previously occupied by the dividing nuclei.

Septa were formed in groups of 1 to 9 in apical compartments of germ-tubes of *Aspergillus nidulans*; as in leading hyphae a septal group was completed in about 9 min. Similarly, the interval (1.8 h) between successive septation cycles in germ-tubes of *A. nidulans* was of approximately the same duration as the doubling time of the organism (2.1 h), suggesting that in this organism septa are formed at a fixed point in the duplication cycle, presumably soon after nuclear division. Groups of septa formed in germ tubes of

Aspergillus nidulans (Fiddy & Trinci, 1976a) and *Polyporus arcularius* (Valla, 1973b), unlike those in leading hyphae, divide apical compartments unequally (Figs. 8.4 and 8.5). This unequal division of apical compartments by septation persists until the germ tubes attain a linear growth rate.

Septation cycles in different germ-tubes of a single spore of *Aspergillus nidulans* were observed to be out of phase by as much as 53 min, indicating that septation cycles (and hence duplication cycles) in the different compartments of the same mycelium are not synchronized. Thus septation like nuclear division (Niederpruem & Jersild, 1972), is synchronized within a compartment but not necessarily between compartments of the same mycelium.

8.5 Branch Initiation in Leading Hyphae and Undifferentiated Mycelia

Branch initiation in fungi is an event which is physiologically analogous to cell division in unicellular micro-organisms.

Branch initiation associated with septation

The variety of branching patterns (Fig. 8.7) produced by hypae of *Neurospora crassa* recovering from an osmotic shock (Trinci & Collinge, 1974) suggests that a branch can potentially be formed from any part of the compartment wall, including the septum. The precise location of a branch is thus probably determined by a cytoplasmic event. Vesicles observed at the tips of fungal hyphae (Grove & Bracker, 1970) may contain wall precursors and/or the enzymes (including enzymes which lyse wall polymers) required for the insertion of precursors into the existing wall (Bartnicki-Garcia, 1973). Such vesicles have been implicated in wall extension (primary wall formation?) during branching (Nolan & Bol, 1974), germ-tube emergence (Bracker, 1971), budding (McCully & Bracker, 1972; Cortat, Matile & Wiemken, 1972), and the formation of intrahyphal branches from septa (Trinci & Collinge, 1974). If it is assumed (Collinge & Trinci, 1974) that vesicles are generated throughout the peripheral growth zone of hyphae, then the imposition of a barrier (*e.g.* a septum) to acropetal vesicle transport will result in an accumulation of vesicles behind the barrier. Vesicle accumulation and their subsequent fusion with the hyphal wall may therefore be the first step in branch initiation. Vesicles have been observed behind plugged septa of *N. crassa* (Trinci & Collinge, 1974) and vesicles containing cellulolytic activity have been implicated in branch initiation in *Achlya ambisexualis*, a fungus which has walls containing cellulose (Nolan & Bal, 1974). If vesicle transport within an intercalary compartment bounded by complete septa continues to be polarized in an acropetal direction then branches would be expected to arise just behind the proximal septum (with respect to the hyphal apex). Branches are in fact initiated from this position (Table 8.3) in *Geotrichum candidum* (Fig. 8.8) and *Basidiobolus ranarum* (Fig. 8.2) but not in *Mucor hiemalis* (Fiddy & Trinci, 1976c) although all three fungi are thought to form complete septa. Branches are also usually initiated immediately behind or close to septa in *Ceratocystis fagacearum* (Aist, 1969), monokaryotic (Volz & Niederpruem, 1968) and dikaryotic (Niederpruem *et al.*, 1971) hyphae of *Schizophyllum commune*, dikaryotic hyphae of *Coprinus sterquilinus* and *Coprinus lagopus* (Buller, 1958) and other fungi (Park & Robinson, 1967). Basidiomycete septa have dolipores

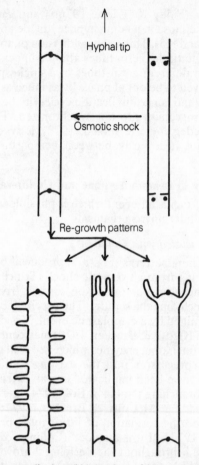

Fig. 8.7 Re-growth patterns displayed by hyphae of *Neurospora crassa* which had been flooded with water. Redrawn from Trinci & Collinge (1974).

which at least initially, allow the transport of cytoplasm, including mitochondria, between adjacent compartments. However, as far as branch initiation is concerned such septa *ultimately* appear to function as if they form barriers to vesicle transport (Table 8.3).

The lag between septation and branch initiation in *Geotrichum candidum* is remarkably constant, being 26 ± 6 min for intercalary compartments of undifferentiated mycelia and 28 ± 9 min for intercalary compartments of leading hyphae (Fiddy & Trinci, 1976b). There was a lag of about 25 min between septation and branch initiation in *Basidiobolus ranarum* (Robinow, 1963) and branches were initiated from intercalary compartments of *Neurospora crassa* about 30 min after they had received an osmotic shock (Collinge & Trinci, 1974). In *N. crassa*, regrowth branches were usually formed behind the proximal septa of intercalary compartments, suggesting that acropetal vesicle transport was maintained during the recovery period. Intrahyphal rather than lateral branches were, however, formed by some compartments (Fig. 8.7) suggesting that the osmotic shock may sometimes

Table 8.3 Location of branches formed from intercalary compartments of leading hyphae

Region of intercalary compartment§	Number of branches in each compartment region expressed as a percentage of the total number of branches				
	Geotrichum candidum†	*Basidiobolus ranarum*/	*Coprinus sterquilinus*¶	*Coprinus lagopus*¶	*Aspergillus nidulans*‡
1	92	94	ca. 100	ca. 100	27
2	5	0	0	0	41
3	3	4	0	0	17
4	1	2	0	0	10
5	0	0	0	0	5

* C. Fiddy & A. P. J. Trinci (unpublished)
† Fiddy & Trinci (1976*b*)
‡Fiddy & Trinci (1976*a*)
§ Each compartment was divided by eye into 5 equal portions, 1 being nearest the hyphal tip and 5 furthest from the tip
¶ Data of Buller (1958)

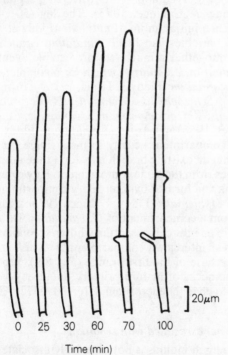

Fig. 8.8 Septation and branch initiation during growth of a hypha of *Geotrichum candidum*. From Fiddy & Trinci (1976*b*).

disrupt normal development. When Buller (1958) damaged hyphae of *Pyromena confluens* he observed that intrahyphal branches were initiated from septa about 30 min after their pores had been plugged. Intrahyphal hyphae have been observed in several fungi under a variety of cultural conditions (Lowry & Sussman, 1966; Chan & Stephen, 1967; Calonge,

1968; Trinci & Righelato, 1970). Buller (1958) suggested that the formation of such branches is a normal event in old colonies. It is not known how intra-hyphal branching is suppressed during normal hyphal morphogenesis. The multilateral branching observed (Fig. 8.7) from some compartments of *N. crassa* recovering from an osmotic shock suggests that this treatment may sometimes disrupt the normal polarity of vesicle transport with the result that vesicles fuse with the lateral hyphae of the compartment along its entire length.

There was a significant correlation between septation and branching in *Aspergillus nidulans* and usually each intercalary compartment initially produced a single branch (Fiddy & Trinci, 1976a). However, these branches, unlike those of *Basidiobolus ranarum* and *Geotrichum candidum*, were formed from all regions of the lateral wall of the compartment (Table 8.3). Intercalary compartments of *Pyronema confluens* also do not appear to be highly polarized with respect to the site of branch initiation (Buller, 1958). This suggests that complete septa or dolipores form an effective barrier to vesicle movement whereas incomplete septa do not. Certainly, vesicles have been observed apparently traversing septal pores of *Neurospora crassa* (Trinci & Collinge, 1973). The lag (50 ± 40 min) between septation and branch initiation in *A. nidulans* (Fiddy & Trinci, 1976a) is longer and more variable than in *G. candidum*. Branch initiation in *A. nidulans* is thus altogether a much less predictable event than in moulds such as *G. candidum* or *B. ranarum* which form complete septa.

In *Geotrichum candidum* (Fiddy & Trinci, 1976b), *Basidiobolus ranarum* (Robinow, 1963), *Schizophyllum commune* (Niederpruem et al., 1971), *Pyronema confluens, Rhizoctonia solani, S. commune, Coprinus lagopus, C. sterquilinus* (Buller, 1958) and *Aspergillus nidulans* (Fiddy & Trinci, 1976a) each intercalary compartment usually forms a single branch. This observation suggests that an existing branch may for a time inhibit the formation of further branches from the same compartment. However, up to four quite short branches may be formed by intercalary compartments of undifferentiated mycelia of *M. hiemalis* (Fiddy & Trinci, 1976c). The formation of a second branch from a compartment of *G. candidum* was usually preceded by the formation of an intercalary septum whilst a second branch was usually only formed from an intercalary compartment of *A. nidulans* after a septum had formed at the base of the first branch (Fig. 8.9). Septa are commonly formed across the necks of mature primary branches where they connect with their parent hyphae (Niederpruem & Jersild, 1972; King & Alexander, 1969).

Branch initiation not associated with septation

Of course branching in moulds is not invariably associated with septation. For example, *Rhizopus stolonifer, Actinomucor repens* and other members of the Mucorales branch normally, but are aseptate. Further, dichotomous branching in *Allomyces* (Emerson, 1955) and *Geotrichum candidum* (Trinci, 1970) is not associated with septation and branches are also frequently formed from the apical compartments of Basidiomycetes (Butler, 1972; Valla, 1973a) and Ascomycetes. For example, *Aspergillus nidulans* branches sub-apically producing two or more branches per tip (Trinci, 1970) and branches are formed just behind the tips of leading hyphae of

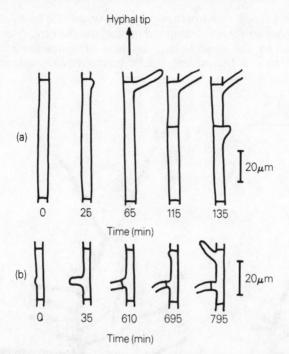

Fig. 8.9 Formation of second branches produced by intercalary compartments in *Geotrichum candidum* (a) from Fiddy & Trinci (1976*b*) and *Aspergillus nidulans* (b) from Fiddy & Trinci (1976*a*).

Neurospora crassa (Zalokar, 1959*b*). In addition Morris (1975) has isolated several temperature-sensitive mutants of *A. nidulans* which branch normally, but do not form septa at the restrictive temperature.

It has been claimed that branches in *Alternaria solani* (King & Alexander, 1969) and *Polyporus arcularius* (Valla, 1973*b*) are only initiated from the apical compartment of hyphae; the apical compartments of these fungi produced branches at all stages of the duplication cycle. Such branches in *A. solani* were 40 to 70 μm long before nuclei migrated into them from the parent compartment. Up to 30 nuclei migrated from an apical compartment of *P. arcularius* into a single branch. Nuclei present in apical compartments and branches of leading hyphae of *P. arcularius* divided synchronously.

I have suggested here and elsewhere (Trinci, 1974) that branches are initiated at locations in mycelia where for one reason or another there is an accumulation of vesicles. The correlation between septation and branching has been discussed above and in addition, it is possible to visualize circumstances under which vesicles may accumulate in certain regions of a hypha in the absence of septation. For example an increased vesicle concentration will occur in an apical compartment if the rate of supply of vesicles to the apex exceeds the rate at which they are incorporated into the tip wall. Due to the polarized nature of hyphal growth (and therefore presumably of vesicle transport) such increases in vesicle concentration are likely to occur at or close to the hyphal tip. These are in positions where hyphae are actually observed to branch dichotomously or sub-apically. The hypothesis that vesicles are not uniformly distributed throughout the cytoplasm of a mould is supported by the observation that new branches are initiated in a mycelium before all the existing tips have attained their maximum rate of extension (Fig. 8.10).

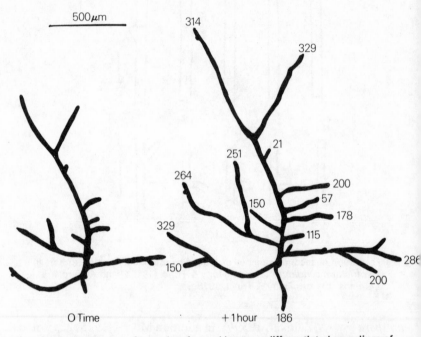

Fig. 8.10 Extension rate of branches formed by an undifferentiated mycelium of *Mucor hiemalis*. Growth of the mycelium during a one hour period is illustrated. The figures give the extension rate of each tip in the previous hour. Redrawn from Trinci (1974).

I have also suggested that the growth of an undifferentiated mycelium may be considered in terms of the duplication of a hypothetical 'growth unit' which consists of a tip associated with a specific length of hypha (Trinci, 1974). As observed by Bull & Trinci (1977) the hyphal growth unit is a physiological but not a morphological entity and clearly differs qualitatively from the cell as the growth unit of unicellular micro-organisms. The observation that the extension rates of hyphae of an undifferentiated mycelium of *Mucor hiemalis* varied from 21 to 329 μm/h suggests that the length of hypha actually associated with each tip probably varies over a wide range (Fig. 8.10). However, branch initiation, like the division of a single cell, may be regulated by changes in cytoplasmic volume which accompany growth. Thus, when the mean hyphal growth unit of a mycelium (volume of cytoplasm per hyphal tip) exceeds a critical length (volume) a new branch is initiated somewhere in the mycelium, its precise location being determined in a manner which has yet to be fully understood.

8.6 Regulation of the Duplication Cycle

Helmstetter, Cooper, Pierucci & Revelas (1968) have proposed that initiation of DNA synthesis in bacteria is governed by a positive control mechanism whereby a hypothetical initiation protein is continuously formed as a constant fraction of the total protein. They suggest that initiation occurs when the protein reaches a critical level and that the protein may be used up in the initiation process so that further initiation only occurs after more protein has been produced. New rounds of replication in bacteria appear to be initiated at a constant cell volume per chromosome origin ratio which is independent of the growth rate for doubling times greater than 70 min (Donachie, 1968; Pritchard, 1968). Thus

$$(m/n)_i = k_i \qquad (8.1)$$

where m = cell mass, n = chromosome origins present and k_i is a constant which has been called the initiation mass. Thus when the condition of eqn 8.1 pertains initiation will occur, leading to a doubling in the number of chromosome origins.

A positive control mechanism has also been proposed by Sachsenmaier, Remy & Plattner-Shobel (1972) to explain the timing of synchronous mitosis in *Physarum polycephalum*. They suggest that nuclei are induced to divide synchronously by a factor which accumulates during interphase up to a threshold level and subsequently disappears or is transformed during mitosis. Sachsenmaier *et al.* (1972) propose that the onset of mitosis is determined by a critical ratio between the number of initiation molecules and the number of nuclei in the cytoplasm and that initiation molecules accumulate during interphase at a rate proportional to plasmodial mass.

The principal features of the duplication cycle in moulds are summarized in Table 8.4. It would appear that mitosis in fungi, as in slime moulds, is initiated when the ratio between cytoplasmic mass to nuclear number attains a critical value. However, Clutterbuck (1969) and Fiddy & Trinci (1976a) found that in diploid strains this ratio had approximately double the value observed in haploid strains of the same species. They thus conclude that the volume of cytoplasm per genome is the critical ratio which regulates the timing of mitosis.

A positive control mechanism which is substantially similar to those proposed to explain the regulation of mitosis in *Physarum polycephalum* or initiation of DNA replication in bacteria may be advanced to explain the timing of mitosis in fungi. A factor produced during interphase in fungi at a rate proportional to cytoplasmic mass may initiate mitosis when its concentration per genome exceeds a certain critical value. The observation of Kessel & Rosenberger (1968) that the duration of mitosis and the (S) period of DNA synthesis in *Aspergillus nidulans* is not altered by growth rate suggests that the mechanism which regulates nuclear division in fungi probably operates, as in bacteria, by influencing the frequency of initiation of DNA replication.

If a cytoplasmic signal does regulate mitosis, the observations that mitosis in adjacent compartments (King & Alexander, 1969) and duplication cycles in compartments of the same mycelium (Fiddy & Trinci, 1976a) are not synchronized suggest that, in contrast to *Physarum polycephalum* and other

Table 8.4 Comparison of the cell cycle of *Schizosaccharomyces pombe* with the duplication cycle in moulds

Organism	Mitosis initiated when ratio, volume of cytoplasm: no. of nuclei attains critical value	Synchronous or near synchronous mitosis in apical compartment	Septation				Branching or cell separation
			Temporal relationship between mitosis and septation	Synchronous or near synchronous mitosis in apical compartment (Spatial relationship between sites of nuclear division and septation)	Spatial relationship between sites of nuclear division and septation	Quantitative relationship between the number of nuclei which divide and the number of septa formed	
Schizosaccharomyces pombe	+	Does not apply	+		+	+	+ +
Basidiobolus ranarum	+	Does not apply	+		+	+	+ +
Schizophyllum commune monokaryon	+	Does not apply	+		+	+	+ but delayed
Schizophyllum commune dikaryon	+	+	+ + +		+	+	+ but delayed
Aspergillus nidulans	+	+	+ +		Apparently not	Apparently not	+ but delayed
Alternaria solani	+	+	+		Apparently not	Apparently not	+ but delayed
Polyporus arcularius	+	+ but some do not divide	+		Apparently not	Apparently not	?
Thanatephorus cucumeris	?	+	+		Apparently not	+	+ but delayed

slime moulds, there is incomplete cytoplasmic mixing in fungal mycelia. Indeed, mitosis may not even be perfectly synchronized within a single apical compartment (King & Alexander, 1969; Clutterbuck, 1970; Robinow & Caten, 1969). Incomplete cytoplasmic mixing is probably an inevitable consequence of the polarized nature of hyphal growth. Although mitosis in adjacent compartments of a mycelium is not usually synchronized, King & Alexander (1969) and Valla (1973b) found that nuclei in apical compartments and their branches divided synchronously. Fiddy & Trinci (1976a) suggested that the observed decay in synchrony in undifferentiated mycelia of *Aspergillus nidulans* was correlated with septation. It is certainly possible that even incomplete septa may interfere with cytoplasmic mixing in a coenocyte. Fiddy & Trinci (1976a) also suggested that mitosis may be initiated in intercalary compartments and their branch initials when, as a result of branch growth, the cytoplasmic volume per nucleus ratio attains a critical value.

Some cytoplasm synthesized within intercalary compartments may be transported via septal pores towards the hypha tip and contribute to apical compartment growth. If this happens it might be expected that in the absence of nuclear migration the duplication cycle in apical compartments would proceed at a faster rate than in intercalary compartments, since the former would attain the critical cytoplasmic volume per nucleus value more rapidly. The finding of Nishi *et al.* (1968) that the generation time of nuclei in apical compartments of *Aspergillus oryzae* is about half the value observed in distal, intercalary compartments is consistent with this notion. Further, several workers have observed that nuclei in intercalary compartments divide more slowly than those in apical compartments (Aist, 1969). In contrast, however, Fiddy & Trinci (1976a) found that the mean interval between successive cycles of septation (and consequently duplication cycle duration) in leading hyphae of *A. nidulans* was identical with the mycelial doubling time of the organism.

The type of negative control mechanisms which have been proposed to explain the regulation of DNA synthesis in bacteria (Pritchard, Barth & Colling, 1969) could also be advanced to explain the timing of mitosis in fungi.

During hyphal growth the positions of nuclei within an apical compartment change in a manner which suggests that they are attached to a cytoplasmic element which stretches or grows uniformly at the same rate as tip extension. However, it is unlikely that this element is the protoplasmic membrane since it is usually assumed that this membrane only increases in surface area at the hyphal tip. Morris (1975) has isolated five temperature-sensitive mutants of *Aspergillus nidulans* which display an abnormal nuclear distribution after 4 h at the restrictive temperature; the nuclei in such hyphae were clustered near septa but were absent from considerable lengths of mycelium. The existence of such mutants confirms that there is a specific mechanism which regulates nuclear distribution in the wild type. The suggestion that nuclei may be self motile (Wilson & Aist, 1967) may also be significant.

The precision with which coenocytic apical compartments of leading hyphae are divided in two by septation (Alexander & King, 1969; Clutterbuck, 1970; Fiddy & Trinci, 1976a) suggests that septal initiation is a well regulated event. In uninucleate cells, *e.g. Schizosaccharomyces pombe* Johnson, Yoo & Calleja, 1973), compartments of monokaryotic (Robinow, 1963; Niederpruem & Jersild, 1972) and dikaryotic mycelia (Niederpruem & Jersild, 1972), and the dikaryotic tips of ascogenous hyphae, septa are formed on cytoplasmic sites previously occupied by dividing nuclei. It is possible that in these fungi, part of the mitotic apparatus may form some kind of template for subsequent septation. In *Basidiobolus ranarum* and *Aspergillus nidulans* (Robinow, 1963; Robinow & Caten, 1969) one daughter nucleus remains close to the mitotic site whilst the other migrates away from it; perhaps only the former is involved in septation. However, in *A. nidulans* (Clutterbuck, 1970), *Alternaria solani* (Alexander & King, 1969), *Fusarium oxysporum* (Koenig & Howard, 1962) and *Polyporus arcularius* (Valla, 1973*a*, *b*) there is a temporal but no obvious quantitative or spatial relationship between dividing nuclei and subsequent septation.

Morris (1975) isolated five temperature sensitive mutants of *Aspergillus nidulans* which failed to form septa at the restrictive temperature. All mutants grew and underwent numerous nuclear division at the restrictive temperature, indicating that growth and nuclear division are not dependent upon septation.

8.7 Hyphal and Colony Extension

The peripheral growth zone

The peripheral growth zone (Trinci, 1971) of a colony has the following characteristics: (i) in the case of colonies growing at a linear rate, its width remains constant; (ii) only growth within the peripheral growth zone contributes to radial expansion of the colony; central regions of the colony increase in biomass at a slow rate but this growth does not contribute to radial expansion of the colony; and (iii) growth within the peripheral growth zone occurs at or close to the maximum specific growth rate of the organism for the prevailing conditions. Trinci (1971) showed that the radial expansion rate (K_r) of a colony is a function of the width (w) of its peripheral growth zone and the specific growth rate (α) of the mould. Thus,

$$K_r = w\alpha \qquad (8.2)$$

However, growth may not occur at the maximum specific growth rate of the organism throughout the peripheral growth zone but may instead decrease within the peripheral growth zone with distance from the margin of the colony (Koch, 1975; Bull & Trinci, 1977).

Since the radial expansion of a colony is a function of the rate of extension of its leading hyphae, eqn 8.2 also defines the growth of such hyphae. Thus leading hyphae have peripheral growth zones of constant length. The peripheral growth zone of hyphae which form complete septa, *e.g. Geotrichum candidum, Basidiobolus ranarum* and *Mucor hiemalis*, must be limited to their apical compartments. Hyphae of *G. candidum* extend at a linear rate although their apical compartments (the putative peripheral growth zone) vary in length from about 250 to 400 μm during periodic cycles of septation. The cytoplasm at the rear of apical compartments of *G. candidum* is usually highly vacuolated and hence may not make an appreciable contribution to tip extension. This study does, however, show

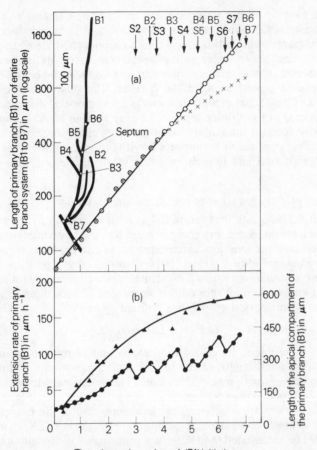

Fig. 8.11 (a) Observed growth of a primary branch, B1 (×) and the whole branching system, B1 to B7 (○) produced by an intercalary compartment of *G. candidum*. The solid line indicates growth as predicted by eqn. (8.3). The inset drawing shows the primary branch (B1) and the whole branching system (B1 to B7) at the end of the observation period. The times at which the septa (S) and the branches (B) were produced are indicated by the vertical arrows. (b) Extension rate (▲) of the primary branch (B1) shown in (a) and length (●) of its apical compartment (peripheral growth zone). From Fiddy & Trinci (1976*b*).

that a hypha can extend at a linear rate, although its peripheral growth zone is *apparently* increasing in length. Deceleration in the rate of extension of hyphae of *B. ranarum* is observed during mitosis (Fig. 8.1), *i.e.* before septation (Robinow, 1963). Clutterbuck & Roper (1966), Lhoas (1968), Nishi *et al.* (1968) and Fiddy & Trinci (1976*a*) suggest that the length of the peripheral growth zones of leading hyphae of various *Aspergillus* species is the same as, or slightly less than, the maximum length of their apical compartments. Leading hyphae of *Aspergillus* species, *Alternaria solani* and the monokaryotic or dikaryotic mycelia of Basidiomycetes (Niederpruem & Jersild, 1972; Butler, 1961; Valla, 1973*b*) continue to extend at a linear rate even after septation has reduced their apical compartments to about half maximum length. This observation suggests that either newly formed inter-calary compartments continue for a time to support tip extension or that, as in *G. candidum*, only distal regions of apical compartments (equivalent to its length just after septation?) contribute to tip extension and hence, constitute the peripheral growth zone of the hyphae. The observation of King & Alexander (1969) that cytoplasm moves in an acropetal direction through newly formed but not older septa of leading hyphae of *A. solani* suggests that newly formed intercalary compartments may contribute to tip ex-tension. This conclusion is supported by the long lag sometimes observed between septation and branch initiation in *A. nidulans* (Fiddy & Trinci, 1976*a*).

Growth of primary branches of undifferentiated mycelia

Equation 8.2 suggests that the initial extension rate of a primary branch produced from an intercalary compartment bounded by complete septa may be a function of the length of the compartment and the specific growth rate of the organism. Fiddy & Trinci (1976*b*) showed that a primary branch of *Geotrichum candidum* and its parent intercalary compartment increased in length at an exponential rate until it was about 325 μm long (Fig. 8.11*a*). Thus during this period growth was defined by eqn 8.3

$$e_n L_1 = e_n L_0 + \alpha (t_1 - t_0) \qquad (8.3)$$

where L_0 = combined length of the branch and compartment at time t_0, L_1 = the combined length of the branch and compartment at time t_1 and α = the specific growth rate of the mould. The extension rate of the primary branch continued to increase until it was about 700 μm long (Fig. 8.11*b*) and there was a correlation between the extension rate of the branch and the length of its apical compartment (peripheral growth zone?). Such a relation-ship would be anticipated (eqn 8.2) since presumably the specific growth rate of the organism is a constant. The total length of the branching system produced by the intercalary compartment, which eventually consisted of 7 hyphae, increased in length exponentially until it was at least 1.5 mm long (Fig. 8.11).

Primary branches of *Aspergillus nidulans* could also be shown to grow exponentially for a time when it was assumed that a short length of parent hypha in addition to the parent intercalary compartment initially contri-buted to branch growth (Fiddy & Trinci, 1976*a*).

The maximum length of the peripheral growth zone of a branch can be determined by measuring its length at the time when it first attains its maximum rate of extension (Trinci, 1974). This method may, however, over-estimate the peripheral growth zone of fungi which form complete septa and its takes no account of the contribution of the parent intercalary compartment to branch growth.

Fig. 8.4 shows several cycles of septation in a germ tube of *Aspergillus nidulans* which had an initial length of about 200 μm. The length of the apical compartment just after a period of septation (*i.e.* its minimum length) and the length just before the next period of septation (*i.e.* its maximum length) increased with hyphal length and hence time. Germ tubes of *A. nidulans* attained their maximum rate of extension when they were about

600 μm long whilst their apical compartments first attained their maximum length when the germ tubes were about 800 μm long (Fiddy & Trinci, 1976a). Thus, as in *Geotrichum candidum*, acceleration in germ tube extension rate appears to be correlated with an increase in apical compartment length. This correlation is indirect evidence of a relationship between apical compartment and peripheral growth zone length.

The extension zone of a hypha may be determined by measuring the length of the tapered portion of its tip (Trinci & Halford, 1975). Steele & Trinci (1975a) compared the morphology and growth kinetics of hyphae of undifferentiated mycelia of *Neurospora crassa* with their leading hyphae at the margin of mature colonies. There may be a direct relationship between the extension rate of a hypha and such morphological parameters as extension zone length, hyphal diameter, apical compartment length and intercalary compartment length. The dimensions of all these parameters will increase until the hypha eventually attains its linear rate of growth. Since the specific growth rate of an organism is a constant it follows from eqn 8.2 that an acceleration in tip extension will be correlated with an increase in peripheral growth zone length. Steele & Trinci (1975b) proposed that the shape and length of the extension zone and hyphal diameter may be functions of (a) the rate of supply of wall and membrane precursors (vesicles?) to the hyphal tip and (b) the rate of rigidification of the primary wall.

Apical branching

I have described elsewhere the growth kinetics of apical branching in *Geotrichum candidum* and *Aspergillus nidulans*; in both species apical branching was a normal feature of hyphal morphogenesis at the colony margin (Trinci, 1970). Apical branching in *A. nidulans*, but not in *G. candidum*, was preceded by a short lived deceleration in growth rate. However, the reorganization of the hyphal tip to form two extending points takes place remarkably rapidly with minimal interference with hyphal extension. Within 14 to 44 min of their initiation the apical branches had attained the linear growth rate previously displayed by the parent tip.

8.8 Formation of Mature Colonies

Undifferentiated mycelia initially increase in total length at an exponential rate (Trinci, 1974); in *Mucor hiemalis* exponential growth continues until the mycelium is at least 15 mm in total hyphal length. Growth of such mycelia is balanced, *i.e.* all extensive properties increase at the same rate (Table 8.2). This kind of result was predicted by Koch (1975). After an initial period the branches produced by a mycelium increase in number exponentially at more or less the same specific rate as the increase in mycelial length (Fig. 8.12). We have called the ratio between mycelial length and the number of branches, the hyphal growth unit (Caldwell & Trinci, 1973; Trinci, 1974). The size of the hyphal growth unit oscillates initially, but tends to a constant value as the mycelium increases in size (Fig. 8.12). Thus growth of an undifferentiated mycelium can be considered in terms of the exponential increase of a hypothetical growth unit consisting of a tip and a specific length of hypha. However, as observed previously, the hyphal growth unit is a physiological but not a morphological entity.

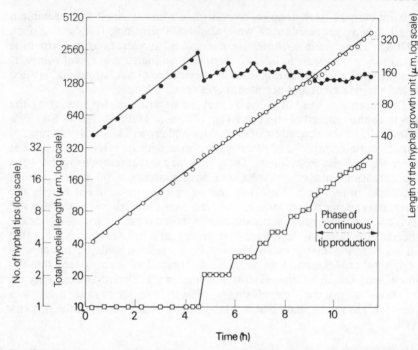

Fig. 8.12 Growth of a mycelium of *Geotrichum candidum*. □, No. of tips; ○, total hyphal length; ●, hyphal growth unit. From Trinci (1974).

Total length of an undifferentiated mycelium increases exponentially until environmental conditions become more growth limiting than when growth commenced. In a homogeneous batch (liquid) culture the whole mycelium is affected at the same time, but where pellets are formed or with colonies on solid media, only the central part of the mycelium may be affected. Growth of most fungi on solid medium probably starts to decelerate from an exponential rate when its mycelium attains a total length of only a few millimetres. There may be a direct relationship between the length of the hyphal growth unit of a fungus and the length of its mycelium at the onset of the deceleration phase of growth (Trinci, 1973; Morrison & Righelato, 1974), *i.e.* mycelial length at the onset of the deceleration phase of growth is less for moulds which branch profusely (*e.g. Penicillium*) than for moulds which are sparsely branched (*e.g. Mucor*). Koch (1975) has come to a similar conclusion. He assumes that mycelia grow logistically, *i.e.* growth is initially exponential but eventually decelerates in proportion to the utilization of some factor in the environment, so that eventually the biomass produced per unit area of substrate attains the 'carrying capacity' (maximum yield) of the environment. Koch developed formulae to describe growth of fungal colonies and concluded that growth laws for colony mass depend critically upon the degree to which hyphae at the margin of the colony initially occupy uncolonized medium, *i.e.* upon the branching pattern of the mould. However, as observed by Righelato (1975) the logistic growth law 'is of limited value because as such it does not test any hypothesis concerning the cause of the slowing growth rate.'

Conditions which are unfavourable for growth develop in the medium as a result of the activities of the mould. Such unfavourable conditions include changes in nutrient concentration, oxygen tension and pH and the accumulation in the medium of products of secondary metabolism. The onset of these changes is almost certainly correlated with the mould decelerating from its previous exponential mode of growth. Such changes in the medium are reflected by differentiation of the mycelium which ultimately results in the formation of the type of mature colony described by Yanagita & Kogane (1962). The first signs of this differentiation process is that branches are formed at the centre of the mycelium which are narrower than those formed earlier and their growth is more meandering and less radially directly (Plomley, 1959; Park & Robinson, 1967). A mycelium which is apparently at this stage of development was drawn (Fig. 8.13) by Buller (1958). This drawing shows the very well ordered nature of a mycelium and illustrates why moulds are so effective and efficient in colonizing solid substrates. At least three mechanisms are involved in regulating the form of a mycelium, *viz.* mechanisms which regulate hyphal polarity, branch initiation and the spatial distribution of hyphae with respect to each other and the substrate. These mechanisms are discussed by Bull & Trinci (1977).

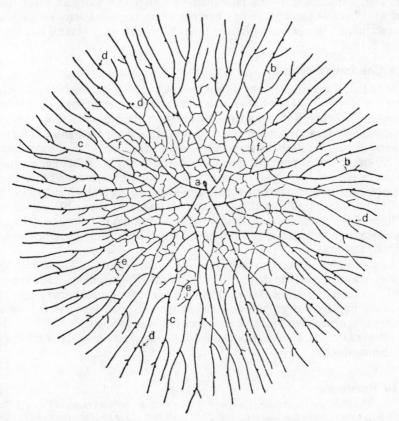

Fig. 8.13 Diagram of a dikaryotic mycelium of *Coprinus sterquilinus*. From Buller (1958).

Leading hyphae differ in diameter, compartment length, extension zone length and extension rate from the primary and secondary branches which they subtend (Table 8.1). It can be assumed that each hypha in a mycelium has the genetic potential to develop into a leading hypha. Realization of this potential by hyphae in a mature colony is probably determined by chance, *i.e.* whether or not the hypha happens to grow into a relatively uncolonized part of the substrate at the margin of the colony. The mean distance between leading hyphae at the margin of a colony remains approximately constant during colony expansion indicating that new leading hyphae arise as the colony increases in diameter. The hyphae are almost certainly formed from primary branches which gradually increase in growth rate and diameter, *etc.* until they assume the position and characteristics (Table 8.1) of new leading hyphae (Trinci, 1973). The transformation of a primary branch into a leading hypha probably occurs as a chance event when such a branch happens to extend into a relatively uncolonized part of the substrate at the margin of the colony. Bull & Trinci (1977) describe some of the morphological and kinetic changes which occur as a mould colonizes a solid substrate.

It is possible that all hyphae in an undifferentiated mycelium which is growing exponentially can potentially develop the morphological and growth characteristics of leading hyphae. However, relatively few hyphae have attained this potential by the time the mycelium is several mm long (Fig. 8.10).

8.9 Concluding Remarks

I hope that the foregoing has shown the relevance of the duplication cycle concept to studies of hyphal growth and morphogenesis. The existence in moulds of mechanisms to regulate and integrate the synthesis of organelles, macromolecules and enzymes is self evident. That these mechanisms may be similar if not identical to those present in yeasts and other unicellular micro-organisms is perhaps less obvious. Hartwell (1974) has shown that valuable information about the cell cycle of *Saccharomyces cerevisiae* may be obtained from studying temperature-sensitive mutants. Morris (1975, 1976) has illustrated that this powerful technique can also be applied to studies of the duplication cycle in moulds. We would of course very much like to know how events in cell or duplication cycles are integrated in such an orderly fashion.

I am sure that some, if not all, of the speculative ideas put forward in this review may at some time in the future be shown to be erroneous. However, I feel that these hypotheses are worth advancing at the present time if only to indicate that hyphal growth and morphogenesis may be viewed from a standpoint other than that traditionally held by mycologists.

I thank Dr. C. F. Thurston for helpful discussion and constructive criticism of the manuscript.

8.10 References

AIST, J. R. (1969). The mitotic apparatus in fungi, *Ceratocystis fagacearum* and *Fusarium oxysporum. Journal of Cell Biology* **40**, 120–35.

AIST, J. R. & WILLIAMS, P. H. (1972). Ultrastructure and time course of mitosis in the fungus *Fusarium oxysporum. Journal of Cell Biology* **55**, 368–89.

AIST, J. R. & WILSON, C. L. (1966). A study of asexual nuclear division in *Ceratocystis fagacearum* with phase-contrast microscopy. *Phytopathology* **56**, 869 (Abstr.).

BARTNICKI-GARCIA, S. (1973). Fundamental aspects of hyphal morphogenesis. *Symposium Society of General Microbiology* **23**, 245–67.

BRACKER, C. E. (1971). Cytoplasmic vesicles in germinating spores of *Gilbertella persicaria*. *Protoplasma* **72**, 381–97.

BULL, A. T. & TRINCI, A. P. J. (1977). The physiology and metabolic control of fungal growth. *Advances in Microbial Physiology* **15**.

BULLER, A. H. R. (1958). Further observations on the Coprini together with some investigations on social organization and sex in the Hymenomycetes. *Researches in Fungi*, Vol. 4. New York: Hafner Publishing Co.

BUTLER, G. M. (1961). Growth of hyphal branching systems in *Coprinus disseminatus*. *Annals of Botany* **25**, 341–52.

BUTLER, G. M. (1972). Nuclear and non-nuclear factors influencing clamp connection formation in *Coprinus disseminatus*. *Annals of Botany* **36**, 263–79.

CALDWELL, I. Y. & TRINCI, A. P. J. (1973). The growth unit of the mould *Geotrichum candidum*. *Archiv für Mikrobiologie* **88**, 1–10.

CALONGE, F. D. (1968). Origin and development of intrahyphal hyphae in *Schlerotinia fructigena*. *Mycologia* **60**, 932–42.

CAMPBELL, A (1957). Synchronization of cell division. *Bacteriological Reviews* **21**, 253–72.

CHAN, S. & STEPHEN, R. C. (1967). Intrahyphal hyphae in the genus *Linderina*. *Canadian Journal of Botany* **45**, 1995–8.

CLUTTERBUCK, A. J. (1969). Cell volume per nucleus in haploid and diploid strains of *Aspergillus nidulans*. *Journal of General Microbiology* **55**, 291–300.

CLUTTERBUCK, A. J. (1970). Synchronous nuclear division and septation in *Aspergillus nidulans*. *Journal of General Microbiology* **60**, 133–5.

CLUTTERBUCK, A. J. & ROPER, J. A. (1966). A direct determination of nuclear distribution in heterokaryons of *Aspergillus nidulans*. *Genetical Research, Cambridge* **7**, 185–94.

COLLINGE, A. J. & TRINCI, A. P. J. (1974). Hyphal tips of wild-type and spreading colonial mutants of *Neurospora crassa*. *Archiv für Mikrobiologie* **99**, 353–68.

CORTAT, M., MATILE, P. & WIEMKEN, A. (1972). Isolation of glucanase-containing vesicles from budding yeasts. *Archiv für Mikrobiologie* **82**, 189–205.

DONACHIE, W. D. (1968). Relationship between cell size and time of initiation of DNA replication. *Nature, London* **219**, 1077–9.

EMERSON, R. (1955). In *Aspects of synthesis and order in growth*, pp. 171–208. Edited by D. Rudrick. Princeton, N.J.: Princeton University Press.

FENCL, Z., MACHEK, F. & NOVAK, M. (1969). Kinetics of product formation in multistage continuous culture. In *Fermentation Advances* pp. 301–24, edited by D. Perlman. New York and London: Academic Press.

FIDDY, C. & TRINCI, A. P. J. (1976a). Mitosis, septation, branching and the duplication cycle in *Aspergillus nidulans*. *Journal of General Microbiology* **97**, 169–84.

FIDDY, C. & TRINCI, A. P. J. (1976b). Nuclei, septation, branching and growth of *Geotrichum candidum*. *Journal of General Microbiology* **97**, 185–92.

FIDDY, C. & TRINCI, A. P. J. (1976c). Septation in mycelia of *Mucor hiemalis* and *Mucor rammanianus*. *Transactions of the British Mycological Society* **68**, 118–20.

FLENTJE, N. T., STRETTON, H. M. & HAWN, E. J. (1963). Nuclear distribution and behaviour throughout the life cycles of *Thanatephorus*, *Waitea* and *Ceratobasidium* species. *Australian Journal of Biological Sciences* **16**, 450–67.

GIRBART, M. (1971). Ultrastructure of the fungal nucleus. II. The kinetochore equivalent (KCE). *Journal of Cell Science* **2**, 453–73.

GOODAY, G. W. (1971). An autoradiographic study of hyphal growth of some fungi. *Journal of General Microbiology* **67**, 123–5.

GREGORY, P. H. (1966). The fungus spore: what is it and what it does. *The fungus spore*, pp. 1–13, edited by M. F. Madelin. London, Butterworths.

GROVE, S. N. & BRACKER, C. E. (1970). Protoplasmic organization of hyphal tips among fungi: vesicles and Spitzenkörper. *Journal of Bacteriology* **104**, 989–1009.

HARTWELL, L. H. (1974). *Saccharomyces cerevisiae* cell cycle. *Bacteriological Reviews* **38**, 164–98.

HELMSTETTER, C. E., COOPER, S., PIERUCCI, O. & REVELAS, E. (1968). On the bacterial life sequences. *Cold Spring Harbor Symposium in Quantitative Biology* **33**, 809–22.

JOHNSON, B. F., YOO, B. Y. & CALLEJA, S. B. (1973). Cell division in yeasts: movement of organelles associated with cell plate growth of Schizosaccheromyces pombe. Journal of Bacteriology 115, 358–66.

KESSEL, M. & ROSENBERGER, R. F. (1968). Regulation and timing of deoxyribonucleic acid synthesis in hyphae of Aspergillus nidulans. Journal of Bacteriology 95, 2275–81.

KING, S. B. & ALEXANDER, L. J. (1969). Nuclear behaviour, septation and hyphal growth of Alternaria solani. American Journal of Botany 56, 249–53.

KOCH, A. L. (1975). The kinetics of mycelial growth. Journal of General Microbiology 89, 209–16.

KOENIG, R. & HOWARD, R. L. (1962). Nuclear division and septum formation in hyphal tips of Fusarium oxysporum. American Journal of Botany 49, 666 (Abstr).

LHOAS, P. (1968). Growth rate haploidization of Aspergillus niger in medium containing p-fluorophenylalanine. Genetical Research, Cambridge 12, 305–15.

LOWRY, R. J. & SUSSMAN, A. S. (1966). Intra-hyphal hyphae in 'clock' mutants of Neurospora. Mycologia 58, 541–8.

MCCULLY, E. K. & BRACKER, C. E. (1972). Apical vesicles in growing bud cells of heterobasidiomycetous yeasts. Journal of Bacteriology 109, 922–6.

MCCULLY, E. K. & ROBINOW, C. F. (1971). Mitosis in the fission yeast Schizosaccharomyces pombe: a comparative study with light and electron microscopy. Journal of Cell Science 9, 475–507.

MITCHISON, J. M. (1971). The biology of the cell cycle. London: Cambridge University Press.

MORRIS, N. R. (1975). Mitotic mutants of Aspergillus nidulans. Genetical Research, Cambridge 26, 237–54.

MORRIS, N. R. (1976). A temperature-sensitive mutant of Aspergillus nidulans reversibly blocked in nuclear division. Experimental Cell Research 98, 204–10.

MORRISON, K. B. & RIGHELATO, R. C. (1974). The relationship between hyphal branching, specific growth rate and colony radial growth rate in Penicillium chrysogenum. Journal of General Microbiology 81, 517–20.

NIEDERPRUEM, D. J. & JERSILD, R. A. (1972). Cellular aspects of morphogenesis in the mushroom Schizophyllum commune. Critical Reviews in Microbiology 1, 545–76.

NIEDERPRUEM, D. J., JERSILD, R. A. & LANE, P. L. (1971). Direct microscopic studies of clamp connection formation in growing hyphae of Schizophyllum commune. I. The dikaryon. Archiv für Mikrobiologie 78, 268–80.

NISHI, A., YANAGITA, T. & MARUYAMA, Y. (1968). Cellular events occurring in growing hyphae of Aspergillus oryze as studied by autoradiography. Journal of General and Applied Microbiology 14, 171–82.

NOLAN, R. A. & BAL, A. K. (1974). Cellulase localization in hyphae of Achlya ambisexualis. Journal of Bacteriology 117, 840–3.

PARK, D. & ROBINSON, P. M. (1967). A fungal hormone controlling internal water distribution normally associated with cell ageing in fungi. Symposium Society of Experimental Biology 21, 323–36.

PLOMLEY, N. J. B. (1959). Formation of the colony in the fungus Chaetomium. Australian Journal of Biological Sciences 12, 53–64.

PRITCHARD, R. H. (1968). Control of DNA synthesis in bacteria. Heredity (Abstract) 23, 472.

PRITCHARD, R. H. (1974). Review lecture on the growth and form of a bacterial cell. Philosophical Transactions of the Royal Society of London Series B. 267, 303–37.

PRITCHARD, R. H., BARTH, P. T. & COLLING, J. (1969). Control of DNA synthesis in bacteria. Symposium Society General Microbiology 19, 263–98.

REES, H. & JINKS, J. L. (1952). The mechanism of variation in Penicillium heterokaryons. Proceedings of the Royal Society of Medicine, Series B. 140, 100–6.

RIECK, V. T., PALUMBO, S. A. & WITTER, L. D. (1973). Glucose availability and the growth rate of colonies of Pseudomonas fluorescens. Journal of General Microbiology 74, 1–8.

RIGHELATO, R. C. (1975). Growth kinetics of mycelial fungi. In The filamentous fungi, Vol. 1, pp. 77–103, edited by J. E. Smith and D. R. Berry. London: Edward Arnold.

ROBINOW, C. F. (1963). Observations on cell growth, mitosis and division in the fungus Basidiobolus ranarum. Journal of Cell Biology 17, 123–52.

ROBINOW, C. F. & CATEN, C. E. (1969). Mitosis in Aspergillus nidulans. Journal of Cell Science 5, 403–31.

ROSENBERGER, R. F. & KESSEL, M. (1967). Synchrony of nuclear replication in individual hyphae of Aspergillus nidulans. Journal of Bacteriology 94, 1464–9.

SACHSENMAIER, W., REMY, V. & PLATTNER-SHOBEL, R. (1972). Initiation of synchronous mitosis in *Physarum polycephalum*. A model of the control of cell division in eukariots. *Experimental Cell Research* **73**, 41–8.

STEELE, G. C. & TRINCI, A. P. J. (1975a). Morphology and growth kinetics of hyphae of differentiated and undifferentiated mycelia of *Neurospora crassa*. *Journal of General Microbiology* **91**, 362–8.

STEELE, G. C. & TRINCI, A. P. J. (1975b). The extension zone of mycelial hyphae. *New Phytologist* **75**, 583–7.

TRINCI, A. P. J. (1970). Kinetics of apical and lateral branching in *Aspergillus nidulans* and *Geotrichum lactis*. *Transactions of the British Mycological Society* **55**, 17–28.

TRINCI, A. P. J. (1971). Influence of the peripheral growth zone on the radial growth rate of fungal colonies. *Journal of General Microbiology* **67**, 325–44.

TRINCI, A. P. J. (1973). The hyphal growth unit of wild type and spreading colonial mutants of *Neurospora crassa*. *Archiv für Mikrobiologie* **91**, 127–36.

TRINCI, A. P. J. (1974). A study of the kinetics of hyphal extension and branch initiation of fungal mycelia. *Journal of General Microbiology* **81**, 225–36.

TRINCI, A. P. J. & COLLINGE, A. J. (1973). Structure and plugging of septa of wild type and spreading colonial mutants of *Neurospora crassa*. *Archiv. für Mikrobiologie* **91**, 355–64.

TRINCI, A. P. J. & COLLINGE, A. J. (1974). Occlusion of the septal pores of damaged hyphae of *Neurospora crassa* by hexagonal crystals. *Protoplasma* **80**, 56–67.

TRINCI, A. P. J. & HALFORD, E. (1975). The extension zone of stage I sporangiophores of *Phycomyces blakesleeanus*. *New Phytologist* **74**, 81–3.

TRINCI, A. P. J. & RIGHELATO, R. C. (1970). Changes in constituents and ultrastructure of hyphal compartments during autolysis of glucose-starved *Penicillium chrysogenum*. *Journal of General Microbiology* **60**, 239–49.

VAN DER VALK, P. & WESSELS, J. G. N. (1973). Mitotic synchrony in multinucleate Schizophyllum protoplasts. *Protoplasma* **78**, 427–32.

VALLA, G. (1973a). Rythmes des divisions nucléaires et de la septation dans les articles apicaux en croissance du mycelium de *Polyporus arcularius*. *C. R. Acad. Sciences, Paris* **279**, 2649–52.

VALLA, G. (1973b). Division nucléaires, septation et ramification chez l'haplonte de *Polyporus arcularius* (Batsch) Ex. Fr. *Naturaliste Can.* **100**, 479–92.

VALLA, G. (1974). Rhythms of nuclear divisions and septation in growing mycelium of a Basidiomycete mushroom, *Polyporus arcularius* (Batsch) ex. Fr.: variation of the rhythm periods in relation to the temperature of cultivation. *Journal of Interdisciplinary Cycle Research*, **5**, 223–30.

VOLZ, P. A. & NIEDERPRUEM, D. J. (1968). Growth, cell division and branching patterns of *Schizophyllum commune* Fr. Single basidiospore germlings. *Archiv für Mikrobiologie* **61**, 232–45.

WILSON, C. L. & AIST, J. R. (1967). Motility of fungal nuclei. *Phytopathology* **57**, 769–71.

YANAGITA, R. & KOGANE, F. (1962). Growth and cytochemical differentiation of mold colonies. *Journal of General and Applied Microbiology* **8**, 201–13.

YANAGITA, T. & KOGANE, F. (1963). Cytochemical and physiological differentiation of mold pellets. *Journal of General and Applied Microbiology* **9**, 179–87.

ZALOKAR, M. (1959a). Enzyme activity and cell differentiation in *Neurospora*. *American Journal of Botany* **46**, 555–9.

ZALOKAR, M. (1959b). Growth and differentiation of *Neurospora* hyphae. *American Journal of Botany* **46**, 602–10.

ZICKLER, D. (1971). Déroulement des mitoses dans les filaments en croissance de quelques Ascomycètes. *C.R. Academie Sciences Paris* **273**, 1687–9.

CHAPTER 9

Fungal Dimorphism: a Particular Expression of Cell Wall Morphogenesis
P. R. STEWART AND P. J. ROGERS

9.1 Introduction

The yeast-mycelium dimorphism, exhibited by a variety of fungal species (Table 9.1), may be described as the process in which the mycelial habit of growth, with cells in hyphal or filamentous form, is transformed by some change in cultural or environmental conditions, so that a yeast-like or unicellular morphology is adopted at the cellular level. The process is generally thought to be reversible, though cyclical processes may account for this. At the colonial level, the effect is one of transformation of rough, filamentous colonies of the mycelial stage to smooth, uniform colonies of the yeast phase.

The change in cell shape (and colonial morphology as a consequence) is the most evident feature, and as in other walled organisms the cell wall determines cell shape, the determinants of growth and morphogenesis of the cell wall have been necessarily emphasized as critical to an understanding of the transformation between yeast and mycelial states. Cytokinesis is a key process in these considerations, since the balance between yeast, pseudohyphal and hyphal cell types may be seen, superficially at least, as some function of the relative rates of axial growth of a cell, and its septation or budding and separation from the parent cell.

Dimorphism is thus a matter of morphogenesis interwoven intricately with cell growth and division. Many of those concerned with the exploration of its basic mechanisms and control see dimorphism as a useful model eukaryotic system in which to examine the fundamental basis of morphogenesis and cellular differentiation. Because of the prevalence of dimorphism amongst the clinically important fungi (Table 9.1), others see an understanding of the phenomenon as being significant in the control of an increasingly important group of mycoses which afflict man and animals. Dimorphism is likewise of considerable relevance to the industrial production of single-cell protein, since the capacity to form pellicles or mats by yeasts used in biomass production may alleviate difficulties inherent in harvesting individual cells from large quantities of culture fluid.

The terms used to describe the dimorphic state—yeast and mycelium, and the word dimorphism—tend to indicate that two exclusive states exist. This is not the case, since up to five different cell types may be discerned in cultures of *Phialophora* (Oujezdsky, Grove & Szaniszlo, 1973) and *Cladosporium* (Hardcastle & Szaniszlo, 1974), and the dividing line between a single, round or ovoid cell (yeast) and filamentous cell (hypha), may sometimes be difficult to discern (Scheer & Weaver, 1953). For example, the germ tube which develops from blastospore or yeast-phase cells in *Candida* has most of the primordial characteristics of a filament, yet if its development is limited to a few times the diameter of the initiating cell, it can hardly be said to have developed as a filament. Likewise, the formation of a pseudomycelium (pseudohypha may be a more accurate term) in which yeast cells fail to form complete septa, or to separate, represents another intermediate state between true yeast and hyphal cell forms.

Hyphae of dimorphic fungi may convert to yeast phase by one of three basic mechanisms: by lateral budding, by terminal budding, or by arthrospore formation resulting from fragmentation. Yeast phase cells convert to hyphae by germ tube formation and elongation.

In this overview and review of dimorphism, we do not intend to catalogue organisms which show the phenomenon, nor to consider some of the earlier physiological data concerning the cultural and other conditions which induce or stimulate morphic transformation. Instead, we intend to consider some of the processes which may be fundamental and integral to the transformation from one cell type to another. We shall limit the discussion in this way because the general and practical aspects of dimorphism have been comprehensively reviewed elsewhere in the past decade (Romano, 1966; Robertson, 1968; Bartnicki-Garcia, 1968, 1973; Smith & Galbraith, 1971; Brody, 1973; Cabib, 1975).

The more recent reviews by Bartnicki-Garcia (1973) and by Brody (1973) are seminal ones in the general area of fungal morphogenesis. They represent the culmination of much of the earlier experimental information into models and catalogues of questions about the molecular and cellular bases of morphogenesis in fungi. We have therefore elected to consider: (i) the status of models and proposals concerning dimorphism as they existed four years ago, using the reviews of Bartnicki-Garcia (1973) and Brody (1973) as starting points; (ii) the relevance to fungal dimorphism of data obtained, and concepts developed, in plant cell wall biogenesis, an area in which significant ultrastructural and biochemical observations have recently been made; (iii) experimental developments of the past four or five years in the fundamental molecular and ultrastructural aspects of dimorphism, particularly in terms of models and hypotheses relevant to (i) and (ii) above; and finally (iv) proposals as to general and specific approaches which might profitably be adopted in future studies of fungal dimorphism.

Table 9.1. Significant species of fungi which exhibit dimorphism

Species	Significance
Aspergillus parasiticus	Not reported to be pathogenic; other species, which are not dimorphic, cause aspergillosis.
Blastomyces dermatitidis	Agent of blastomycosis, or suppurative and granulomatous lesions especially of lungs, skin and bones.
Candida albicans	Agent of candidiasis (together with other species of *Candida*), an infection of the skin, lungs, vagina and mouth, and occasionally causes septicaemia, and infections of the heart or brain. Hyphae are found in tissues.
	Other species of *Candida* which are variably dimorphic are used for biomass or single-cell protein production.
Cladosporium werneckii	Agent of tinea nigris, a superficial infection of the skin characterized by brown to black plaques or patches, especially on the palm of the hands. Hyphae are found in the skin, and the yeast form develops after isolation. The organism is pleomorphic in that up to three types of hyphae and two of yeast-like cells may develop.
Coccidioides immitis	Agent of coccidiomycosis, which may take one of two forms: acute but self-limiting respiratory infection, or chronic, malignant, disseminated disease of skin, bones and internal organs. Endospore stage in tissues grows by internal divisions to form spherules containing many new endospores.
Cryptococcus neoformans	Agent of cryptococcosis, a subacute or chronic infection primarily of the brain and meninges. Yeast phase in tissue, but shows only transient transformation to pseudomycelia (pseudohyphae) in primary isolations.
Geotrichum candidum	Agent of geotrichosis, a rare disease in which lesions occur in the mouth, intestinal tract and lungs. The dimorphic mechanism is unusual: oblong or spherical arthrospores are present in infected tissue, and in culture these form buds which elongate into germ tubes then form septate hyphae. The hyphae then segment into arthrospores.
Histoplasma capsulatum	Agent of histoplasmosis, which usually takes the form of a primary, acute but benign pulmonary infection, and more rarely, a progressive, chronic, malignant infection of blood cells, intestines, and upper respiratory tract. Hyphae present in tissues.
Mucor spp.	Agent of mucormycosis and (with other Phycomycetes) phycomycosis, in which the walls and lumina of blood vessels are invaded, or chronic subcutaneous infections occur. Occurs particularly in patients with uncontrolled diabetes mellitus. Coenocytic hyphae present in tissues.
Mycotypha microspora, M. africana	Saprophytes found in air and soil.
Paracoccidioides brasiliensis	Agent of paracoccidiomycosis, a chronic granulomatous disease of the skin, mucous membranes, lymph nodes and other internal organs. Yeast phase in tissues.

Species	Significance
Phialophora dermatitidis	One of a number of agents of chromoblastomycosis, in which warty, cutaneous nodules develop slowly, usually on the feet and legs. Forms large, brown, septate bodies (not filaments) in tissues. Yeast-like cells develop in primary culture which later turn to hyphae.
Aureobasidium (Pullularia) pullulans	Common as mildew of painted surfaces or associated with spoiled, stored fruit.
Sporothrix schenkii	Agent of sporotrichosis, a chronic infection with nodular lesions of lymph nodes and skin. Yeast phase occurs in tissues.

This table is compiled primarily from data of Conant, Smith, Baker & Callaway (1971).

9.2 Current Models and Concepts in Fungal Dimorphism

An ultimate description of dimorphism will embody mechanisms which account for altered cell shape in terms of: (i) differential rates of growth of different regions of the cell wall; or (ii) assembly of wall elements in different arrangements to generate different shapes; or (iii) modulation of the timing of septation and separation following division; or some combination of all three.

The available evidence (Bartnicki-Garcia, 1973, and the present review) tends to emphasize and favour option (i) above. As Smith (1975) points out 'morphological development within the fungi may be reduced to a question of cell wall morphogenesis', and thus models of cell wall growth should be noted in the search for an understanding of dimorphism. Consequently, we have considered and taken into account studies carried out with either fungi which are not dimorphic, or plant and algal cells, the other major group of walled cells. In particular, we have emphasized the observations made with morphological mutants of *Neurospora* (reviewed by Scott & Mahoney, 1976), and the chrysophycean algae (reviewed by Gunning & Steer, 1975).

We have also sought to bring together recent data on the physiological status of the dimorphic fungal cell and how this may, through effects mediated by cyclic AMP in particular, bring about altered morphological states. A review of note in this respect is that by Willingham (1976), although it primarily considers animal cells.

Two other recent reviews of parallel interest, concerning shape determination in bacteria, are those by Henning (1975) and Giesbrecht, Wecke & Reinicke (1976).

Differential tip and wall growth as an explanation of morphogenesis

In his review, Bartnicki-Garcia (1973) drew together the relevant threads of evidence relating to: (i) the organization of the cytoplasm in the growing region of the cell (with special emphasis on the apical tip); (ii) the possible importance of a balance between polymer synthesis and hydrolysis in growth regions; (iii) the probable location of wall synthesizing and lysing enzymes;

and (iv) autoradiographic identification of sites at which cell growth occurs most actively.

The issues raised by Bartnicki-Garcia (1973) about which something more can now be said are: (i) the nature of the vesicles which underly the plasma membrane and cell wall in actively growing regions of the cell; (ii) the origins and movement of vesicles to these sites; (iii) the discharge of the material of the vesicle, including its membrane; and (iv) the localization and activity of enzymes concerned with the synthesis of the polysaccharides of the cell wall.

General issues which relate to cell wall morphogenesis in fungi and yeasts

Brody (1973) carefully and usefully distinguishes between differentiation (changes in the selective and specific expression of particular genes) and morphogenesis (the shaping of three-dimensional structures from gene products). He lists a series of questions, posed in general terms, relating biochemical and genetic processes to an expression of shape, primarily through formation of a cell wall. The questions concern: (i) the role of enzymes in the shape-determining process; (ii) the possibility that shape determination can be completely described by chemical analysis; (iii) the genetic control of shape determination; (iv) the primary biochemical products of the genes directly involved in these processes; and (v) the co-ordination of shape-determining processes at the level of the gene or gene products.

Brody (1973) outlines the desirable features of an experimental system which may permit answers to these questions to be found. At the single cell level, *Neurospora* is indisputably the best characterized model organism in these respects, and a summary is given of findings with that organism. The genetic approach has been successfully applied to *Neurospora*, and this subject is discussed further below.

The question nevertheless remains for *Neurospora*: how do hyphal branches form? Brody considers that the answer lies in the co-ordination of synthesis and assembly of wall polymers, and the effect of regulatory perturbations on the kinetics of apical growth and branching. In passing, he notes that the transport and assembly of cell wall precursors are areas of potential importance, about which little is known.

An analogy for fungal morphogenesis: cell wall morphogenesis in unicellular plants

The Golgi apparatus and its component dictyosomes, cisternae and vesicles are likely to be of significance in any model which seeks to explain the ultrastructural basis of dimorphism in fungi. The involvement of these organelles in cell wall and scale biogenesis in unicellular plants has been impressively demonstrated (reviewed by Gunning & Steer, 1975), and it is worth examining this evidence briefly before moving to a consideration of ultrastructure in the dimorphic fungal cell.

Single-celled planktonic flagellates such as *Chrysochromulina* cover their protoplasts with sculptured scales of two or more morphologically different types (Gunning & Steer, 1975). These scales are composed primarily of polysaccharide, and they thus provide a useful chemical analogy for plant or fungal cell wall formation.

The scales are readily recognized not only as they finally ensheath the cell, but also in various stages of development within the cisternae of the Golgi apparatus of the cells. The micrographs of Fig. 9.1 clearly demonstrate the polarity, or dorsiventral character, of the cisternal stacks (dictyosomes) in the Golgi apparatus. Endoplasmic reticulum or nuclear envelope is located at the 'bottom' or proximal section of the stack, adjacent to tightly packed, narrow cisternae, which merge into a series of less regular cisternae which at the 'top' or distal end of the stack contain scales indistinguishable in structure from those deposited outside the cell. Less mature scales are present in intermediate cisternae in the stack. In places, cisternae bearing mature scales may be seen to have fused with the plasma membrane, presumably in preparation for the release of the scales to the external surface of the cell.

The physical sequence, ranging from endoplasmic reticulum or nuclear envelope at the bottom, through tightly packed to loosely packed cisternae containing scales in contact with the plasma membrane, is assumed to reflect the temporal sequence of scale formation and deposition outside the cell.

Whether the nascent scales are passed from one cisterna to another, or whether the cisternae plus contents move as units through the stack, cannot be determined from these studies. The latter possibility is favoured (Gunning & Steer, 1975) and new cisternae appear to be formed at the bottom of the stack to compensate for those lost from the top to the plasma membrane. There is thus also a requirement for the return, direct or indirect, of the membrane components from the cisternae or vesicles fusing with the plasma membrane, to the bottom of the dictyosome, and this may occur via the endoplasmic reticulum.

Two other aspects of scale production in *Chrysochromulina* require comment. Firstly, scales may be of two to four different types. Each cisterna of the Golgi apparatus normally produces a particular type of scale, and each cisterna behaves as an individual unit during the assembly process, yet with an overall control exercised to make products in the correct proportions. An understanding of this individuality would be a most significant contribution to the total question of cell wall morphogenesis. The use of morphogenetic mutants would facilitate such studies, and indeed in *Chrysochromulina*, analysis of morphogenetic variants indicates that the expression of scale variation is first seen in the shape and proportions of scales synthesized by the Golgi apparatus. It is interesting to speculate whether altered states of gene expression, rather than mutation, would also be reflected in altered morphology and synthesis of the scales. If so, the relevance of these studies to fungal dimorphism becomes even greater.

Pleurochrysis, another unicellular alga, represents an even more useful approximation to the fungal cell. This organism has a continuous wall, generally like that of higher plants and fungi. The biogenesis of this wall has been studied extensively by Brown (1969, 1973, 1974) and Brown *et al.* (1970). In section the wall appears as a laminar arrangement with interspersed amorphous material (Fig. 9.2). The unit laminae are flat and disc-like, approximately 1 μm in diameter, and are synthesized by the Golgi

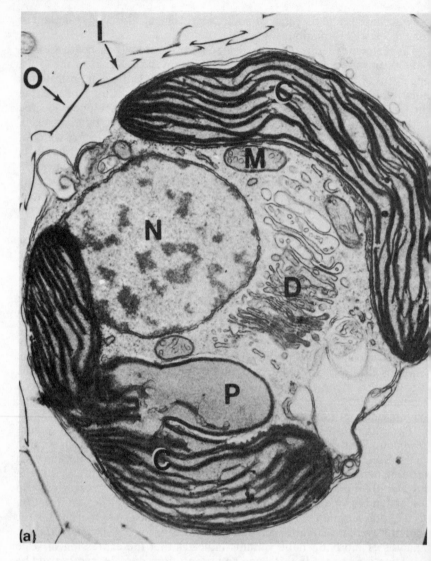

Fig. 9.1(a) Section of *Chrysochromulina chiton*, showing the two types of scale which ensheath the cell. O, outer scale; I, inner scale; C, chloroplast; D, dictyosome of Golgi apparatus; M, mitochondrion; N, nucleus; P, pyrenoid. ×10,000. From Manton (1967) and Gunning & Steer (1975).
(b) Dictyosome containing partially and completely formed inner (I) and outer (O) discs in *C. chiton*. The proximal face lies adjacent to endoplasmic reticulum (ER), and the distal face is above this, near the plasma membrane (PM). Each type of scale is found within an individual cisterna. ×30,000. From Manton (unpublished) and Gunning & Steer (1975).

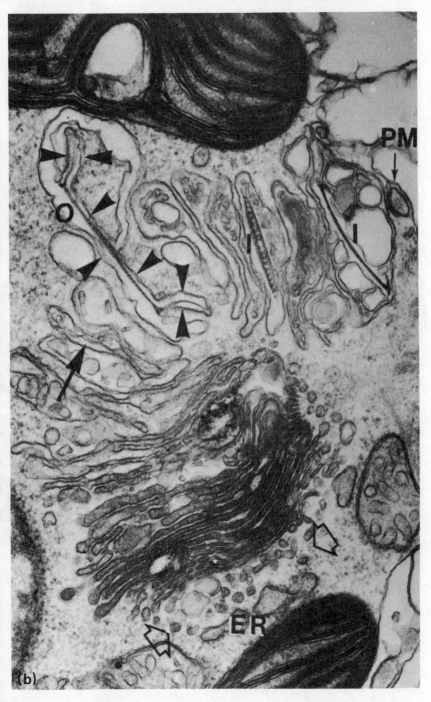

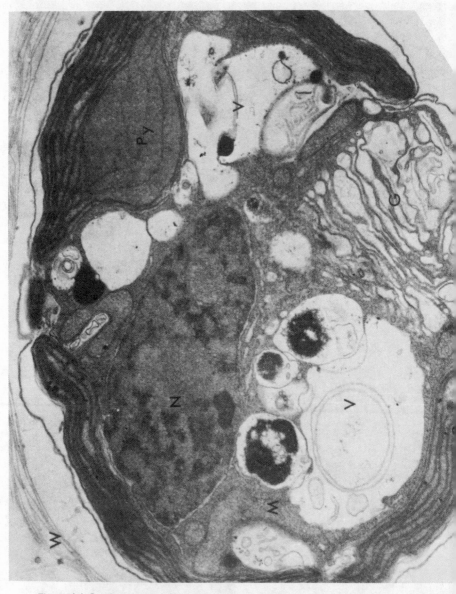

Fig. 9.2(a) Section through *Pleurochrysis scherfelii*, showing a dictyosome of the Golgi apparatus (G) adjacent to the plasma membrane, and the laminate cell wall (W). C, chloroplast; M, mitochondrion; N, nucleus; Py, pyrenoid; V, vacuole. ×15,000. From Brown *et al*. (1970).

(b) Fully mature, distal cisternae (left), containing cell wall (W) discs discharging at the plasma membrane (PM) in *P. scherfelii*. ×55,000. From Brown *et al*. (1970).

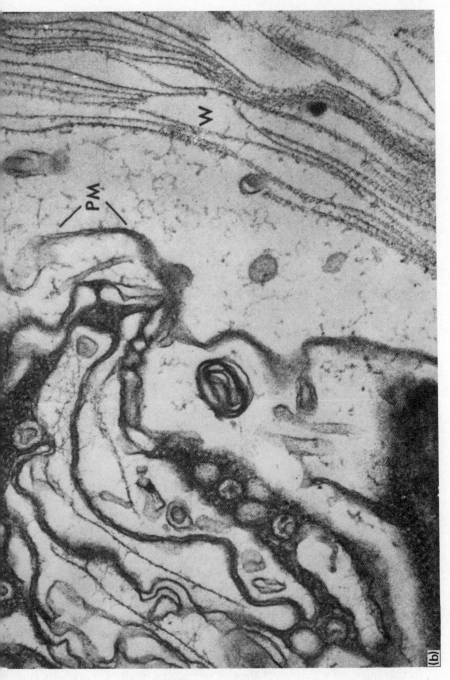

apparatus in much the same way as scales are produced in *Chryso-chromulina*. The discs of *Pleurochrysis* consist of three components (Brown, 1973): radial polysaccharide microfibrils, spiral cellulose–peptide microfibrils, and amorphous coating polysaccharide.

Pools of precursors for synthesis of these components can be located cytochemically, and are found in specific regions of rough endoplasmic reticulum associated with, and possibly formed from the outer nuclear envelope (*cf.* Plate 27 in Gunning & Steer, 1975). Precursor pools bleb from the rough endoplasmic reticulum and condense to form cisternae at the proximal face of the Golgi apparatus. Cisternal membranes then control the directed, sequential synthesis of radial microfibrils, initially folded, later unfolded in configuration. Spiral microfibrils are synthesized and added later, followed by amorphous material.

The flow of discs, from the Golgi apparatus through the plasma membrane to the cell wall may involve bundles of microtubules, which remove cisternae blocking the approach of Golgi vesicles and expose specific receptors for the vesicles on the plasma membrane. Colchicine depolymerizes the microtubules, and prevents the uncovering of receptors which is necessary for vesicle fusion with plasma membrane (Brown, 1974).

Control of scale deposition over the inner surface of the cell wall is thought to be achieved by rotation of the protoplast within the cell wall, thus directing the distal face of the Golgi apparatus towards different regions of the cell wall and liberating wall laminae in succession to different parts of the growing wall. The origin of the motile force for this movement is clearly of significance in understanding differential wall growth and its morphogenetic implications. Finger-like projections associated with a large parietal sac are thought by Brown (1973) to be part of this motile system.

Caution is obviously necessary in attempting to transpose a model based on one type of cell to another quite different type. However, the importance of the *Pleurochrysis* studies to cell wall morphogenesis is the emphasis they give to: (i) the processing of cell wall elements through the Golgi cisternal system; (ii) the directedness of this process, including the apparent importance of cytoplasmic or protoplasmic motility, and a possible role for microtubules; and (iii) the likelihood that cisternal or vesicle membranes and plasma membranes are continuous, after fusion, and thus may be alike in certain functional terms.

The relevance of these studies to fungal dimorphism is clear. Mechanisms which regulate a directed synthesis of wall material in cells contain the elements of control which may determine alternative morphic states. In particular, the question of the presence and physical state of preassembled wall elements in the vesicles of the apical region or of the Spitzenkörper (Bartnicki-Garcia, 1973) is one which, with current cytochemical and immunochemical methods, ought to be resolvable. An answer should indicate whether the apical vesicles and Spitzenkörper represent, functionally at least, the fungal equivalent of the distal face of the Golgi apparatus. A further question which might then be considered is whether crystallization or polymerization of polysaccharide elements to form wall matrix or microfibrils takes place before, during, or after exocytosis from apical or Spitzenkörper vesicles (*cf.* BroQet al., 1970).

Moreover, the tentative identification of a role for microtubules, albeit by way of effects of an inhibitor (colchicine), in deposition of cell wall elements in *Pleurochrysis* (Brown, 1974), suggests directions for new experimental approaches with fungi. Before considering these, it would be profitable to ask whether there is already ultrastructural evidence available from dimorphic fungi which fits with this general model of wall growth and morphogenesis.

.3 Ultrastructural Changes during Transformation in Dimorphic Fungi

Though it is generally assumed to be the case (for example, see Smith, 1975), the evidence for an involvement of the Golgi apparatus or its equivalent in fungi, together with vesicles and cisternae, microtubules and microfilaments, in guiding and delivering precursors to the site of cell wall growth and development, is not great. Clear demonstrations of a typical Golgi structure, or portions thereof, have been reported infrequently for the fungi (for references see Johnson, Yoo & Calleja, 1973, and Mahvi, Spicer & Wright, 1974) and possibly not at all in the case of the dimorphic fungi (*e.g.*, see Carbonell; 1967, Cassone, Simonetti & Strippoli, 1973; Grove, Oujezdsky & Staniszlo, 1973).

Likewise, the identification of microfilaments and microtubules in fungal cells is not common; the significant observations in this respect are summarized by Gull (1975). The root cause of this absence of data is the persistent difficulty of preserving structure in hyphal and yeast cells for electron microscopy. As a consequence, there has been an emphasis on the nature of the cell wall and the more obvious cytoplasmic structures, in searching for differences, similarities, and the ultrastructural dynamics of cell wall formation in different morphic types of fungi.

The study of Grove & Bracker (1970) exemplifies both the possibilities and difficulties of detailed descriptions of the organization and distribution of subcellular structures in the apical tip of growing hyphae. These workers examined in detail the interesting general similarities in the basic pattern of hyphal tip organization among a range of taxonomically diverse fungi. They observed an apical zone of vesicles, as found in other cells which grow by apical activity (*e.g.* pollen tubes, root hairs, algal rhizoids). Within this zone of vesicles is a concentrated region of smaller vesicles, known as the Spitzenkörper, a structure frequently observed by light microscopy (for ultrastructural details see Chapter 3). The Spitzenkörper (which is not present in Zygomycetes and Oömycetes) is visible only during active extension growth of the hyphal tip. As growth stops, for example due to changing culture conditions, the Spitzenkörper fades rather than moves away (McClure, Park & Robinson, 1968). When growth resumes the Spitzenkörper reappears before growth again becomes evident. Girbhardt (cited by McClure *et al.*, 1968)) had earlier noted that the position of the Spitzenkörper in the apical dome is related to the subsequent direction of extension of the hypha, an eccentric position for example preceding a turning of the tip.

Similarities are evident in size and electron opacity of vesicles at the site of wall growth with those further back along the filament axis, and nearer the 'Golgi-like' structures of the hyphal cell. Grove & Bracker (1970) also observed the apparent fusion of cytoplasmic vesicles with the plasma membrane, with the concomitant transfer of contents of the vesicles to the periplasmic space, to the inner surface of cell wall, or to some combination of these. In most respects, apart from the lack of identification of a classical Golgi apparatus, the process of transfer of new wall material to the growing wall is the same as that thought to occur in plant cells (Fig. 9.1 and 9.2 and Gunning & Steer, 1975). The failure to identify a Golgi system of cisternae raises the possibility that the Spitzenkörper may functionally represent the Golgi apparatus and that the larger vesicles of the apical region are 'mature' vesicles which fuse with the plasma membrane and deposit their partly pre-assembled cell wall material at the inner boundary of the cell wall. The vesicles constituting the Spitzenkörper may arise other than from the nuclear membrane or endoplasmic reticulum, or perhaps are present but not readily identified as they migrate through the cytoplasm from their remote sites of formation to the hyphal tip.

Alternatively, the different sized vesicles of the apical cone and Spitzenkörper may represent vesicles with different functions, one group perhaps containing lytic enzymes for wall softening, the other containing cell wall precursors for growth (see 9.4).

Analysis of the origin of cytoplasmic vesicles, and their relationship to the plasma membrane and cell wall, is available only to a limited degree for dimorphic fungi. Two aspects of these studies will now be considered in particular: (i) the structure of the plasma membrane and local regions of the cytoplasm; and (ii) the structure and development of the cell wall.

Local cytoplasmic and plasma membrane organisation in regions of local wall growth

Takeo & Nishiura (1974) used freeze-etching to examine differences between hyphal ultrastructure and that of arthrospores (ovoid or rectangular cells formed by hyphal fragmentation) in *Mucor javanicus*. The plasma membranes of hyphae are smooth and do not possess rod-like invaginations characteristic of lower fungi. The inner surface sometimes carries microfibrils, which could be continuations or extensions of wall fibrils. Vesicles and large tubular membrane structures are abundant in the cytoplasm. There was no evidence of microtubules or microfilaments at the resolution achieved in this study.

In the mature arthrospore, many round depressions about 50 nm in size are present in the cell membrane, and vesicles and small vacuoles are more abundant than in hyphae. These vesicles and vacuoles are concentrated in the outer zones of the cytoplasm, near the plasma membrane.

Takeo (1974) subsequently examined the ultrastructure of yeast phase cells of species of *Mucor*, by freeze-etching. These cells have a loose network of microfibrils on the outer wall surface, a feature which is not seen in the higher, ascomycetous yeasts. The cell membrane, as in hyphae and arthrospores, is not invaginated, though the plasma membrane possesses many round depressions, 50 nm or more in size as in arthrospores. Yeast-phase cells are large, multinucleate and lacking in endoplasmic reticulum.

Phialophora dermatitidis is an unusual fungus which can differentiate into yeast-like or hyphal cells, as well as into variants of these, suggesting that it is polymorphic, rather than dimorphic (Oujezdsky *et al.*, 1973). The thick-walled yeast cell, formed from thin-walled yeast at a resting stage, gives rise to pseudohyphae and thence hyphae, whereas thin-walled yeast cells arise by budding, either from other thin-walled yeast cells or laterally from hyphae. To the experimentalist, *Phialophora* offers certain advantages, particularly for ultrastructural studies, since morphological changes between a number of different types occur in a systematic fashion within the same culture. For example, membranous vesicles appear to be prominent in buds emerging from hyphae or thin-walled cells (Grove *et al.*, 1973) but not in pseudohyphae developing from thick-walled yeast cells (Oujezdsky *et al.*, 1973). Oujezdsky *et al.* (1973) also present one of the few micrographs which show microtubules in a dimorphic fungus.

Cell wall structure in dimorphic fungi

Ultrastructural events related to the interconversion of yeast and mycelial phases in *Sporothrix schenckii* have been studied by Lane & Garrison (1970) and Garrison, Boyd & Mariat (1975). Two alternative modes of development of yeast cells occur in this organism: by direct lateral budding of hyphae, and by oidial cell formation at the tips of hyphae.

Yeast cells formed by either mechanism are characterized by an outer cell wall layer composed of rough, microfibrillar material which reacts strongly with dialyzed iron or permanganate. The hyphal wall bears a much thinner layer of this material. Despite these differences in wall structure between yeast and hyphal cells, no particular differences in the structure of plasma membrane or underlying cytoplasm were evident, except that there may be greater numbers of mitochondria in the hyphae.

In *Mucor rouxii*, Lara & Bartnicki-Garcia (1974) noted that the newly forming cell wall of budding yeast-phase cells is apparently a continuation of the innermost layers of the parent cell wall, a phenomenon observed in other yeast phases of dimorphic fungi and in other yeasts (summarized by Grove *et al.*, 1973). The bud wall remains much thinner than that of the parent cell wall until the bud is about half the diameter of the parent cell. A septum then begins to form centripetally, by invagination of the plasmalemma. Accumulation of endoplasmic reticulum and vesicle elements occurs in this region. Two layers of electron-dense wall material appear within the space delimited by the invaginating plasmalemma creating a double cross-wall. This occurs before the plasmalemma has completely sealed off the newly budded cell at the neck region. The synthesis of microfibrillar material of each cross wall occurs as a continuation of the inner layer of the neck wall. The two newly forming cross walls are thus firmly and independently anchored to the neck wall (Fig. 9.3). The two cells remain attached through the neck wall, which breaks during abscission, presumably as a consequence of specific lytic action on the wall at that time. These structural aspects of budding are similar to those seen in the ascomycetous yeasts. The similarity is striking considering the considerable biochemical differences between cell wall material of *Mucor* and *Saccharomyces*, reflecting as it must substantial differences in the pathways of synthesis of the elements of the cell wall. The similarities may indicate that the basic structural mechanisms of cell wall biogenesis have been conserved in eukaryotes during phylogeny.

Although yeast cell budding in *Mucor* (which gives rise to a new yeast cell) does not relate directly to the dimorphic process in this organism, the fact that three morphologically distinct cell wall types (parent, bud, septum) are present during budding, that the two newly formed walls (bud, septum) represent outgrowths of only part of the fabric of the inner layer of the original or parent wall, and that cell wall growth is centripetal and localized in the septum while generalized in the expanding bud, makes this budding process potentially a useful model system in which to examine oriented cell wall synthesis. An understanding of the guiding role of the plasma membrane in septum wall development would be valuable in an understanding of the role of this membrane in wall morphogenesis in general. Cabib (1975) takes a similar view about septum biogenesis in *Saccharomyces cerevisiae*, and his review of recent studies in this field contains much of relevance to fungal dimorphism.

Cassone *et al.* (1973) observed striking changes to the organization of amorphous, granular and fibrous layers of blastospores of *Candida albicans* as germ tube formation began. An electron-transparent layer, possibly rich in chitin, develops from inside the blastospore wall to form an early germ tube, accompanied apparently by degradative (mechanical and lytic)

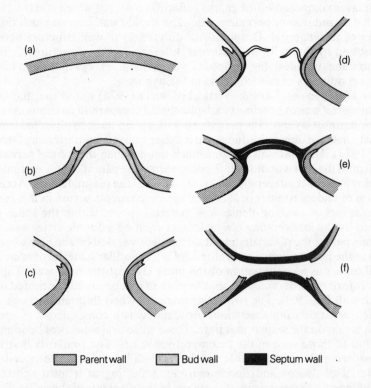

Fig. 9.3 Schematic representation of wall biogenesis during budding of yeast-phase cells in *Mucor rouxii*. (a) intact parent cell wall. (b) incipient bud formation. (c) bud before septation begins. (d) septum formation begins. (e) septum is completed as new wall material is laid down over ingrown plasma membrane. (f) abscission. From Lara & Bartnicki-Garcia (1974).

changes to the blastospore wall structure overlaying the electron-transparent layer. As the germ tube elongates, cell wall layers corresponding to those of the blastospore wall develop around the maturing tube.

An intimate structural relationship between plasma membrane and cell wall is suggested by the freeze-fracture studies of the yeast form of *Aureobasidium* (*Pullularia*) *pullulans* by Ramos & Acha (1975). Cell wall degradation by lytic enzymes yielded protoplasts which when examined by freeze-etching show a smooth external face to the plasma membrane, and a particle-covered internal face. The external face is seen when protoplasts are used, but rarely when intact cells are fractured, indicating strong adherence and intimate contact between the external surface of the cell membrane and the cell wall.

Mechanisms for directing precursors to the site of cell wall synthesis

If it is accepted that the rate of growth and the development of shape of the cell wall in fungi are determined by the rate of flow of vesicles or cisternae from the polysaccharide–polypeptide synthesizing and assembling organelle to the growing point(s) of the hyphal cell wall, then a critical question to be answered is: how do the vesicles or cisternae containing precursors or pre-fabricated components find their way to the specific growth points? It may well be that the regulation of cell wall growth is determined not by the rate of arrival of cell wall elements at the plasma membrane, but by the rate at which they are transported across or further processed by this membrane, or the rate at which they are deposited and assembled in the wall itself. It seems useful, however, to look to the earlier events in the sequence for information on the control of rate and direction. The physical transfer of vesicles from origin in the Golgi apparatus or Spitzenkörper region to the site of wall growth could be a starting point.

Several possibilities suggest themselves as to how vesicles may find their way from Golgi apparatus to specific sites in the cell wall. The prospect that they may simply diffuse through the cytoplasm may be dismissed since it would be a random process, and therefore unlikely to result in vesicles arriving in quantity at any specific part of the internal surface of the cell.

Cytoplasmic streaming (cyclosis) could deliver vesicles rapidly to particular points in the cell. However, unless such streams are themselves specifically directed, they may provide little more direction than does the process of diffusion, though random collisions between vesicles and sites of deposition would occur with higher frequency. If this is the case, then specific mutual recognition sites on vesicles and plasma membrane or other target sites may be important in the delivery and deposition of products at the site of cell wall growth.

The endoplasmic reticulum and microtubules have been implicated in guiding the movement of Golgi vesicles to sites in the cell (Gunning & Steer, 1975). In plants, the endoplasmic reticulum is thought to be an important mediating as well as synthesizing organelle in the deposition of wall substances (Cronshaw, 1974). In budding cells of *Saccharomyces*, the endoplasmic reticulum may guide or compartmentalize wall-softening enzymes (glucanases) into regions where outgrowth is to occur (Matile, Cortat, Wiemken & Frey-Wyssling, 1971). The vesicles containing glucanase may themselves be elaborated by the endoplasmic reticulum. The isolation and identification of glucanase-containing vesicles from *Saccharomyces* is described by Cortat, Matile & Wiemkin (1972), though these data may need to be interpreted cautiously (Cabib, 1975).

Byers & Goetsch (1976) speculate on the possibility that an ordered ring of filaments (10 nm thick) associated with the plasma membrane of budding cells of *Saccharomyces cerevisiae* may be part of a control mechanism which limits vesicle fusion with the membrane at the neck of the budding cell until such time in division when septum formation begins. The possibility that microtubules may carry out similar sequestering or protective roles in guiding vesicles to specific sites on the plasma membrane is explored in more detail in the next section.

Microtubules and microfilaments in cell wall synthesis

Early studies indicated that microtubules (diameter about 25 nm) in walled (plant) cells could influence cell shape, not by their own rigidity, but by determining sites of deposition of cellulose microfibrils in the growing cell wall. It is a general observation, in a wide variety of cell types, that microtubules at the cytoplasmic face of the plasma membrane lie with the same orientation as the cellulose fibrils in the adjacent cell wall (Gunning & Steer, 1975). Exceptions to this generalization exist, for example in *Chlorella*, and in cells in which wall deposition occurs in the presence of colchicine, which depolymerizes microtubules into tubulin subunits. Moreover, in root apex cells and in *Nitella*, exposure to colchicine leads to production of swollen, almost spherical cells, as though cellulose fibril deposition has become random and turgor pressure induces cells to form an optimal, spherical shape. Fibril orientation returns to normal when the drug is removed.

What mechanisms might be postulated to involve microtubules with the oriented deposition of fibrils in cell walls? Microtubules may position and stabilize polysaccharide-synthesizing enzymes attached to the plasma membrane. Alternatively the drug may affect the distribution of proteins or particles in membrane faces, as is the case in *Tetrahymena* (Wunderlich, Muller & Speth, 1973). Microtubules could also specify sites of access of vesicles or cisternae to the plasma membrane, either by presenting a barrier or by generating a guiding structure, thus regulating the supply of precursors for cell wall synthesis. This may be the role for the collar of tubular material seen in the neck of budding *Saccharomyces* cells (Byers & Goetsch, 1976).

Microfilaments appear in thin sections of many animal and some plant cells as lightly staining structures, 5–8 nm in diameter. They are most conspicuous when associated together in bundles. In plant cells, their elusiveness has been variously attributed to their absence from many cell types, or their general sparsity in other than animal cells. Gull (1975) summarizes observations of microfilaments in fungi; they have not so far been reported in dimorphic fungi.

Microfilaments are thought to represent polymerized actin, which may be part of contractile systems in the cytoplasm of cells. However, that the filaments are in fact actin requires a specific criterion, such as binding of anti-actin antibody or heavy meromyosin, to be satisfied. In animal cells, cytochalasin B has been used to perturb microfilaments in cells in various stages of division, resulting in reversible blockages at specific stages in cytokinesis. Plant cells showing tip growth respond to cytochalasin B by cessation of cytoplasmic streaming (cyclosis) and growth. Cyclosis is thought to involve cytoplasmic contractile mechanisms which are actin-dependent, and the inhibition of growth may be a consequence of inhibition of movement of vesicles or cisternae to growing points beneath the cell wall. Cytoplasmic motility is commonly observed in fungi.

It is possible that microtubules and microfilaments act in concert, as 'skeleton' and 'muscle' of cells respectively. It would clearly be of considerable interest to know what effects inhibitors such as colchicine and cytochalasin B have on dimorphic transformation, and more particularly, whether microtubules and microfilaments are present and how they are organized in the dimorphic fungal cell. Once again, the lessons learnt by plant microscopists (Parthasarathy & Muhlethaler, 1972; Hepler & Palevitz, 1974) could profitably be noted by mycologists beginning a search for these often elusive structures.

9.4 Biochemical Aspects of Cell Wall Structure and Biogenesis

Lysis and synthesis

A frequent subject of conjecture is that the integrity of the cell wall may be important as a determinant of extensibility and shape change during growth of the cell (Fig. 9.4). A wall which is weakened at many points may, under the influence of the turgor pressure of the protoplast, expand or grow in a number of directions if deposition of new material occurs randomly at weakened sites. That is, the cells would grow approximately as spheres. Likewise, apical or polar growth of filaments could be accounted for by deposition of new material at weakened sites which were restricted to one region of the cell wall.

The experimental evidence suggests that for hyphae at least, weaker regions do exist at the apical tip, since when lysis is induced, it tends to occur at the tips of hyphae in *Mucor rouxii* (Bartnicki-Garcia & Lippman, 1972; Dow & Rubery, 1975). Whether this is a consequence of activation of enzymes in the apical tip, enzymes which otherwise remain sequestered or latent in some way (Bartnicki-Garcia & Lippman, 1972), or whether it results from destabilization of certain polymers of the cell wall (Dow & Rubery, 1975), is a question which is not yet resolved.

Bartnicki-Garcia & Lippman (1972) propose that hyphal tip bursting induced by osmotic shock, lowered pH, or heating, involves an active chemical process of apical wall weakening. The temperature coefficient of osmotically induced lysis is in the region 1.3–2.1, rather than 1.0 which would be expected if the process were purely osmotic. They therefore suggest that enzymic processes, which weaken the wall, are involved in tip bursting. In intact growing cells, the process of lysis is counterbalanced by the addition of new cell wall material (Fig. 9.4).

By contrast, Dow & Rubery (1975) suggest that macromolecular or microfibrillar stability, rather than activation of latent lytic enzymes, accounts for the opposing effects of Ca^{2+} and H^+ which they observed on hyphal tip bursting in *Mucor rouxii*. They point to the importance of cations to plant cell wall structures as illustrated also by the effects of Ca^{2+} and H^+

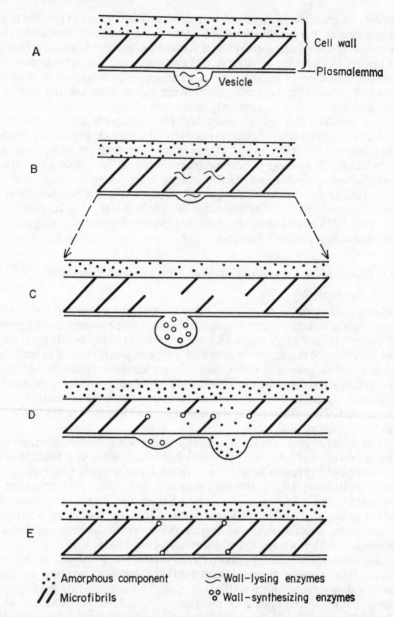

Fig. 9.4 Hypothetical representation of the processes of wall lysis and weakening (a, b), wall expansion (c), and addition of new cell wall units (d) to form an extended cell wall (e). From Bartnicki-Garcia (1973).

on cell growth. Acidic pectic polysaccharides exist as Ca^{2+} salts and account for a major fraction of the total matrix (non-fibrillar) material of the plant cell wall. Elevated external Ca^{2+} concentrations result in hardening of the cell wall and inhibition of extension (Ray & Baker, 1965). Replacement of stabilizing Ca^{2+} with H^+ results in greater extensibility of the cell wall

(Davies, 1973). These consequences may be explained by the dissimilar effects of Ca^{2+} and H^+ on the viscosity of pectic acid solutions: similar effects are seen on mucoran, an acidic polysaccharide which is a matrix component of *Mucor* cell walls (Dow & Rubery, 1975). These workers do not exclude a role for synthesis of fibrils and matrix material in maintaining tip integrity, but point out that regulation of ionic conditions in the growing regions of the cell wall is a further element to consider in the control of wall growth and morphogenesis.

Dabbagh, Conant, Nielsen & Burns (1974*b*) examined cell wall and protoplast growth in saprophytic species of the cryptococci. *Cryptococcus diffluens*, *C. laurentii* and *C. albidus* grow normally at 27°C but lyse when shifted to 37°C, in contrast to the pathogenic *C. neoformans*, which is not restricted for growth until the temperature exceeds 39°C. With osmotic stabilization, lysis of the saprophytes does not occur at 37°C, and protoplasmic growth continues. Microscopic examination under these conditions indicates aberrant wall biosynthesis, and a pore or fissure develops where normally a bud initial would appear. Lysis in unbuffered medium is thus apparently a consequence of an imbalance between protoplasmic growth, and wall plasticization without new wall synthesis at the site of bud formation.

These workers also found that the UDPglucose, a precursor for glucan biosynthesis, accumulates and is excreted by *Cryptococcus diffluens* at 37°C but not at 27°C, indicating that at the restrictive temperature precursor biosynthesis continues, though its effective utilization apparently does not (Dabbagh, Conant & Burns, 1974*a*). No differences were noted in the sulphydryl content of the cell surface at the two temperatures, measured by binding of labelled N-ethylmaleimide, nor of the p-1,3-exoglucanase activity of intact cells, cell homogenates or culture filtrates of cultures of *C. diffluens* grown at the two temperatures. Both sulphydryl content of cell walls, and glucanase activity, have been implicated in cell wall synthesis and morphogenesis in fungi (Nickerson & Falcone, 1956; Cortat *et al.*, 1972). However, the enzyme protein disulphide reductase (Nickerson & Falcone, 1956), which is proposed to regulate the sulphydryl content of cell walls by cleaving inter-protein disulphide crosslinks, thus weakening wall structure, is not present in all fungi (Matile *et al.*, 1971), and its presence has not been specifically demonstrated in *Cryptococcus diffluens*. Similarly, the glucanase which has been proposed as involved in yeast cell wall biogenesis is an endoglucanase, which was not measured in this study.

The question of wall rigidity and softening in the growing apical region of hyphae of *Geotrichum candidum* has been examined by Trinci & Collinge (1975), using longitudinal and serial transverse sectioning methods. They propose a tentative model involving: (i) at the growing tip, a primary wall which is relatively extensible either because cross-linkage has not developed between the microfibrils, and/or because the region is exposed to hydrolytic enzymes which limit or decrease interaction between microfibrils; (ii) in the next region back from the growing tip, the wall has the same thickness as the apex, but has become inelastic either because lytic vesicles are no longer interacting with it, and/or because stable cross-linkages have been formed between microfibrils. This region is cylindrical, compared with the conical apical region; and (iii) further down the hypha, secondary thickening and maturation take place.

Direct measurements of vesicle concentration in the growing region provides quantitative support for the now commonplace observation that vesicles occur in substantially greater numbers at the site of the growing wall than in regions of maturation or where there is no growth. In *Geotrichum*, the number of vesicles range from about 700 per μm^3 at the leading edge of the apical zone, to about 50 per μm^3 at the base of the apex and in mature regions of hyphae (Trinci & Collinge, 1975).

Studies such as these, highlighting as they do the concentration of active growth and morphogenetic processes into small regions of the cell, point out the difficulty that users of a gross biochemical approach may have in detecting the subtle organizational and functional changes at the hyphal tip, which undoubtedly characterize the morphogenetic processes underlying dimorphism.

Location of biosynthetic enzymes

There is little direct evidence bearing on the location of the enzyme systems responsible for the synthesis of the components of the fungal or yeast cell wall, though it is often assumed that they are part of the plasma membrane or of the growing cell wall. On the other hand, in the unicellular algae already discussed, an essentially complete mechanism for polymerization of precursors, and their assembly into intact scales or wall units exists within the vesicles and cisternae of the Golgi stack or dictyosome. These two models may thus appear to predict different localizations for some of the enzymes involved in cell wall biogenesis. These differences may not be contradictory, however, since it is possible that enzymes for cell wall polymer synthesis found in or on the plasma membrane originate in the membranes of the Golgi apparatus, becoming part of the plasma membrane as a consequence of fusion of Golgi vesicles or cisternae with that membrane. Nevertheless, important differences in the apparent involvement of the plasma membrane in cell wall biogenesis may be evident experimentally. These differences should be noted, since they may indicate significant variation from one cell type to another in the origin, and rates and location of processing and maturation of cell wall precursors and units.

Marriott's (1975) measurements of the enzymic activity of purified plasma membranes from yeast and mycelial forms of *Candida albicans* may be interpreted in these terms. Mannan synthase was enriched 3.7 times into a partly purified plasma membrane fraction from yeast cells, compared with the protoplast lysate. When prepared from mycelial cells, the enrichment was 7.5 times. While these levels of purification may not be great enough to exclude the possibility that Golgi apparatus or other secretory vesicles or membranes are being purified into the same fraction, the difference between them may indicate differences between the two cell types in the rate or mode of delivery of vesicles from Golgi apparatus (or equivalent) to plasma membrane.

The production of precursor carbohydrates for cell wall synthesis, and the activity of initial enzymes for synthesis of chitin and mannan were measured in yeast and mycelial forms of *Candida albicans* by Chattaway *et al.* (1973). The activity of phosphomannose isomerase, the branch-point enzyme for mannose synthesis and thus for mannan synthesis, is twice as high in yeast

phase cultures as in mycelial cultures at the time of maximum transformation (4–6 h cultures). The cell walls of yeast phase cells, however, contain
only marginally more polymerized mannose than mycelial cells (Chattaway,
Holmes & Barlow, 1968). Glutamine: fructose-6-phosphate aminotransferase, the branch point enzyme for N-acetyl glucosamine synthesis and thus
for chitin synthesis, is also generally greater in yeast phase cells. The chitin
content of yeast cells, however, is substantially *lower* than that of mycelial
phase cells. These results, indicating that first enzyme activities for pathways
leading to cell wall polymers do not correspond to the amount of polymer
formed by the cells, suggest that measurement of enzyme activities as such in
cell fractions or homogenates may not be of great value to an understanding
of the dimorphic process. It is the activity of the pathways *in situ* which will
probably repay careful measurement and analysis; endogenous regulatory
processes will clearly be significant in this respect. Wright's (1973) proposals
on the control of morphogenesis in the cellular moulds would be of more
than passing interest to an investigator embarking upon such studies in the
dimorphic fungi.

The potential importance of metabolic flux to cell wall morphogenesis is
highlighted by the morphological (colonial) mutants of *Neurospora* (Scott &
Mahoney, 1976). Colonial mutants in *Neurospora* are characterized by
increased frequency of hyphal branching, which results in a compact colonial
morphology, and decreased growth rates. The specific enzyme defect is
manifested as a decreased activity of glucose-6-phosphate dehydrogenase.
This in turn is reflected in lowered concentrations of pyridine nucleotides,
though the proportion of these in the reduced state remains approximately
the same as in wild-type cells. It is notable that Chattaway *et al.* (1973) found
that an increased flux through the hexose monophosphate shunt (and thus
NADPH generation) corresponds with a period of minimum mycelial
development in temperature transformed cultures of *Candida albicans*.
Whether this can be explained simply as an effect on wall weakening by way
of increased reductive capacity, and thus of protein disulphide reductase
(Nickerson & Falcone, 1956), remains to be established.

Possibly related to the altered NADPH concentrations found in colonial
mutants of *Neurospora* is the lowered proportion of linolenic acid in the
neutral lipids and phospholipids (Scott & Mahoney, 1976). In revertants,
and t_s mutants grown at the permissive temperature, the wild type concentrations of NADPH and linolenic acid are re-established. Moreover, the
growth rate and morphology of a particular semi-colonial mutant (*fr*) can be
made identical to that of wild type cells by supplementation of the growth
medium with linolenic acid, though not with mono-unsaturated and saturated fatty acids. It should be noted, however, that this simple nutritional
reversal of phenotype has not been achieved with a number of other colonial
mutants. A somewhat similar observation on altered fatty acid composition
in yeast and mycelial phases of *Mucor genevensis* has been noted by Gordon,
Stewart & Clark-Walker (1971): both cell types were grown aerobically, and
the yeast phase was induced using phenethyl alcohol; yeast cells contained
lower concentrations of palmitoleic and higher concentrations of oleic acids,
compared with mycelial phase cells.

Whether the effects of mutation on the hexose monophosphate shunt in
Neurospora are ultimately reflected, through deficiency of a particular lipid,

in the specific function of membranes in the cell can only be a subject for conjecture at this stage. The plasma membrane suggests itself as an obvious candidate in this respect because of its direct involvement with the secretory and synthetic processes of cell wall biogenesis. A more direct indication that plasma membrane function is affected is given by the semi-colonial mutant *fr*, which in liquid culture synthesizes only 10% of the cAMP of wild type cells (Scott, Mishra & Tatum, 1973). This effect appears to be a consequence of altered activity of the adenyl cyclase, an enzyme which generally is thought to be attached to the plasma membrane of eukaryotic cells.

The question of how cAMP may be related to morphogenesis is considered in greater detail in the next section.

9.5 Relationship of Metabolic Status of the Cell to Dimorphic Transformation

Dimorphism and respiration: breaking the nexus

A causal relationship between the induction of mycelial growth and the development of respiratory or oxidative metabolism has been strongly implied in much of the earlier literature on metabolic aspects of fungal dimorphism (reviewed by Smith & Galbraith, 1971; Brody, 1973). For instance, fermentative, yeast-like cells generally form when sporangiospores of *Mucor* are germinated and grown under an atmosphere of CO_2, while an actively respiring mycelium develops under aerobic conditions. Yeast-like cells form hyphae when anaerobic cultures are aerated (Bartnicki-Garcia & Nickerson, 1962; Bartnicki-Garcia, 1968).

Moreover, this dimorphic transformation is inhibited by acriflavin or chloramphenicol (Haidle & Storck, 1966), phenethyl alcohol (Terenzi & Storck, 1969), sodium fluoride, and by elevated glucose (Bartnicki-Garcia, 1968) and carbon dioxide concentrations (Bartnicki-Garcia & Nickerson, 1962). An unidentified agent in peptone also favours the development of yeast-like cells (Nickerson & Bartnicki-Garcia, 1962; Elmer & Nickerson, 1970). All of these factors are thought to inhibit the yeast to mycelium transition by impairing respiratory adaptation and oxidative growth. This view was strongly supported when Storck & Morrill (1971) discovered that respiratory deficiency in *Mucor bacilliformis* introduced by spontaneous mutation results in an obligate yeast form.

More recent studies with *Mucor rouxii* (Friedenthal, Epstein & Passeron, 1974) and *Mycotypha* (Schulz, Kraepelin & Hinkelmann, 1974) have stressed the relationship between mycelial growth and oxidative metabolism. Schulz *et al.* (1974), on the basis of inhibitor studies with antimycin A and KCN, concluded that a causal relationship exists between the predominance of fermentation or respiration and the yeast or mycelial phenotypes respectively. But other factors, such as temperature and pH, that could not obviously be tied to inhibition of oxidative metabolism, have dramatic effects on morphological expression. Hall & Kolankaya (1974), for example, reported that lowered pH, high temperature, dense inoculum and high hexose level increased the proportion of the yeast form in cultures of *Mycotypha*. Furthermore, strain to strain variations are observed. Friedenthal *et al.* (1974) found that anaerobiosis does not prevent the elongation of preformed hyphae obtained under aerobic conditions, provided that cellular integrity is maintained. The anaerobic mycelia had fermentative metabolism, judged by their content of respiratory enzyme complexes, and the production of ethanol. These data strongly argue against the nexus between respiratory metabolism and mycelial morphology.

Chloramphenicol specifically inhibits *in vivo* mitochondrial protein synthesis in *Saccharomyces cerevisiae* resulting in cells that no longer respire (Clark-Walker & Linnane, 1967). Growth of *Mucor genevensis* in high concentrations of chloramphenicol increases the proportion of cells with yeast-like morphology, but small club-like mycelial forms are still present, even though cyanide-sensitive respiration is lost (Clark-Walker, 1973). The ability of *Mucor* to grow small mycelial projections in aerobic conditions without normal mitochondrial function indicates that oxidative phosphorylation is not essential for apical growth (Clark-Walker, 1973).

To separate the effects of respiration and catabolite concentration on dimorphism, Rogers, Clark-Walker & Stewart (1974) studied the behaviour, both metabolic and morphological, of *Mucor genevensis* grown in continuous culture with glucose as the limiting nutrient. When growth is limited by glucose, so that the steady state glucose concentration is less than $25 \mu M$, the yeast form exists exclusively, under anaerobic conditions. In the transition from anaerobic to aerobic cultures, respiratory adaptation and metabolism parallels the yeast/mycelium conversion. However, with chloramphenicol present and respiratory development completely inhibited, hyphal development still occurs at these very low glucose concentrations. In aerobic conditions, the presence of excess glucose induces the yeast form with active oxidative metabolism. Interestingly, mycelial cultures equilibrated in low steady state glucose concentrations under aerobic conditions could only be reverted to the yeast form by increasing the steady state glucose concentration. The conclusions are that oxidative metabolism is not an essential feature of mycelial morphology, and that glucose is the most important cultural factor determining morphology in *Mucor genevensis*. Similar conclusions were drawn from a study of another dimorphic member of the Mucorales, *Cokeromyces poitrasii* (Rogers & Gleason, 1974). More recent studies with continuous cultures of *Candida albicans* (Shepherd & Sullivan, 1976) have also shown that under aerobic conditions, morphology depends not on oxidative metabolism, but on the nature of the carbon source.

The clear lack of dependence of mycelial development in *Mucor* upon oxidative metabolism is well demonstrated by a series of important experiments by Paznokas & Sypherd (1975). If *Mucor racemosus* is grown in air and in the presence of dibutyryl cAMP, it develops into yeast-like form, but with a high functional respiratory capacity. However, when grown with continuous flushing by nitrogen the mycelial form develops with a low respiratory capacity. Obviously mycelial growth is not dependent on oxidative metabolism, and in the search for a primary effector to initiate a dimorphic response, cAMP appears to be a key substance.

Effects of cAMP on dimorphism

Cyclic AMP was first implicated as an effector of dimorphism in *Mucor* by Larsen & Sypherd (1974) who showed that the addition of dibutyryl cAMP to yeast-like cultures of *Mucor racemosus* inhibits transformation to hyphae after exposure to air. Endogenous cAMP levels in yeast-form cells decline about fourfold prior to the appearance of hyphal germ cells. It had also been

shown at this time that cAMP relieves repression of respiration in aerobically adapting *Escherichia coli* (Okinawaka & Dobroqosz, 1967; Abound &
Burger, 1972) and yeast cells (Sy & Richter, 1972; Tsuboi & Yanagishima,
1973). The involvement of cAMP in the reversal of catabolite repression by
glucose in particular suggests that this nucleotide might play a role in the
regulation of dimorphism of fungi. When yeast-phase cells of *Mucor* are
aerated a decline in the cAMP level is evident before any morphological
changes towards hyphae are observed (Larsen & Sypherd, 1974). Paznokas
& Sypherd (1975) concluded from their studies, referred to earlier, that
low intracellular cAMP concentrations correlate closely, though not invariably, with the mycelial state of the organism. In some way, the transformation process is also influenced by the nature of the carbon source, as
mentioned earlier.

In a related study, Orlowski & Sypherd (1976) found that cyclic
guanosine $3',5'$-monophosphate (cGMP) levels in *Mucor racemosus*
remained relatively constant during the conversion of the yeast form to
mycelia; added exogenous cGMP or the dibutyryl derivative had no effect
on morphology and did not alter the effect that cAMP has on cell
morphology.

In all living systems that have so far been studied, cAMP is synthesized by
the reaction catalyzed by adenyl cyclase, and is hydrolysed by cAMP
phosphodiesterase. In principle, any alteration in the amount of activity of
either of these two enzymes could affect intracellular cAMP concentration.
Paveto, Epstein & Passeron (1975) showed that in the case of *Mucor rouxii*
the decline in intracellular cAMP levels during the yeast to mycelium
transformation corresponds in particular to the rapid increase in phosphodiesterase activity. There is at most a slight increase in the activity of the
cyclase. EDTA (10 mM) inhibits the phosphodiesterase activity by 70%,
but theophylline, a powerful phosphodiesterase inhibitor in most systems, is
only slightly inhibitory. Mg^{2+} and Ca^{2+} have only a slight affect on activity.
Adenyl cyclase from *Mucor rouxii* requires Mn^{2+} for activity and Mg^{2+} will
not replace Mn^{2+}.

Since the activities of adenyl cyclase and phosphodiesterase are sensitive
to divalent cations, it may be argued that the effects of EDTA and high
concentrations of cations such as Zn^{2+} in causing a dimorphic response in
fungi are a direct result of changing the intracellular cAMP levels via
modulation of the activity of these enzymes. For instance in an earlier report
(Zorzopulus, Jobbagy & Terenzi, 1973), it was shown that low concentrations of EDTA (0.1 mM) added to anaerobic yeast-like cultures of
Mucor rouxii induce mycelial growth. In *Aspergillus parasiticus* it has been
reported that the dimorphic response may follow one of two patterns of
morphogenesis depending on the presence or absence of Mn^{2+} ions in the
medium (Garrison & Boyd, 1974). Dimorphism in the fungus *Phialophora
verrucosa* is also affected by the presence or absence of Mn^{2+} in defined
medium (Reiss & Nickerson, 1974). The response is specific towards Mn^{2+}
and not towards other cations. Moreover, it was concluded that part of the
Mn^{2+} is bound to the cell walls, as well as being present in the cytoplasm. Its
association with cell walls might suggest a role in wall biosynthesis and/or a
regulatory role for adenyl cyclase.

Zinc is also implicated in the morphogenesis of a wide variety of micro-
organisms including the dimorphic fungi, *Histoplasma capsulatum* (Pine &

Peacock, 1958), *Mucor rouxii* (Bartnicki-Garcia & Nickerson, 1962) and *Candida albicans* (Yamaguchi, 1975). Since metals such as Mn^{2+} and Zn^{2+} have multi-functional roles in biosynthesis (Evans, Morhenn, Jones & Tomkins, 1974) including cell wall, DNA, RNA and fatty acid biosynthesis, it may be too simplistic to interpret the prime effect as one upon cAMP in the face of these pleiotropic alterations to cellular metabolism. Nevertheless, taken together with the observations of Dow & Rubery (1975) on Ca^{2+} protection against hyphal tip-bursting, the need for a systematic examination of the effect of metal ions on biosynthetic and structural features of the apical tip in fungi becomes evident.

Although the site of action of cAMP in altering patterns of morphogenesis is as yet unclear, the results with dimorphic organisms and differentiation in other fungi (Uno & Ishikawa, 1974; Schwalb, 1974; Silvermann & Epstein, 1975; Torres, Flawia, Terenzi & Tellez-Inon, 1975) present an interesting parallel to studies with the morphology mutants of *Neurospora* referred to earlier. As was mentioned, the mutations concerned invariably affect the pentose phosphate pathway, especially the amount of glucose-6-phosphate dehydrogenase present, and cause a colonial mode of growth (Brody, 1973; Scott & Solomon, 1975; Scott & Mahoney, 1976). Furthermore, mutant-like morphology is observed after inhibition of adenyl cyclase activity in wild-type cells (Scott *et al.*, 1973). Scott & Solomon (1975) examined the effects of a range of unrelated drugs (including atropine, theophylline, histamine and chloroquine) on the morphology and intracellular cAMP levels of wild type *Neurospora*. Each of the drugs decreases the endogenous cAMP concentration of mycelia and the cells assume a colonial or semicolonial growth morphology similar to that of known morphology mutants. While this evidence may indicate a relationship between morphological abnormalities and reduction in cAMP concentration, a simple relationship between cAMP and morphogenesis should not be assumed, if only because the cAMP content of most *Neurospora* morphology mutants is normal (Scott & Solomon, 1975).

Studies with another *Neurospora* mutant (*fr*) may be of significance to an understanding of dimorphism in fungi (Scott *et al.*, 1973). On plates it exhibits a semicolonial phenotype. The intracellular cAMP levels are 10% of the wild type in liquid cultures and 50–69% on solid media. If the cultures are supplemented with linolenic acid, however, a wild type growth morphology and normal cAMP levels are restored. This suggests that cAMP deficiency may be a consequence of linolenic acid deficiency. In fact this supposition is supported by the finding that the adenyl cyclase, which is a plasma membrane-bound enzyme, is more thermolabile in the *fr* mutant than in the wild type cell, and has altered physical characteristics (Scott *et al.*, 1973; Scott & Solomon, 1975). In particular, the sedimentation properties are altered and this could be construed as being due to an altered lipid–protein interaction. Effects of temperature on adenyl cyclase or on phosphodiesterase, and thus on intracellular cAMP concentrations, may provide a basis for explaining transformation in the temperature-sensitive dimorphic fungi.

The phosphodiesterase of the *fr* mutant is also membrane-bound (Scott & Solomon, 1973), but there appears to be no difference between wild-type and mutant enzymes. When theophylline is administered at millimolar levels, the growth rate is stimulated and a more spreading phenotype is produced. This effect could be due to an inhibition of phosphodiesterase in the mutant cells resulting in a higher intracellular cAMP concentration.

It is believed that there is an involvement of contractile and structural proteins attached to membranes during hormonal activation of adenyl cyclase in animal cells (Willingham, 1976). The effect of agents which perturb microtubule and microfilament structure (for example cytochalasins, colchicine), may therefore have their effects mediated or modulated by effects on the localization or activity of adenyl cylase. This possibility adds a further dimension to the role of the motile and contractile systems in the morphogenesis of the fungal cell. Betina, Mičeková & Nemec (1972) have reported that cytochalasins induce branching and swelling of fungal hyphae.

From the preceding description of the relationship between cAMP levels and dimorphism it seems that the physiological factors that are capable of inducing a dimorphic response may in many cases be rationalized as substances or conditions that affect the activity of adenyl cyclase and phosphodiesterase, and consequently alter the intracellular cAMP concentration. However, all the physiological factors involved in producing dimorphic responses cannot easily be incorporated into this scheme. Moreover, the question of how cAMP ultimately generates a morphological effect is left suspended. In this regard there are three areas of interest that deserve comment (Fig. 9.5). Firstly, three different DNA-dependent RNA polymerases have been characterized in *Mucor rouxii* by Young & Whiteley (1975*a*, *b*) and although all three are present in both yeast and mycelial cultures, changes occur in the relative activities of the three enzymes during the yeast to mycelial transition. The authors have suggested that the increased specific activity of polymerase II may be responsible for the increased transcription rate during the yeast to mycelium conversion, and that cAMP may play a role in these primary events, perhaps in a fashion similar to that in the regulation of the *lac* operon in *Escherichia coli* and gene regulation during spore development in bacteria (Dickson, Abelson,

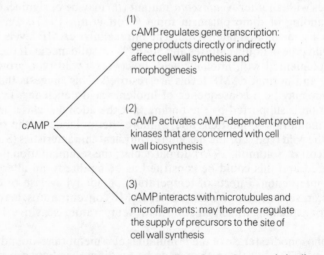

Fig. 9.5 Possible roles for cAMP in the regulation of morphogenesis in dimorphic fungi.

Barnes & Reznikoff, 1975). The altered products of gene expression then presumably affect cell wall biogenesis and morphogenesis either directly or indirectly.

The second possible connection between cAMP and morphogenesis stems from the known property of cAMP to regulate protein kinase activities (Simantov & Sachs, 1975). Formation of a complex between the kinase, cAMP and cAMP-binding protein releases the catalytic subunit of the kinase so that it can transfer phosphate groups to a variety of proteins including membrane proteins and histones. In a study of the regulation of growth and differentiation in neuroblastoma cells, Simantov & Sachs (1975) showed that mutations could be induced which affect the temperature sensitivity of cAMP-binding proteins. The temperature sensitivity is transmitted in turn to cAMP-dependent kinases that are involved in the regulation of the growth and differentiation of the cells. Once more, the fact that temperature induces morphological changes in a number of dimorphic fungi suggests a parallel with the above experiments, and perhaps the central mechanisms involved in triggering morphological changes are common to a wide range of eukaryotic organisms.

The third possibility is that cAMP interacts directly with the microtubules and microfilaments which are proposed to regulate shape and motility, at least in animal cells (Willingham, 1976). As in the examples with fungi given above, altered intracellular concentrations of cAMP are associated with morphological changes in cultured animal cells. Moreover, agents which interfere with microtubular function prevent most of the gross morphological changes induced by elevated cAMP levels (Willingham, 1976). Willingham & Pastan (1975) adduce data to show that cAMP stimulates microtubule assembly, increases cellular adhesion to the substratum, and inhibits microfilament (putatively actin) mediated contraction in cultured animal cells. The significance of these observations to dimorphism in fungi is not the effect on cell shape as such, since the shape of the fungal protoplast is determined by the external wall in the intact cell, but rather the possible control by microtubules and microfilaments of delivery of precursor material through the cytoplasm to the site of cell wall synthesis, as discussed in 9.3.

9.6 Conclusions and Outlook

Three matters seem clear from recent studies of fungal dimorphism. Firstly, an understanding of the dimorphic process will come hand in hand with the fundamental understanding of cell wall growth and morphogenesis in filamentous fungi and yeasts. Secondly, information from studies of plant cell ultrastructure will be useful in designing new approaches to the study of hyphal and yeast cell ultrastructure. Thirdly, ultrastructural methods need to be devised which permit the specific chemical identification of key components in the flow of precursors from possibly remote regions of the cytoplasm to the growing, inner regions of the cell wall adjacent to the plasma membrane.

The methods will thus necessarily emphasize the microscopic approach, with closely integrated chemical, immunochemical and radioautographic procedures which will permit the specific identification of: (i) macromolecules constituting intracellular precursors of cell wall fibrils and matrix components; (ii) fusion of specific types of vesicles with each other and with other membranes such as the plasma membrane so that exocytotic and membrane circulation patterns may be identified; and (iii) microtubules and microfilaments, and subunits of these structures as they may exist as unpolymerized pools in the cells.

Inhibitors which act on certain of the potential underlying regulatory processes may also be usefully employed, particularly in association with the specific ultrastructural approaches referred to above. In particular, the effects of: (i) inhibitors of microtubule assembly, such as the vinca alkaloids (vinblastine, vincristine) and colchicine; (ii) putative inhibitors of actin polymerization and depolymerization, such as the cytochalasins and phalloidin; and (iii) inhibitors of cAMP synthesis and degradation, such as theophylline, caffeine, atropine, histamine and quinolines, should be examined systematically for effects on dimorphic transformation, on ultrastructural changes in yeast, mycelial and intermediate or aberrant forms, and on specific biochemical processes, such as metabolic flux rates and concentrations of likely key molecules, ions, and enzymes.

Finally, little is known of the underlying molecular processes of differentiation associated with fungal dimorphism, since there is practically nothing known of the role of particular genes, or of control exerted at the levels of transcription and translation. Greater attention might profitably be given to the isolation of morphogenetic mutants or variants as working material. Although it is true that genetic mapping is not possible for many dimorphic fungi, this may well change soon (for example, see Kwon-Chung, 1974). In any case the approach has validity if only as a means to disconnect interlinked metabolic and assembly processes without resort to physiological perturbation or the use of inhibitors.

9.7 References

ABOUND, M. & BURGER, M. (1972). The effect of anaerobiosis and uncouplers on the stimulation of β-galactosidase synthesis by cyclic $3',5'$-adenosine monophosphate in *Eschericia coli*. *Journal of General Microbiology* **71**, 311–18.

BARTNICKI-GARCIA, S. (1968). Cell wall chemistry, morphogenesis, and taxonomy of fungi. *Annual Review of Microbiology* **22**, 87–108.

BARTNICKI-GARCIA, S. (1973). Fundamental aspects of hyphal morphogenesis. *Symposium Society General Microbiology* **23**, 245–67.

BARTNICKI-GARCIA, S. & LIPPMAN, E. (1972). The bursting tendency of hyphal tips of fungi: presumptive evidence for a delicate balance between wall synthesis and wall lysis in apical growth. *Journal of General Microbiology* **73**, 487–500.

BARTNICKI-GARCIA, S. & NICKERSON, W. J. (1962). Induction of yeast-like development in *Mucor* by carbon dioxide. *Journal of Bacteriology* **84**, 829–40.

BETINA, V., MIČEKOVÁ, D. & NEMEC, P. (1972). Antimicrobial properties of cytochalasins and their alteration of fungal morphology. *Journal of General Microbiology* **71**, 343–9.

BRODY, S. (1973). Metabolism, cell walls, and morphogenesis. In *Developmental regulation: aspects of cell differentiation*, pp. 107–154. Edited by S. J. Coward. New York: Academic Press.

BROWN, R. M. (1969). Observations on the relationship of the Golgi apparatus to wall formation in the marine chrysophycean alga, *Pleurochrysis scherffelii* Pringsheim. *Journal of Cell Biology* **41**, 109–23.

BROWN, R. M. (1973). The role of the Golgi apparatus in scale and coccolith biogenesis. *Journal of Cell Biology* **59**, 35a.

BROWN, R. M. (1974). New observations on membrane flow and product transport in the Golgi apparatus of *Pleurochrysis*. *Journal of Cell Biology* **63**, 39a.

BROWN, R. M., FRANKE, W. W., KLEINIG, H., FALK, H. & SITTE, P. (1970). Scale formation in chrysophycean algae. I. Cellulosic and noncellulosic wall components made by the Golgi apparatus. *Journal of Cell Biology* **45**, 246–71.

BYERS, B. & GOETSCH, L. (1976). A highly ordered ring of membrane-associated filaments in budding yeast. *Journal of Cell Biology* **69**, 717–21.

CABIB, E. (1975). Molecular aspects of yeast morphogenesis. *Annual Review of Microbiology* **29**, 191–214.

CARBONELL, L. M. (1967). Cell wall changes during the budding process of *Paracoccidioides brasiliensis* and *Blastomyces dermatitidis*. *Journal of Bacteriology* **94**, 213–23.

CASSONE, A., SIMONETTI, N. & STRIPPOLI, V. (1973). Ultrastructural changes in the wall during germ-tube formation from blastospores of *Candida albicans*. *Journal of General Microbiology* **77**, 417–26.

CHATTAWAY, F. W., HOLMES, M. R. & BARLOW, A. J. E. (1968). Cell wall composition of the mycelial and blastospore forms of *Candida albicans*. *Journal of General Microbiology* **51**, 367–76.

CHATTAWAY, F. W., BISHOP, F., HOLMES, M. R., ODDS, F. C. & BARLOW, A. J. E. (1973). Enzyme activities associated with carbohydrate synthesis and breakdown in the yeast and mycelial forms of *Candida albicans*. *Journal of General Microbiology* **75**, 97–109.

CLARK-WALKER, G. D. (1973). Relationship between dimorphology and respiration in *Mucor genevensis* studied with chloramphenicol. *Journal of Bacteriology* **116**, 972–80.

CLARK-WALKER, G. D. & LINNANE, A. W. (1967). The biogenesis of mitochondria in *Saccharomyces cerevisiae*. A comparison between cytoplasmic respiratory deficient mutant yeast and chloramphenicol inhibited wild-type cells. *Journal of Cell Biology* **34**, 1–14.

CONANT, N. F., SMITH, D. T., BAKER, R. D. & CALLAWAY, J. L. (1971). *Manual of clinical mycology*, 3rd Edition. Philadelphia: W. B. Saunders Co.

CORTAT, M., MATILE, P. & WIEMKIN, A. (1972). Isolation of glucanase-containing vesicles from budding yeast. *Archiv für Mikrobiologie* **82**, 189–205.

CRONSHAW, J. (1974). Phloem differentiation and development. In *Dynamic aspects of plant ultrastructure*, pp. 391–413. Edited by A. W. Robards. London: McGraw-Hill.

DABBAGH, R., CONANT, N. F. & BURNS, R. O. (1974*a*). Effect of temperature on saprophytic cryptococci: observations relating to wall biosynthesis at non-permissive growth temperatures. *Journal of General Microbiology* **85**, 190–202.

DABBAGH, R., CONANT, N. F., NIELSEN, H. S. & BURNS, R. O. (1974*b*). Effect of temperature on saprophytic cryptococci: temperature-induced lysis and protoplast formation. *Journal of General Microbiology* **85**, 177–89.

DAVIES, P. J. (1973). Current theories on the mode of action of auxin. *Botanical Review* **39**, 139–71.

DICKSON, R. C., ABELSON, J., BARNES, W. M. & REZNIKOFF, W. S. (1975). Genetic regulation: the *lac* control region. *Science* **187**, 27–35.

DOW, J. M. & RUBERY, P. M. (1975). Hyphal tip bursting in *Mucor rouxii*: antagonistic effects of calcium ions and acid. *Journal of General Microbiology* **91**, 425–8.

ELMER, A. W. & NICKERSON, W. J. (1970). Nutritional requirements for growth and yeast-like development of *Mucor rouxii* under carbon dioxide. *Journal of Bacteriology* **101**, 593–602.

EVANS, R. B., MORHENN, V., JONES, A. L. & TOMKINS, G. M. (1974). Concomitant effects of insulin on surface membrane conformation and polysome profiles of serum-starved BALB/C 3T3 fibroblasts. *Journal of Cell Biology* **61**, 95–106.

FRIEDENTHAL, M., EPSTEIN, A. & PASSERON, S. (1974). Effect of potassium cyanide, glucose and anaerobiosis on morphogenesis of *Mucor rouxii*. *Journal of General Microbiology* **82**, 15–24.

GARRISON, R. G. & BOYD, K. S. (1974). Ultrastructural studies of induced morphogenesis by *Aspergillus parasiticus*. *Sabouraudia* **12**, 179–87.

GARRISON, R. G., BOYD, K. S. & MARIAT, F. (1975). Ultrastructural studies of the mycelium-to-yeast transformation of *Sporothrix schenkii*. *Journal of Bacteriology* **124**, 959–68.

GIESBRECHT, P., WECKE, J. & REINICKE, B. (1976). On the morphogenesis of the cell wall of staphylococci. *International Review of Cytology* **44**, 225–318.

GORDON, P. A., STEWART, P. R. & CLARK-WALKER, G. D. (1971). Fatty acid and sterol composition of *Mucor genevensis* in relation to dimorphism and anaerobic growth. *Journal of Barcteriology* **107**, 114–20.

GROVE, S. N. & BRACKER, C. E. (1970). Protoplasmic organization of hyphal tips among fungi: vesicles and Spitzenkörper. *Journal of Bacteriology* **104**, 989–1009.

GROVE, S. N., OUJEZDSKY, K. B. & STANISZLO, P. J. (1973). Budding in the dimorphic fungus *Phialophora dermatitidis. Journal of Bacteriology* **115**, 323–9.

GULL, K. (1975). Cytoplasmic microfilament organization in two basidiomycete fungi. *Journal of Ultrastructure Research* **50**, 226–32.

GUNNING, B. E. S. & STEER, M. W. (1975). *Ultrastructure and the biology of plant cells.* London: Edward Arnold.

HAIDLE, C. W. & STORCK, R. (1966). Control of dimorphism in *Mucor rouxii. Journal of Bacteriology* **92**, 1236–44.

HALL, M. J. & KOLANKAYA, N. (1974). The physiology of mould-yeast dimorphism in the genus *Mycotypha* (Mucorales). *Journal of General Microbiology* **82**, 25–34.

HARDCASTLE, R. V. & SZANISZLO, P. J. (1974). Characterization of dimorphism in *Cladosporium werneckii. Journal of Bacteriology* **119**, 294–302.

HENNING, U. (1975). Determination of cell shape in bacteria. *Annual Review of Microbiology* **29**, 45–60.

HEPLER, P. K. & PALEVITZ, B. A. (1974). Microtubules and microfilaments. *Annual Review of Plant Physiology* **25**, 309–62.

JOHNSON, B. F., YOO, B. Y. & CALLEJA, G. B. (1973). Cell division in yeasts: movement of organelles associated with cell plate growth of *Schizosaccharomyces pombe. Journal of Bacteriology* **115**, 358–66.

KWON-CHUNG, K. J. (1974). Genetics of fungi pathogenic for man. *Critical Reviews in Microbiology* **3**, 115–33.

LANE, J. W. & GARRISON, R. G. (1970). Electron microscopy of the yeast to mycelial phase conversion of *Sporotrichum schenckii* (1970). *Canadian Journal of Microbiology* **16**, 747–9.

LARA, S. L. & BARTNICKI-GARCIA, S. (1974). Cytology of budding in *Mucor rouxii*: wall ontogeny. *Archives of Microbiology* **97**, 1–16.

LARSEN, A. D. & SYPHERD, P. S. (1974). Cyclic adenosine 3′,5′-monophosphate and morphogenesis in *Mucor racemosus. Journal of Bacteriology* **117**, 432–8.

McCLURE, W. K., PARK, D. & ROBINSON, P. M. (1968). Apical organization in the somatic hyphae of fungi. *Journal of General Microbiology* **50**, 177–82.

MAHVI, T. A., SPICER, S. S. & WRIGHT, N. J. (1974). Cytochemistry of acid mucosubstances and acid phosphatase in

Crytococcus neoformans. Canadian Journal of Microbiology **20**, 833–8.

MANTON, I. (1967). Further observations on scale formation in *Chrysochromulina chiton. Journal of Cell Science* **2**, 411–18.

MARRIOTT, M. S. (1975). Enzymic activity of purified plasma membranes from the yeast and mycelial forms of *Candida albicans. Journal of General Microbiology* **89**, 345–52.

MATILE, P., CORTAT, M., WIEMKEN, A. & FREY-WYSSLING, A. (1971). Isolation of glucanase-containing particles from budding *Saccharomyces cerevisiae. Proceedings of the National Academy of Sciences of the United States of America* **68**, 636–40.

NICKERSON, W. J. & BARTNICKI-GARCIA, S. (1962). Nutrition, growth and morphogenesis of *Mucor rouxii. Journal of Bacteriology* **84**, 841–58.

NICKERSON, W. J. & FALCONE, G. (1956). Identification of protein disulphide reductase as a cellular division enzyme in yeast. *Science* **124**, 722–3.

OUJEZDSKY, K. B., GROVE, S. N. & SZANISZLO, P. J. (1973). Morphological and structural changes during the yeast-to-mould conversion of *Phialophora dermatitidis. Journal of Bacteriology* **113**, 468–77.

OKINAKA, R. T. & DOBROQOSZ, W. J. (1967). Catabolite repression and the Pasteur effect in *Escherichia coli. Archives of Biochemistry and Biophysics* **120**, 451–3.

ORLOWSKI, M. & SYPHERD, P. S. (1976). Cyclic guanosine 3′,5′-monophosphate in the dimorphic fungus *Mucor racemosus. Journal of Bacteriology* **125**, 1226–8.

PARTHASARATHY, M. V. & MUHLETHALER, K. (1972). Cytoplasmic microfilaments in plant cells. *Journal of Ultrastructure Research* **38**, 46–62.

PAZNOKAS, J. L. & SYPHERD, P. S. (1975). Respiratory capacity, cyclic adenosine 3′,5′-monophosphate and morphogenesis of *Mucor racemosus. Journal of Bacteriology* **124**, 134–9

PAVETO, C., EPSTEIN, A. & PASSERON, A. (1975). Studies on cyclic adenosine 3′,5′-monophosphate levels and adenylate cyclase and phosphodiesterase activities in the dimorphic fungus *Mucor rouxii. Archives of Biochemistry and Biophysics* **169**, 449–57.

PINE, L. & PEACOCK, C. L. (1958). Studies on the growth of *Histoplasma capsulatum*. IV. Factors influencing conversion of the mycelial phase to the yeast phase. *Journal of Bacteriology* **75**, 167–74.

RAMOS, S. & ACHA, I. G. (1975). Cell wall enzymatic lysis of the yeast form of *Pullularia pullulans* and wall regeneration by protoplasts. *Archives of Microbiology* **104**, 271-7.

RAY, P. M. & BAKER, D. B. (1965). The effect of auxin on synthesis of oat coleoptile wall constituents. *Plant Physiology* **40**, 353-68.

REISS, E. & NICKERSON, W. J. (1974). Control of dimorphism in *Phialophora verrucosa. Sabouraudia* **12**, 202-13.

ROBERTSON, N. F. (1968). The growth process in fungi. *Annual Review of Phytopathology* **6**, 115-36.

ROGERS, P. J. & GLEASON, F. H. (1974). Metabolism of *Cokeromyces poitrasii* grown in glucose-limited continuous culture at controlled oxygen concentrations. *Mycologia* **66**, 919-25.

ROGERS, P. J., CLARK-WALKER, G. D. & STEWART, P. R. (1974). Effects of oxygen and glucose on energy metabolism and dimorphism of *Mucor genevensis* grown in continuous culture: reversibility of yeast-mycelium conversion. *Journal of Bacteriology* **119**, 282-93.

ROMANO, A. H. (1966). Dimorphism. In *The Fungi*, vol. 2, pp. 181-209. Edited by G. C. Ainsworth and A. S. Sussman. New York: Academic Press.

SCHERR, G. H. & WEAVER, R. H. (1953). The dimorphism phenomenon in yeasts. *Bacteriological Reviews* **17**, 51-92.

SCHWALB, M. N. (1974). Effect of adenosine 3',5'-cyclic monophosphate on the morphogenesis of fruit-bodies of *Schizophyllum commune. Archives of Microbiology* **96**, 17-20.

SCHULZ, B. E., KRAEPELIN, G. & HINKELMANN, W. (1974). Factors affecting dimorphism in *Mycotypha* (Mucorales): a correlation with the fermentation/respiration equilibrium. *Journal of General Microbiology* **82**, 1-13.

SCOTT, W. A. & MAHONEY, E. (1976). Defects of glucose-6-phosphate and 6-phosphogluconate dehydrogenases in *Neurospora* and their pleiotropic effects. *Current Topics in Cellular Regulation* **10**, 205-36.

SCOTT, W. A. & SOLOMON, B. (1975). Adenosine 3',5'-cyclic monophosphate and morphology in *Neurospora crassa*. Drug induced alterations. *Journal of Bacteriology* **122**, 454-63.

SCOTT, W. A., MISHRA, N. C. & TATUM, E. L. (1973). Biochemical genetics of morphogenesis in *Neurospora. Brookhaven Symposia in Biology* **25**, 1-18.

SHEPHERD, M. G. & SULLIVAN, P. A. (1976). The production and growth characteristics of yeast and mycelial forms of *Candida albicans* in continuous culture. *Journal of General Microbiology* **93**, 361-70.

SILVERMANN, P. M. & EPSTEIN, P. M. (1975). Cyclic nucleotide metabolism coupled to cytodifferentiation of *Blastocladiella emersonii. Proceedings of National Academy of Sciences of the United States of America* **72**, 442-6.

SIMANTOV, R. & SACHS, L. (1975). Temperature sensitivity of cyclic adenosine 3',5'-monophosphate-binding proteins and the regulation of growth and differentiation in neuroblastoma cells. *Journal of Biological Chemistry* **250**, 3236-42.

SMITH, J. E. (1975). The structure and development of filamentous fungi. In *The Filamentous Fungi*. vol. 1, pp. 1-15, Edited by J. E. Smith and D. E. Berry. London: Edward Arnold.

SMITH, J. E. & GALBRAITH, J. C. (1971). Biochemical and physiological aspects of differentiation in the fungi. *Advances in Microbial Physiology* **5**, 45-124.

STORCK, R. & MORRILL, R. C. (1971). Respiratory-deficient, yeast-like mutant of *Mucor. Biochemical Genetics* **5**, 467-79.

SY, J. & RICHTER, D. (1972). Content of cyclic 3',5'-adenosine monophosphate and adenyl cyclase in yeast at various growth conditions. *Biochemistry* **11**, 2788-91.

TAKEO, K. (1974). Ultrastructure of polymorphic *Mucor* as observed by means of freeze-etching. II. Vegetative yeast forms grown under anaerobic conditions. *Archives of Microbiology* **99**, 91-8.

TAKEO, K. & NISHIURA, M. (1974). Ultrastructure of polymorphic *Mucor* as observed by means of freeze-etching. I. Vegetative growth of mycelium and arthrospore formation in submerged and aerated cultures. *Archives of Microbiology* **98**, 175-85.

TERENZI, H. F. & STORCK, R. (1969). Stimulation of dimorphism in *Mucor rouxii. Journal of Bacteriology* **92**, 1236-44.

TORRES, H. N., FLAWIA, M. M., TERENZI, H. F. & TELLEZ-INON, M. F. (1975). Adenylate cyclase activity in *Neurospora crassa*. In *Advances in cyclic nucleotide research*. vol. 5. pp. 67-86. Edited by G. I. Drummond, P. Greangurd and G. A. Robinson. New York: Raven Press.

TRINCI, A. R. J. & COLLINGE, A. J. (1975). Hyphal wall growth in *Neurospora crassa* and *Geotrichum candidum*. *Journal of General Microbiology* **91**, 355–61.

TSUBOI, M. & YANAGISHIMA, N. (1973). Effect of cAMP, theophylline and caffeine on the glucose repression of sporulation in *Saccharomyces cerevisiae*. *Archiv für Mikrobiologie* **93**, 1–12.

UNO, I. & ISHIKAWA, T. (1974). Effect of glucose on the fruiting body formation and adenosine 3′,5′-cyclic monophosphate levels in *Coprinus macrorhizus*. *Journal of Bacteriology* **120**, 96–100.

WILLINGHAM, M. C. (1976). Cyclic AMP and cell behavior in cultured cells. *International Review of Cytology* **44**, 319–57.

WILLINGHAM, M. C. & PASTAN, I. (1975). Cyclic AMP and cell morphology in cultured fibroblasts. Effects on cell shape, microfilament and microtubule distribution and orientation to substratum. *Journal of Cell Biology* **67**, 146–59.

WRIGHT, B. E. (1973). *Critical variables in differentiation*. Englewood Cliffs, U.S.A.: Prentice Hall, Inc.

WUNDERLICH, F., MULLER, R. & SPETH, V. (1973). Direct evidence for a colchicine-induced impairment in the mobility of membrane components. *Science* **182**, 1136–8.

YAMAGUCHI, H. (1975). Control of dimorphism in *Candida albicans* by zinc. Effect on cell morphology and composition. *Journal of General Microbiology* **86**, 370–2.

YOUNG, H. A. & WHITELEY, H. R. (1975*a*). DNA-dependent ribonucleic acid polymerases in the dimorphic fungus *Mucor rouxii*. *Journal of Biological Chemistry* **250**, 479–87.

YOUNG, H. A. & WHITELEY, H. R. (1975*b*). Changes in the levels of DNA-dependent RNA polymerases during the transition of the dimorphic fungus *Mucor rouxii* from yeast-like to mycelial growth. *Experimental Cell Research* **91**, 216–22.

ZORZOPULUS, J., JOBBAGY, A. J. & TERENZI, H. F. (1973). Effects of ethylenediaminetetraacetate and chloramphenicol on mitochondrial activity and morphogenesis in *Mucor rouxii*. *Journal of Bacteriology* **115**, 1198–204.

CHAPTER 10

Sclerotium Formation

H. J. WILLETTS

10.1 Introduction

In mycological literature many different fungal structures have been referred to as sclerotia and all have in common the ability to survive unfavourable conditions for varying periods as independent bodies. Sometimes the term 'sclerotium' is used in a restrictive sense to include resting vegetative bodies that germinate to form reproductive structures or clearly related forms. Other sclerotium-like structures are known by terms such as resting mycelia, bulbils, microsclerotia, and pseudorhizae. Whetzel (1945) distinguished between a 'sclerotial stroma' and a 'substratal stroma', the former when the structure consisted of hyphal material only and the latter (sometimes referred to as a 'pseudosclerotium') when host tissue or portions of the substratum, which served as a food reserve, were enclosed in the medulla. In the present context the term 'sclerotium' is used for hyphal aggregates which are alike functionally in that they are able to survive periods of adverse conditions too severe for ordinary vegetative mycelia. When conditions become favourable, germination may be by means of hyphae (*myceliogenic*), by the production of asexual spores (*sporogenic*), or by the internal or external development of fruit bodies (*carpogenic*). Renewed growth may be by more than one of these types of germination.

Early mycologists soon realised that sclerotia constituted a resting stage in the life-cycles of fungi and that these structures had the potential to produce enormous amounts of inoculum. Many studies have been carried out on sclerotia during the past 150 years and a large literature has accumulated. Recently, some sclerotium-forming fungi, *e.g. Sclerotinia sclerotiorum* and related species, have attracted greater interest from plant pathologists owing to an increase in crop losses caused by these pathogens. This is probably associated with the use of new systemic fungicides that control many fungal pathogens but not some of those that form sclerotia. Considerable interest has been centred on the uses of metabolites of the sclerotia (ergots) of *Claviceps* spp. in modern medicine although the morphogenesis of the structure is not well understood. Possibly related products, useful to man, may be found in sclerotial tissues of other fungi. Apart from their economic importance, morphogenetic studies on sclerotia could contribute to an understanding of the development of fungal structures such as sporocarps and other vegetative structures, since most multi-hyphal bodies are organized in a similar manner. It has been suggested that sclerotia provide an example of convergent evolution and that they have originated from different fungal bodies, including aborted sexual fruiting bodies, asexual sporogenous tissue and aggregates of vegetative mycelia (Willetts, 1972).

Butler (1966) published a comprehensive account of the structure and development of fungal vegetative structures with a detailed section on sclerotia. Subsequent reviews on sclerotium morphogenesis have been published by Willetts (1972) and Chet and Henis (1975). Survival of sclerotia under adverse environmental conditions has been discussed by Coley-Smith and Cooke (1971), who also included details of germination, and Willetts (1971).

Three stages of sclerotium development can be distinguished: *initiation* (*i.e.* the aggregation of hyphae to form a small sclerotial primordium); *increase in size* (*i.e.* branching, growth and compaction of hyphae so that the sclerotium attains its maximum size); and *maturation* (*i.e.* final differentiation of the sclerotium involving the formation, in most instances, of a rind, changes in the composition and distribution of materials within the sclerotial hyphae and its isolation from the parent colony and surroundings). It is convenient to separate morphological aspects of sclerotial development into these stages but difficulties arise when attempts are made to confine physiological and biochemical processes, particularly those taking place after small hyphal aggregates have been formed, within arbitrary divisions.

In this chapter some morphological aspects of development will be considered first, followed by a discussion of initiation and then selected physiological and metabolic processes associated with sclerotial morphogenesis.

10.2 Morphological Aspects of Development

Types of sclerotium development

Some pioneer studies on sclerotium development were carried out towards the end of the last century but most of our detailed information has been obtained in the last two decades. Townsend and Willetts (1954) investigated the morphological features of sclerotium genesis and they distinguished three main types of development—*terminal, strand* and *loose*.

In the *terminal* type (Fig. 10.1), the sclerotium initial is formed by active branching at the tip of one hypha or the tips of several adjacent hyphae. Frequent septa are formed across the branches and often there is fusion of several branches. Further branching and growth give a compact knot of hyphae and several such initials may coalesce during the development of a mature sclerotium. Some examples of fungi whose sclerotia are formed in this way are *Botrytis cinerea, B. allii, Pyronema domesticum* and *Sclerotinia sclerotiorum.*

The *strand* type of sclerotium development is by the branching of hyphal strands. For example, an initial of *Sclerotium rolfsii* develops by multiple intercalary branching of the parallel hyphae of one strand or of several strands in the region of intersection. Frequent septation takes place and the

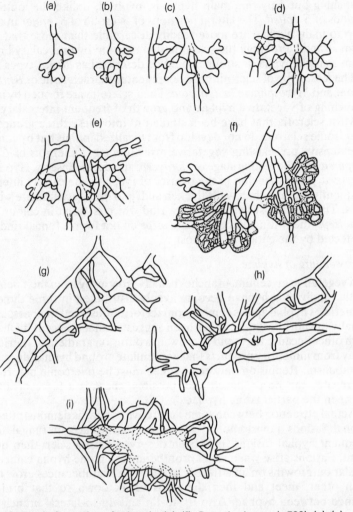

Fig. 10.1 Stages in formation of sclerotia. (a)–(f), *Botrytis cinerea* (×500). (a)–(e) drawn at intervals of 4 h to show terminal sclerotial initial formed by dichotomous branching and septation of the hyphae; (f) later stage of same initial showing coalescence of hyphae, and pigmentation of the walls. (g)–(i), *Sclerotinia gladioli* (×500). Stages in development of sclerotial initial, showing branching of a simple mycelial strand and septation and fusion of branches. (a)–(f), after Willetts, H. J. (1949). *Thesis*. University of Bristol. (g)–(i), after Townsend, B. B. and Willetts, H. J. (1954).

branches become interwoven to form a compact hyphal aggregate. Usually a rind differentiates around the interwoven hyphal mass and an outer pseudoparenchymatous cortex can be distinguished from a prosenchymatous medulla. The sclerotia are formed in positions that are lateral to main hyphae. Other fungi whose sclerotia are formed from strands are *Phymatotrichum omnivorum* and *Typhula* spp. The sclerotia of some species *e.g. Sclerotinia gladioli* (Fig. 10.1) and *S. minor* develop by intercalary branching but only one main hypha is involved, unlike the multi-hyphal strands of *S. rolfsii*. The lateral positions of sclerotia of *S. minor* in relationship to main hyphae are more readily discernible than those that develop from strands. Probably the sclerotia that develop by intercalary branching from a single hypha or strand should be described as *lateral* types.

The *loose* type of sclerotium development is characteristic of *Rhizoctonia solani* and *Mycosphaerella ligulicola*. Their sclerotia are formed by irregular branching of vegetative hyphae and growth of frequent intercalary septa.

Most sclerotia that have been studied fit into these three groups. However, some sclerotia do not develop from localized initials but by conversion of extensive pre-existing vegetative mycelium *e.g.* the ergots of *Claviceps purpurea* and the resting stage of *Cordyceps* in infected insects. Hyphae of *C. purpurea*, which develop in the ovaries of rye, become closely interwoven, frequently septate, their walls thicken and lipid drops accumulate within the cells. The outermost hyphae form a rind which is violet in colour. Unlike *Claviceps*, the resting bodies of *Cordyceps* do not form a fungal rind but are protected by the cuticle of the host.

Interweaving of hyphae

In a vegetative mycelium, branches usually grow away from their neighbours so that individual hyphae become evenly distributed in three dimensions. Sometimes hyphal apices come in contact fortuitously but their association is usually of short duration. It has been suggested that adjacent hyphae repel each other because they tend to grow up a diffusion gradient of nutrients and away from inhibitory products that accumulate around hyphae during active metabolism. Repulsion between hyphae must be overcome before hyphal aggregates can be produced and, possibly, there is a force of attraction between the participating hyphae.

Actual attraction between some vegetative hyphae is demonstrated by the hyphal fusions (anastomoses) that often take place in fungal mycelia. Frequent hyphal fusions have been observed during sclerotium development. Fusions arise when small protuberances on one hypha coincide with similar outgrowths on an adjacent hypha. The protuberances grow towards each other, meet and then the walls break down so that bridges are formed between hyphae. Also tip-to-tip and tip-to-lateral branch fusions take place. Anastomoses provide permanent unions between hyphae, increased mechanical strength to tissues and routes of transfer of nutrients and organelles.

A feature of many sclerotium-forming fungi is the production of abundant mucilage that accumulates as a layer on the outer surfaces of hyphal walls. Often large amounts of mucilage accumulate between sclerotial hyphae so that they are suspended in a mucilaginous matrix. Sometimes a mucilaginous layer is formed also on the surfaces of sclerotia. Several recent studies on the chemical composition of mucilage in stromata of the brown rot fungi (*Monilinia* spp.) indicate that mucilages produced by them are predominantly polysaccharide. Willetts (1972) suggested that mucilage could act as an adhesive that holds sclerotial hyphae loosely together. Enzymes may also be involved as in the adhesion of cells of *Dictyostelium discoideum* (Gerisch, Malchow, Wilhelms & Luderitz, 1969) but these proteins would be difficult to detect owing to the small amounts involved. Apart from loosely cementing hyphae together, mucilage could have a modifying effect on factors that affect the spatial dispersal of hyphae in vegetative mycelia, perhaps nullifying those that normally inhibit hyphal association and contact.

Structure of mature sclerotia

In general, the tissues of sclerotia are arranged in concentric zones but there is considerable variation in structural detail. Many sclerotia have an outer layer (rind) and medullary region that comprises the main part of the structure. The rind and medulla may be homogeneous or further concentric layers may be discernible.

A *rind* delimits the outer surfaces of most sclerotia although sometimes a continuous skin or crust of dried-up, aerial hyphae is formed outside the rind. Gaps in the crust reveal swollen tips of rind cells packed together to form a continuous layer. Septa are laid down close to the apices of peripheral hyphae and the terminal cells become swollen and globose while their walls become thickened, agglutinated and often pigmented. This type of rind develops after the sclerotium has attained its maximum size and does not become distorted by pressures exerted by the growth of medullary hyphae within *e.g. Botrytis cinerea* and *Sclerotinia sclerotiorum*. The rind of *Sclerotium rolfsii* develops before the sclerotium has become fully differentiated and the internal pressures which are produced subsequently tend to stretch the layer. Early rind development is more common among the lateral type sclerotia while late rind formation is normally a characteristic of sclerotia that form terminally. With only a few exceptions, a new layer of thickened, pigmented cells differentiates when the rind of a mature sclerotium is cut away. Pigmentation develops by the accumulation of melanins in and between rind cells. Usually there is a decrease in pigmentation from the surface, presumably because of the reduction of oxygen concentrations inwards and only infrequently are more than a few layers of cells pigmented. Polyphenoloxidases are active during melanization.

The *medullary region* of many sclerotia is divided into an outer cortex and a central medulla. The cortex usually consists of several rows of cells that form a pseudoparenchymatous tissue; frequently the hyphae are thin-walled and have dense contents. In the large central medulla, the filamentous nature of the hyphae can be clearly seen. A variety of different types of hypha have been observed in the prosenchymatous tissue of the medulla, including large, thick-walled, storage hyphae and smaller hyphae with numerous organelles. Often medullary hyphae are embedded in a mucilaginous matrix. The composition of walls of sclerotial tissue differs from that of vegetative hyphae.

10.3 Initiation of Sclerotia

Although there have been numerous studies on the physical factors affecting the initiation of sclerotia most of the results remain as a mass of uninterpreted empirical data that have not contributed significantly to a general understanding of the processes involved. In many instances, environmental factors most favourable for vegetative growth are those that have been reported as optimal for the production of sclerotia. However, very few workers in this field have attempted to study initiation and subsequent development separately, although the conditions required for these processes seem to be different. The available evidence suggests that initiation results from a perturbation of the environment that alters metabolism of the mycelium in the regions of the colony where sclerotia develop later. Post-initiation development depends on hyphal growth and therefore growth of young sclerotia would be expected to respond in a similar manner as vegetative mycelia when subjected to changes of the environment. Corner (1950) suggested that initiation is the only stage of sclerotium development by *Typhula* spp. that is co-ordinated, and subsequent growth takes place in an haphazard manner without the usual form determining factors which control the differentiation of other multi-hyphal fungal structures such as sporocarps. Corner concluded that the sclerotia of *Typhula* spp. are abortive fruit bodies that have become adapted to survive adverse environmental conditions, particularly low temperatures. However, in most species, there is some degree of co-ordination of growth of a sclerotium after a primordium has been produced.

General factors

When mycelia are cut or torn there is often stimulation of sclerotium production. Restricitions on linear or radial growth by mechanical or chemical barriers can also induce sclerotial primordia. Thus, when marginal hyphae grow against the sides of the culture dish, sclerotia are initiated and differentiate close to the edges of the colony or some distance in from the margin. Bedi (1958) found that sclerotia of *Sclerotinia sclerotiorum* did not develop when staling products were removed or prevented from forming in culture media. Presumably the staling substances accumulated in the medium around the colony and, when certain threshold concentrations were reached, formed a chemical barrier that arrested marginal growth but stimulated sclerotium production. The addition of staling products from other cultures of the fungus resulted in significant increases in number, size and rate of formation of sclerotia compared with control plates. Environmental factors such as light, temperature, humidity and aeration also influence sclerotium production (Chet & Henis, 1975) but these will not be discussed further in this chapter.

Geiger & Goujon (1970) postulated that a protein was responsible for sclerotium development by *Sclerotium rolfsii*. They suggested that the morphogenetic substance was produced in mature mycelium and transported to hyphal tips at the margin of the colony where terminal growth was inhibited. Subsequently, sclerotia developed in positions lateral to the main hyphae.

Unfortunately, none of these studies gives any indication of the actual substances involved in sclerotial morphogenesis and none can be used to explain initiation in biochemical terms. Three groups of substances (sulphur-containing compounds, polyphenols and polyphenoloxidases) have been investigated because they seem to influence sclerotium production. The work carried out on them provides some insight into the metabolic processes involved in sclerotium development but interpretation of the results is still highly speculative.

Sulphur-containing compounds

External application of sulphur-containing amino acids (methionine and cysteine) to cultures inhibits the formation of sclerotia by *Sclerotium rolfsii* whereas the addition of iodoacetate, metal chelates or potassium iodate, which are sulphydryl group (-SH) antagonists, induces larger sclerotia, arranged in concentric rings around the inoculum (Chet, Henis & Mitchell, 1966). It was suggested that the -SH groups of the cell walls or of certain enzymes with a role in morphogenesis were the sites where these substances acted. From later work with the disodium salt of ethylene diaminetetraacetic acid, Chet & Henis (1968) concluded that a copper-linked sulphydryl-containing entity repressed sclerotial development but that iodoacetic acid, chelating agents and potassium iodate modified the effect of the entity resulting in sclerotial production. Trevethick & Cooke (1971), who carried out similar experiments, postulated that the action of these chemicals was more indirect and that there was no specific action on a repressor of sclerotial morphogenesis. Support for this hypothesis was given by Wong & Willetts (1974) who suggested that the stimulation of sclerotial production was a direct result of blockage of a major biochemical pathway. According to Wong (1975), the -SH protein repressor of sclerotium morphogenesis referred to by Chet & Henis (1968) could be the -SH enzyme, glyceraldehyde-3-phosphate dehydrogenase (Gly-3-PDH).

In most instances, the results obtained from experiments with sulphur-containing compounds and -SH antagonistic agents have been considered in relationship to wall plasticity and branch formation. Obviously, active branching of hyphae is essential for sclerotium development and this can be achieved only if areas of the cell walls of parent hyphae are plastic. When added to the culture media substances, such as ethanol and acetate, which inhibit lateral branch development in mycelium of *Sclerotium rolfsii*, suppress sclerotium production, while threonine and lactose, which induce lateral branches, result in an increase in the number of sclerotia formed (Henis, Okon & Chet, 1973). However, enzymes containing -SH groups are involved in many metabolic processes in normal mycelial development, including energy providing pathways. Thus, it would be expected that a range of activities would be affected by -SH group modifiers such as iodoacetate or heavy metals. Cysteine is a precursor in the formation of coenzyme A, which affects the entry of carbohydrates into the Krebs cycle. Blockage of the -SH groups would therefore inhibit this metabolic cycle. If there is also inhibition of Gly-3-PDH then both the glycolytic and Krebs cycles will be blocked while the pentose phosphate pathway could become more active. Possibly, a shift in energy balance and/or reducing potential (NADPH or NADH) could be the trigger to sclerotium initiation. Alternatively, since these pathways supply precursors for other metabolic processes, changes in the availability of a secondary metabolite may be responsible for the aggregation of hyphae and subsequent sclerotium development.

Polyphenols and polyphenoloxidases

Polyphenoloxidases, both tyrosinase and laccase, have been found to be active at the time of sporocarp initiation. Wong & Willetts (1974) used polyacrylamide gel electrophoresis to study the activities of these enzymes at different stages of mycelium and sclerotium development by *Sclerotinia sclerotiorum*. Tyrosinase was not detected in aerial or submerged vegetative mycelium but the enzyme was moderately active in sclerotium initials, while during the rapid growth stage it became very active. Throughout the maturation stage the enzyme continued to be active. Laccase was detected in vegetative mycelium and at all stages of sclerotium development, but its greatest activity was detected in initials and young sclerotia. The failure to detect tyrosinase activity in young, vegetative mycelia agrees with the result obtained from studies on other fungi. Horowitz *et al.* (1970) postulated that activity is not detected in rapidly growing vegetative mycelium of *Neurospora* because the enzyme is repressed by a rapidly turning-over or an unstable protein. Inhibition of the repressor could result in greater activity of tyrosinase. Wong & Willetts (1974) found that the highest activities of polyphenoloxidases were during those stages when there was no pigmentation of hyphae, suggesting that the enzymes are involved in more than just melanization of the rind.

Recently, Marukawa, Funakawa & Satomura (1975) concluded that a secondary phenolic metabolite produced by *Sclerotinia sclerotiorum* and *S. libertiana* has a role in the induction of sclerotium initials and melanogenesis of sclerotia. The metabolite, which was named sclerin, has been isolated and was found to be a monohydric phenol of formula $C_{13}H_{14}O_4$ with three methyl groups on the aromatic ring and a methyl group linked to a secondary carbon atom (Kubota, Tokoroyama, Kamikawa & Satomura, 1966). Sclerin was produced at the mycelial stage, increased in concentration during sclerotium initiation and development, reached maximum concentration soon after sclerotia were fully grown and then its concentration decreased. When sclerin was added to cultures of *S. libertiana* that did not normally produce sclerotia, initials developed although they did not differentiate into mature structures. Possibly this suggests a sequence of events controlled by different factors. The activities of polyphenoloxidase and peroxidase were stimulated by the addition of sclerin to culture media while *p*-aminobenzoic acid, which inactivates polyphenoloxidases, inhibited the production of both sclerin and sclerotia.

Thus there is evidence that phenolics accumulate and polyphenoloxidases are active during the development of sclerotia and, possibly, these substances could have a morphogenetic role. Apart from accumulating in actively growing colonies, phenolics are produced when mycelia are mechanically damaged or become autolysed after contact with physical or chemical barriers. These are conditions that stimulate sclerotium production. When concentrations of phenolics become toxic, marginal growth of colonies is arrested. The way most mycelia develop suggests that hyphal apices have a restraining effect on growth of lateral branches although the mechanism of control is not known. When growth of the hyphal apex is inhibited, this restraint seems to be removed and more active branching takes place in sub-apical and/or lateral positions. Small hyphal aggregates formed in this way could become sclerotium initials, and develop into mature sclerotia under favourable conditions. A terminal initial develops from active branching near a hyphal apex. In many instances growth of the whole margin of the colony is arrested and sclerotia are produced later in a ring corresponding approximately to the position occupied by the edge of the colony at the time of growth inhibition. If there is renewed radial growth after primordia have been initiated, and linear extension is arrested again later, owing to a build-up of toxic materials in the substratum, a second ring of sclerotia may be produced. This could be the explanation for the production of sclerotia in one or more concentric rings by some fungi, *e.g. Sclerotinia sclerotiorum*. Most sclerotia are produced in aerial positions and this, also, can be accounted for on an inhibition hypothesis. The hyphae in and on the substratum are in more direct contact with toxins that accumulate in the medium while aerial hyphae are less affected and more able to branch and form sclerotium initials. However, sclerotia of some fungi, *e.g. S. minor*, are arranged irregularly amongst the vegetative mycelium. Sclerotia of *S. minor* are formed from short, aerial hyphae, arising in positions lateral to main hyphae, while those of *S. sclerotiorium*, which form in concentric circles, develop terminally on primary branches (Willetts & Wong, 1971).

The possible mechanism of action of polyphenols and polyphenoloxidases in fungal morphogenesis, including sclerotium initiation, is a matter of speculation. Wilson (1968), who obtained a correlation between tyrosinase activity and the formation of large numbers of perithecial primordia by *Hypomyces* spp., suggested several ways that tyrosinase could affect morphogenesis of fungi. These included: changes in the permeability and organization of cellular membranes, as found by Cory (1967) with blood cells; inhibition of synthesis of certain proteins; an effect on DNA by oxidation of nucleohistone tyrosine; and inactivation of enzymes essential for normal vegetative growth by oxidation of their tyrosyl residues. Solti & Telegdi (1972) found that tyrosinase inactivated Gly-3-PDH and they suggested that the site of action was the tyrosyl residues of proteins. Inactivation of Gly-3-PDH could suppress glycolysis and possibly cause a stimulation of the pentose phosphate pathway.

Thus, phenolics may inhibit enzymes involved in mycelial development, *e.g.* wall synthesis, and stop hyphal growth at the margin of the colony. However, aerial hyphae will be less affected by the accumulation of toxins in the medium than hyphae submerged in or growing on the surface of the substrate. With growth inhibition of the latter, additional nutrients could become available to aerial hyphae. Usually linear extension of hyphae into the air is restricted owing to desiccation and if active sub-apical branching of the aerial hyphae takes place, sclerotial initials can be produced.

10.4 Physiology and Metabolism

After the aggregation of hyphae to form initials, growth of sclerotia to their maximum sizes takes place relatively quickly. Often sclerotia are fully mature within 7–10 days of initiation. In recent years, some of the physiological and metabolic changes that take place during rapid increase in size have been studied by several workers, most of whom have used *Sclerotinia sclerotiorum* and to a lesser extent *S. trifoliorum* and *Sclerotium rolfsii* for their investigations. Some of the important physiological and metabolic processes associated with growth of sclerotia are discussed below, mainly with reference to *S. sclerotiorum*. Probably many of the findings and conclusions obtained for sclerotium formation by this species can be applied, in general terms, to morphogenesis of sclerotia by other fungi.

Sources and utilization of carbohydrates

Nutritional studies on the growth of sclerotia have shown that there is a rapid movement of nutrients into young primordia. Cooke (1971) found that 88% of the dry weight of fully developed sclerotia of *Sclerotinia sclerotiorum* was attained within 3 days of sclerotium initiation, indicating the relatively short duration of the active growth phase. Most of the materials which move into sclerotia are in the form of carbon compounds. The nutrients must come originally from the surrounding substratum but considerable amounts are probably obtained at the time of sclerotium enlargement and compaction from reserve and structural materials that have accumulated in the mycelium during early vegetative growth of the colony.

Mycelia and sclerotia grow well on a large number of carbon sources. Wang & LeTourneau (1971) found that sclerotia were readily produced by *Sclerotinia sclerotiorum* on D-glucose, D-mannose, D-fructose, L-arabinose sucrose, raffinose, maltose and lactose, while trehalose, L-sorbose, D-arabinose and several sugar alcohols (polyols), including D-mannitol and L-arabitol, were unsatisfactory sources of carbon. However when glucose was combined with trehalose or polyols there was an increase in mycelia growth and sclerotium production. Organic acids were not good sources of carbon for development of either mycelia or sclerotia. Presumably, there are differences in carbon requirements between different fungi and even between isolates of the same species since the hosts on which they are found will affect their utilization of carbon.

Cooke (1971) fed (^{14}C)-glucose to mycelium of *Sclerotinia sclerotiorum* in an investigation of carbon physiology during sclerotium development by this fungus. He found that in the early stages of sclerotium growth there was active respiration and exudation of carbohydrates into liquid droplets that formed on the surface of young sclerotia. After 36 h of development there was a decline in ($^{14}CO_2$) evolution and exudation ceased after about 3–4 days. Apart from the need for glucose as a substrate for respiration there must be a diversion of carbohydrates into walls from the beginning of hyphal proliferation and branching, since new walls are formed whenever growth occurs. Glucose will also be converted to other hexoses and storage products and the extent of these conversions must depend upon the rate of glucose uptake in relation to its utilization in respiration and wall formation. Presumably, later in development, an increasing proportion of the glucose taken up is used in the formation of ethanol-insoluble storage materials and for thickening of hyphal walls. The amounts of polyols that accumulate inside the sclerotium and in the droplets on its surface will be dependent on the availability of carbohydrates after the requirements of essential growth processes have been met. If uptake of glucose exceeds its utilization in other metabolic pathways then soluble carbohydrates will accumulate within the sclerotial tissues and, since these are in the soluble form, some will be lost from the hyphae, either by active exudation or passive leakage. An explanation for reduced exudation after about 4 days could be that pathways concerned with the synthesis of insoluble storage compounds become active at this stage resulting in increased use of and competition for nutrients and consequently glucose is used as soon as it enters the mycelium. Thus, the amounts of soluble sugars are reduced and there is a decline in the concentration of carbohydrates in liquid droplets which may continue, for a time however, to increase in size because of the exudation of water.

There have been several studies on the composition of sclerotia of *Sclerotinia sclerotiorum* particularly with reference to ethanol-soluble carbohydrates. Trehalose, mannitol and arabitol have been found in quantities of 4.5, 4.8 and about 2% respectively of oven-dry weight of mature sclerotia. Also small quantities of glucose, mannose and fructose have been reported. Glycogen appears to be the major insoluble carbohydrate reserve. However, the detection of soluble carbohydrates in sclerotial tissues may not be an indication of their importance in metabolism. Thus, some carbohydrates may not accumulate in hyphae because they are rapidly utilized in essential growth and morphogenetic processes. Nevertheless, the reserve materials that build up in mature tissues of resting structures are sources of energy and nutrients for subsequent growth. It has been shown that trehalose disappears when sclerotia of *S. sclerotiorum* and *Phymatotrichum omnivorum* germinate and it was concluded that the sugar is used up in the production of fruit bodies or mycelia, respectively.

Lewis & Smith (1967) discussed in detail the occurrence and roles of sugar alcohols, particularly mannitol, in fungi. Mannitol is a frequent storage compound in fungi and is commonly found with trehalose and glycogen. Both mannitol and trehalose can be elaborated into insoluble polymers and the two are interconvertible. Mannitol has several advantages over sugars as a storage product. Its molecule is more reduced than that of hexoses and stores correspondingly more reducing power. As Lewis & Smith (1967) pointed out, this is of importance in fungi compared with green plants since, in the latter, reduced coenzymes are provided during the photosynthetic process. They suggested that the frequent occurrence of the pentose phosphate pathway in fungi may represent a substitute mechanism for the provision of reducing power, with polyols acting as compounds for its storage. Polyols are synthesised by the reduction of polysaccharides by polyol dehydrogenases. Consequently, the rapid synthesis of mannitol or other polyols in sclerotial initials may promote the flow of carbohydrates to them.

Sources and utilization of nitrogen

Information on the utilization of nitrogen by sclerotium-forming fungi is limited compared with that on carbon sources. Several studies have been made on the most suitable sources of nitrogen for sclerotium formation. It has been found that sclerotium formation by *Botrytis cinerea* and *Verticillium albo-atrum* is dependent on the carbon: nitrogen ratio of the growth medium and this probably applies to the sclerotia of other fungi.

Most sclerotium-forming fungi utilize inorganic sources of nitrogen and no specific differential effects have been reported when ammonium and nitrate salts were used to supply nitrogen. Most of the available evidence indicates that organic nitrogen sources are more suitable for sclerotium growth than inorganic forms. Wang & LeTourneau (1972) used many different amino acids as nitrogen sources in nutritional studies on sclerotium growth by *Sclerotinia sclerotiorum*. They found that compounds closely related to the Krebs cycle (*e.g.* aspartic and glutamic acids, alanine and serine) were the most suitable for this purpose. They suggested that the production of organic acids of or related to the Krebs cycle could be associated with sclerotium formation. Both aspartic and glutamic acids are central to amino acid metabolism. Transamination of α-ketoacids with glutamic acid as the amino group donor represents the major pathway (in animals) for the introduction of α-amino groups in the biosynthesis of most amino acids.

Chet & Henis (1975) referred to abundant sclerotium formation by *Sclerotium rolfsii* when grown on L-threonine and they attributed this to the role of the latter in the glyoxylic acid cycle. The inhibitory effect of sulphur-containing amino acids, especially cysteine, has already been discussed in the section on sclerotium initiation.

Translocation

A primordium will differentiate into a sclerotium only if it is able to obtain supplies of water and nutrients during the active growth phase. Many initials remain as small, loose, fluffy, hyphal aggregates owing to their inability to compete for available nutrients with other growth centres in the colony and, probably, also because of inhibition by more mature sclerotia in the regions of hyphal concrescence.

Cooke (1970) found that the bulk of nutrients had passed into sclerotial primordia of *Sclerotinia sclerotiorum* within about 48 h of their initiation and movement of materials had stopped after about 72 h. An efficient method of transport of soluble carbohydrates, other nutrients and water is needed for rapid growth of sclerotium initials. Wilcoxson & Sudia (1968) discussed the movement of nutrients into sclerotia in a general review of translocation in fungi. Nutrients usually pass from regions of high concentration along a decreasing concentration gradient to metabolically active sites. Although cytoplasmic streaming has been shown to assist transport of nutrients through mycelia, movement can take place without streaming. Probably expenditure of energy is needed to bring about translocation in many instances. Transport to sclerotia is through a few translocatory hyphae that are functional from early stages of development. Studies with (^{14}C)-glucose have shown that a physiological connection between sclerotia and vegetative mycelia by translocatory hyphae is maintained throughout the active growth phase and until the sclerotia begin to mature. The high metabolic activities of initials during development use up the translocated nutrients and thus a concentration gradient is maintained along which carbohydrates and other materials move. Translocation is aided also by exudation of sugars into liquid droplets. The accumulation of exudation droplets on the surfaces of sclerotia is discussed separately below.

Translocation may be terminated by exhaustion of nutrients, a reduction in the metabolic activities of the sclerotium so that the concentration gradient is lost, or severence of translocatory hyphae. It has been found that damage to translocatory hyphae results in loss of ability of young sclerotia to obtain nutrients, even when bathed in a nutrient-rich medium.

Exudation

The exudation of liquid droplets on the surfaces of fungal structures is frequently encountered, especially when there is an abundance of sugars in the substratum. Exudation droplets form on the surfaces of sclerotia both in the laboratory and in the field. Remsberg (1940) obtained a crystalline residue when droplets taken from the surfaces of sclerotia of some *Typhula* species were dried on a glass slide. There have been several recent studies on xudation from sclerotia of *Sclerotinia sclerotiorum* to determine the physiological and biochemical properties of the exudates (Cooke, 1969, 971; Colotelo, Sumner & Voegelin, 1971; Colotelo, 1973).

Very small liquid droplets, often visible only under magnification, develop n young sclerotial primordia of *Sclerotinia sclerotiorum*. The small droplets oalesce to form larger ones which can be seen with the naked eye. A film or kin encloses the liquid exudate. The droplets often become pale yellow in olour probably owing to the accumulation of phenolic substances. When clerotia are fully grown and a rind has differentiated, the droplets slowly lecrease in size and there is a loss in their dry matter. Eventually all the iquid is lost from within the skin, which remains on the surface of the clerotium as a small, wrinkled, deflated sac. Colotelo (1973) concluded that he decrease in size is caused by evaporation of water from the droplet while he constituents are reabsorbed into the sclerotium and further metabolized.

The exudates are complex in composition and contain inorganic ions and organic compounds, some of which have enzymatic properties. Cooke 1969) found that the exudates on sclerotia of *Sclerotinia sclerotiorum* contain trehalose, mannitol, and inositol and traces of glucose. Mannitol was absent from exudates of *S. trifoliorum* despite the presence of this polyol within the sclerotia. Other constituents reported in exudation droplets include potassium, magnesium and sodium cations, free amino acids, free ammonia, fatty acids and a variety of enzymes such as catalase, peroxidase, polyphenoloxidase, and β-glucosidase. Trehalase activity has not been letected in the exudates but pectolytic and cellulolytic properties have been lemonstrated.

It seems that movement of carbohydrates from sclerotia into droplets is not a simple, indiscriminate leakage process but that some selective mechanism operates during exudation. Probably the phenomenon keeps internal osmotic balance within tissues of young sclerotia and maintains a concentration gradient so that nutrients will continue to move into the structure. Also the conversion of carbohydrates into insoluble compounds contributes to physiological balance. Cooke (1971) considered that the role of exudation is partly to bring about rapid dehydration of developing sclerotia. Colotelo (1973) suggested that liquid droplets form on young primordia at the time of their initiation and subsequently; as the amount of free space within the sclerotia is reduced by active growth and multiplication of medullary hyphae, the tiny droplets coalesce and are forced to the surface, where they form large droplets. Probably some of the constituents in the droplets come from peripheral hyphae which rupture owing to a build up of internal pressures or after autolysis.

Formation of exudation droplets could result from either an increase in permeability of membranes of sclerotial hyphae or the operation of specific membrane carrier systems such as those which operate in sugar transport in yeast cells. In any case, accumulation of materials, particularly ions, in the exudation drops would be expected to alter the permeability of hyphal membranes in localized areas. Any factor that affects membrane permeability could influence the loss of soluble carbohydrates and thus translocation to primordia. The work carried out by Allaway & Jennings (1970) is of interest on this context. They found that the balance of cations in the medium affected the ability of the fungus *Dendryphiella salina* to retain mannitol, arabitol and some unidentified compounds inside the mycelium. Increase in sodium ion concentration of the external solution in the absence of calcium increased permeability. Colotelo (1973) studied changes in concentrations of cations in exudation droplets on sclerotia of *Sclerotinia sclerotiorum* and his observations indicated that sodium, potassium and magnesium ions were exuded up to 6 days after initiation. The exudation of cations during the rapid growth phase would contribute to a reduction in osmotic potential of sclerotial cells. This is desirable for the maintenance of flow of nutrients to the structure. The concentrations of potassium ions in the exudation droplets decreased very quickly after maximum sclerotium size had been attained and Colotelo suggested that potassium ions were reabsorbed. In mature sclerotia high osmotic potentials are desirable so that excessive dehydration or mechanical damage to the protoplasts is minimized under conditions of desiccation, and high and low temperatures. A detailed study of the effects of selected cations on the permeability of membranes of sclerotial hyphae, carbohydrate uptake by initials and resistance of sclerotia to adverse environmental conditions could prove rewarding. Also the fate of the carbohydrates in the droplets is of interest.

Enzymology

Corsini & LeTourneau (1973) found that a typical Krebs cycle existed in *Sclerotinia sclerotiorum* during mycelial development. After 5 days of growth the specific activities of malate dehydrogenase and citrate synthase were more than double their values after 10 and 15 days incubation. Increased activities of these selected enzymes at 5 days were associated with the high rate of cellular synthesis at that period. Organic acids accumulate in the culture medium, an observation made previously by several workers and the pH of the medium was reduced owing to the accumulation in it of succinate, malate, fumate and oxalate. The high production of organic acids is probably related to the need for essential synthetic intermediates and catalysts during energy metabolism. It has been postulated that changes in pH of culture media could be one of the morphogenetic factors leading to sclerotium initiation.

Chet, Retig & Henis (1972) studied patterns of soluble proteins and enzymes in vegetative mycelium and at three stages of sclerotium formation in the fungus *Sclerotium rolfsii*. They used gel isoelectric focusing for their investigation and found that there were distinct enzyme patterns for each stage of development. However, the enzymes were randomly selected, and this limited the usefulness of the results in the interpretation of metabolic events associated with sclerotium differentiation.

Wong & Willetts (1974) studied nine key enzymes of major metabolic pathways during the formation of sclerotia by *Sclerotinia sclerotiorum*. They made protein extracts from aerial and submerged mycelia and from five different stages of sclerotium development; enzyme activities were determined by means of polyacrylamide gel electrophoresis. Succinate dehydrogenase (SDH) and glucose-6-phosphate dehydrogenase (Glu-6-PDH) were moderately active in the submerged mycelium while in the non-sclerotial aerial mycelium arylesterase and acid phosphatase were very active. Probably SDH and Glu-6-PDH are involved in the absorption of materials from the liquid media by submerged mycelium, thereby providing the nutrients required for active mycelial growth. After vegetative colonies had developed, sclerotium initials formed on the surfaces of the cultures. Presumably, much of the energy needed for sclerotium differentiation is obtained from structural and reserve materials in the aerial mycelium by the degrading activities of enzymes such as esterases and acid phosphatase. Almost all the enzyme systems included in the study were active in sites of sclerotium initiation. High levels of aerobic respiration were suggested by the increased activities of glycolytic (Gly-3-PDH) and Krebs cycle respiratory (SDH) enzymes. Moderate activities of Glu-6-PDH and phosphogluconate dehydrogenase (PGDH), which are associated with biosynthetic processes, were detected. Tyrosinase and laccase were moderately and highly active, respectively. The pentose phosphate shunt enzymes (Glu-6-PDH and PGDH) were most active in young compacting sclerotia and their increased activities coincided with a reduction in the activities of the enzymes of the glycolytic and Krebs cycle pathways. Probably oxygen levels are reduced in sclerotial tissues during the compacting process and lack of oxygen affects the activities of the enzymes of the Krebs cycle while the pentose phosphate pathway, although an aerobic route, requires less oxygen and is affected to a lesser extent. Metabolism of hexoses by means of the pentose phosphate pathway would generate the NADH essential for the biosynthetic reactions associated with development.

Suppression of the glycolytic and Krebs cycle pathways and the stimulation of the pentose phosphate pathways seem important during the compaction stage of sclerotium differentiation. Support for this has been given by Marukawa *et al.* (1975) who suggested that the role of sclerin in sclerotium formation is associated with suppression of the glycolytic pathway and stimulation of the pentose phosphate pathway.

Once sclerotia have reached their maximum size they become fully mature within a short time, often only a few days. During this period there is a decrease in hydration of the tissues and an increase in the deposition of structural and storage materials. When the rind is fully differentiated, the cortical and medullary hyphae become isolated from the parent mycelia and the external environment forming a persistent resting stage. Sclerotia have many adaptations that enable them to survive adverse physical and nutritional conditions and to resist biological degradation in a balanced and efficient manner for long periods. Presumably, some of the large amounts of endogenous reserves are used up during the resting period, since it seems unlikely that metabolic activities will be completely suspended, but there will be an abundance of available nutrient for the production of sporocarps, asexual spores and/or mycelia when conditions become favourable for germination. Thus sclerotia serve an important role in the survival and propagation of many fungi.

10.5 References

ALLAWAY, A. E., & JENNINGS, D. H. (1970). The influence of cations on glucose transport and metabolism by, and the loss of sugar alcohols from, the fungus *Dendryphiella salina*. *New Phytologist* **69**, 581–93.

BEDI, K. S. (1958). The role of stale products in the formation of sclerotia of *Sclerotinia sclerotiorum* (Lib.) de Bary. *Indian Phytopathology* **11**, 29–34.

BUTLER, G. M. (1966). Vegetative structures. In *The fungi*, Vol. 1, pp. 83–112.

Edited by G. C. Ainsworth and A. S. Sussman. New York and London: Academic Press.

CHET, I. & HENIS, Y. (1968). The control mechanism of sclerotial formation in *Sclerotium rolfsii*. Sacc. *Journal of General Microbiology* **54**, 231–6.

CHET, I. & HENIS, Y. (1975). Sclerotial morphogenesis in fungi. *Annual Review of Phytopathology* **13**, 169–92.

CHET, I., HENIS, Y. & MITCHELL, R. (1966). The morphogenetic effect of sulphur-containing amino acids, glutathione and iodoacetic acid on *Sclerotium rolfsii* Sacc. *Journal of General Microbiology* **45**, 541–6.

CHET, I., RETIG, N. & HENIS, Y. (1972). Changes in total soluble protein and in some enzymes during morphogenesis of *Sclerotium rolfsii* as detected by the gel isoelectric focusing technique. *Journal of General Microbiology* **72**, 451–6.

COLEY-SMITH, J. R. & COOKE, R. C. (1971). Survival and germination of fungal sclerotia. *Annual Review of Phytopathology* **9**, 65–92.

COLOTELO, N. (1973). Physiological and biochemical properties of the exudate associated with developing sclerotia of *Sclerotinia sclerotiorum* (Lib.) De Bary. *Canadian Journal of Microbiology* **19**, 73–9.

COLOTELO, N., SUMNER, J. L. & VOEGELIN, W. S. (1971). Chemical studies on the exudate and developing sclerotia of *Sclerotinia sclerotiorum* (Lib.) De Bary. *Canadian Journal of Microbiology* **17**, 1189–94.

COOKE, R. C. (1969). Changes in soluble carbohydrates during sclerotium formation by *Sclerotinia sclerotiorum* and *S. trifoliorum*. *Transactions of the British Mycological Society* **53**, 77–86.

COOKE, R. C. (1970). Physiological aspects of sclerotium growth in *Sclerotinia sclerotiorum*. *Transactions of the British Mycological Society* **54**, 361–5.

COOKE, R. C. (1971). Physiology of sclerotia of *Sclerotinia sclerotiorum* during growth and maturation. *Transactions of the British Mycological Society* **56**, 51–9.

CORNER, E. J. H. (1950). *A monograph of Clavaria and allied genera*. London and New York: Oxford University Press.

CORSINI, D. L. & LETOURNEAU, D. (1973). Organic acid metabolism in *Sclerotinia sclerotiorum*. *Archiv für Mikrobiologie* **90**, 59–64.

CORY, J. (1967). Evidence for a role of tyrosyl residues in cell membrane permeability. *Journal of Biological Chemistry* **242**, 218–21.

GEIGER, J. P. & GOUJON, M. (1970). Mise en évidence dans les extracts de thalle d'un facteur morphogène responsable de l'apparition des sclérotes du *Corticium rolfsii* (Sacc.) Curzi. *Comptes rendus hebdomadaires des seances, Academie des Sciences, Paris*, (Serie D) **271**, 41–4.

GERISCH, G., MALCHOW, D., WILHELMS, H. & LUDERRITZ, I. (1969). The new species-specificity of cell membranes containing polysaccharide-antigens of *Dictyostelium discoideum*. *European Journal of Biochemistry* **9**, 229–36.

HENIS, Y., OKON, Y. & CHET, I. (1973). The relationship between early hyphal branching and sclerotial formation in *Sclerotium rolfsii*. *Journal of General Microbiology* **79**, 147–50.

HOROWITZ, N. H., FLING, M., FELDMAN, H. M., PALL, M. L., & FROEHNER, S. C. (1970). Derepression of tyrosinase synthesis in *Neurospora* by amino acid analogs. *Developmental Biology* **21**, 147–56.

KUBOTA, T., TOKOROYAMA, T., KAMIKAWA, T. AND SATOMURA, Y. (1966). The structures of sclerin and sclerolide, metabolites of *Sclerotinia libertiana*. *Tetrahedron Letters* **42**, 5205–10.

LEWIS, D. H. & SMITH, D. C. (1967). Sugar alcohols (polyols) in fungi and green plants. I. Distribution, physiology and metabolism. *New Phytologist* **66**, 143–84.

MARUKAWA, S., FUNAKAWA, S. & SATOMURA, Y. (1975). Role of sclerin on morphogenesis in *Sclerotinia sclerotiorum* de Bary (including *S. libertiana* Fuckel). *Agricultural and Biological Chemistry Bulletin* **39**, 645–50.

REMSBERG, R. E. (1940). Studies in the genus *Typhula*. *Mycologia* **32**, 52–96.

SOLTI, M. & TELEGDI, M. (1972). The effect of tyrosinase on glyceraldehyde-3-phosphate dehydrogenase and glycerophosphate dehydrogenase. *Acta Biochemica et Biophysica Academiae Scientiarum Hungaricae* **3**, 227–32.

TOWNSEND, B. B. & WILLETTS, H. J. (1954). The development of sclerotia of certain fungi. *Transactions of the British Mycological Society* **37**, 213–21.

TREVETHICK, J. & COOKE, R. C. (1971). Effects of some metabolic inhibitors and sulphur-containing amino acids on sclerotium formation in *Sclerotium rolfsii*, *S. delphinii* and *Sclerotinia sclerotiorum*. *Transactions of the British Mycological Society* **57**, 340–2.

WANG, S. Y. & LETOURNEAU, D. (1971). Carbon sources, growth, sclerotium formation and carbohydrate composition of *Sclerotinia sclerotiorum*. *Archiv für Mikrobiologie* **80**, 219–33.

WANG, S. Y. & LETOURNEAU, D. (1972). Amino acids as nitrogen sources for growth and sclerotium formation in *Sclerotinia sclerotiorum*. *Transactions of the British Mycological Society* **59**, 509–12.

WHETZEL, H. H. (1945). A synopsis of the genera and species of the Sclerotiniaceae, a family of stromatic inoperculate Discomycetes. *Mycologia* **37**, 648–714.

WILCOXSON, R. D. & SUDIA, T. W. (1968). Translocation in fungi. *Botanical Reviews* **34**, 32–52.

WILLETTS, H. J. (1949). *Hons. Thesis. University of Bristol.*

WILLETTS, H. J. (1971). The survival of fungal sclerotia under adverse environmental conditions. *Biological Reviews* **46**, 387–407.

WILLETTS, H. J. (1972). The morphogenesis and possible evolutionary origins of fungal sclerotia. *Biological Reviews* **47**, 515–36.

WILLETTS, H. J. & WONG, A. L. (1971). Ontogenetic diversity of sclerotia of *Sclerotinia sclerotiorum* and related species. *Transactions of the British Mycological Society* **57**, 515–24.

WILSON, J. K. (1968). Physiology of sexual reproduction in *Hypomyces solani* f. sp. *cucurbitae*. V. Influence of tyrosine on perithecial primordium formation. *Phytopathology* **58**, 1697–9.

WONG, A. L. (1975). *Sclerotial morphogenesis and taxonomy of Sclerotinia sclerotiorum and related species.* Unpublished Ph.D. thesis, University of New South Wales, Australia.

WONG, A. L. & WILLETTS, H. J. (1974). Polyacrylamide-gel electrophoresis of enzymes during morphogenesis of sclerotia of *Sclerotinia sclerotiorum*. *Journal of General Microbiology* **81**, 101–9.

CHAPTER 11

Asexual Sporulation in Filamentous Fungi

J. E. SMITH

11.1 Introduction

Within the developmental cycle of filamentous fungi, the asexual spore can claim a unique duality of function being integrally involved both at the beginning and at the end of the cycle.

At genesis spores are almost always single, spherical or ellipsoidal in shape, hyaline and smooth-walled. Although many will remain in this form others will develop septa, surface appendages, complex shapes and often become pigmented (Hawker & Madelin, 1976). In contrast to the vegetative mycelium the spore is normally delimited from the thallus and is characterized by a minimal metabolic turnover, low water content, and lack of cytoplasmic movement (Gregory, 1966). Coming at the end of the growth cycle when the mycelium is normally experiencing conditions of stress (Smith & Galbraith, 1971), the specialized spore represents a means of propagation, survival and dispersal. Spores have evolved in many ways to combat deleterious environmental pressures, *e.g.* extremes of temperature, drying and wetting, radiation, and other exogenous physico-chemical influences and of course animal and microbial digestion. Furthermore, they have also adapted to gain maximum assistance from air and water currents and other means of achieving dispersal or transfer to new areas of exploitation.

The initiation of the vegetative cycle is a universal function of the spore. Under balanced environmental conditions, the spore will undergo germination and subsequent development of the mycelial state. However, most spores exhibit some variable degree of dormancy by which the germination process is delayed until conditions are appropriate for successul rather than abortive germination and growth. Dormancy has been considered as a reversible interruption of the phenotypic development of the fungus (Sussman, 1966) and must therefore represent a means of regulating spore dispersal *in time* (Sussman, 1976). The alleviation of dormancy and subsequent initiation of germination has been the subject of intense study and the

:oncepts of germination activators have been lucidly expressed by Sussman 1976). Macko, Staples, Yaniv & Granados (1976) have further examined he role of endogenous molecules that regulate spore germination.

Fungal spores differ widely in their ontogeny and biological functions. Within the filamentous fungi it is possible to find motile zoospores, sessile :porangiospores, oidia, conidia, spermatia, aecidiospores, urediospores, eleutospores and chlamydospores (for refs. see Weber & Hess, 1976).

Sporulation in fungi must be considered as an integration of several biological aspects. Thus the ability of a fungus to change from a vegetative orm to a reproductive form is the end result of genetic competence esponding to specific physico-chemical environmental changes which in urn creates the internal stimulus to regulate the cellular activity, *viz.* primary and secondary metabolic pathways (Turian, 1974). The net result of :his interplay is the establishment of a new developmental pattern with its unique metabolism and morphology.

This chapter will be concerned with the ontogeny of fungal spores and the environmental and biochemical regulation of their formation. Several reviews on sporulation in filamentous fungi have appeared in recent years and these should be consulted for a deeper insight into this complex area of differentiation (Smith & Galbraith, 1971; Turian & Bianchi, 1972; Smith & Anderson, 1973; Smith & Berry, 1974; Turian, 1974, 1975, 1976; Lovett, 1975).

11.2 Spore Ontogeny

Filamentous fungi have evolved a bewildering array of spore types and mechanisms for spore formation. Such morphological mechanisms should not be considered as being separate and distinct but rather as parts of a continuously, intergrading and overlapping spectrum of morphological developments to achieve the formation and release of propagating units (Carmichael, 1971). Spore ontogeny has long been studied using ordinary light and phase contrast microscopy but more recently several distinct but complementary new approaches have been emerging using scanning and transmission electron microscopy, time-lapse cinemicroscopy, fluorescent antibody staining and biochemical analyses (for refs. see Kendrick 1971*a*; Talbot, 1971; Weber & Hess, 1976).

The formation of asexual spores by filamentous fungi can be broadly subdivided into four main categories, *viz.* fragmentation, fission, extrusion and cleavage (Fig. 11.1; Carmichael, 1971). In *fragmentation* the cytoplasm gradually becomes concentrated into particular cells within the mycelium while the remaining cells become empty of cytoplasmic material. Formation by *fission* occurs when the cells within the mycelium break apart at double-walled septa. *Extrusion* formation occurs when the spores are produced as extrusions from the ends or the sides of the sporogenous cells. Finally, there is *cleavage* in which the protoplasmic contents of a cell splits or divides into fragments, each of which becomes surrounded by a wall. Spore formation by fragmentation or fission are examples of thallic sporogenesis. In holothallic spore ontogeny, all wall layers of the sporogenous cell are involved in the formation of the spore wall while in enterothallic spore formation the outer wall of the sporogenous cell does not become part of the outer wall layer of the mature spore (Kendrick, 1971*b*). Extrusion or blastic development is

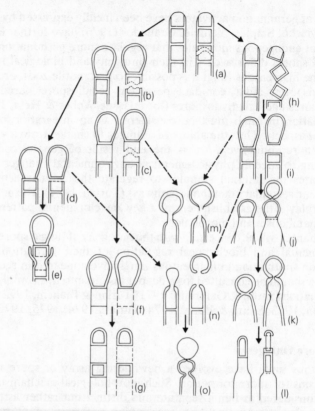

Fig. 11.1 Conidia classified by dehiscence. (a) An undifferentiated hypha releasing a propagule by fracture; (b)–(g) conidia released by sacrifice of a supporting cell; (h)–(l) conidia released by fission of a double septum; (m)–(o), conidia released by fracture of a fine connective (from Carmichael, 1971).

essentially the *de novo* growth of a spore initial from the fertile locus of a sporogenous cell (Cole, 1975*a*). Spore formation by cleavage is an example of endospore formation.

Three distinct methods of releasing the conidial spore type from the conidiogenic cells of the mycelium can be distinguished. In certain fungi, spore release occurs by having a fragile intercalary cell which becomes sacrificed. This may be by a single thin-walled cell (Fig. 11.1c), by a cell with a thin abscission ring (Fig. 11.1d) or by a cell whose wall is endogenously digested enzymatically (Fig. 11.1f), g). The second method utilizes a double septum at the base of the spore, separation being achieved by splitting of the double septum (Fig. 11.1h, l). In the third method the spore is produced on such a fine connection that no further separation mechanism is required (Fig. 11.1m, n) (Carmichael, 1971; Kendrick, 1971*a*).

Asexual spores in Basidiomycotina, Ascomycotina and Deuteromycotina

Within these divisions there are many different types of spores and mechanisms for their generation. In particular, the Deuteromycotina or

Fungi Imperfecti demonstrate an immense array of spore forming mechanisms which may be solitary (Moniliales), variously aggregated (Melanconials), or enclosed (Sphaeropsidales). Furthermore, the hyaline or coloured conidia display an astonishing variety of size, shape and configuration, with or without septation (Talbot, 1971).

The characteristic asexual spore in these groups is the conidium. Deriving from the Kananaskis Conference (Kendrick, 1971a) the conidium has been defined as 'a specialized, nonmotile, asexual propagule, usually caducous, not developing by cytoplasmic cleavage or free-cell formation'. Conidia arise from conidiophores which may be a simple cell or a system of conidiogenic cells with or without a differentiated supporting structure. Conidiophores function to position the developing conidium away from the medium and parent mycelium. Only rarely does a single cell function as a conidiophore (see p. 222) but the conidiophore can be a single hypha, branched or unbranched, or it may be a complex of hyphae (Talbot, 1971; Kendrick 1971a).

Pirozynski (1971) has attempted to classify the diverse types of conidiophores into groups based on characteristics of conidiogenic cells and supporting structures (Fig. 11.2).

Some of the main types of conidia will be briefly examined although for a detailed and authoritative study reference should be made to Hughes (1953, 1970, 1971), Talbot (1971), Kendrick (1971a) and Weber & Hess (1976).

BLASTOSPORE, RADULASPORE, POROSPORE AND ARTHROSPORE
Conidiospores of this type are characterized by a simple mechanism of formation and, with the exception of the arthrospore, by having at maturation a narrow point of attachment with the conidiogenic cell. Blastospores arise by budding from either somatic hyphae or conidiophores and are not born on stipules. Characteristic blastospore formers are *Cladosporium*, *Phymatotrichum* and the microconidia of *Neurospora*. When blastospores are abstricted leaving behind rasp-like denticiles on the conidiophore they are termed radulaspores (Talbot, 1971), *e.g. Cercospora*, *Botrytis* and *Aureobasidium pullulans*.

With the porospore or poroconidium (Ellis, 1971) the conidium develops through a pore in the wall of the conidiophore. For the proconidium to develop, a channel is induced enzymatically in the outer wall of the conidiogenic cell and the inner wall can then develop a protrusion. Examples of poroconidia are *Alternaria*, *Stemphyllium*, *Curvularia* and *Helminthosporium*.

In contrast to the three previous spore types the arthrospore does not form by budding but rather by the apex of the conidiogenic cell being expanded and delimited by a septum early in development (meristem arthroconidium) or by the conversion of the hyphae in the absence of a meristem (arthroconidium). Conidia may be formed in a basipetal sequence or irregularly. A revised concept of thallic conidiogenesis has been proposed by Cole (1975b). Meristem arthroconidia occur in *Erysiphe* and the dikaryotic aecidiospores of the Uredinales while arthrospores can occur in many fungi, *i.e. Oidiodendron*, *Sporendonema*, *Briosia*, *Geotrichum* and *Thielaviopsis* (Kendrick, 1971b).

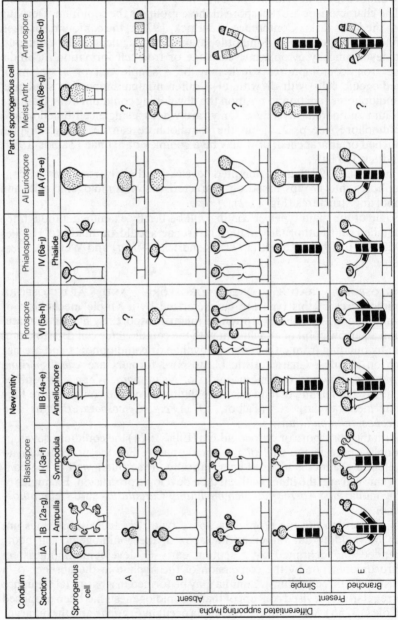

Fig. 11.2 Classification of conidiophores (from Pirozynski, 1971).

PHIALOCONIDIA   When conidia are produced in basipetal sequence from a special structure, the phialide, they are termed phialospores. Phialides are conidiogenic cells wherein the first conidium initial develops within an apical extension of the cell and is liberated by rupture or dissolution of the upper wall of the parent cell. The ruptured parent cell may or may not form a distinctive collarette around the developing phialoconidium. Thereafter, a basipetal succession of enteroblastic conidia may be produced in which the conidium wall does not derive from the conidiogenic cell wall. Successive production of conidia from a fixed conidiogenous locus is another feature of the phialide. The length of the phialide does not change during the production of successive conidia. Typically, phialoconidia often do not show clear attachment scars. A phialide may proliferate percurrently and produce conidiogenic loci at higher levels or a succession of phialides may sometimes be produced in a sympodial manner. Examples of phialoconidia are numerous and include such useful and well documented fungi as *Aspergillus*, *Penicillium* and *Trichoderma* (Subramanian, 1971; Cole & Kendrick, 1969). The spermatia of the Uredinales are typical phialoconidia (Hughes, 1970).

The complete understanding of the initiation of phialoconidium development and the genesis of the conidial wall has only been deduced by careful studies of the ultrastructure of the complete process (Beckett, Heath & McLaughlin, 1974).

Each conidium successively produced from a phialide may be shed from the phialide as a separate unit, and these conidia may then either slime down to form gloeoid masses at the tip of the phialide or form loose or fragile chains. However, in some species, the successively produced conidia are held together by an outer wall common to all of them as in species of *Aspergillus*, *Penicillium* and *Paecilomyces*; the conidia in these fungi are seen to form persistent chains, in contrast to the loose, fragile chains seen in other fungi.

CHLAMYDOSPORE   A chlamydospore is a terminal or intercalary cell which is characterized by the production of thick protective walls and a lack of any mechanism for liberation (Fig. 11.1a, b). These spores are firmly attached to the mycelium and unlike the previously listed spore types do not possess a clearly differentiated mechanism to achieve separation from the wall of the adjacent cell or cells. Liberation is only achieved by indirect breakdown or by enzymic lysis by other organisms. Chlamydospore formation can be considered to have biological significance since it allows the fungus to form a resistant structure under conditions which limit or inhibit macromolecular synthesis (Cochrane & Cochrane, 1970). Chlamydospores occur widely throughout the fungi.

Asexual spores in the Oömycetes and Zygomycotina

The asexual or imperfect state of the majority of the Oömycetes and Zygomycotina is characterized by the cleavage of the cellular contents of special cells (sporangia) into motile zoospores or non-motile sporangiospores. In this type of spore formation the protoplasm of the sporangium is reorganised internally to form a number of smaller cells. There is a considerable increase in the area of cell surface with concomitant increase in plasmalemma. Cytoplasmic division occurs separately from nuclear division and cell wall formation. In zoospore formation, naked, flagellated cells are produced, whereas in sporangiospore formation, a rigid spore wall develops before release from the sporangium (Hughes, 1971; Bartnicki-Garcia & Hemmes, 1976).

In some species of the Oömycetes and Zygomycotina sporangia have acquired some or all of the characteristics of conidia. In the Oömycetes, zoosporogenesis or indirect germination (Bartnicki-Garcia & Hemmes, 1976) is typically found in aquatic environments with such well studied genera as *Pythium, Saprolegnia, Phytophthora* and *Asphanomyces*. Some members of the Oömycetes have adopted a definite terrestrial habitat and in so doing have developed deciduous sporangia that germinate directly. Accompanying this change to a conidia-like spore has been the development of a much-branched sporangiophore (conidiophore) bearing terminal deciduous sporangia, *e.g. Peronospora*, sympodial development of successive conidia on a simple conidiophore and finally the development of successive conidia from percurrent proliferations through conidium scars (Hughes, 1971). Conidia-like sporulation has thus been achieved by delaying or suppressing the zoosporangial state.

In the Mucorales of the Zygomycotina typical sporangiospore formation is a characteristic of the group and is regularly observable in *Mucor, Rhizopus* and *Absidia*. However, again in this group there is clear evidence of an evolutionary development towards a direct germination of the sporangium. This is achieved by the reduction of the number of sporangiospores in the sporangium so that in the most developed forms the sporangia are only 1-spored. As such the sporangial wall may sometimes only be differentiated with great difficulty from the wall of the spore, *e.g. Cunninghamella elegans*. The decrease in the number of sporangiospores in such species is paralleled by an increase in the production of sporangioles, merosporangia and 1-spored sporangia (Hughes, 1971).

Arthrospore and chlamydospore formation occurs throughout the Phycomycetes.

Microcycle sporulation

Microcycle sporulation has been defined in bacteria as the immediate recapitulation of sporogenesis following spore germination (Vinter & Slepecky, 1965). A similar phenomenon is now known to exist in fungi and has been variously termed microcycle conidiation in *Aspergillus niger* (Anderson & Smith, 1971*a*, 1972), *Neurospora crassa* (Cortat & Turian, 1974), *Penicillium digitatum* (Zeidler & Margalith, 1972, 1973), *P. urticae* (Sekiguchi, Gaucher & Costerton, 1975*a*), and *Paecilomyces varioti* (J. G. Anderson and J. E. Smith, unpublished results); iterative germination in *Helminthosporium spiciferum* (Mangenot & Reisinger, 1976); precocious sporulation in *H. sativum* (Boosalis, 1962); and lag-phase sporogenesis in *Blastocladiella emersoni* (Hennessy & Cantino, 1972).

A feature common to most forms of microcycle conidiation is the increase in size of the spore by spherical growth during the period of elevated temperature. The role of temperature in regulating this process is discussed in detail in Chapter 18. However, it would appear that in the thermic controlled microcycle conidiation the spherical cells are unable to germinate at the elevated temperatures due probably to the inhibitory effect of temperature on the apical growth process. Growth is not stopped but occurs by uniform wall deposition resulting in spherical cells of variable size and thickness. Only when cells are removed to a lower temperature regime can apical outgrowth occur. Why a reproductive form should then occur is not yet understood.

Ultrastructural studies on spherical cell formation and early outgrowth of the conidiophore have been carried out for *Aspergillus niger* (Smith, Gull, Anderson & Deans, 1976) and *Penicillium urticae* (Sekiguchi *et al.*, 1975*a, b*) and for conidia formation in *P. urticae* (Sekiguchi *et al.*, 1975*c*). Phialoconidia formation by the microcycle method in *P. urticae* was indistinguishable from normal subaerial formation. The conidia-producing apparatus in microcycle conidiation may be considerably modified or reduced as in *P. urticae*, becoming simply a branched or unbranched, septate hypha bearing terminal phialides and conidia or being little modified, as in *A. niger* where the conidiophore is similar in morphology but reduced in size from the normal subaerial form (Fig. 11.3*a, b*).

The undoubted value of microcycle conidiation (Fig. 11.4) for synchronous spore production and subsequent biochemical analysis of conidiation

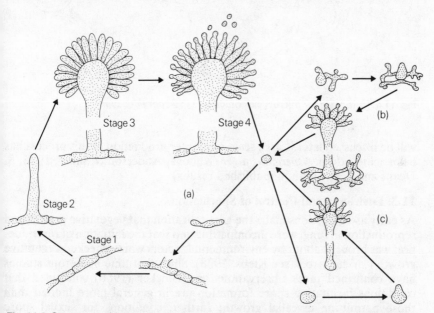

Fig. 11.3 Summary of induced morphogenetic sequences leading to conidiophore development in *Aspergillus niger* under submerged agitated conditions. (a) Sequence of morphological changes in replacement fermenter culture. Stage 1, conidiophore initiation; Stage 2, conidiophore elongation; Stage 3, vesicle and phialide formation; Stage 4, conidiospore production. (b, c) Forms of microcycle conidiation. In (b) a branched mycelial system and a mature conidiophore are produced from an enlarged conidium. Treatment consists of incubation at 41° for 15 h followed by 30° for 12 h. In (c) a mature conidiophore is produced from an enlarged conidium in the complete absence of vegetative development. Treatment consists of incubation at 44° for 48 h followed by 30° for 15 h (From Smith & Anderson, 1973).

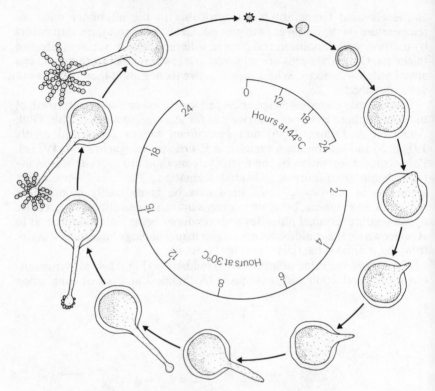

Fig. 11.4 Time scale for microcycle conidiation in *Aspergillus niger*.

will be discussed later. Large scale fermenter production of this process has been achieved for *Aspergillus niger* (Kuboya, Anderson & Smith, 1976; S. Deans and J. E. Smith, unpublished results).

11.3 Environmental Control of Sporulation

As early as 1898 Klebs made the generalization that vegetative growth and reproduction in fungi were incompatible and that the initiation of reproduction was brought about by environmental factors which checked vegetative growth (for early refs. see Klebs, 1928). Since that time numerous studies have confirmed these observations and Hawker (1966) concluded that conditions facilitating spore formation are in general more limited than those permitting mycelial growth; further, conditions for sexual spore formation are also invariably more exacting than for asexual spore formation. Almost all environmental parameters have been investigated and most show some influence on sporulation. How can so many intrinsically different factors have such a formulative effect on a developmental process?

Microbial sporulation commonly occurs following a variable period of vegetative growth. Within the eubacteria, the actinomycetes, yeasts and lower fungi there is a diverse range of sporulation forms but all have in common the fact that they are intracellular processes of differentiation during which the cell becomes subdivided by neomembrane formation;

while vegetative growth continues such differentiation processes will normally remain unexpressed. Although vegetative growth and sporulation are not necessarily completely incompatible, some antagonism is to be expected if it is postulated that alternative metabolic pathways are in operation. Perhaps vegetative growth and sporulation should be considered as cellular processes which are competing for limiting metabolic intermediates rather than as mutually exclusive phenomena. If vegetative growth and sporulation do not normally occur simultaneously, and are separated by definite metabolic shifts, the point of change may be associated with the limitation of vegetative growth due to nutrient exhaustion or to some other environmentally controlled limiting process. Many examples exist in the fungi in which sporulation is associated with or follows from some degree of limitation of vegetative expression. The implications of these observations will be discussed later.

At this point it is pertinent to consider the natural evolution of cultivation techniques which have been used to examine this type of developmental system and, in particular, to show how controlled fermenter conditions have given a new insight to the whole problem.

The induction of sporulation in fungi

Sporulation in most filamentous fungi can be a complex form of development and if meaningful interpretations are to be made it is imperative that the experimental conditions should impose rigorous control over the developmental patterns. Only then will it be possible to gain a fuller understanding of the interrelationships that exist between the inducing environmental conditions and the genetic potential of the fungus for sporulation (Turian, 1974, 1976; Smith & Berry, 1974).

SURFACE CULTIVATION   Many studies have made use of surface growing cultures and these have yielded much useful information on the influence of the environment on the sporulation process together with reliable genetic interpretations (see Chapter 12). However, the very nature of the growth morphology precludes any meaningful information at the biochemical level. The mycelial mat formation characteristic of surface cultures represents a grossly heterogeneous state which is rarely amenable to worthwhile biochemical investigation (Smith & Anderson, 1973). It must always be remembered that this is the 'natural' condition for growth of mycelial fungi in nature and by imposing the obviously unnatural conditions of submerged culture (see p. 224) important surface contact phenomena may be overlooked.

Recently, however, surface culture studies have made valuable contributions to our understanding of the timing of conidiation. Spore-derived surface colonies of *Aspergillus nidulans* initiate conidiophore and spore development only after a 24 h period of vegetative growth (Axelrod, 1972). This period of vegetative growth preceding initiation of conidiophore development is not related to nutrient supply and depletion of nutrients, in particular carbon and nitrogen, is not necessary to initiate conidiophore development (Pastushok & Axelrod, 1976). It is now considered that this period of vegetative growth is related to the known gene control of conidiophore initiation (Axelrod, Gealt & Pastushok, 1973; Gealt & Axelrod,

1974), and indicates the time required to complete this endogenous gene-controlled activity (Pastushok & Axelrod, 1976).

Although the main vegetative mass of mycelium in surface cultures is heterogeneous in form and function, synchronous development of sporulation can be achieved (Stine & Clark, 1967; Zentmyer & Erwin, 1970; Tan, 1976). Under conditions of imposed synchronous development it becomes possible to gain a better understanding of causally related biochemical systems.

SUBMERGED CULTIVATION   Submerged cultivation of fungi can lead to a more homogeneous mycelium particularly if growth is in the filamentous rather than the pellet form. Furthermore, most filamentous fungi will remain entirely vegetative in submerged culture. This fact is consistent with the observation that most differentiated structures are characteristic of the aerial mycelium in surface cultivation. The cause of the inhibition of sporulation in submerged culture is complex and most probably involves a combination of many factors including oxygen tension (Turian & Bianchi, 1972), changes in the physical nature of the hyphal wall associated with submergence (Morton, 1961), and the direct contact of the sporogenous cells with nutrients and inhibitory factors in the medium (Galbraith & Smith, 1969a; Smith & Galbraith, 1971). More recently, sporulation in submerged culture has been routinely achieved by manipulation of the medium components and related environmental parameters. The most general condition for induction of sporulation is the reduction or exhaustion of assimilable nitrogen while carbohydrate is still available (Morton, 1961; Turian & Bianchi, 1972; Smith & Anderson, 1973; Vezina & Singh, 1975). By varying the nitrogen source of a simple liquid synthetic medium it has been possible to regulate the transition between the mycelial state and con-idiogenic state in *Neurospora crassa* (Turian & Bianchi, 1972) and *Aspergillus niger* (Galbraith & Smith, 1969a). Thus it becomes possible to compare the biochemical events in differentiating and non-differentiating mycelium by utilizing culture media which differ in one factor only.

Many of the early studies and some current studies involve simple flask culture which although ensuring adequate contact between mycelium and medium does create intrinsic disadvantages: in particular, poor conditions of aeration, variation between samples and inoculum size and difficulties in regulating pH. For this reason and others there is an overwhelming necessity for developmental mycologists to study many aspects of filamentous mycelial differentiation using more sophisticated fermenters where it becomes feasible to grow within one vessel quantities of relatively similar mycelium under more controlled experimental conditions of aeration, temperature, agitation, pH control, incremental feeding and replacement culture (Bull & Bushell, 1976). Regrettably, with the exception of industrial studies, the use of fermenter techniques in basic filamentous studies, let alone complex developmental studies such as sporulation, is still the exception rather than the rule.

Growth and sporulation in filamentous fungi are subject to a multitude of internal and external regulatory controls. Within the confines of a monitored fermenter it becomes possible to gain a deeper insight into the role of the external regulatory factors. Of particular interest is the study of the transfer of O_2 from air bubbles to cell reaction sites since an aerobic stimulus seems to be important to induce a Pasteur effect switching the predominantly alcoholic fermentation of the vegetative hyphae into an oxidative, conidiogenic type of metabolism (Turian, 1973, 1975; Anderson & Smith, 1971b). One of the main physical factors affecting O_2 transfer is the physical structure or morphology of the organism. Thus as the organism undergoes developmental changes, the mycelial form and growth pattern will consistently interplay with O_2 transfer, necessitating regular reassessment.

More information is required on the fundamental responses of filamentous fungi to changes in ionic strength, pH, adsorbed gas concentrations, shear and other factors. Furthermore, not enough is known about the rheology of mycelial fermentation systems, known to be non-Newtonian in some cases, which must have a marked effect on the circulatory patterns of solids, liquids and gases within the system. Although many of the answers will be subjective they will undoubtedly shed some light on the complex interrelationship between external and internal regulatory control of vegetative growth and sporulation. Further advances in the knowledge and availability of on-line *in vivo* sensors will allow determination of substrates (C, N and mineral ions), intermediates (NAD, RNA, proteins) and cell wall structure without involving the removal of the mycelium (and subsequent alteration of activity) from the fermenter. Developmental mycology will advance with the advance in fermentation technology and analytical techniques.

A disadvantage of most submerged fermentation systems is the heterogeneous nature of mycelial form, and subsequent biochemical analysis is merely a summation of many physiological states. The ability to control the process of development leading to synchronous attainment of each phase is a condition which can be achieved in part with sporulation in *Aspergillus niger* (Anderson & Smith, 1971b), *Achlya* sp. (Griffin & Breuker, 1969) and *Penicillium digitatum* (Zeidler & Margalith, 1972) by either nutrient changes or cultural manipulations. Studies with *A. niger* have been particularly worthwhile. By means of a replacement medium fermenter technique, it has been possible to study individually the various developmental phases of conidiogenesis. The first morphological event of conidiation, foot cell formation (Stage 1), is induced by growth in a medium in which N is the limiting nutrient (LN medium) (Fig. 11.3a). It is not possible to induce foot cells in a high nitrogen medium in which both the N and C sources become exhausted simultaneously or in a low oxygen medium. Conidiophore elongation (Stage 2) occurs in LN medium after the exhaustion of exogenous N although the continued presence of the carbon source (glucose) is essential. When the mycelium is left indefinitely in this medium no further morphological development will occur. However, replacement of the culture to a new medium containing a N source and a TCA cycle intermediate (citrate) as the carbon source will induce vesicle and phialide formation (Stage 3). At this stage no conidia will develop from the phialides and indeed prolonged exposure to this medium can cause dedifferentiation with the formation of conidiophores from the phialides. Multiple dedifferentiations are not uncommon. Conidia formation (Stage 4) is most effectively induced by transfer into a medium with glucose as the C source and nitrate as the N source. Thus by means of nutrient replacement in the controlled environment of the fermenter it has been possible to follow the synchronous maturation of conidiophores. Subsequently, this has permitted a detailed examination of the biochemical changes associated with each stage of development (Smith & Ng, 1972; Ng, Smith & Anderson, 1972; Lloyd, Anderson, Smith & Morris, 1972, Ng, Smith & McIntosh, 1973a).

Microcycle conidiation of *Aspergillus niger* (Fig. 11.4) and *Paecilomyces varioti* can now be routinely obtained in fermenter cultivation (Kuboye *et al.*, 1976; Aryee, 1974). This system undoubtedly offers many advantages over conventional mycelial sporulation because of the limited amount of vegetative growth present before induction of the conidiophore. Furthermore, the entire conidiating apparatus arises from a single cell—the giant cell (Anderson & Smith, 1971a, 1972) which will permit more accurate measurement of relevant changes.

The transient nature of the medium environmental conditions during growth within a batch fermenter, *i.e.* nutrient uptake and waste product release into a closed system, can complicate the interpretation of developmental responses. Thus it cannot be deduced from a batch culture system whether sporulation results from nutrient limitation or from limitation of growth rate imposed by these conditions. However, chemostat culture does permit a study of fungi at various growth rates and under various metabolic steady states. Studies with *Penicillium chrysogenum* (Righelato, Trinci, Pirt & Peat, 1968) and *Aspergillus niger* (Ng, Smith & McIntosh, 1973b) have emphasized the importance of growth rate in determining the nature and the extent of sporulation. Conidiation in *A. niger* appears to be determined by an interaction between growth rate and the ratio of the carbon and nitrogen sources in the medium. This will be discussed in detail later.

An interesting feature of the chemostat culture studies with *Penicillium chrysogenum* (Righelato *et al.*, 1968) and with *Aspergillus niger* (Ng *et al.*, 1973b) is the occurrence of a considerable reduction in the complexity of the conidiophore. Studies with *A. niger* in batch culture have previously shown that conidiophores produced in submerged culture were essentially similar to, although smaller than, the normal subaerial structures (Anderson & Smith, 1971a, b). In chemostat culture conidiophores were characterized by possessing small vesicles with few phialides and occasionally conidia were observed to develop from modified hyphal tips. These modifications in conidiophore complexity may represent only a partial switch-on of the conidiophore mechanism. It also may indicate that under certain conditions the morphological and biochemical events of conidiophore development which precede conidia formation can be by-passed. However, recently, normal conidiation of *A. niger* has been achieved in chemostat culture (S. Cave, D. Pitt and J. E. Smith, unpublished results).

Growth rate and sporulation

When the growth of a fungus is being controlled by the intrinsic characteristics of the organism rather than the environment, *i.e.* the rate of nutrient uptake into the mycelium and rate of utilization are constant, the organism is considered to be exhibiting balanced growth. The rate at which maximum or optimal growth rate declines is both genetically specific and environmentally dependent and may be regarded as a direct indication of the degree of imbalance in the various metabolic functions. The nature of such an imbalance will determine the rates at which essential intermediates accumulate or disappear together with the sequence of change; the net effect will be some form of control over which pathways are initiated and the sequence of initiation. In ideal growth conditions the whole of the biosynthetic capacity of the cell can be expected to be required to maintain the high rate of growth and cell division. Growth becomes unbalanced when the growth rate is limited by the reduction in availability of an essential nutrient factor or by the influence of other environmental factors known to influence cellular activity *viz.* temperature, O_2 concentration, pH, *etc.*

In filamentous fungi, vegetative growth is normally manifested by apical tip extension. The factors regulating apical tip growth are enumerated in Chapter 4 and will not be considered here in detail. In unicellular fungi, growth can be easily monitored by the production of new daughter cells. However, in filamentous fungi, there is no separation of cells during growth, but rather growth is by apical extension and consequently there exists a gradient of age along the filament. Organelles multiply in subapical zones while further back the cellular constituents are characterized by progressive autolysis. Consequently, in any one hyphal filament there will be a distinction between areas of active replication and high metabolic activity and areas of little or no replication and low metabolic activity (see also Chapters 8 and 20). Much confusion exists concerning the true nature of growth particularly in conditions of N limitation and excess C supply. In this case the organisms can continue to take up C compounds and an increase in biomass will result, however, this may not be considered a true example of growth. The concepts of autocatalytic and assimilatory growth have been considered by Bu'Lock (1975).

In autocatalytic growth the basic regulatory systems of the organism, *viz.* nucleic acid and protein synthesis, increase and this represents a true indication of the growth potential of the organism. Conversely, assimilatory growth involving the active uptake of nutrients from the environment may not always be integrated with autocatalytic growth since fungi not actively multiplying may continue to take up carbohydrate molecules and convert them into polymeric form, *e.g.* fats and lipids. Although the total weight of the fungus increases during assimilatory growth, there is no increase in the replicatory systems. The difference between autocatalytic and assimilatory growth can be readily observed in fungal cultures when N becomes limiting while the medium contains carbon molecules in excess. While little replicatory growth is occurring, biomass increases due to continued uptake of C compounds, with subsequent polymerization to form storage molecules. Tempest (1975) considers that the continued uptake of C compounds during N limitation reflects inadequate modulation of uptake processes, while the formation of 'storage' polymers is primarily to prevent the overproduction of key compounds such as ATP, reduced pyridine nucleotides and acetyl-CoA.

Thus balanced growth and autocatalytic growth are synonomous and whenever an essential environmental factor becomes limiting, growth becomes unbalanced causing major metabolic alterations, including alternative metabolic pathways, excretion of intermediary metabolites, synthesis of storage polymers and morphological differentiation.

The relationship between balanced and unbalanced growth has been analysed by Bu'lock and others (for refs. see Bu'lock, 1975) with particular emphasis on secondary metabolite production. Bu'lock (1975) considers that secondary metabolism is an aspect of differentiation implied by limited growth. Secondary metabolism is related not only to the existence of nutrient limitation but also to the intensity, type and character of the limitation. Not all types of limitation to growth produce the same type of secondary metabolite. Within a single fungus (*e.g. Gibberella fujikuroi*) variable limitations will evoke different intensities and types of secondary metabolite formation (Bu'lock, Detroy, Hostalek & Minim-Al-Shakarchi, 1974).

The relationship between growth limitation and sporulation in fungi has long been implied but only recently understood. Many studies have shown that sporulation in fungi follows on from nutrient exhaustion (for refs. see Hawker, 1966; Turian & Bianci, 1972; Smith & Berry, 1974). In all such examples sporulation occurred after the fungus had been exposed to conditions that severely limited balanced growth. However, it does not always follow that unbalanced growth will result in sporulation; in many cases the exhaustion of C will rapidly lead to autolysis. Clearly, a movement away from balanced growth is a prerequisite to sporulation, but the manner of achieving the unbalanced state must also be critical. Because most studies of fungal sporulation have involved batch cultivation where environmental parameters such as nutrient levels are transitory, it has never been possible to relate growth rate accurately to developmental changes. However, by using continuous cultivation techniques with *Penicillium chrysogenum* (Righelato *et al.*, 1968), *Aspergillus niger* (Ng *et al.*, 1973*b*), and *Geotrichum candidum* (Robinson & Smith, 1976) it has been possible to analyse growth rate and the factors limiting growth and their relationship with sporulation. Indeed, the conditions of growth occurring in a typical continuous system are probably very close to conditions in nature where one or more nutrients may be present at extremely low, growth-limiting concentrations whilst others are present in excess of requirements. Furthermore, studies on the control of sporulation in bacteria have demonstrated the suitability of continuous systems for examining the effects on sporulation of changing growth rates since growth rate can be made to depend on the concentration of a single component of the medium and so avoid effects due to qualitative differences (Dawes & Mandelstam, 1968).

With *Penicillium chrysogenum*, Righelato *et al.* (1968) obtained limited conidiation when the glucose feed rate was lowered to the maintenance ration, (the amount required to sustain the viability of the fungus) or a little above this level. At high growth rates in glucose limitation the mycelium remained vegetative. El Kotry (1970) has shown that conidiation occurred in maltose-limited chemostat culture of *Aspergillus oryzae* at a dilution rate of $0.15 \, h^{-1}$ when the concentration of maltose was decreased to 0.25% (w/v). Ng *et al.*, (1973*b*) have examined the effect of growth rate and the nature of several substrate limitations on conidiation in *A. niger*. Under citrate limitation, conidiation intensity varied inversely with dilution rate. Conidiation was difficult to achieve under glucose limitation; at low dilution rates which allowed very limited conidiation, steady state could not be maintained due to the onset of autolysis. At high dilution rates when steady state was readily obtained conidiation did not occur.

The difficulties of obtaining good conidiation with glucose limitation may be related to catabolite repression at high dilution rates while at lower dilution rates (below concentrations of glucose that will cause catabolite repression) the substrate is supplied at too low a level and subsequently autolysis sets in before conidiation can be fully established. Under ammonium limitation with citrate as the carbon source there was no conidiation, while NO_3 limitation with citrate gave conidiophore initiation at low dilution rates but also early wash out. Recently S. Cave and D. Pitt (unpublished results) have shown good conidiation in a glucose limited NO_3 medium.

Thus, although the experiments with varying dilution rate under citrate limitation demonstrate that conidiation can be controlled by growth rate, the studies with glucose limitation also show that the composition of the medium must also be considered. Similar relationships can be seen with the nitrogen limitation studies and also the C/N relationship. These findings indicate that conidiation in *Aspergillus niger* is determined by the concentration and nature of the carbon and nitrogen sources in the culture medium. Limitation of growth brought about by the control of dilution rate/growth rate demonstrates that the intensity of the limitation promotes conidiation. However, the type and character of this limitation is also critical. Studies with light-induced conidiation of *Aspergillus ornatus* (Hill, 1976) have inferred that the internal mycelial glucose concentration is an important factor in regulating conidiation. Under the conditions of continuous light inhibition of glucose uptake and phosphorylation precedes conidiation.

Studies with *Geotrichum candidum* have shown that arthrospores are produced at growth rates up to $0.10 \, h^{-1}$ although higher spore production occurred at lower specific growth rates (Robinson & Smith, 1976). Thus whereas spore induction may not be entirely dependent on low specific growth rates there is clearly a definite correlation between increasing sporulation and decreasing specific growth rate.

Sporulation in filamentous fungi can be considered as an aspect of differentiation that results from specific types of limitation to the normal unlimited pattern of vegetative growth. Although nutrient limitation is one of the more obvious means of limiting growth of a fungus other environmental factors such as light, temperature, *etc.* which have been shown to influence sporulation may do so indirectly by influencing essential growth limiting reactions. Thus perhaps growth rate is the central factor determining the nature and extent of sporulation in filamentous fungi.

11.4 Biochemical Considerations of Sporulation

To understand the biochemical basis of fungal sporulation it is necessary to know which constituents of the cell are involved and more particularly which of these are critical and limit the rate of specific reactions at various points in the orderly sequence of developmental progression (Killik & Wright, 1974). It is now widely accepted that the primary critical factor(s) controlling cellular differentiation is differential gene activation. Thus qualitative and/or quantitative changes in rates of enzyme synthesis are believed to be a major variable limiting differentiation and it follows that the rate of synthesis of mRNA will limit the rate of enzyme synthesis.

Clearly, enzyme activity must play a critical role in determining morphogenetic change since few macro-changes can occur without considerable molecular synthesis. However, it is difficult if not impossible to conclude from *in vitro* enzyme studies alone that (a) an enzyme does not exist or occurs in another form or in a different place *in vivo*; or (b) what level of enzymes are normally present in great excess *in vivo* compared to their substrates (Gustafson & Wright, 1972). It is now also recognized that mechanisms other than the regulation of enzyme synthesis can profoundly influence developmental patterns and include enzyme stabilization, activation and modulation of enzyme activity and concentration by substrate availability, and allosteric control (for refs. see Killick & Wright, 1974; Wright, 1973; Smith & Berry, 1974). Undoubtedly, a key critical variable in the control of enzyme activity will be the availability of low molecular weight precursors or substrates and the flux in the levels of these metabolites will be of central importance in regulating developmental patterns (for refs. see Gustafson & Wright, 1972; Smith & Anderson, 1973).

Yet another critical factor in regulating enzyme activity is the complex aspect of cellular compartmentalization (Svere & Moshbach, 1974). Compartmentalization is a necessary structural feature of all living cells and cell systems and as such will be involved in cellular regulation and differentiation. Within individual cells, compartmentalization is achieved with organelles (nuclei, mitochondria, lysosomes, glyoxosomes and vacuoles) separated from each other by enclosing membranes. Thus we find that glycolytic enzymes are cytosolic, TCA cycle enzymes are mitochondrial and glyoxolate enzymes are glyoxosomal. The widely used practice in studies of fungal differentiation of crude cell smashing and analysis of the resultant 'enzyme brew' must surely be misleading and future analysis of enzyme systems should be by *in vivo* measurements or by careful extraction of the various components of the cell by controlled rupturing and digestion techniques in isotonic conditions. Only in this way will it be possible to appreciate the complexity and interrelationships of cellular organelles and metabolic pathways.

It may also be conceivable that many specific enzyme changes may also occur as a result of alterations in the degradative phase of enzyme turnover and not as a response to changes in the rate of synthesis.

This section will not attempt to exhaustively examine the many biochemical studies that have been carried out during sporulation in filamentous fungi but rather, arbitrarily, select a few examples of each type. A fuller appreciation will be gained from the reviews by Turian & Bianchi, 1972; Turian, 1974, 1975, 1976; Smith & Galbraith, 1971; Smith & Anderson, 1973; Ashworth & Smith, 1973; Smith & Berry, 1974; Lovett, 1975).

Cell wall formation and sporulation

The cellular shape and form of fungi is mainly determined by the presence of the rigid outer cell wall. The fungal cell wall is a structured and complex assembly of polymers in which the chemical nature of the polymers and the spatial relationship of the polymers to each other can alter during growth and differentiation (Bartnicki-Garcia, 1968, 1973; Rosenberger, 1976). Most studies on wall formation have been concerned with vegetative growth and the enzymology of this process is considered in Chapter 4. Only a few attempts have been made to analyse the chemical and physical changes of wall structure related to sporulation in filamentous fungi and even here it is not yet possible to identify whether the changes are causal or secondary to the process of sporulation.

In certain Oömycete fungi the β-glucans occupy a major role in cell wall structure (Bartnicki-Garcia & Hemmes, 1976). The β-glucans have both a quantitative and qualitative role comprising a major portion of the cell mass and also being the principal cell wall constituents. During sporulation in *Phytophthora*, massive quantities of cytoplasm move from the vegetative hyphae into the expanding sporangia (Zentmyer & Erwin, 1970). The reserves of mycelial mycolaminarum represent the main source of glucose for sporangial wall synthesis and zoospore genesis. Within the sporangium the relative proportions of cytoplasmic to wall glucan of the mycelium is reversed and the thicker sporangial wall has a greater amount of glucan than the vegetative wall.

Zonneveld (1974) has demonstrated a relationship between the α-1, 3 glucan and the α-1, 3 glucanase system in the regulation of cleistothecium and conidiophore formation in *Aspergillus nidulans*. When α-1, 3 glucan and α-1, 3 glucanase were synthesized in adequate amounts cleistothecia were formed—the more α-1,3 glucan available the more cleistothecia were formed. High quantities of glucose in the medium lead to a greater formation of α-1,3 glucan with a corresponding decrease in conidiation. It was conjectured that the rapid synthesis of α-1,3 glucan prevented conidiophore initials from breaking through the cell wall which thickened at an early stage.

Membranes and sporulation

Fungal hyphae possess a complex membranous system including plasma, nuclear, mitochondrial and vesicular membranes interconnected to some extent by an endoplasmic reticulum (Rose, 1976). As the outer living surface of the cell the plasma membrane may be the primary receptor of all external stimuli which can be transmitted internally promoting cellular reaction and change. Sussman (1976) views membranes as ordered and continuous arrays of interacting molecules in which sequential perturbations of adjacent molecules could be caused by the change in a single molecule, such that a cascade effect could occur. Since membranes control both the entry and the exit of molecules to cells changes in membrane permeability may be a vital factor in understanding morphogenesis. In some differentiating systems a period of isolation from the environment allowing more accurate control of cellular intermediate flux has been postulated to occur due to changes in membrane structure and permeability (Wright, 1967).

Undoubtedly, the qualitative and quantitative composition of membranes can be induced to change by a variety of external physical and metabolic influences. The compositional and topographical flexibility of bacterial membranes is well exemplified in the membrane biosynthesis and differentiation which accompanies sporulation and subsequent spore formation in certain bacteria. During such sporulation distinct qualitative changes in membrane proteins and lipids have been measured.

Although many studies have been carried out to analyse the complex molecular architecture of fungal cell walls, 'no one class of membrane from filamentous fungi has been isolated and purified to the point at which extensive chemical analyses are justified' (Rose, 1976). The challenge is out and surely must be accepted. The answers to so many aspects of fungal differentiation in particular, sporulation, must surely be related to changes in structure and permeability of membranes.

Energy pathways and sporulation

GLYCOLYSIS   Glycolysis may be considered broadly as the metabolic sequence by which sugars are converted to small molecules which can then enter the pathways of terminal oxidation or biosynthesis. The two principal pathways of glycolysis *viz.* Embden–Meyerhof (EM) and Pentose Phosphate (PP) pathways are almost of universal occurrence in filamentous fungi (Cochrane, 1976). Of particular current interest is the elucidation of the relative contributions of each pathway to carbohydrate metabolism and further to consider whether the balance of the pathways has any causal relationship in developmental processes such as sporulation in filamentous fungi. Many studies have sought to establish the relative contributions of these pathways during sporulation involving selective inhibition, measurement of specific enzyme activities, isotope labelling techniques, measurement of glycolytic intermediate flux, and also using specific genetic mutants (for ref. see Turian & Bianchi, 1972; Smith & Anderson, 1973; Turian, 1976).

There has been extensive examination of carbon catabolism during conidiation of *Aspergillus niger* (Valenzuela-Perez & Smith, 1971; Smith, Valenzuela-Perez & Ng, 1971; Smith & Valenzuela-Perez, 1971; Ng *et al.*, 1972). In these experiments *in vitro* enzyme determinations were coupled with radiorespirometric analysis of glucose metabolism *in vivo* in order to give a more reliable estimation of *in vivo* changes occurring during glucose catabolism. During conidiophore development the PP pathway enzymes had higher specific activities than the EM pathway enzymes and when taken together with the radiorespirometric analysis strongly implied that the direct oxidation of glucose through the PP pathway is important during conidiophore development. High EM pathway activity occurred in media that did not support conidiophore initiation. These results are in agreement with other studies which have demonstrated that the PP pathway is the major glycolytic route during sporulation in *Neurospora crassa* (Turian & Bianchi, 1972), *Aspergillus nidulans* (Carter & Bull, 1969), and *Puccinia carthami* (Daly, Sayre & Pazur, 1957). In *N crassa* the PP pathway apparently achieves a peak of activity at conidiophore induction before the appearance of the counteracting activity of NADPase at the end of the blastic phase of development. There is a progressive move from catabolism to anabolism leading to the development of the gluconeogenic pathway necessary for the synthesis of structural polysaccharides during the arthric phase *Neurospora* conidiation (for refs. see Turian, 1976).

OXIDATIVE METABOLISM   Oxidative metabolism appears to be essential to sporulation in most fungi and environmental conditions which favour glycolysis inhibit conidial development. There have been many reports concerning the relative roles of the tricarboxylic acid cycle (TCA) and the glyoxylate cycle in fungal sporulation. High activity of key enzymes of the TCA cycle during conidiophore development in replacement cultures of *Aspergillus niger* (Ng *et al.*, 1973*a*) together with the stimulation of conidiation by acids oxidised by the TCA cycle (Galbraith & Smith, 1969*a*) lend support to the hypothesis that oxidative metabolism is essential for normal conidiation. The involvement of the TCA cycle in *Neurospora* conidiation has been clearly demonstrated (Urey, 1971; Stine, 1967; Turian, Oulevey-Matikian & Cortat, 1973; Ojha, Velmelage & Turian, 1969). It has been considered that an alternate cyanide-insensitive and antimycin A resistant pathway operates during conidiation in many filamentous fungi (Oulevey-Matikian & Turian, 1968).

Isocitrate lyase has been shown to increase in specific activity during sporulation in several fungi. In most cases this increase in activity is not paralleled by a similar increase in the other important glyoxylate cycle enzyme malate synthase. With *Aspergillus niger* high activity of isocitrate lyase has been shown to be associated with conidiating mycelium in flask culture (Galbraith & Smith, 1969*b*), replacement culture (Ng *et al.*, 1973*a*) and in continuous culture (Ng, Smith & McIntosh, 1974). Thus several environmental variations inducing conidiation also induce activity of isocitrate lyase. Similar observations have been made for other fungi. The low activity of malate synthase in all these systems may further suggest that the glyoxylate formed by isocitrate lyase activity is transaminated with alanine to form glycine—an important precursor of RNA synthesis. It is possible that a source of glyoxylate rather than an active isocitrate lyase is vital for conidia development since good conidiation occurs when glyoxylate is present in the media (Galbraith & Smith, 1969*a*).

The isolation and biochemical characterization of mitochondria and glyoxosomes during the developmental process of sporulation is currently being studied with microcycle conidiation of *Aspergillus niger*. Functional mitochondria have been isolated from the giant cells by a novel cell smash technique (B. Davis & J. E. Smith, unpublished results) or by enzymic digestion of the thick cell walls (Davis, D'Avillez Paixae, Deans & Smith, 1977). By these methods it is hoped to follow developmental changes primarily at the organelle level and thus avoiding some of the problems of interpretation previously discussed. Microcycle conidiation will undoubtedly allow a clearer understanding of causally related changes since the system does not have the normal large amount of masking vegetative mycelium.

INTERMEDIATE FLUX   The essential properties of a fungal cell will be determined largely by the rhythmic pattern of variation in constituent levels of metabolites which result from the functioning of innumerable metabolic control circuits. In this way cellular differentiation may well result from alterations in these periodicities in such a way that changes in intermediate levels may control not only enzyme activity but also the sequential release of nuclear information necessary for progressive differentiation (Hanson, 1975). Gustafson & Wright (1972) consider that changes in the concentration of key metabolites during differentiation are in themselves sufficient to cause the qualitative and quantitative changes in metabolite flux that occur during morphogenesis. Thus it becomes increasingly obvious that there is more to cell differentiation than the regulation of transcription and translation.

The extensive biochemical studies on sporulation in *Dictyostelium discoideum* have lead to ingenious mathematical models for predicting biochemical changes associated with or essential to differentiation. Unfortunately, little attempt has been made to emulate these elegant studies with other developmental systems including filamentous fungi. However, limited studies with *Aspergillus niger* using mycelia from flask culture (Smith, Valenzuela-Perez, 1971) and replacement fermenter culture (Smith & Ng, 1972) have shown that major changes occur in the levels of many intermediates of glycolysis during differentiation in each system analysed. The fact that there was no obvious correlation between the measured levels of intermediates in the two examples studied may imply that critical changes in the concentration of metabolites essential to differentiation may be met and balanced by compensatory mechanisms within the cell. Thus different types of initiators of differentiation or sporulation may achieve the same end result from different biochemical approaches. In this way the PP pathway has been shown to be essential during the initiation of sporulation in many fungi while in media containing glutamate (Smith & Valenzuela-Perez, 1971), wherein there is a lowered demand for reducing power, the PP pathway does not have the same importance. However, in each case the mycelium freely sporulates.

Macromolecular synthesis and sporulation

There have been relatively few studies relating nuclear expression to sporulation in filamentous fungi, due mainly to the paucity of basic nucleic acid metabolism studies in these organisms. However, the limited studies that have been carried out do indicate that sporulation is under nuclear control.

With *Trichoderma viride*, a fungus which requires light stimulation for sporulation, there is evidence for continuous synthesis of RNA during critical periods of photo-induced sporulation. It is possible that new species of RNA are being transcribed at these periods (Galun & Gressel, 1966). By means of DNA/RNA hybridization techniques it was possible to reveal differences in RNA species during early vegetative growth but not during the critical period following photo-induction. It was considered that this apparent lack of photo-induced transcription was related to the following: the transition is not regulated by transcriptional control but may be regulated by translational control; only a small part of the genome is transcribed and this cannot be detected by the hybridization techniques; and the change in RNA transcription is considerable but is restricted to the relatively few aerial hyphae which become conidiophores so that it is masked by the massive vegetative contribution (Stavey, Galun & Gressel, 1972).

RNA and protein synthesis have been studied during sporulation of *Achlya bisexualis* (Griffin & Breuker, 1969; Timberlake, McDowell, Cheney & Griffin, 1973) and in *Allomyces arbuscula* (Burke, Seale & McCarthy, 1972; Fahnrich, 1974a, b; Stumm & Croes, 1975). Sporulation in *A. bisexualis* seems to be induced by starvation with concomitant turnover of both RNA and protein. By means of incorporation rates, pool specific activities and the loss of radioactivity from pre-labelled protein they calculated a turnover rate of 75%/h. Clearly, a large proportion of the amino acids released by degradation are being re-utilized at all stages of development (Timberlake *et al.*, 1973). Griffin & Breuker (1969) have further shown that rRNA and tRNA synthesis continues during sporulation.

O'Day & Horgen (1974) have studied acid phosphatase activity during sporulation of *Achlya* and by fractionation experiments and inhibition studies have shown that both transcription and translation are required for the increased phosphatase level and its maintenance during sporulation.

There have been few studies of the effect of selective antimetabolites on the process of sporulation in filamentous fungi. Actinomycin D blocks sporogenesis in *Allomyces arbuscula* (Burke & Germond, 1976) and in *Achlya ambisexualis* (Griffin & Breuker, 1969) but not in *A. arbuscula* (Burke, Seale & McCarthy, 1972). In general, a higher concentration of inhibitor is required to block sporulation than is required to block germination. This effect may well be due to a change in permeability during vegetative growth and sporulation (Burke & Germond, 1976).

The foregoing studies must raise the question as to whether sporulation is controlled by definite time steps for transcription and translation (enzyme control) or by substrate and product concentrations involving feedback inhibition or stimulation (intermediate control). Both concepts are valid and the real solution lies in the integration of the control systems. There can be no one key controlling step in sporulation but rather an ordered sequence(s) of control mechanisms (see also Chapter 1).

11.5 References

ANDERSON, J. G. & SMITH, J. E. (1971a). The production of conidiophores and conidia by newly germinated conidia of *Aspergillus niger* (microcycle conidiation). *Journal of General Microbiology* **69**, 185–97.

ANDERSON, J. G. & SMITH, J. E. (1971b). Synchronous initiation and maturation of *Aspergillus niger* conidiophores. *Transactions of the British Mycological Society* **56**, 9–29.

ANDERSON, J. G. & SMITH, J. E. (1972). The effects of elevated temperatures on spore swelling and germination in *Aspergillus niger*. *Canadian Journal of Microbiology* **18**, 289–97.

ARYEE, V. (1974). Microcycle conidiation in *Paecilomyces varioti*. *M.Sc. Thesis University of Strathclyde, Glasgow.*

ASHWORTH, J. M. & SMITH, J. E. (1973). Eds. *Microbial differentiation*. London: Cambridge University Press.

AXELROD, D. E. (1972). Kinetics of differentiation of conidiophores and conidia by colonies of *Aspergillus nidulans*. *Journal of General Microbiology* **73**, 181–4.

AXELROD, D. E., GEALT, M. & PASTUSHOK, M. (1973). Gene control of developmental competence in *Aspergillus nidulans*. *Developmental Biology* **34**, 9–15.

BARTNICKI-GARCIA, S. (1968). Cell wall chemistry, morphogenesis and taxonomy in fungi. *Annual Review of Microbiology* **22**, 87–108.

BARTNICKI-GARCIA, S. (1973). Fundamental aspects of hyphal morphogenesis. *Symposium Society of General Microbiology* **23**, 245–67.

BARTNICKI-GARCIA, S. & HEMMES, D. E. (1976). Some aspects of the form and function of Oömycete spores. In *The fungal spore: form and function*. pp. 593–641, edited by D. J. Weber and W. M. Hess. New York: John Wiley & Sons.

BECLETT, A., HEALTH, I. B. & MCLAUGHLIN, D. J. (1974). *An atlas of fungal ultrastructure*. London: Longman.

BOOSALIS. M. G. (1962). Precocious sporulation and longevity of conidia of *Helminthosporium sativum* in soil. *Phytopathology* **52**, 1172–7.

BULL, A. T. & BUSHELL, M. E. (1976). Environmental control of fungal growth. In *The filamentous fungi*, Vol. II, pp. 1–31, edited by J. E. Smith and D. R. Berry. London: Edward Arnold.

BU'LOCK, J. D. (1975). Secondary metabolism in fungi and its relationship to growth and development. In *The filamentous fungi*, Vol. I, pp. 33–58, edited by J. E. Smith and D. R. Berry. London: Edward Arnold.

BU'LOCK, J. D., DETROY, R. W., HOSTALEK, Z. & MINIM-AL-SHAKARCHI, A. (1974). Regulation of secondary biosynthesis in *Gibberella fujikuroi*. *Transations* of the British Mycological Society 62, 377–89.

BURKE, D. D. & GERMOND, M. (1976). Effect of inhibitors on the morphology and growth of Allomyces macrogynus. Journal of General Microbiology 95, 213–9.

BURKE, D. D., SEALE, T. W. & McCARTHY, B. J. (1972). Protein and ribonucleic acid synthesis during the diploid life cycle of Allomyces arbuscula. Journal of Bacteriology 110, 1965–72.

CARMICHAEL, J. W. (1971). Blastospores, aleuriospores, chlamydospores. In Taxonomy of fungi imperfecti, pp. 50–70, edited by B. Kendrick. Toronto and Buffalo: University of Toronto press.

CARTER, B. L. A. & BULL, A. T. (1969). Studies of fungal growth and intermediary metabolism under steady and non-steady conditions. Biotechnology and Bioengineering 11, 785–804.

COCHRANE, V. W. (1976). Glycolysis. In The filamentous fungi Vol. II, pp. 65–91, edited by J. E. Smith and D. R. Berry. London: Edward Arnold.

COCHRANE, V. W. & COCHRANE, J. C. (1970). Chlamydospore development in the absence of protein synthesis in Fusarium solani. Developmental Biology 23, 345–54.

COLE, G. T. (1975a). Conidiation ontogeny in Hyphomycetes: a review of current developmental concepts. Proceedings of University of Madras Symposium on Taxonomy of Fungi.

COLE, G. T. (1975b). The thallic model of conidiogenesis in the Fungi Imperfecti. Canadian Journal of Botany 53, 2983–3001.

COLE, G. T. & KENDRICK, B. (1969). Conidium ontogeny in Hyphomycetes: the phialides of Phialophora, Penicillium and Ceratocystis. Canadian Journal of Botany 47, 779–89.

CORTAT, M. & TURIAN, G. (1974). Conidiation of Neurospora crassa in submerged culture without mycelial phase. Archives of Microbiology 95, 305–9.

DALY, J. M., SAYRE, R. M. & PAZUR, J. H. (1957). The hexosemonophosphate shunt as the major respiratory pathway during sporulation of rust of safflower. Plant Physiology 32, 44–8.

DAVIS, G., D'AVILLEZ PAIXAE, M. T., DÉANS, S. G. & SMITH, J. E. (1977). Protoplast formation from giant cells of Aspergillus niger. Transactions of the British Mycology Society in press.

DAWES, J. W. & MANDELSTAM, J. (1968). Sporulation of Bacillus subtilis in continuous culture. Journal of Bacteriology 103, 529–35.

EL KOTRY, R. A. R. (1970). Physiological studies on the production of amylase by Aspergillus oryzae in batch and continuous cultivation. Ph.D. Thesis, University of Strathclyde, Glasgow.

ELLIS, M. B. (1971). Porospores. In Taxonomy of fungi imperfecti. pp. 71–4. Edited by B. Kendrick. Toronto and Buffalo: University of Toronto Press.

FAHNRICH, P. (1974a). Untersuchunger zur Entwicklung des Phycomyceten Allomyces arbuscula. 1. Einfluss von inhibitoren der Protein und Nucleinsauresynthese auf die differenzierung von gametagien. Archives of Microbiology 98, 85–92.

FAHNRICH, P. (1974b). Untersuchungen zur Entwicklung des Phycomyceten Allomyces arbuscula. II. Enfluss von Inhibitoren der Protein und Nucleinsauresythose auf die Gametogenise. Archives of Microbiology 99, 147–53.

GALBRAITH, J. C. & SMITH, J. E. (1969a). Sporulation of Aspergillus niger in submerged liquid culture. Journal of General Microbiology 59, 31–45.

GALBRAITH, J. C. & SMITH, J. E. (1969b). Changes in activity of certain enzymes of the tricarboxylic acid cycle and the glyoxylate cycle during the initiation of conidiation of Aspergillus niger. Canadian Journal of Microbiology 15, 1207–12.

GALUN, F. & GRESSEL, J. (1966). Morphogenesis in Trichoderma: suppression of photoinduction by 5-fluorouracil. Science 151, 696–8.

GEALT, M. & AXELROD, D. E. (1974). Coordinate regulation of enzyme inducibility and developmental competence in Aspergillus nidulans. Developmental Biology 41, 224–32.

GREGORY, P. H. (1966). The fungus spore: what it is and what it does. In The fungus spore, pp. 1–13, edited by M. F. Madelin. London: Butterworths.

GRIFFIN, D. H. & BREUKER, C. (1969). Ribonucleic acid synthesis during the differentiation of sporangia in the water mold Achlya. Journal of Bacteriology 98, 689–96.

GUSTAFSON, S. L. WRIGHT, B. E. (1972). Analyses of approaches used in studying differentiation of the cellular slime mould. Critical Reviews in Microbiology 1, 453–78.

HANSON, R. S. (1975). Role of small molecules in regulation of gene expression and sporogenesis in bacilli. In Spores, Vol. 6, pp. 318–26, edited by P.

Gerherdt, R. N. Coustilou and H. L. Sedoff. Washington D.C.: American Society for Microbiology.

HAWKER, L. E. (1966). Environmental influences on reproduction. In *The fungi*, Vol. 2, pp. 435–469, edited by G. C. Ainsworth and A. S. Sussman, New York, London: Academic Press.

HAWKER, L. E. & MADELIN, M. F. (1976). The dormant spore. In *The fungal spore: form and function*, pp. 1–72, edited by D. J. Weber and W. M. Hess, New York: John Wiley & Sons.

HENNESSY, S. W. & CANTINO, E. C. (1972). Lag-phase sporogenesis in *Blastocladiella emersonii*: induced formation of unispored plantlets. *Mycologia* **64**, 1066–87.

HILL, E. P. (1976). Effect of light on growth and sporulation of *Aspergillus ornatus*. *Journal of General Microbiology* **95**, 39–44.

HUGHES, S. J. (1953). Conidiophores, conidia and classification. *Canadian Journal of Botany* **31**, 577–659.

HUGHES, S. J. (1970). Ontogeny of spore forms in Uredinales. *Canadian Journal of Botany* **48**, 2147–57.

HUGHES, S. J. (1971). Phycomycetes, Basidiomycetes and Ascomycetes as Fungi Imperfecti. In *Taxonomy of fungi imperfecti*, pp. 7–36, edited by B. Kendrick, Toronto and Buffalo: University of Toronto Press.

KENDRICK, B. (1971a). Ed. *Taxonomy of fungi imperfecti*. Toronto and Buffalo: University of Toronto Press.

KENDRICK, B. (1971b). Arthroconidia and meristem arthroconidia. In *Taxonomy of fungi imperfecti*, pp. 160–75, edited by B. Kendrick, Toronto and Buffalo: University of Toronto Press.

KILLICK, K. A. & WRIGHT, B. E. (1974). Regulation of enzyme activity during differentiation in *Dictyostelium discoideum*. *Annual Review of Microbiology* **28**, 139–66:

KLEBS, G. (1928). *Die Bedingungen der Fortpflanzung Bei einigen Algen und Pilzen*. Berlin: Fischer-Verlag.

KUBOYE, A. O., ANDERSON, J. G. & SMITH, J. E. (1976). Control of autolysis of a spherical cell form of *Aspergillus niger*. *Transactions of the British Mycological Society* **67**, 27–31.

LLOYD, G., ANDERSON, J. G., SMITH, J. E. & MORRIS, E. O. (1972). The effect of medium nitrogen and of conidiation on esterase synthesis in *Aspergillus niger*. *Transactions of the British Mycological Society* **59**, 63–70.

LOVETT, J. S. (1975). Growth and differentiation in the water mold *Blastocladiella emersonii*: cytodifferentiation and the role of ribonucleic acid and protein synthesis. *Bacteriological Reviews* **39**, 345–404.

MACKO, V., STAPLES, R. C., YANIV, Z. & GRANADOS, R. R. (1976). Self-inhibitors of fungal spore germination. In *The fungal spore: form and function*, pp. 73–98, edited by D. J. Weber and W. M. Hess. New York: John Wiley & Sons.

MANGENOT, F. & REISINGER, O. (1976). Form and function of conidia as related to their development. In *The fungal spore: form and function*, pp. 789–847, edited by D. J. Weber and W. H. Hess.

MORTON, A. G. (1961). The induction of sporulation in mould fungi. *Proceedings of the Royal Society B* **153**, 548–69.

NG, A. M. L., SMITH, J. E. & McINTOSH, A. F. (1973a). Changes in activity of tricarboxylic acid cycle and glyoxylate cycle enzymes during synchronous development of *Aspergillus niger*. *Transactions of the British Mycological Society* **61**, 13–20.

NG, A. M. L., SMITH, J. E. & McINTOSH, A. F. (1973b). Conidiation of *Aspergillus niger* in continuous culture. *Archiv für Mikrobiologie* **88**, 119–26.

NG, A. M. L., SMITH, J. E. & McINTOSH, A. F. (1974). Influence of dilution rate on enzyme synthesis in *Aspergillus niger* in continuous culture. *Journal of General Microbiology* **81**, 425–34.

NG, W. S., SMITH, J. E. & ANDERSON, F. G. (1972). Changes in carbon catabolic pathways during synchronous development of *Aspergillus niger*. *Journal of General Microbiology* **71**, 495–504.

O'DAY, D. A. & HORGEN, P. A. (1974). The developmental patterns of lysosomal enzyme activities during Ca^{2+}-induced sporangium formation in *Achlya bisexualis*. 1. Acid phosphatase. *Developmental Biology* **39**, 116–24.

OJHA, M. N., VELMELAGE, R. & TURIAN, G. (1969). Malonate metabolism, stimulation of conidiation and succinic dehydrogenase activity in *Neurospora crassa*. *Physiologia Plantarum* **22**, 819–26.

OULEVEY-MATIKIAN, N. & TURIAN, G. (1968). Controle metabolique et aspects ultrastructuraux de la conidiation (macromicroconidies) de *Neurospora crassa*. *Archiv für Mikrobiologie* **60**, 35–58.

PASTUSHOK, M. & ALEXROD, D. E. (1976). Effect of glucose, ammonium and media maintenance of the time of conidophore initiation by surface colonies of

Aspergillus nidulans. Journal of General Microbiology **94**, 221-4.

PIROZYNSKI, K. A. (1971). Characters of conidiophores as taxonomic criteria. In *Taxonomy of fungi imperfecti*, pp. 37-49, edited by B. Kendrick. Toronto and Buffalo: University of Toronto Press.

RIGHELATO, R. C., TRINCI, A. P. J., PIRT, S. J. & PEAT, A. (1968). The influence of maintenance energy and growth rate on the metabolic activity, morphology and conidiation of *Penicillium chrysogenum. Journal of General Microbiology* **50**, 399-412.

ROBINSON, P. M. & SMITH, J. H. (1976). Morphogenesis and growth kinetics of *Geotrichum candidum* in continuous culture. *Transactions of the British Mycological Society* **66**, 413-20.

ROSE, A. H. (1976). Chemical nature of membrane components. In *The filamentous fungi* Vol. II, pp. 308-327, edited by J. E. Smith and D. R. Berry. London: Edward Arnold.

ROSENBERGER, R. F. (1976). The cell wall. In *The filamentous fungi* Vol. II, pp. 328-44, edited by J. E. Smith and D. R. Berry. London: Edward Arnold.

SEKIGUCHI, J., GAUCHER, G. M. & COSTERTON, J. W. (1975a). Microcycle conidiation in *Penicillium urticae*: an ultrastructural investigation of spherical spore growth. *Canadian Journal of Microbiology* **21**, 2048-58.

SEKIGUCHI, J., GAUCHER, G. M. & COSTERTON, J. W. (1975b). Microcycle conidiation in *Penicillium urticae*: an ultrastructural investigation of conidial germentation and outgrowth. *Canadian Journal of Microbiology* **21**, 2059-68.

SEKIGUCHI, J., GAUCHER, G. M. & COSTERTON, J. W. (1975c). Microcycle conidiation in *Penicillium urticae*: an ultrastructural investigation of conidiogenesis. *Canadian Journal of Microbiology* **21**, 2069-83.

SMITH, J. E. & ANDERSON, J. G. (1973). Differentiation in the Aspergilli. *Symposium Society of General Microbiology* **23**, 295-337.

SMITH, J. E. & BERRY, D. R. (1974). *An introduction to biochemistry of fungal development*. London: Academic Press.

SMITH, J. E. & GALBRAITH, J. C. (1971). Biochemical and physiological aspects of differentiation in the fungi. *Advances in Microbial Physiology* **5**, 45-134.

SMITH, J. E. & NG, W. S. (1972). Fluorometric determination of glycolytic intermediates and adenylates during sequential changes in replacement culture of *Aspergillus*

niger. Canadian Journal of Microbiology **18**, 1657-64.

SMITH, J. E. & VALENZUELA-PEREZ, J (1971). Changes in intracellular concentrations of glycolytic intermediates and adenosine phosphates during the growth cycle of *Aspergillus niger. Transactions o, the British Mycological Society* **57**, 103-10.

SMITH, J. E., VALENZUELA-PEREZ, J. & NG, W. S. (1971). Changes in activities of the Embden-Meyerhof-Parnas and pentose phosphate pathways during the growth cycle of *Aspergillus niger. Transactions of the British Mycological Society* **57**, 93-101.

SMITH, J. E., GULL, K., ANDERSON, J. G. & DEANS, S. G. (1976). Organelle changes during fungal spore germination. In *The fungal spore: form and function*, pp. 301-54, edited by D. J. Weber and W. M. Hess. New York: John Wiley & Sons.

SRERE, P. A. & MOSBACH, K. (1974). Metabolic compartmentation: symbiotic, organellar, multienzymic and microenvironmental. *Annual Review of Microbiology* **28**, 61-83.

STAVY, R., GALUN, R. & GRESSEL, J. (1972). Morphogenesis in *Trichoderma*: RNA/DNA hybridization studies. *Biochemica et Biophysica Acta* **259**, 321-9.

STINE, G. J. (1967). Enzyme activities during the asexual cycle of *Neurospora crassa*. 1. Succinic dehydrogenase. *Canadian Journal of Microbiology* **13**, 1203-10.

STINE, G. J. & CLARK, A. H. (1967). Synchronous production of conidiophores and conidia of *Neurospora crassa*. *Canadian Journal of Microbiology* **13**, 447-53.

STUMM, C. & CROES, A. F. (1975). Polyribosomes in different stages of the life cycle of the water mold *Allomyces arbuscula. Archives of Microbiology* **102**, 117-22.

SUBRAMANIAN, C. V. (1971). The phialide. In *Taxonomy of fungi imperfecti*, pp. 93-119, edited by B. Kendrick. Toronto and Buffalo: University of Toronto Press.

SUSSMAN, A. S. (1966). Dormancy and spore germination. In *The fungi*, Vol. 2, pp. 733-64, edited by G. C. Ainsworth and A. S. Sussman. New York: Academic Press.

SUSSMAN, A. S. (1976). Activators of fungal spore germination. In *The fungal spore: form and function*, pp. 101-39, edited by D. J. Weber and W. M. Hess. New York: John Wiley & Sons.

TALBOT, P. H. B. (1971). *Principles of fungal taxonomy*. London: Macmillan.

TAN, K. K. (1976). Light-induced synchronous conidiation in the fungus *Botrytis cinerea*. *Journal of General Microbiology* **93**, 278–82.

TEMPEST, D. W. (1975). The application of continuous culture to studies of microbial adaptation to low nutrient environments. *6th International Symposium on Continuous Culture of Microorganisms, Oxford* (Abstract). In press.

TIMBERLAKE, W. E., McDOWELL, L., CHENEY, J. & GRIFFIN, D. H. (1973). Protein synthesis during differentiation of sporangia in the water mold *Achlya*. *Journal of Bacteriology* **116**, 67–73.

TURIAN, G. (1973). Induction of conidium formation in *Neurospora* by lifting of catabolite repression. *Journal of General Microbiology* **79**, 347–50.

TURIAN, G. (1974). Sporogenesis in fungi. *Annual Review of Phytopathology* **12**, 129–37.

TURIAN, G. (1975). Differentiation in *Allomyces* and *Neurospora*. *Transactions of the British Mycological Society* **64**, 367–80.

TURIAN, G. (1976). Spores in Ascomycetes, their controlled differentiation. In *The fungal spore: form and function*, pp. 715–88, edited by D. J. Weber and W. M. Hess. New York: John Wiley & Sons.

TURIAN, G. & BIANCHI, D. E. (1972). Conidiation in *Neurospora*. *Botanical Reviews* **38**, 119–54.

TURIAN, G. OULEVEY-MATIKIAN, N. & CORTAT, M. (1973). Recherches sur la differenciation conidienne de *Neurospora crassa*. V. Ultrastructure de la sequence macroconidiogene. *Annales Microbiologie* **124A**, 443–58.

UREY, J. C. (1971). Enzyme patterns and protein synthesis during synchronous conidiation in *Neurospora crassa*. *Developmental Biology* **26**, 17–27.

VALENZUELA-PEREZ, J. & SMITH, J. E. (1971). Role of glycolysis in sporulation of *Aspergillus niger* in submerged culture. *Transactions of the British Mycological Society* **57**, 111–19.

VEZINA, C. & SINGH, K. (1975). Transformation of organic compounds by fungal spores. In *The filamentous fungi*, Vol. I, pp. 158–92, edited by J. E. Smith and D. R. Berry. London: Edward Arnold.

VINTER, V. & SLEPECKY, R. A. (1965). Direct transition of outgrowing bacterial spores to new sporangia without intermediate cell division. *Journal of Bacteriology* **90**, 803–7.

WEBER, D. J. & HESS, W. M. (1976). Eds. *The fungal spore: form and function*. New York: John Wiley & Sons.

WRIGHT, B. E. (1967). On the evolution of differentiation. *Archiv für Mikrobiologie* **59**, 335–44.

WRIGHT, B. E. (1973). *Critical variables in differentiation*. Englewood Cliffs: Prentice-Hall.

ZEIDLER, G. & MARGALITH, P. (1972). Synchronized sporulation in *Penicillium digitatum* (Secc.). *Canadian Journal of Microbiology* **18**, 1685–90.

ZEIDLER, G. & MARGALITH, P. (1973). Modification of the sporulation cycle in *Penicillium digitatum*. *Canadian Journal of Microbiology* **19**, 481–3.

ZENTMYER, G. A. & ERWIN, D. C. (1970). Development and reproduction of *Phytophthora*. *Phytopathology* **60**, 1120–7.

ZONNEVELD, B. J. M. (1974). α-1,3 glucan synthesis correlated with α-1,3 glucanase synthesis, conidiation and fructification in morphogenetic mutants of *Aspergillus nidulans*. *Journal of General Microbiology* **81**, 445–51.

CHAPTER 12

Genetics of Vegetative Growth and Asexual Reproduction

A. J. CLUTTERBUCK

12.1 Introduction

The main contribution of genetics to studies of development lies in the provision of morphological mutants. Examination of the phenotypes of these may point to subdivisions of the developmental process not necessarily suggested merely by inspection of the developing wild-type organism. Secondly, if mutations are used in conjunction with biochemical studies it may be possible to demonstrate parallel morphological and biochemical changes and hence establish a causal relationship. As with physiological studies of development, where the starting point of an investigation is an alteration of the environment rather than the induction of a mutation, it is a problem to decide whether an observed morphological alteration is a primary or secondary effect of the input change, but the advantage of the genetical approach is that this input change is confined to a single gene, whereas an environmental disturbance may attack an organism at a number of unrelated points.

12.2 Vegetative Growth

Hyphal growth rates and branching

While mutants defective in vegetative growth have been used as genetic markers in many fungi, the most intensive use of such mutants in investigations of morphogenesis has been made in *Neurospora crassa* (Garnjobst & Tatum, 1967). *Neurospora* normally grows as a diffuse, rapidly spreading mycelium and morphological mutants studied are generally characterized by a slower hyphal extension rate as well as distinctive culture appearance (Fig. 12.1). Slow growth alone, however, is not necessarily accompanied by abnormal morphology; this is illustrated by auxotrophic mutants, which grow slowly on unsupplemented medium, but among which only choline and inositol auxotrophs give rise to a distinctive 'colonial' morphology (Fuller & Tatum, 1956; Crocken & Nyc, 1964).

A precise morphological analysis of mutants originally characterized in terms of gross culture appearance is often not as simple as might be supposed

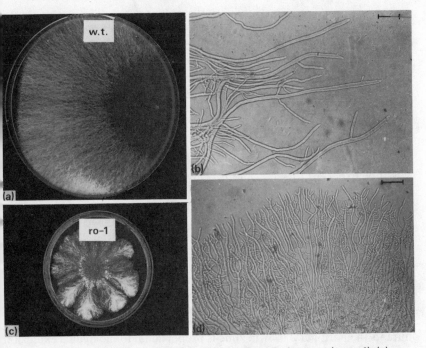

Fig. 12.1 *Neurospora crassa* wild-type and a morphological mutant (*ropy-1*). (a) Wild-type colony on a 14 cm dish; (b) wild-type hyphae; (c) *ropy-1* grown on a 9 cm dish for the same time as a.; and (d) *ropy-1* hyphae: note frequent branching, narrow, slow-growing hyphae with irregular outlines. Scale marker in (b) and (d) = 100 μm.

since it may depend on the erratic behaviour of different areas of the mycelium rather than a regular discrepancy in one parameter. Such irregularities of behaviour may in some instances suggest environmental interactions, for instance *spray* and *frost* mutants of *Neurospora* both produce relatively normal hyphae on the surface of agar, but are only able to produce slow-growing, highly branched submerged hyphae.

The term 'colonial' is applied to *Neurospora* mutants which form discrete colonies with slower, denser hyphal growth than the wild-type. Trinci (1973a, b) has shown that for a number of spreading colonial (*spco*) mutants there is no difference from the wild-type in their specific growth rate (a measure of exponential growth in mass in shaken culture), but only in the elongation rate of individual hyphae. This is also true for one morphological mutant of *Aspergillus nidulans* (Bainbridge & Trinci, 1969). Decreased hyphal elongation rates are only consistent with unaltered total growth if the colony density is increased, as observed in colonial mutants. This could come about either by increased hyphal branching or by more persistent growth of existing branches and Trinci (1973a, b) in fact found that while all *spco* mutants had increased branching frequencies, the correlation of this with growth rate was very imperfect, so it seems likely that both mechanisms of increasing hyphal density operate.

Ryan, Beadle & Tatum (1943) showed that colony growth on agar required the activities of only a limited length of hypha behind the growing tips. If a cut is made across the mycelium behind this zone, growth of the hyphal tips at the colony edge is unaffected. Trinci (1973a) found that the size of this 'peripheral growth zone' is closely correlated with the hyphal extension rate of the mutants. In some, but not all, cases the septa behind the peripheral growth zone have plugged pores (Trinci & Collinge, 1973), but there is no direct evidence as to whether this is the cause or effect of material failing to pass through the pores towards the hyphal tips.

The primary defect in these mutants might be an increase in hyphal branching: the overload of hyphal tips competing for the same growth in mass of the cytoplasm would result in reduced extension rates for all of them. However, in this case a close correlation between growth rate and branching would be expected. It seems more likely that the basic defect is in hyphal extension, leaving spore growth of cytoplasmic mass to be taken up by a combination, characteristic of different mutants, of extra branching and of extra growth of branches which might otherwise slow down or stop.

Defective tip growth might itself have more than one cause. There is clearly some fairly complex machinery concerned with wall synthesis at the hyphal tip (see Chapters 3 and 4) and a hypha which in *Neurospora* at its fastest can extend by three or four hyphal diameters per minute will also require an effective transport system for wall and membrane building materials to the growth point (see Volume II, Chapter 2). These provide obvious points susceptible to failure in mutants. The fact that in the *spco* mutants the peripheral growth zone is proportional to the hyphal extension rate could be interpreted as an indication that the transport pathway is normal, in which case it is fair to look for biochemical defects in these mutants in the fields of polysaccharide and membrane synthesis.

While mutants with excessive hyphal branching have been given considerable attention it is perhaps worth asking why mutants deficient in branching ability have not been isolated. One answer is that a totally branchless hypha might be difficult to catch! However, while fungal mutants with reduced branch frequencies are not uncommon, those studied further prove to be auxotrophs which behave in this manner only when deprived of the required supplement. Striking cases are mutants of *Aspergillus nidulans* defective in utilization of carbon or nitrogen sources which may have normal growth rates for colony diameter, but drastically reduced colony densities. This may readily be seen as an adaptive measure, allowing the colony to search for scarce nutrients, but it is only switched on by certain deficiencies. For instance, a putrescine auxotroph of *A. nidulans* grows as a small, tight colony on partially supplemented medium, but can also be switched to a large spidery form, probably of similar total mass, if starved for carbon or nitrogen. This behaviour has been used as the basis for a method for isolating mutants defective in utilization of particular carbon or nitrogen sources (Herman & Clutterbuck, 1966).

Hyphal walls and morphological mutants

Trinci (1973a) observed that there were good correlations between colony growth rate and both hyphal compartment length and hyphal diameter. The correlation between extension rate and hyphal diameter for individual hyphae of the *Neurospora* wild-type is shown in Fig. 12.2. Main and branch hyphae fit the same curve, unlike the situation in *Coprinus*, in which the

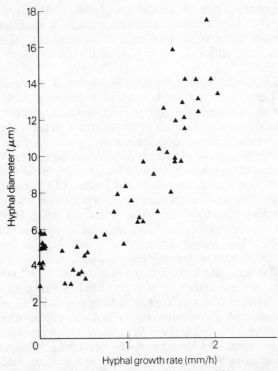

Fig. 12.2 Diameter and growth rate of individual hyphae of *Neurospora crassa* wild-wild-type RL3-8A on Vogel's agar. Each point represents one hypha.

Table 12.1 Mean hyphal diameter and growth rate of *Neurospora crassa* morphological mutants grown on agar. All measurements are means for ten hyphae

Mutant	Diameter μm	Growth rate mm/h	Within wild-type range?
ol-4	9.2	0.53	No
pco-4	5.9	0.46	Yes
e-1	11.4	0.39	No
o-1	4.6	0.32	Yes
g-1	6.7	<0.3	No
alloon	4.4	<0.3	Yes
ol-6	8.4	<0.3	No
ol-8	7.9	<0.3	No
ol-2	6.2 (20)*	<0.3	No
mco-9	6.3	<0.3	?
pco-3	6.1	<0.3	?
cumbo	5.8	<0.3	Yes
pco-5	5.7	<0.3	Yes
ol-5	5.7	<0.3	Yes
ol-10	5.6	<0.3	Yes
rost	5.6	<0.3	Yes
ol-3	5.5	<0.3	Yes
ol-7	5.3	<0.3	Yes

* Hyphae of *col-2* swell with age to 20 μm diameter.

branches do not arise from the hyphal tips (Butler, 1961). Table 12.1 show that while the data for many morphological mutants fit on the same curve as the wild-type—as those for the *spco* mutants measured by Trinci (1973*a*)— some mutants have strikingly fat hyphae indicative, perhaps, of weak c extensible walls.

Wall composition has been correlated with colonial growth in mutants b de Terra & Tatum (1963) and Mahadevan & Tatum (1965). The forme work showed an increase in the ratio of glucosamine to glucose in a numbe of colonial mutants, while the latter authors noticed, in particular, changes i a fraction containing polypeptide, glucan and galactosamine. Wrathall & Tatum (1974) further showed that all colonial mutants tested had less wa polypeptide than normal which is a surprising finding in a diverse collectio of mutants unless it is a result rather than a cause of colonial morphology The *doily* mutant was unusual in that its polypeptides did not bind to DEA cellulose and may therefore have been unusually basic. Edson & Brod (1976) found that this mutant had very little galactosamine in its walls an had defects in galactosamine synthesis although the *doily* gene did nc appear to be the structural gene of the enzymes tested.

In *Aspergillus nidulans* there are two mutants with 'ballooning' hypha due to defects in the metabolism of sugars used as wall components: on requires glucosamine (Cohen, Katz & Rosenberger, 1969) and the othe mannose (Valentine & Bainbridge, 1975). Both are repaired by addition o the appropriate sugar to the medium. Septa have a high glucosamine conten (Mahadevan & Tatum, 1967) so it would be interesting to know whether th septa of the glucosamine deficient mutant are normal, and conversel whether aseptate mutants in *Neurospora* (Garnjobst & Tatum, 1967) and *Aspergillus* (Morris, 1975) are deficient in the synthesis of chitin or othe particular polysaccharides.

Enzyme deficiencies in Neurospora colonial mutants

The next step in the analysis of *Neurospora* colonial mutants, the identifica tion of some of them with specific enzyme deficiencies, has been amply reviewed (Brody, 1973; Scott, Mishra & Tatum, 1973; Mishra, 1977). These mutants and enzymes include the ragged (*rg*) locus in *N. crassa* and *rg-1* anc *rg-2* in *N. sitophila* (phosphoglucomutase, Brody & Tatum, 1967; Mishra & Tatum, 1970), *col-2*, *frost* and *balloon* (glucose-6-phosphate dehydrogen- ase, G6PD, Brody & Tatum, 1966; Scott & Tatum, 1970), *col-3* and *col-10* (6-phosphogluconic acid dehydrogenase, 6GPD, Lechner, Fuscaldo & Bazinet, 1971; Scott & Abramsky, 1971) and *pgi* (glucose phosphate isomerase, Muryama & Ishikawa, 1975).

These mutants present a number of problems. Firstly, there is the surpris- ing number of loci concerned with the same enzymes. The G6PD mutants are an extreme example in which all three mutants have relatively subtle alterations in the enzyme which are most readily interpreted as indications that all three loci are structural genes. Furthermore, there is a fourth such locus: *suC*, identified as a suppressor of *col-2* (Scott & Brody, 1973). The enzyme is a tetramer (Scott, 1971), so four structural loci are possible, although one is forced to wonder how many more loci might be turned up by a deliberate search for G6PD mutants! It is perhaps more realistic to suppose that some of the loci may be responsible for post-translational modifications of the enzyme or may be concerned with coding for, or regulating, alternative subunits of the enzyme (Scott, 1971). The *cot-2* (temperature-sensitive colonial) mutant of *Neurospora* has altered invertase and it has been suggested (Eggerding, Randall & Sargent, 1975) that post-translational modification might in this case act on a carbohydrate moeity of the enzyme while the morphological effect of *cot-2* would result from modifications of wall carbohydrates.

The second problem with morphological mutants which have been identified with specific enzymes is that these are all concerned with central carbon metabolism rather than, for instance, polysaccharide synthesis itself. While the significant feature of the *pgi* mutants may be their failure to provide fructose-6-phosphate for the synthesis of glucosamine or mannose, the importance of the carbohydrate products of enzymes which are defective in other mutants is less clear. For this reason a number of pleiotropic effects of these genes have been considered as the root of morphological abnormality. Firstly, there is the possibility that sugar phosphates accumulated by *col-2, ragged* and possibly other mutants may have allosteric effects on the enzymes of wall polysaccharide synthesis in the same way that glucose-6-phosphate is known to affect glycogen synthase. Secondly, since the pentose phosphate pathway is the main source of reduced NADPH deficiencies in this pathway would be expected to lead to low levels of all the pyridine nucleotides and of reduced fatty acids as found by Brody (1970) and Brody & Nyc (1970) in the G6PD mutants. The consequences of these deficiencies could, in turn be membrane defects leading to morphological effects similar to those already noted in choline and inositol auxotrophs. Evidence for this route to colonial morphology is provided by the fact that the morphology of the *frost* mutant on solid medium is greatly improved by the addition of linolenic acid (Scott *et al.*, 1973).

A further clue to the basis of colonial morphology was found in experiments by Scott & Solomon (1975) which showed that this morphology is produced in wild-type strains by inhibitors of cyclic AMP metabolism. Following this lead, Scott *et al.* (1973) demonstrated low cyclic AMP levels and a thermolabile adenyl cyclase in the *frost* mutant. Addition of linolenic acid reversed these effects. They reason that this second enzyme defect in the *frost* mutant is probably the result of abnormality of the membranes to which it is bound. Even here, however, the relevance of these findings to morphology is obscure, the nearest parallel being known effects of cyclic AMP on glycogen metabolism in *Neurospora* (Tellez-Iñón & Torres, 1970) and multiple effects on carbohydrate metabolism in other organisms.

Conclusions

Considerable difficulties, both conceptual and experimental, have been encountered in attempting to use mutants to define a morphogenetic pathway for vegetative growth. It is evident that wall synthesis is a crucial component: the wall-less *slime* mutant of *Neurospora* (Emerson, 1963) is the perfect demonstration of this, but is unfortunately, genetically very complex. However colony morphology may be even more complex than hyphal morphology since it depends primarily on hyphal branching patterns, the basis for which is virtually unexplored. It is probable that the relationship between mass growth and hyphal extension rates is an important influence on branching and it may be through these components that wider areas of cell organization such as intracellular transport, membranes, *etc.* become involved.

12.3 Asexual Reproduction

Categories of mutants and numbers of loci

Collections of mutants defective in asexual reproduction have been made in a variety of organisms. Those concerned with Ascomycetes or Fungi Imperfecti which will be discussed here have generally included such a wide variety of mutant types that they are only manageable if subdivided in some way.

Martinelli & Clutterbuck (1971) attempted to estimate the number of genes involved in conidiation in *Aspergillus nidulans* by comparing frequencies of conidiation mutants with frequencies of auxotrophs or other known mutants. The conclusion that there are between 45 and 150 genes specifically concerned with conidiation could not be very exact, but the exercise did serve to show that the total might not be so great as to inhibit further investigation and it also provided a collection of mutants suitable for classification according to the stage of development affected. By far the largest class of mutants (85% of the total) also suffers from reduced colony growth rate. These mutants are best separated from specifically aconidial mutants since they must carry a general defect rather than one confined to the developmental process. For this reason such mutants were excluded from the estimates of numbers of loci quoted above. Many of the *Neurospora* colonial mutants conidiate poorly and therefore belong to this group as do auxotrophs which cannot conidiate at supplement levels which allow normal growth. These mutants demonstrate that hyphal growth and conidiation share many processes and although the missing functions may not be specific to asexual reproduction, they may provide clues to the metabolic steps most essential to it.

Mutants blocked before the start of development form a second large and complex group. It is apparent that before conidation is initiated the fungus has a number of competing developmental pathways open to it. Therefore in a collection of aconidial mutants, besides those actually defective in conidiation itself, there are likely to be some diverted away from conidiation by excessive sexual reproduction or aerial mycelium formation. Such mutants can be regarded as defective at *strategic* loci (Clutterbuck, 1977).

Using a similar military analogy, mutants blocked during the course of development, if they result in a diversion of the developmental programme may be regarded as mutants at *tactical* loci, while others which show evidence of failing in development due to insufficiency of metabolic supplies might be allocated to *support* loci. This class would include both those lacking materials specific to development and also those, discussed above which lack materials common to hyphal and conidial growth. Finally, there are *auxiliary* loci which code for functions uniquely or otherwise brought into play during development, but which cannot be considered as essential to it.

The order of action of developmental genes

Analysis of the order of action of genes in development may be decided in most cases merely by comparing the final stage achieved by a mutant with the sequence of stages through which the wild-type passes. A more exacting procedure is to construct double mutants between those of different types with the expectation that the double mutant will show the phenotype of the mutant whose developmental blockage was the earlier. Barbata, Valdes & Sermonti (1973) considered a third possible method involving unidirectional complementation in heterokaryons between pairs of aconidial mutants of *Aspergillus nidulans*. Only for some classes of mutants is the basis of this method clear: in *A. nidulans* the mycelium and conidiophores are multinucleate and therefore may be heterokaryotic, but the sterigmata and conidia are uninucleate, so a heterokaryon between an early mutant and a mutant blocked in these later stages would show complementation of the first mutant, but not of the second. The result would be that the heterokaryon would produce conidia only of the mutant with the early block. Complementation studies with conidiation mutants of *Trichoderma viride* (Weinman-Greenshpan & Galun, 1969) and *A. niger* (Hannan, 1975) suggest that complementation patterns in heterokaryons may be complicated by false negative results resulting from unequal nuclear ratios in the mycelium, dominance of some mutants or variations in nuclear migration. Where complementation can be studied in diploids, the picture with respect to allelism of mutants is likely to be clearer (*e.g.* Clutterbuck, 1969), but this method gives no information about the order of action of the genes involved.

Strategic mutants

Developmental mutants of *Penicillium baarnense* well illustrate the choice between alternative developmental pathways. The wild-type does not conidiate profusely but does produce numerous sexual fruiting bodies. Spontaneous and induced mutants of three types have been studied (Bouvier, 1967, 1968): those producing sterile aerial mycelium only, profusely conidiating mutants, and abortive sexual mutants. The phenotypes of double mutants (Bouvier, 1969) show that the sequence of defects in these mutants is in the order given above: for instance, a double mutant derived from a mycelial and a conidial mutant forms only sterile mycelium and no conidia or cleistothecia. Similarly, profuse conidiation pre-empts sexual reproduction. Mutants of *Aspergillus nidulans* examined by Zonneveld (1974) suggest that conidiation may absorb carbohydrate resources which would otherwise be stored in the form of α-1,3,glucan for utilization during sexual reproduction.

While the induced mutants of *Penicillium baarnense* behaved as nuclear mutations, some of the spontaneous mutants gave ratios in crosses which suggested cytoplasmic inheritance. This finding is not unique: many markers involved in studies of cytoplasmic inheritance in filamentous fungi concern morphological differences and these include switches between mycelial, conidial and sexual phenotypes. In particular, variation in the balance between conidiation and sexual reproduction in wild strains of *Aspergillus* and *Penicillium* (see Chapter 21 and also Jinks, 1966) has frequently been shown to have a cytoplasmic basis. Jinks (1959) suggested that the 'dual phenomenon' (the conidial–mycelial switch evident in wild isolates) might also be due to cytoplasmic inheritance, and Roper (1958) and Dorn, Martin & Purnell (1967) have both described 'mycelial' or 'fluffy' mutants which could transfer their mutant properties to other strains in heterokaryons. However, in both cases chromosomal genes necessary for the mutant phenotype could be mapped and transmission in heterokaryons was sporadic rather than free. Furthermore, Roper (1958) noticed that transmission frequency declined with time after isolation of the original mutants. He suggested that an alternative explanation in terms of a transduction-like process was possible and further parallels now suggest themselves with mutator phage or with transfer of insertion sequences in bacteria (*cf*. Peterson, 1970). Dorn (1970) noticed two other unusual properties of his mutants: they were phenotypically repaired by certain nucleic acid and protein synthesis inhibitors, and they invaded other colonies. The latter property may merely reflect the ability of aerial myclium to invade other colonies from above, or it might, more interestingly, suggest that unregulated mycelial growth is a result of insensitivity to a natural mechanism of self-inhibition possessed by normal colonies.

In *Neurospora*, mutants producing microconidia rather than macroconidia (Barratt & Garnjobst, 1949) might be considered as examples of strategic mutants, although they may not reflect a balance between developmental pathways as much as the difficulty of observing one form of development in the presence of the other.

Initiation of conidiation

The conditions necessary for conidiation in *Neurospora* have been reviewed by Turian & Bianchi (1972) and Nelson, Selitrennikoff & Siegel (1975*b*). Nelson *et al.* (1975*b*) have demonstrated properties of aerial mycelium which make it clear that it should be regarded as the first step in conidiogenesis and they have also isolated mutants defective in this step. While, in *Neurospora*, aerial mycelia and conidia can be induced at any time by starvation and exposure to light (Siegel, Matsuyuma & Urey, 1968), in *Aspergillus* competence to respond to exposure to air by initiation of conidiation is not possible until cultures, grown either on agar surfaces or in shaken culture, have reached a specific age (Axelrod, 1972). At 36°C this age is 20 h, but a mutant has been isolated which achieves competence two hours earlier (Axelrod, Gealt & Pastushok, 1973). The nature of this stimulus and of competence to respond are unknown, but the ability to respond is accompanied by reduction in the induced levels of some enzymes apparently unrelated to conidiation (Gealt & Axelrod, 1974) and a fall in the cellular level of cyclic AMP (Clutterbuck, 1975) which suggest a general shift in metabolism at this time. Starvation can also induce conidiation in submerged cultures (see Chapter 11, Saxena & Sinha, 1973), but need not accompany the more effective stimulation by exposure to air (Pastushok & Axelrod, 1976).

Cyclic AMP is also relevant to conidiation in *Neurospora*: the *cr-1* mutants (one of three *crisp* loci) are slow growing and conidiate prematurely on short aerial hyphae (Garnjobst & Tatum, 1970). They also have defective adenyl cyclase activity and low intracellular cyclic AMP (Terenzi, Flawiá & Torres, 1974; Terenzi, Flawiá, Tellez-Iñón & Torres, 1976). Moreover, normal growth and morphology are restored by high exogenous cyclic AMP levels. The role of cyclic AMP in conidiation may not be simple, however, since the wild-type shows only a very transient increase in cyclic AMP at the start of conidiation. The *frost* mutant, which is also short of this compound, as discussed earlier, does not conidiate prematurely, but it is possible that the aerial hyphae of this mutant are less abnormal than the submerged mycelium.

Tactical or morphogenetic mutants

Once development has reached a stage at which some recognizable structure is formed, it is easier to distinguish different types of mutation affecting further development. Tactical or morphogenetic mutants are those which have noticeably abnormal morphology due to failure of the developmental process to move on to the next stage, so that growth continues by maintenance or repetition of the previous step. The *fluffy* mutant of *Neurospora* and *bristle* mutants of *Aspergillus* (see Fig. 12.3d) are examples of mutants producing conidiophores which develop no further but continue to grow as elongated aerial structures. In both cases these aerial hyphae can be recognized as conidiophores by the formation in them of characteristic enzymes (see below) and in the case of the *bristle* mutants by their thickened walls (Oliver, 1972). The *bristle* mutants were found in the course of an intensive hunt for mutants which revealed only one other locus giving

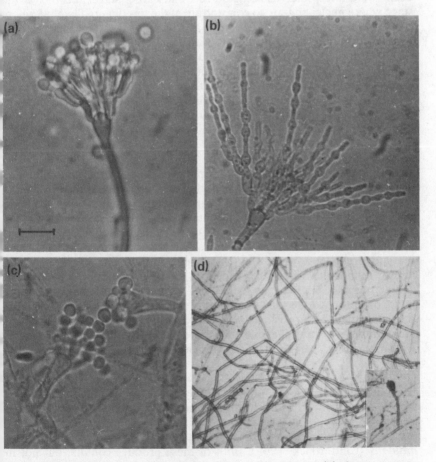

Fig. 12.3 *Aspergillus nidulans* conidiation mutants: (a) wild-type; (b) *abacus* mutant; (c) *stunted* mutant; and (d) *bristle* mutant (with wild-type conidial head at same scale inset). Scale marker for (a), (b) & (c) = 10 μm, for (d) scale marker = 100 μm.

mutants of this type. Mutants at this second locus fail at the stage of the switch from apical growth of sterigmata to the subapical growth characteristic of the extrusion of a series of conidia from a phialide. The resulting *abacus* phenotype (see Fig. 12.3b) appears to be due to repetition of the phialide stage. Similar mutants have been described as *'Cladosarum olivaceum'* in *A. niger* (Yuill & Yuill, 1938), *'fuzzy'* in *A. fonsaceous* (Raper & Fennell, 1953) and mutant 22B in *A. aureolatus* (Vujičić & Muntanjola-Cvetković, 1973).

Neurospora conidia can be described as blasto-arthrospores (Turian & Bianchi, 1972) in which no switch to supapical growth is required but disarticulation of the spores does require the lysis of septa between adjacent spores of a chain and a search for morphogenetic mutants of *Neurospora* affecting this stage has yielded *csp* (conidial separation) mutants, which apparently lack a wall lytic enzyme activity, mapping at two loci (Selitrennikoff, Nelson and Siegal, 1974). Mutants at two other loci, *acon-2* and *acon-3* are interpreted as being defective at earlier stages: *acon-2* at the stage where nuclei migrate to concentrate at the tips of aerial hyphae where conidia will later form, and *acon-3* at the following stage where these hyphae form spore-like swellings (Matsuyama, Nelson & Siegel, 1974). Again a parallel to the first of these mutants exists in *Aspergillus* where the aps (anucleate primary sterigmata) mutants have primary sterigmata which do not develop further because nuclei fail to move into them from the vesicle (Clutterbuck, 1977). In these mutants, however, the defect is not confined to conidiation since hyphal nuclei are abnormally irregular in their distribution and hyphal growth rate is also somewhat reduced.

Mutations at support loci

During a search for specifically morphogenetic or tactical mutants it is usually apparent that there are many other mutants with less readily defined effects on conidiation. Many of these are mutants at support loci which fail not only in the development but also in growth of those structures already formed. These mutants tend to fail at ill-defined stages as if the conidiophore was gradually running out of resources. The fact that the majority of such mutants are also slow-growing (Martinelli & Clutterbuck, 1971) would appear to mean that many cell functions are operative in both hyphal and conidiating states rather than there being a separate set of genes governing metabolic processes in the two phases of the life cycle. On the other hand, the fact that a surprisingly large proportion of these mutants are temperature-sensitive suggests that there may be a separate set of genes whose functions are only required at high temperatures.

Auxiliary loci

Partially defective conidiation, which may result from mutation at support loci, is also likely to result from mutation of loci whose function is the auxiliary one of merely increasing the efficiency of conidiation. Two examples of such mutants are the *medusa* and *stunted* mutants of *Aspergillus nidulans* (Clutterbuck, 1969). Both of these produce conidia, but not in normal numbers because their conidial apparatus is abnormal: *medusa* mutants bear conidia on top of multiseriate sterigmata instead of the normal two tiers, while *stunted* mutants (Fig. 12.3c) have diminutive conidiophores lacking both elongation and wall thickening. Curiously enough, all mutants of both types, like the *acon-3* mutants of *Neurospora* are also defective in sexual reproduction, which once more stresses that relatively few genes are confined in their functions to one phase of the life cycle.

An auxiliary biochemical character investigated in *Neurospora*, NADase, provides an instructive example of the use of mutants to determine the importance of physiological characteristics to an organism. This enzyme is normally found only in the aerial hyphae and conidia and this has led to suggestions that it plays an essential part in conidiation (Urey, 1971). Its phase specificity is confirmed by its presence in *fluffy*, *acon* and *csp* mutants, all of which are blocked after the induction of aerial hyphae, and its absence from mutants failing to make aerial hyphae. NADase is also found in conidiating mycelium of *Aspergillus nidulans*, including that of the conidiophore-blocked *bristle* mutants (A. J. Clutterbuck, unpublished results). However, the crucial experiment of isolating, in *Neurospora*, mutants specifically defective in the structural gene for this enzyme has proved that this is a 'luxury molecule' (Nelson *et al.*, 1975*a*, *b*) whose absence does not produce any obvious defect in the conidia.

In fact, the phase specificity of this enzyme is not complete since it can be induced in substrate mycelium by zinc deficiency (Kaplan, Colowick & Nason, 1951). This finding could be interpreted as suggesting that the enzyme may have a role in hyphal growth under some conditions, but it could equally well be argued that zinc starvation is merely mimicking the induction process occurring during conidiation.

The search for conidiation-phase specific enzymes in *Neurospora* has also led to the examination of various extracellular hydrolytic enzymes which are autogenously induced during conidiation: trehalase, invertase, cellobiase, aryl-β-glucosidase, α-amylase, laminarinase and aryl sulphatase (see Schmit & Brody, 1976, for review). None of these enzymes, however, is confined to the conidiation phase for its induction and where mutants deficient for a specific enzyme are available, their presence does not prevent the formation of otherwise normal conidia. It seems likely, therefore, that conidiation normally occurs under circumstances which coincidentally produce derepression of catabolite-repressed and other enzymes.

Similar conclusions can be drawn from studies of some hydrolytic enzymes in *Aspergillus nidulans* (A. J. Clutterbuck, unpublished results). While increases in phosphatases, proteases and esterases accompany conidiation (see Chapter 11), similar changes occur in an aconidial *bristle* mutant, and in any case most of these changes are not apparent until after most of the conidia have been formed and they can be delayed further, without affecting conidiation, if the surface mat of mycelium is periodically shifted to new medium.

In *Penicillium baarnense*, Parisot (1972*a*, *b*) has measured enzymes of the tricarboxylic acid (TCA) cycle, two glycolytic pathways and glutamate metabolism, during development. Activities of these enzymes do change during conidiogenesis and in some cases quantitative changes are also found in various developmental mutants. However, many of the differences between conidial and mycelial mutants develop only in senescent cultures (post-conidiation) and there is little consistency between different mutants with similar phenotypes. The conclusion is that the mutations are probably not in the structural genes for any of the enzymes measured, but that the enzyme differences are pleiotropic effects of these mutations and may be coincidental to their effects on conidiation.

Where the obvious function of an enzyme is clearly concerned with a secondary characteristic of development it is possible to be more confident in predicting phase specificity. This proves to be the case with two phenol oxidases in *Aspergillus nidulans*. Each is concerned with the formation of a separate pigment, one (cresolase) in the conidiophores and sterigmata (Clutterbuck, 1977), and the second (laccase) in conidia (Clutterbuck, 1972), and each has a substrate specificity distinct from the other and from the mycelial phenol oxidases. The enzymes are produced only at the appropriate stage of development (Fig. 12.4) and neither is formed in *bristle* mutants that fail to reach these stages. The auxiliary nature of these enzymes is demonstrated by the *ivory* conidiophore mutants and *yellow* spore mutants which are morphologically normal, although lacking, in each case, one of these enzymes.

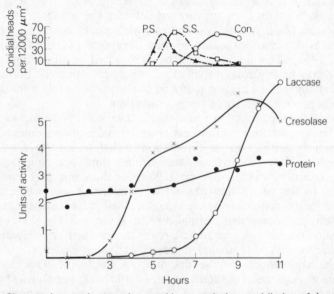

Fig. 12.4 Changes in protein, cresolase and laccase during conidiation of *Aspergillus nidulans* grown on agar. Conidiation was initiated at 0 h by removal of a cellophane membrane under which the mycelium had been grown for 24 h. Protein and enzyme activities were assayed spectrophotometrically and are expressed in arbitrary units. The upper graph shows the timing of the stages of conidiation: P.S. = primary sterigmata, S.S. = secondary sterigmata, Con. = conidia. The first appearance of conidiophores would have preceded the P.S. stage 1–2 h.

Regulation of developmental genes

The majority of conidiation mutants affect support or auxiliary functions and of the very few strictly morphogenetic mutants only the *csp* mutants of *Neurospora* can even tentatively be assigned an enzymatic function. It is therefore a possibility, at least, that even these morphogenetic loci are not structural genes for developmentally specific proteins, but regulatory loci which co-ordinate, in patterns unique to the developmental situation, the activities of enzymes or other proteins which may be used in other combinations elsewhere. If both regulatory loci and structural loci were specific to each developmental situation, it should not be difficult to pick up mutants of both types unless the regulatory genes were very small or refractory to mutation. In *Aspergillus nidulans*, where an intensive search was made for mutants affecting one well-defined area of development, all 75 morphogenetic mutants obtained mapped at the two loci: *abacus* and *bristle* (Clutterbuck, 1969). If this result is taken at face value, either these two loci are likely to be regulatory for structural genes essential to hyphal growth as well as conidiation or else the sequence of development is in some sense autoregulatory, in which case the gene products of morphogenetic loci would again have regulatory functions, but they would be in addition to structural or enzymatic functions.

The sequence of regulation, if not its mechanism, is strikingly illustrated by the two phenol oxidasés described above, in particular the cresolase. This enzyme, although not present in non-leaky *bristle* mutants, is found, apparently at the wild-type level, in any *bristle* mutants which appear at all leaky as judged by their morphology. Thus even the least activity of the *bristle* gene product, whatever that may be, is sufficient to switch on cresolase synthesis. None of the *bristle* mutants, on the other hand, produces any laccase unless the mutant is so leaky as to produce conidia, so in this case it seems that the *bristle* gene may have completed its activity before laccase is synthesized. These two enzymes can be taken as examples of many other unknown proteins whose regulation may follow the same pattern, for instance, the *abacus* activity must be switched on at the same time as the laccase, and presumably numerous other genes concerned with spore function will come into play either at the same time, or possibly when the *abacus* gene in turn has completed its action.

Conclusions

In summary, we may say that the use of mutants in studies of vegetative reproduction has shown that very few loci are specific to this phase of morphogenesis. The large numbers of mutants affecting both hyphal growth and conidiation show that these stages of the life cycle share many functions. Even when mutants with normal hyphae but abnormal conidiation are obtained, some of these are sexually sterile. Finally, there is as yet no evidence of separate regulatory and structural loci for each step of morphogenesis so the whole picture is one, if not of economy in numbers of genes involved, at least of opportunism in the use of existing genes in a variety of conditions.

12.4 References

AXELROD, D. E. (1972). Kinetics of differentiation of conidiophores and conidia by colonies of *Aspergillus nidulans*. *Journal of General Microbiology* **73**, 181–4.

AXELROD, D. E., GEALT, M. & PASTUSHOK, M. (1973). Gene control of developmental competence in *Aspergillus nidulans*. *Developmental Biology* **34**, 9–15.

BAINBRIDGE, B. W. & TRINCI, A. P. J. (1969). Colony and specific growth rates of normal and mutant strains of *Aspergillus nidulans*. *Transactions of the British Mycological Society* **53**, 473–5.

BARBATA, G., VALDES, L. & SERMONTI, G. (1973). Complementation among developmental mutants in *Aspergillus nidulans*. *Molecular and General Genetics* **126**, 227–32.

BARRATT, R. W. & GARNJOBST, L. (1949). Genetics of a colonial microconidiating mutant strain of *Neurospora crassa*. *Genetics* **34**, 351–69.

BOUVIER, J. (1967). Un nouvel organisme, le *Penicillium baarnense* Van Beyma, pour l'étude des phénomènes de développement. *Comptes Rendus de l'Académie des Sciences, Paris, Serie D* **265**, 1305–8.

BOUVIER, J. (1968). Induction d'un hétérothallisme *non balancé* chez le *Penicillium baarnense van Beyma*, Ascomycète homothallique. *Comptes rendus de l'Académie des Sciences, Paris, Serie D* **266**, 220–3.

BOUVIER, J. (1969). Ordre d'action des gènes et nombre de noyaux impliqués dans l'initiation et le développement des fructifications du *Penicillium baarnense* Van Beyma. *Comptes rendus de l'Académie des Sciences, Paris, Series D* **269**, 171–4.

BRODY, S. (1970). Correlation between reduced nicotinamide adenine dinucleotide phosphate levels and morphological changes in *Neurospora crassa*. *Journal of Bacteriology* **101**, 802–7.

BRODY, S. (1973). Metabolism, cell walls, and morphogenesis. In *Developmental regulation: aspects of cell differentiation*, pp. 107–154, edited by S. J. Coward. New York & London: Academic Press.

BRODY, S. &. NYC, J. F. (1970). Altered fatty acid distribution in mutants of *Neurospora crassa*. *Journal of Bacteriology* **104**, 780–6.

BRODY, S. &. TATUM, E. L. (1966). The primary biochemical effect of a morphological mutation in *Neurospora crassa*. *Proceedings of the National Academy of Sciences, U.S.A.* **56**, 1290–7.

BRODY, S. &. TATUM, E. L. (1967). Phosphoglucomutase mutants and morphological changes in *Neurospora crassa*. *Proceedings of the National Academy of Sciences, U.S.A.* **58**, 923–30.

BUTLER, G. M. (1961). Growth and branching systems in *Coprinus disseminatus* *Annals of Botany* **25**, 341–52.

CLUTTERBUCK, A. J. (1969). A mutational analysis of conidial development in *Aspergillus nidulans*. *Genetics* **63**, 317–27.

CLUTTERBUCK, A. J. (1972). Absence of laccase from yellow-spored mutants of *Aspergillus nidulans*. *Journal of General Microbiology* **70**, 423–35.

CLUTTERBUCK, A. J. (1975). Cyclic AMP levels during growth and conidiation. *Aspergillus New Letter* **12**, 13–15.

CLUTTERBUCK, A. J. (1977). The genetics of conidiation of *Aspergillus nidulans*. In *The physiology and Genetics of* Aspergillus. Edited by J. E. Smith and J. Pateman. In press.

COHEN, J., KATZ, D. & ROSENBERGER, R. F. (1969). Temperature sensitive mutant of *Aspergillus nidulans* lacking amino sugars in its cell wall. *Nature, London* **224**, 713–15.

CROCKEN, B. J. & NYC, J. F. (1964). Phospholipid variations in mutant strains of *Neurospora crassa*. *Journal of Biological Chemistry* **239**, 1727–30.

DORN, G. L. (1970). Genetic and morphological properties of undifferentiated and invasive variants of *Aspergillus nidulans*. *Genetics* **66**, 267–79.

DORN, G. L., MARTIN, G. M. & PURNELL, D. M. (1967). Genetic and cytoplasmic control of undifferentiated growth in *Aspergillus nidulans*. *Life Sciences* **6**, 629–33.

EDSON, C. M. & BRODY, S. (1976). Biochemical and genetic studies of galactosamine metabolism in *Neurospora crassa*. *Journal of Bacteriology* **126**, 799–805.

EGGERDING, C., RANDALL, J. A. & SARGENT, M. L. (1975). An altered invertase in the *cot*-2 mutant of *Neurospora crassa*. *Journal of General Microbiology* **89**, 102–12.

EMERSON, S. (1963). Slime: a plasmodioid variant of *Neurospora crassa*. *Genetica* **34**, 162–82.

FULLER, R. C. & TATUM, E. L. (1956). Inositol-phospholipid in *Neurospora* and its relationship to morphology. *American Journal of Botany* **43**, 361–5.

GARNJOBST, L. & TATUM, E. L. (1967). A survey of new morphological mutants in *Neurospora crassa*. *Genetics* **57**, 579–604.

GARNJOBST, L. & TATUM, E. L. (1970). New crisp genes and crisp-modifiers in *Neurospora crassa*. *Genetics* **66**, 281–90.

GEALT, M. A. & AXELROD, D. E. (1974). Coordinate regulation of enzyme inducibility and development competence in *Aspergillus nidulans*. *Developmental Biology* **41**, 224–32.

HANNAN, M. A. (1975). Isolation and analysis of conidiation defective mutants of *Aspergillus niger*. *Molecular and General Genetics* **142**, 333–40.

HERMAN, Č. & CLUTTERBUCK, A. J. (1966). A method for selection of auxotrophs by means of 'spidery' growth. *Aspergillus News Letter* **7**, 13–14.

JINKS, J. L. (1959). The genetic basis of 'duality' in imperfect fungi. *Heredity* **13**, 525–8.

JINKS, J. L. (1966). Extranuclear inheritance. In *The fungi*, Vol. II, pp. 619–60, edited by G. C. Ainsworth & A. S. Sussman. New York and London: Academic Press.

KAPLAN, N. O., COLOWICK, S. P. & NASON, A. (1951). *Neurospora* diphosphopyridine nucleotidase. *The Journal of Biological Chemistry* **191**, 473–83.

LECHNER, J. F., FUSCALDO, K. E. & BAZINET, G. (1971). Genetical and biochemical studies of hexose monophosphate shunt in *Neurospora crassa*. II. Characterization of biochemical defects of the morphological mutants colonial-2 and colonial -3. *Canadian Journal of Microbiology* **17**, 789–94.

MAHADEVAN, P. R. & TATUM, E. L. (1965). Relationship of the major constituents of the *Neurospora crassa* cell wall to wild-type and colonial morphology. *Journal of Bacteriology* **90**, 1073–81.

MAHADEVAN, P. R. & TATUM, E. L. (1967). Localization of structural polymers in the cell wall of *Neurospora crassa*. *Journal of Cell Biology* **35**, 295–302.

MARTINELLI, S. D. & CLUTTERBUCK, A. J. (1971). A quantitative survey of conidiation mutants in *Aspergillus nidulans*. *Journal of General Microbiology* **69**, 261–8.

MATSUYAMA, S. S., NELSON, R. E. & SIEGEL, R. W. (1974). Mutations specifically blocking differentiation of macroconidia in *Neurospora crassa*. *Developmental Biology* **41**, 278–87.

MISHRA, N. C. (1977). Genetics and biochemistry of morphogenesis in *Neurospora*. *Advances in Genetics* **19**, 341–405.

MISHRA, N. C. & TATUM, E. L. (1970). Phosphoglucomutase mutants of *Neurospora sitophila* and their relation to morphology. *Proceedings of the National Academy of Sciences, U.S.A.* **66**, 638–45.

MORRIS, N. R. (1975). Mitotic mutants of *Aspergillus nidulans*. *Genetical Research* **26**, 237–54.

MURAYAMA, T. & ISHIKAWA, T. (1975). Characterization of *Neurospora crassa* mutants deficient in glucosephosphate isomerase. *Journal of Bacteriology* **122**, 54–8.

NELSON, R. E., SELITRENNIKOFF, C. P. & SIEGEL, R. W. (1975*a*). Mutants of *Neurospora* deficient in nicotinamide adenine nucleotide (phosphate) glycohydrolase. *Journal of Bacteriology* **122**, 695–709.

NELSON, R. E., SELITRENNIKOV; C. P. & SIEGEL, R. W. (1975*b*). Cell changes in *Neurospora*. In *Cell cycle and cell differentiation*, pp. 291–310, edited by J. Reinert and H. Holtzer. Berlin, Heidelberg & New York: Springer-Verlag.

OLIVER, P. T. P. (1972). Conidiophore and spore development in *Aspergillus nidulans*. *Journal of General Microbiology* **73**, 45–54.

PARISOT, D. (1972*a*). Quelque conséquences physiologiques de mutations affectant la reproduction du *Penicillium baarnense* Van Beyma. *Mycopathologia at Mycologia Applicata* **46**, 33–52.

PARISOT, D. (1972*b*). Effets pléiotropes de gènes contrôlant les processus de reproduction chez *Penicllium baarnense* Van Beyma. *Annales des Sciences Naturelles, Botanique, Paris* 12e *Série* **13**, 1–18.

PASTUSHOK, M. & AXELROD, D. E. (1976). Effect of glucose, ammonium and media maintainance on the time of conidiophore initiation by surface colonies of *Aspergillus nidulans*. *Journal of General Microbiology* **94**, 204–10.

PETERSON, P. A. (1970). Controlling elements and mutable loci in maize: their relationship to bacterial episomes. *Genetica* **41**, 33–56.

RAPER, K. B. & FENNELL, D. I. (1953). Heterokaryosis in *Aspergillus*. *Journal of the Elisha Mitchell Scientific Society* **69**, 1–29.

ROPER, J. A. (1958). Nucleo-cytoplasmic interactions in *Aspergillus nidulans*. *Cold Spring Harbor Symposia on Quantitative Biology* **23**, 141–53.

RYAN, F. J., BEADLE, G. W. & TATUM, E. L. (1943). The tube method of growth measurement of *Neurospora*. *American Journal of Botany* **30**, 784–99.

SAXENA, R. K. & SINHA, U. (1973). Conidiation of *Aspergillus nidulans* in submerged liquid culture. *Journal of General* and Applied Microbiology 19, 141–6.

SCHMIT, J. C. & BRODY, S. (1976). Biochemical genetics of Neurospora crassa conidial germination. Bacteriological Reviews 40, 1–41.

SCOTT, W. A. (1971). Physical properties of glucose-6-phosphate dehydrogenase from Neurospora crassa. Journal of Biological Chemistry 246, 6353–9.

SCOTT, W. A. & ABRAMSKY, T. (1973). Neurospora 6-phosphogluconate dehydrogenase. II. Properties of two purified mutant enzymes. Journal of Biological Chemistry 248, 3542–5.

SCOTT, W. A. & BRODY, S. (1973). Effects of suppressor mutations on nonallelic glucose-6-phosphate dehydrogenase mutants of Neurospora crassa. Biochemical Genetics 10, 285–95.

SCOTT, W. A. & SOLOMON, B. (1975). Adenosine 3',5'-cyclic monophosphate and morphology in Neurospora crassa: drug-induced alterations. Journal of Bacteriology 122, 454–63.

SCOTT, W. A. & TATUM, E. L. (1970). Glucose-6-phosphate dehydrogenase and Neurospora morphology. Proceedings of the National Academy of Sciences U.S.A. 66, 515–22.

SCOTT, W. A., MISHRA, N. C. & TATUM, E. L. (1973). Biochemical genetics of morphogenesis in Neurospora. Brookhaven Symposia in Biology 25, 1–18.

SELITRENNIKOFF, C. P., NELSON, R. E. & SIEGEL, R. W. (1974). Phase-specific genes for macroconidiation in Neurospora crassa. Genetics 78, 679–90.

SIEGEL, R. W., MATSUYAMA, S. S. & UREY, J. C. (1968). Induced macroconidia formation in Neurospora crassa. Experientia 24, 1179–81.

TÉLLEZ-IÑÓN, M. T. & TORRES, H. N. (1970). Interconvertible forms of glycogen phosphorylase in Neurospora crassa. Proceedings of the National Academy of Sciences, U.S.A. 66, 459–63.

TERENZI, H. F., FLAWIÁ, M. M. & TORRES, H. N. (1974). A Neurospora crassa mutant showing reduced adenylate cyclase activity. Biochemical and Biophysical Research Coomunications 58, 990–6.

TERENZI, H. F., FLAWIÁ, M. M., TELLEZ-IÑÓN, M. T. & TORRES, H. N. (1976). Control of Neurospora crassa morphology by cyclic adenosine 3',5'-monophosphate and dibutyryl adenosine 3',5'-monophosphate. Journal of Bacteriology 126, 91–9.

DE TERRA, N. & TATUM, E. L. (1963). A relationship between cell wall structure and colonial growth in Neurospora crassa. American Journal of Botany 50, 669–77.

TRINCI, A. P. J. (1973a). Growth of wild type and colonial mutants of Neurospora crassa in batch culture and on agar medium. Archiv für Mikrobiologie 91, 113–26.

TRINCI, A. P. J. (1973b). The hyphal growth unit of wild type and spreading colonial mutants of Neurospora crassa. Archiv. für Mikrobiologie 91, 127–36.

TRINCI, A. P. J. & COLLINGE, A. J. (1973). Structure and plugging of septa of wild type and spreading colonial mutants of Neurospora crassa. Archiv für Mikrobiologie 91, 355–64.

TURIAN, G. & BIANCHI, D. E. (1972). Conidiation in Neurospora. Botanical Review 38, 119–54.

UREY, J. C. (1971). Enzyme patterns and protein synthesis during synchronous conidiation in Neurospora crassa. Developmental Biology 26, 17–27.

VALENTINE, B. P. & BAINBRIDGE, B. W. (1975). Properties and chromosomal locations of two mannose mutants. Aspergillus News Letter 12, 31.

VUJIČIĆ, R. & MUNTANJOLA-CVETKOVIĆ (1973). A comparative ultrastructural study of conidium differentiation in the Cladosarum-like mutant 22B of Aspergillus aureolatus. Journal of General Microbiology 79, 45–51.

WEINMAN-GREENSHPAN, D. & GALUN, E. (1969). Complementation in non-conidiating mutants of Trichoderma. Journal of Bacteriology 99, 802–6.

WRATHALL, C. R. & TATUM, E. L. (1974). Hyphal wall peptides amd colonial morphology in Neurospora crassa. Biochemical Genetics 12, 59–68.

YUILL, E. & YUILL, J. L. (1938). Cladosarum olivaceum, a new Hyphomycete. Transactions of the British Mycological Society 22, 194–200.

ZONNEVELD, B. J. M. (1974). α-1,3 glucan synthesis correlated with α-1,3 glucanase synthesis, conidiation and fructification in morphogenetic mutants of Aspergillus nidulans. Journal of General Microbiology 81, 445–51.

CHAPTER 13

Sexual Morphogenesis in the Phycomycetes

H. VAN DEN ENDE

13.1 Introduction

For taxonomists, sex is the chief connecting principle for the fungi of the order Mucorales (Zygomycetes). The morphological aspects of zygospore formation which takes place by fusion of two isogamous gametangia are very similar throughout the group (*cf.* Zycha, Siepmann & Linneman, 1969). Sexual reproduction in these fungi has proved to be an appealing field of research for developmental biologists, biochemists, and students of secondary metabolism. In this chapter, I will review what is known of the physiology of sexual development in the Mucorales. In Volume II of this series, Bu'Lock (1976) has briefly exposed the subject in a different context. The reader is also referred to some valuable review articles written by Gooday (1973), Bu'Lock, Jones & Winskill (1976), and Sutter (1976).

13.2 The Sexual System

Although by the beginning of this century zygospores in mucoraceous fungi were considered to be the result of sexual reproduction, the mechanism of their formation was a matter of much confusion and conflict. That it is much better understood nowadays is largely the work of two outstanding mycologists, Blakeslee and Burgeff.

Blakeslee (1904) demonstrated that in many species, two strains or races can be distinguished which, when grown apart, produce only sporangia, but produce zygospores when grown together. Blakeslee called fungi which exhibited this type of conjugation (the interaction of two differing thalli), *heterothallic*. Many other members of the Mucorales, which reproduced sexually under suitable conditions within one thallus he called *homothallic*. Being convinced that both strains of a heterothallic species were of different sex, though morphologically indistinguishable, he designated them *plus* and *minus* mating types. Most important, of course, was his finding that this bipolarity could be extended to all known heterothallic forms, and that by interspecific matings the *plus* or *minus* characteristic could be assigned to most if not all sexually competent isolates. Even in the absence of genetic evidence he believed that sex was genotypically determined because his strains could be subcultured indefinitely without a change in the qualitative sense (Blakeslee, Welch & Berger, 1927; Blakeslee & Cartledge, 1927; *cf.* also Kniep, 1929). The fact that all available strains could be classified as *plus* or *minus*, without sex intergrades, was reason to suspect that there was something fundamental common to *plus* strains on one hand, and *minus* strains on the other.

Also in homothallic strains Blakeslee found sexual differences which were reminiscent of the *plus/minus* bipolarity: some strains showed incomplete interspecific reactions with heterothallic *plus* or *minus* strains only, others with both or with none. Apparently, even to homothallic strains a *minus*, *plus*, or neutral 'tendency' could often be attributed, though as a rule the distinction was not as clear-cut as in heterothallic forms (Satina & Blakeslee, 1930).

13.3 Sexual Morphogenesis

Basically, the sexual process encompasses the following events. When two compatible hyphae touch, they both may (but do not always) produce a swelling (progametangium) at the site of contact. The progametangia stick together and become delimited by a septum. The resulting anterior cell is the gametangium. Fusion of a pair of gametangia, leading to zygospore formation, takes place by dissolution of the connecting walls. At the posterior side, the supporting hyphae (the suspensors) swell considerably and become yellow coloured. Mature zygospores generally have a dark, thick, crenated or warty cell wall, which is extremely resistant to environmental stress. The developmental course may differ in detail from species to species, and it may be worthwhile to consider some of them more closely.

A *plus* and a *minus* mycelium of *Mucor mucedo* (*Mucoraceae*), growing into proximity with each other, each produce zygophores along the mutual borderline (even without cell-to-cell contact). Zygophores are stout aerial hyphae, differing in appearance from vegetative hyphae. Frequently they are curved in the direction of the partner (zygotropism). The situation is completely symmetrical, though *plus* zygophores may, in general, be somewhat longer. Only on zygophores which have come into sexual contact with each other are gametangia formed.

The fact that zygophores are typical sexual structures, produced only by an encounter with the sexual partner, has made *Mucor mucedo* a popular species for the study of how the mating types interact. In other species, like *M. hiemalis* and *Absidia glauca*, no clearly discernable zygophores are formed. Nevertheless, we must assume that, during sexual reproduction, specialized hyphae are produced which have the ability to fuse with partner cells, because cell fusion (anastomosis) does not occur in vegetative mycelia of the Mucorales (Burnett, 1968). In *Blakeslea trispora* (Choanephoraceae) and *Philobolus crystallinus* (*Pilobolaceae*) the formation of gametangia takes place within a knot of entwined hyphae under the surface of the substrate (Weber & Wolf, 1927; Krafzcyk, 1935). Also in *Phycomyces blakesleeanus* (*Mucoraceae*) the primary functional contact is made under the substrate surface, between partner zygophores which, as in *M. mucedo*, are distinguishable from vegetative hyphae and are formed prior to mycelial contact. They are short, stubby, multibranched hyphae which show oriented growth towards the opposite mating type. After pairing, the zygophores enlarge, grow upward and produce the zygospore in the air (see Sutter, 1975).

Thus sexual morphogenesis in heterothallic Mucorales may be characterized by the following features: (1) specialized hyphae are produced by distance interaction between the mates; (2) these hyphae often show positive chemotropic behaviour to their respective partners, and (3) upon contact, produce swollen protuberances which after delimitation become gametangia.

Also in homothallics, zygotropism and contact induction of gametangia formation are regular phenomena. In *Zygorhynchus moelleri* (*Mucoraceae*) the process starts with the formation of a septum in an erect, aerial hypha. Immediately below the septum, a side branch is produced, which grows curving upward towards the filament delimited by the septum just formed. At the place of contact, a perpendicular outgrowth is produced, which becomes one of the gametangia. The other gametangium is produced by the side branch (*cf.* Green, 1927). From this behaviour, a sexual difference between the main branch and the side branch seems evident. Also, the cytoplasm of the side branch differs from that of the terminal, delimited, hypha in that it contains many oil globules. In *Zygorhynchus exponens*, conjugation takes place in a similar way between gametangia arising on branches of the same filament which, however, are only rarely separated by a septum. Often a side branch 'misses' the main branch it originated from, and subsequently produces a secondary side branch with which it fuses, as in *Absidia spinosa* (*Mucoraceae*, Burgeff, 1924). Apparently, sexual differentiation in these species is quite flexible. In *Mucor genevensis*, the gametangia are borne on two completely individual hyphae (Ling-Young, 1930).

Zygospore formation in homothallic species is to a large extent determined by external conditions like substrate content, atmospheric humidity, *etc.* (Hawker, 1957). In particular, *Syzigites megalocarpus* (*Mucoraceae*) has been the object of many studies (*e.g.* Blakeslee, 1904; Wenger & Lilly, 1966). Interestingly, this species may produce zygospores in abundance in the neighbourhood of heterothallic species, like *Mucor mucedo* (Werkman & van den Ende, 1974). Other species, like *Cunninghamella echinulata* (*Choanephoraceae*) and an agamic strain of *Zygorhynchus moelleri* have been reported to be stimulated by the proximity of heterothallic species (Blakeslee, 1920; Schipper, 1971). I will comment upon this phenomenon in a later section (p. 268).

13.4 Sex Hormones in the Mucorales: the Trisporic Acids and their Precursors

From the foregoing discussion it will be evident that zygospore production is preceded by cellular interactions at different stages of the sexual process. In many heterothallic species the mating partners exert some influence upon each other without cell-to-cell contact, which leads to the formation of sexual structures, the zygophores. These structures in turn react tropically to each other, enhancing the chances of sexual contact, a phenomenon also displayed by homothallic species. Finally, physical contact between the sexual structures seems to trigger the formation of gametangia. Whatever the means may be by which the mates interact, they do not appear to have a high degree of species specificity, because imperfect matings, *i.e.* sexual reactions without zygospore formation, between different species or even genera, are a common phenomenon.

In 1924, Burgeff proved that zygophore formation in *Mucor mucedo* is governed by low molecular weight substances which diffuse through the substrate and the atmosphere. He covered the mycelium of one mating type with a collodion membrane and placed a piece of agar with a mycelium of the opposite mating type on top of it, in an inverted position. The latter mycelium, growing over the membrane, produced numerous zygophores, which evidently had been induced by emanations of the lower mycelium passing through the membrane. This result was observed irrespective whether the *plus* or the *minus* mating type was underneath. Also, the zygophores were seen to curve downward to the membrane. Burgeff made similar observations with other heterothallic species like *Mucor hiemalis* and *Phycomyces blakesleeanus*. In *M. hiemalis* he demonstrated that zygophore-inducing substances could be transferred through the air, because sexual development occurred when the opposite mating types were in close proximity on separate substrates. Hepden & Hawker (1961) later demonstrated the existence of volatile material, produced by the homothallic *Rhizopus sexualis*, stimulating sexual morphogenesis in this species.

Burgeff (1924) concluded from his study, which also involved many interspecific combinations, that the agents responsible for sex-specific zygophore induction and their orientation were common to many different species and genera of the Mucorales, but final proof, of course, had to await the structural elucidation of these substances.

The first steps in that direction were made by Banbury (1954) and Plempel & Braunitzer (1958) who showed that *plus* and *minus* mycelia of *Mucor mucedo*, grown together in a liquid culture, accumulate substances in the medium which induce zygophores in both mating types of the same fungus. Plempel (1963) was able to a large extent to isolate and purify these substances and to determine many of their properties. The assessment of their structures, however, was the result of a different type of study concerning another fungus, *Blakeslea trispora*, which was carried out by an Italian group (Caglioti *et al.*, 1967; Cainelli, Grasselli & Selva, 1967). These workers sought to identify a soluble factor in mated cultures of *Blakeslea trispora* which dramatically enhanced carotenoid production in this fungus. Two compounds appeared to be responsible, which were named trisporic acids B and C. Trisporic acid A was biologically inactive and has been gnored since. Its chemical structure is uncertain (Bu'Lock, Drake & Winstanley, 1972). Subsequent studies proved that the same trisporic acids

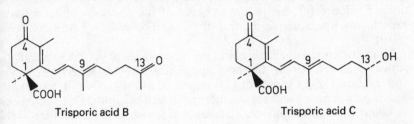

Trisporic acid B   Trisporic acid C

(TA) are active as zygophore-inducing agents in *Mucor mucedo* and, in fact, are also produced by mated cultures of this and other fungi, although in much lower amounts (van den Ende, 1968; Gooday, 1968; Austin, Bu'Lock & Gooday, 1969; Sutter, 1975).

The absolute stereochemistry of TA was investigated by Reschke (1969) and Bu'Lock, Austin, Snatzke & Hruban (1970). On the basis of the circular dichroism of the hydrogenation product with a saturated side chain, it was concluded that the carboxymethyl group at C1 has the axial position, which implies that C1 has the S-configuration. The second centre of asymmetry, in TA C, at C13, has the R-configuration, as was shown by degradation of TA C to the lactone of D-(-)-γ-hydroxy valeric acid. *Cis/trans* isomerization at C9 is possible by light without much effect upon the biological activity (Reschke, 1969; see p. 266). Both isomers could be separated as the methyl esters by thin-layer chromatography (Bu'Lock *et al.*, 1972).

Total synthesis of TA was accomplished by Edwards *et al.* (1971), Isoe, Hayase & Sakan (1971) and White & Sung (1974). The synthetic compounds appeared to have the same biological activity as the natural products (Edwards *et al.*, 1971).

13.5 Biosynthesis of the Trisporic Acids

As can be deduced from their structures, the TA probably are degradation products of the cyclic carotenoids, β- and γ-carotene, which are ubiquitous in heterothallic and many homothallic Mucorales. Tracer studies by Bu'Lock and associates support this suggestion. ^{14}C-labelled β-carotene, retinol and a number of C20 and C18-derivatives are efficiently incorporated into TA (Bu'Lock *et al.*, 1974a). Particularly, a 'β-C18-ketone' and its 4-hydroxyl derivative appeared to serve as good precursors for TA in mated cultures. Probably, a series of cleavage reactions leads from the carotenoids

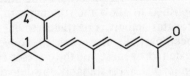

β-C18-Ketone

to the TA skeleton. The hydroxylation at C4 might be an early step in the biosynthesis. The exact nature of these early intermediates is not known however. One of the problems is their extremely low steady state concentrations. Evidence for this is that in some of the above-mentioned trace experiments practically no dilution of the label occurred (*cf.* Bu'Lock *et al.* 1976).

More evidence is available about the latter stages of the TA biosynthetic pathway. Sutter (1970), Sutter, Capage, Harrison & Keen (1973) and Sutter, Harrison & Galasko (1974) were able to explain why TA synthesis occurs only in mated and not in single cultures. They found that *plus* as well as *minus* mycelia secrete small amounts of sex-specific metabolites which are converted to TA by the mating partner only. They suggested that the complete biosynthesis of TA is only accomplished by the co-operative action of both mating types. This was substantiated by the subsequent identification of these metabolites (Bu'Lock, Winskill & Jones, 1974b; Nieuwenhuis & van den Ende, 1975). These compounds and their relationships are shown in Fig. 13.1. From *plus* cultures of *Blakeslea trispora* and *Mucor mucedo* the methyl 4-dihydrotrisporates (IV), and from the corresponding *minus* cultures the trisporins (V) were identified as the major TA precursors. In addition, the trisporols (III) were found in extracts of *minus* cultures of *B trispora* (Austin, Bu'Lock & Drake, 1970).

The *plus*-derived methyl 4-dihydrotrisporates (IV) are converted to TA by *minus* cultures only. This implies that the *minus* strain is able to oxidize the hydroxyl group at C4 to a carbonyl group. Hydrolysis of the carboxymethyl group at C1 is also confined, though less exclusively, to the *minus* strain (Bu'Lock *et al.*, 1972). The *minus*-derived trisporins (V) are converted to TA only in *plus* strains, probably via the trisporols (Nieuwenhuis & van den Ende, 1975; Austin *et al.*, 1970). Apparently, the *plus* possesses the exclusive ability to oxidize the axial methyl group at C1. Thus TA synthesis can proceed in mated cultures as the result of the combination of two incomplete but complementary synthetic pathways, which, at least partly, exhibit strong sex specificity. It is noteworthy that these results were obtained with *Mucor mucedo* as well as with *Blakeslea trispora*, which confirms that the observed specificities are sex-linked, and not the result of strain differences.

Some of the mating-type specific reactions in the *minus* strain of *Mucor mucedo* were more closely investigated by Werkman (1976). The oxidation of methyl 4-dihydrotrisporates (IV) to methyl trisporates (II) is mediated by a NADP-specific dehydrogenase, which is present in the *minus* strain only. Experimentally, this means that a dialysed 120 000 g supernatant of a *minus* homogenate converts methyl 4-dihydrotrisporates to methyl trisporates with concomitant reduction of NADP. In a corresponding *plus* preparation this reaction is virtually absent. However, enzyme purification studies have not yet been accomplished to allow the statement that the proteinaceous species responsible for this activity is restricted to one mating type only.

An esterase has been partially characterized, converting methyl trisporates to TA. Both the dehydrogenase and the esterose enzymes have a K_m of 1 mM and a pH optimum of 7.8. Presumably, the dehydrogenation precedes the demethylation because methyl 4-dihydrotrisporate esterase activity is hardly observed *in vitro*, in contrast to methyl trisporate esterase activity;

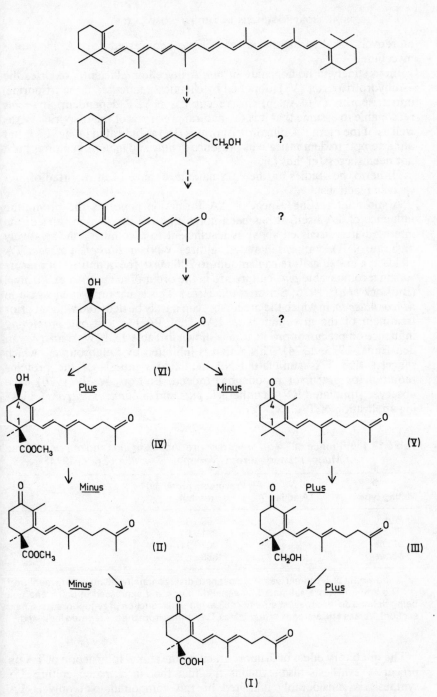

Fig. 13.1 Biosynthetic pathway leading to trisporic acid B. I: trisporic acid B; II: methyl trisporate B; III: trisporol B; IV: methyl 4-dihydrotrisporate B; V: trisporin B; VI: 4-dihydrotrisporin B. The biosynthesis of trisporic acid C is assumed to proceed in a similar way (replace —COCH₃ at C12 by —CHOHCH₃, although interconversion between intermediates of the B- and C-series also might be possible.

moreover, 4-dihydrotrisporic acid is not converted to TA by *minus* mycelium.

Interestingly, a homogenate of *minus* mycelium efficiently oxidizes the 4-dihydrotrisporins (VI) (obtained by chemical reduction of the trisporins) into trisporins (V). Again, this reaction is NADP dependent. It seems reasonable to assume that it is the natural precursor of the trisporins (V) as well as of the methyl 4-dihydrotrisporates (IV) as indicated in Fig. 13.1, but our attempts to demonstrate its presence in *plus* and in *minus* cultures have not been successful thus far.

Hitherto, no studies on the enzymatic level have been reported of *plus* specific reactions.

A most interesting aspect of TA formation concerns the promoting influence of TA itself. It has been noted that the production rates of the intermediates discussed above is insufficient to account for the relatively high rate of TA synthesis in mated cultures. While in *Blakeslea trispora* TA yields in a mated culture can amount to 200 mg/l, the quantities of precursors in a comparable *plus* culture are in the order of magnitude of 0.2 mg/l (Bu'Lock *et al.*, 1976; Sutter *et al.*, 1973). This situation is even worse in *Mucor mucedo*, in which the precursors can hardly be detected without prior treatment of the mycelium with TA. That TA has a strong promoting influence on precursor production is shown in Table 13.1 (Werkman & van den Ende, 1973, 1974). This action is inhibited by 5-fluorouracil, which suggests that TAs stimulate RNA-mediated synthesis of the proteins required for precursor production. Indeed, Bu'Lock *et al.* (1976) have observed stimulated RNA synthesis in *plus* and in *minus B. trispora* following application of TA.

Table 13.1 Influence of TA on precursor production in *plus* and *minus* cultures of *Mucor mucedo* (From Werkman & van den Ende, 1974)

Mating type	TA added[a]?	Precursors produced (units)[b]
Plus	no	69 ± 39
Plus	yes	891 ± 81
Minus	no	207 ± 117
Minus	yes	1863 ± 1053

[a] The amount of TA added was 54 μg per petri dish, containing 20 ml culture medium;
[b] The amounts were estimated by serial dilutions and expressed in units, one unit being defined as the smallest quantity still exhibiting induction of zygophores (see next section). Values are averages of duplicates. Deviations from the average are indicated.

The inhibitory effect of 5-fluorouracil on the stimulating action of TA on precursor synthesis also explains the fact that in a mated culture TA synthesis is considerably inhibited by this compound. Not only is TA synthesis affected when 5-fluorouracil is added to a mated culture (van den Ende, Werkman, Reyngoud & Hendriks, 1970), but also when one of either mycelia is pretreated with this inhibitor (Bu'Lock & Winstanley, 1971; van den Ende, Werkman & Briel, 1972), which indicates that high

TA production in a mated culture cultures is largely the result of the stimulatory action of the end product.

While this is evidence favouring the view that enzymes involved in precursor *synthesis* are more abundant after TA treatment, Werkman (1976) has found that also the dehydrogenase and the esterase in the *minus* mating type, *transforming* precursors to TA have higher activity in TA-treated mycelia of *Mucor mucedo* (Table 13.2). Thus TAs have a profound influence at different sites in the TA-biosynthetic pathway. It is tempting to speculate that the dehydrogenase, acting to produce TA by oxidizing the C4 hydroxyl group in the methyl 4-dihydrotrisporates (IV), is also involved in the conversion of the 4-dihydrotrisporins (VI) to trisporins (V), which in fact concerns the same reaction at C4. In the same fashion, the reaction from the 4-dihydrotrisporins (VI) to the methyl 4-dihydrotrisporates (IV) in the *plus* strain, could be catalyzed by the same enzyme system which is involved in the conversion of the trisporins (V) to TA, because both reactions transform the axial methyl group at C1 to a carboxymethyl group. Confirmation of this could be brought about by testing the substrate specificities of extensively purified enzymes. If true, it would greatly simplify the system, and reduce the difference between *plus* and *minus* mating types to the presence or absence of one of the enzyme systems.

Table 13.2 Stimulation by TA of the enzyme activities involved in the conversion of methyl 4-dihydrotrisporates and methyl trisporates in *Mucor mucedo* (From Werkman, 1976).

Mating type	TA added?	TA (μmoles/h/mg protein) formed from	
		methyltrisporates	methyl 4-dihydrotrisporates
Plus	no	1.7 ± 0.8	0
Plus	yes	2.0 ± 1.4	0
Minus	no	95.7 ± 3.3	12.9 ± 0.2
Minus	yes	195.3 ± 5.8	24.1 ± 0.6

Four-day-old mycelia of *Mucor mucedo* were pretreated with 1 mg TA per 100 ml submerged culture. After 18 h the mycelia were homogenized and centrifuged at 50 000 g. Supernatants were incubated with 5.4 μmoles methyl 4-dihydrotrisporates, 0.5 μmoles NADP and 0.1 μmoles FAD or with 9.2 μmoles methyl trisporates for 16 h. TAs were isolated by extraction and chromatography on DEAE Sephadex columns.

It should be stressed that in these studies no evidence was found that TA could affect the sex specificity of the reactions discussed above. They do not evoke the ability to perform a reaction which is otherwise lacking in a particularly mating type; in other words, they do not confer upon a single heterothallic strain the ability to produce TA by itself. In no case, where ^{14}C-labelled mycelia were incubated with unlabelled TA, could the production of radioactive TA be detected. Also back reactions (which might explain increased intermediate production from added TA) were not observed (B. A. Werkman and H. van den Ende, unpublished results).

13.6 Biological Properties of Trisporic Acids and their Precursors

Zygophore induction

The most remarkable property of TAs is their ability to induce zygophore formation in *both* mating types of *Mucor mucedo* when applied in a well in front of the mycelium in a solid substrate (van den Ende, 1968; Gooday, 1968; Bu'Lock *et al.*, 1972). The *minus* strain generally reacts more vigorously than the *plus* strain, as is shown in Table 13.3, but the difference

Table 13.3 Comparative bioassays on *plus* and *minus* strains of *Mucor mucedo* (From Bu'Lock *et al.*, 1972).

| Compounds | Zygophores per μg | |
	Plus	*Minus*
9-*cis*-trisporic acid C	150	550
9-*trans*-triporic acid C	150	420
9-*cis*-trisporic acid B	260	910
9-*trans*-trisporic acid B	160	450

is seen to be variable when sets of data are compared. The most active isomer, however, is always 9-*cis*-TA B. A typical dose–response curve for a mixture of TA B and TA C is presented in Fig. 13.2 (Wurtz & Jockusch, 1975).

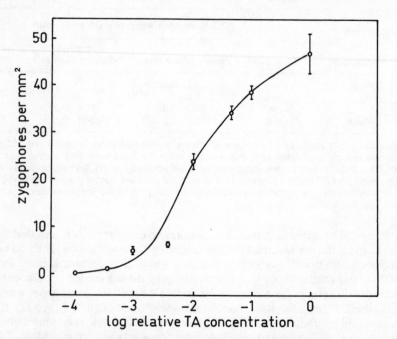

Fig. 13.2 Dose response relationship for the zygophore-inducing action of a mixture of trisporic acids B and C on *Mucor mucedo*. Concentration 1 corresponds to approx. 30 mM in TA; 50 μl-samples of solution were applied (From Wurtz & Jockusch, 1975).

Various precursors induce zygophores, but they display a much stronger sex specificity than TA. While the activity of the methyl trisporates are comparable with those of TA, the trisporols and trisporins evoke zygophores only on the *plus* strain, and the methyl 4-dihydrotrisporates only on the *minus* strain of *Mucor mucedo* (Bu'Lock *et al.*, 1972; Nieuwenhuis & van den Ende, 1975). The activities are equal or slightly lower in comparison with those of TA (T. Wurtz, personal communication). This is compatible with the idea that these precursors derive their activity and specificity from the fact that they are transformed to TA by enzymes which are present in one *or* the other mating type (see preceding section). However, there are no solid arguments against the alternative view that these compounds might have zygophore-inducing activity *per se*, which would imply a different perception mechanism for TA and their precursors. Wurtz & Jockusch (1975) described a mutant of *M. mucedo-minus* which responded to methyl 4-dihydrotrisporates but not to TA. This mutant might have a reduced sensitivity towards TA compared with the methyl 4-dihydrotrisporates; a more plausible explanation, however, provided by the authors, is that the mutant might be blocked in the uptake of TA but not of less polar precursors, which intracellularly could be converted to TA.

Wurtz & Jockusch (1975) found that this and other mutants, not responsive to TA, still showed unimpaired mating behaviour with a wild-type partner. This suggests that TAs do not play an indispensable role in the communication process between the mating mycelia. Also the findings of Mesland, Huisman & van den Ende (1974) and Nieuwenhuis & van den Ende (1975) point in this direction. They found that the sex-specific precursors, which are rather volatile, also induce zygophores via the air. This could be demonstrated by placing a glass vial with an aqueous solution of these precursors in front of the appropriate mycelium of *Mucor mucedo*. Two vegetative mycelia of opposite mating type (producing precursors, not TA), positioned on either side of a slit in the supporting agar, responded to each other's presence by producing zygophores. Thus it seems that TA precursors also have an important role in sexual interaction between the mates. One may envisage that the primary role of TA (formed intracellularly from cross-diffusing intermediates) is the induction of differentiated structures in the mycelia in which they were formed (*cf.* Gooday, 1968; Sutter *et al.*, 1973). In those cases where relatively large amounts of TA are secreted into the medium, as in mated *Blakeslea trispora*, their secondary role could be the overall stimulation of sexual activity in both mating types present in the culture.

The TAs are not only active in *Mucor mucedo* but also in other, related, fungi, like *Phycomyces blakesleeanus* (Sutter, 1975). In homothallic organisms like *Syzigites megalocarpus* (Werkman & van den Ende, 1974) and *Mucor genevensis* (personal observation), they strongly stimulate zygospore production. In an agamic strain of *Zygorhynchus moelleri* no zygospores are produced without added TA.

Stimulation of TA synthesis

In the preceding section we saw that TAs promote the synthesis of TA precursors but also have a stimulating action on the enzymatic activities involved in the sex-specific conversion of these precursors to TAs. The fact that inhibitors of protein synthesis block this action suggests that TA have a controlling action on the synthesis of enzymes involved in TA synthesis. There is reason to presume that this is part of the primary role which I attributed to TA above, namely the induction of sexual structures. This follows from the fact that at least part of the enzymatic machinery involved in TA synthesis (the 4-dihydrotrisporin and methyl 4-dihydrotrisporate dehydrogenase in *Mucor mucedo-minus*) is predominantly localized in the zygophores (Werkman, 1976). Thus, the overall enzyme activity might be directly dependent on the zygophore density in a given mycelium.

Stimulation of carotenoid production

Many authors have noted that sexual reproduction in numerous species is accompanied by increased carotenoid synthesis (*e.g.* Hesseltine, 1961; Barnett, Lilly & Krause, 1956). Particularly in sexual structures, the zygophores, suspensors, and gametangia, carotenoids are abundant. In *Blakeslea trispora* it was shown that the TA are largely responsible for this phenomenon (Sutter & Rafelson, 1968; van den Ende, 1968). Thomas & Goodwin (1967) and Thomas, Harris, Kirk & Goodwin (1967) demonstrated that this action of TA is inhibited by cycloheximide, which suggests that the synthesis of one or more enzymes, limiting overall carotenogenesis, is affected. These enzymes probably function early in the isoprenoid pathway, since also the production of sterols, ubiquinones and prenols is stimulated by TA (Bu'Lock & Osagie, 1973), although not in the same proportions. Bu'Lock (1973) reported that inhibiting sterol biosynthesis with the compound SKF 3301–A markedly affected the formation of zygophores. This suggests that enhancement of sterol biosynthesis is related to the primary action of TA, the induction of sexual morphogenesis. The same might be true for the increase of carotenogenesis, but apart from the fact that a functional carotenogenesis is a prerequisite for TA biosynthesis (Sutter, 1975), the only indication about the role of high carotene levels in sexual structures comes from Gooday, Fawcett, Green & Shaw (1973), who found that β-carotene is oxidatively polymerized to give sporopollenin, a component of the zygospore wall, where its function is to protect the cell against chemical and biological attack.

13.7 The Hormonal System in Homothallic Mucorales

There are several lines of evidence indicating that in homothallic species, sex hormones are operative which are similar, if not identical, to those found in heterothallic species. At interspecific pairings between homo- and heterothallic mycelia, both frequently exhibit sexual morphogenesis, in which the heterothallic zygophores are often oriented towards the homothallic mycelium. Some homothallics produce zygospores only in the presence of a heterothallic strain, or after addition of biologically active preparations derived from heterothallic cultures (Werkman & van den Ende, 1974). These authors were able to demonstrate trace amounts of trisporins and TA in *Zygorhynchus moelleri* and TA in *Z. heterogamus*. Moreover, enzyme activities by which TA were produced from their precursors were found in several homothallic species. Of these, the *minus*-specific dehydrogenase activities which convert the 4-dihydrotrisporins and methyl 4-dihydrotrisporates to trisporins and methyl trisporates, respectively, were found to be localized intracellularly in a rather specific fashion. As mentioned above, zygospore formation in *Z. moelleri* proceeds by the production of a lateral branch which copulates with a higher, and delimited, part of the main hypha. It is in this terminal cell that the dehydrogenase activities are predominantly localized, as appeared from a cytochemical study by Werkman (1976). This is quite significant, because this part of the hypha not only attracts the lateral filament, but, in conjunction with a heterothallic species, also attracts *plus* zygophores. Thus its *minus* character is expressed by two means: by its action on *plus* zygophores, and by the fact that it harbours a *minus*-specific enzyme system. In a sexually incompetent strain, this enzyme system was found to be completely absent (personal observation). No biochemical data are available for the copulating side branch, which does not attract zygophores of any kind in interspecific matings.

In *Zygorhynchus heterogamus*, it seems that the situation is the reverse. While the terminal hypha attracts *minus* zygophores (although less conspicuously than *Z. moelleri* attracts *plus* zygophores) and thus exhibits a *plus* character, it does not contain the above-mentioned dehydrogenase.

The foregoing shows that in the species studied, the terminal cell becomes differentiated from the rest of the mycelium; in *Zygorhynchus moelleri* it attains a *minus*, and in *Z. heterogamus* a *plus* characteristic. This differentiation can be reversed. This follows from the fact that in *Z. exponens* and related species a lateral branch frequently becomes a terminal one (with concomitant change in sexual properties), after producing a secondary lateral hypha. The situation in homothallics is in contrast with heterothallic Mucors where the *plus* or *minus* genotypes are continuously expressed in the whole mycelium, although TA, formed by co-operative action of both mating types, is required for sexual morphogenesis. Of course, more species will have to be studied before this concept about the physiological difference between homothallic and heterothallic strains can be generalized.

All homothallic species that have been dealt with thus far are homokaryotic (though not necessarily isogenic). In several heterothallic species, however, heterokaryotic mycelia arise from germinated zygospores, which contain *plus* as well as *minus* nuclei (Blakeslee, 1906; Orban, 1919). Their origin may be explained by the packaging of nuclei of both mating types into one germ spore (*cf.* Eslava, Alvarez & Delbrück, 1975). Generally, these heterokaryons are not very stable, but in some cases sexual development has been demonstrated. In *Pilobolus crystallinus*, for instance, these mycelia display continuous and uncontrolled sexual differentiation, as one would expect from two heterothallic and continuously expressed genotypes, combined in one thallus (Krafczyk, 1935). In *Phycomyces blakesleeanus*, these strains do not produce zygospores, but 'pseudophores', which resemble the suspensors of zygotes. Their sexual nature is uncertain. Sex heterokaryons of this species, exhibiting the same aberrant phenotype were also prepared artificially by Burgeff (1914, 1924) by microinjection of cytoplasm of one strain into sporangiophores of another strain. More recently, methods to obtain heterokaryons were developed by Heisenberg & Cerda-Olmedo (1968) and Ootaki (1973), by regenerating mixtures of cytoplasm or grafting sporangiophores of two strains to one another. Such heterokaryons yielded valuable information about the distribution and behaviour of genetically distinct nuclei in a highly multinucleate and aseptate mycelium, but about sexual heterokaryons very little was reported. In the hypothetical case that *plus* and *minus* nuclei were equally distributed in one mycelium, one could predict that they would complement each other and produce TA. By testing this, one would not only confirm the ideas presented above about the course of TA biosynthesis, but also yield information about the regulation of this process at the genetic level.

13.8 Zygotropism

One of the properties of zygophores in many Mucorales is the ability to become oriented towards each other, evidently as a response to sex-specific gaseous stimuli, emanated by the partners. Probably the observed curvatures are the result of directed growth, but there are no experimental data to support this. Several authors have observed that zygotropism is accompanied by an increase of the growth rate of zygophores (*e.g.* Gooday, 1975). Plempel (1960, 1962, 1963) and Plempel & Dawid (1961) reported experiments aimed at elucidating the nature of the agents responsible for zygotropism. They concluded that sex-specific, volatile, oxygen-sensitive and short-lived substances are involved. That we do not know more about these hormones might be due to the fact that it seems very difficult to design a proper bioassay with such fragile structures as zygophores.

Mesland *et al.* (1974) observed that zygophores are not only attracted by zygophores, but also, though to a lesser degree, by vegetative mycelium and sporangiophores. These authors suggested that volatile, sex-specific TA precursors, which effect zygophore induction through the air, might be identical to the zygotropic agents (*cf.* also Gooday, 1975; Bu'Lock *et al.*, 1976). Although this view was not substantiated experimentally, it has some appeal for two reasons. In the first place, the methyl 4-dihydrotrisporates and trisporins stimulate zygophore growth just as has been reported for the zygotropic hormones (Mesland *et al.*, 1974). In the second place, one may expect that if zygotropic hormones are in fact TA precursors (or their direct derivatives), they are mainly produced by those parts of the mycelium in which the TA-producing system is concentrated. As we saw above, the *minus* zygophores in *Mucor mucedo* and the terminal copulating cell in *Zygorhynchus moelleri* have been recognized as such. These are exactly the structures which exert an attracting influence on *plus* zygophores. Unfortunately, we can not extend this speculation to structures attracting *minus* zygophores, because no *plus*-specific enzymes involved in TA biosynthesis have been described.

13.9 Is there Sexual Communication by Intercellular Contacts?

The last stage of sexual morphogenesis, the formation of gametangia, takes place after zygophores of different mating type have made cellular contact. It has been speculated that these events are induced by cell contact ('thigmotropism'). Bu'Lock (1974), for instance, has suggested that cellular interaction is mediated by species- and sex-specific macromolecules, similar to the agglutinative glycoproteins of the two mating types in *Hansenula wingei* (Crandall, Lawrence & Saunders, 1974). It is equally feasible, however, that this interaction takes place by means of transmissable substances. Burgeff (1924) has even hypothesized that the same type of substances are involved in typical distance interactions, like zygophore nduction and zygotropism, and also in those reactions which seem to be riggered by cell-to-cell contact. Blakeslee (1904) and Gooday (1975) have tressed the point that asymmetric bulging of the zygophore just before contact can be considered as the beginning of progametangium formation. n fact, the formation of subapical branches or 'pegs' is a common phenomenon observed in chemotropically active systems (*cf.* van den Ende, 1976). In *Pilobolus crystallinus*, Krafczyk (1935) has noted that the delimitation of gametangia, which normally proceeds after cellular contact between sexual hyphae, can also be seen now and then without prior contact with the partner cell. He concludes from his observations that there is not an essential but rather a gradual difference between interaction-at-a-distance and interactions which proceed after cell-to-cell contact. Though this evidence is at best only suggestive, it seems to favor Burgeff's view that at all stages of sexual morphogenesis the same hormones are involved.

13.10 Prospects

From the evidence, presented in this chapter, the following concept may be formulated. In heterothallic Mucorales, the biosynthetic pathway leading to TA has a central function in sexual communication and morphogenesis. Intermediates in this pathway may be the signalling substances in all stages of interaction between the mates, while the end-products are responsible for the induction of sexual morphogenesis. One essential difference between the *plus* and the *minus* mating types on the biochemical level is the ability to carry out differential parts of this pathway, so that both mating types produce different biologically active intermediates, and at the same time are complementary with respect to TA biosynthesis. This system is also operative in homothallic fungi. The locally differentiated sexual structures presumably harbour *plus* or *minus* specific parts of the TA biosynthetic pathway. Probably also the induction of sexual morphogenesis in these fungi is governed by substances derived from this pathway or by TA itself.

However, much of our knowledge is fragmentary. In order to understand more fully the nature of the mating system of homothallic and heterothallic Mucorales, the mode of action of TA and its intermediates will have to be clarified. Also, very little is known of the genetics of the group. The production of sexual mutants and heterokaryons by several workers, and of course the development of methods to germinate zygospores and to produce sexual progeny (*e.g.* Eslava *et al.*, 1975; Cerdá-Olmedo, 1975), have opened new perspectives for future research in this intriguing group of fungi.

13.11 References

AUSTIN, D. J., BU'LOCK, J. D. & GOODAY, G. W. (1969). Trisporic acids: sexual hormones from *Mucor mucedo* and *Blakeslea trispora*. *Nature, London* **223**, 1178–9.

AUSTIN, D. J., BU'LOCK, J. D. & DRAKE, D. (1970). The biosynthesis of trisporic acids from β-carotene via retinal and trisporol. *Experimentia* **26**, 348–9.

BANBURY, G. H. (1954). Processes controlling zygophore formation and zygotropism in *Mucor mucedo* Brefeld. *Nature, London* **173**, 499–500.

BARNETT, H. L., LILLY, V. G. & KRAUSE, R. F. (1956). Increased production of carotene by mixed (+) and (−) cultures of *Choanephora cucurbitarum*. *Science* **123**, 141.

BLAKESLEE, A. F. (1904). Sexual reproduction in the *Mucorineae*. *Proceedings of the National Academy of Sciences, U.S.A.* **40**, 205–319.

BLAKESLEE, A. F. (1906). Zygospore germinations in the *Mucorineae*. *Annals of Mycology* **4**, 1–28.

BLAKESLEE, A. F. (1920). Sexuality in Mucors. *Science* **51**, 375–82.

BLAKESLEE, A. F. & CARTLEDGE, J. L. (1927). Sexual dimorphism in Mucorales. II. Interspecific reactions. *Botanical Gazette* **84**, 51–8.

BLAKESLEE, A. F., WELCH, D. S. & BERGNER, A. D. (1927). Sexual dimorphism in Mucorales. I. Intraspecific reactions. *Botanical Gazette* **84**, 27–50.

BU'LOCK, J. D. (1973). Comparative and functional spects of the isoprene pathway in fungi. *Pure and Applied Chemistry* **34**, 435–61.

BU'LOCK, J. D. (1974). Cascade expression of the mating-type locus in Mucorales. In *Proceedings of the second international symposium of genetics of industrial microorganism's* edited by K. D. Macdonald, pp. 497–509. London and New York: Academic Press.

BU'LOCK, J. D. (1976). Hormones in fungi. In *The filamentous fungi*, Vol. 2, pp. 345–368. Edited by J. E. Smith and D. R. Bessy. London: Edward Arnold (Publishers) Ltd.

BU'LOCK, J. D. & OSAGIE, A. U. (1973). Prenols and ubiquinones in single-strain and mated cultures of *Blakeslea trispora*. *Journal of General Microbiology* **76**, 77–83.

BU'LOCK, J. D. & WINSTANLEY, D. J. (1971). Carotenoid metabolism and sexuality in Mucorales. *Journal of General Microbiology* **68**, xvi–xvii.

BU'LOCK, J. D., AUSTIN, D. J., SNATZKE, G. & HRUBAN, L. (1970). Absolute configuration of trisporic acids and the stereochemistry of cyclization in β-carotene biosynthesis. *Chemical Communnications* 255–6.

BU'LOCK, J. D., DRAKE, D. & WINSTANLEY, D. J. (1972). Specificity and transformations of the trisporic acid series of fungal sex hormones. *Phytochemistry* **11**, 2011–18.

BU'LOCK, J. D., JONES, B. E., TAYLOR, D., WINSKILL, N. & QUARRIE, S. A. (1974*a*). Sex hormones in Mucorales. The incorporation of C20 and C18 precursors into trisporic acids. *Journal of General Microbiology* **80**, 301–6.

BU'LOCK, J. D., WINSKILL, N. & JONES, B. E. (1974*b*). Structures of the mating type specific prohormones of Mucorales. *Chemical Communications* 708–10.

BU'LOCK, J. D., JONES, B. E. & WINSKILL, N. (1976). The apocarotenoid system of sex hormones and prohormones in Mucorales. *Pure and Applied Chemistry*, in press.

BURGEFF, H. (1914). Untersuchungen über Variabilität, Sexualität und Erblichkeit bei *Phycomyces nitens* Kunze I. *Flora* **107**, 259–316.

BURGEFF, H. (1924). Untersuchungen über Sexualität und Parasitismus bei Mucorineen. *Botanische Abhandlungen* (K. Goebel, ed.) **4**, 1–135.

BURNETT, J. H. (1968). *Fundamentals of mycology*. London: Edward Arnold.

CAGLIOTI, L., CAINELLI, G., CAMERINO, B., MONDELLI, R., PRIETO, A., QUILICO, A., SALVATORI, T. & SELVA, A. (1967). The structure of trisporic acid-C acid. *Tetrahedron Supplement* **7**, 175–87.

CAINELLI, G., GRASSELLI, P. & SELVA, A. (1967). Struttura dell' acido trisporico B. *Chimica e l'Industria* (*Milan*) **49**, 628–9.

CERDÁ-OLMEDO, E. (1975). The genetics of *Phycomyces blakesleeanus*. *Genetical Research, Cambridge* **25**, 285–96.

CRANDALL, M., LAWRENCE, L. M. & SAUNDERS, R. M. (1974). Molecular complementarity of yeast glycoprotein mating factors. *Proceedings of the National Academy of Sciences, U.S.A.* **71**, 26–9.

EDWARDS, J. A., SCHWARTZ, V., FAJKOS, J., MADDOX, M. L. & FRIED, J. H. (1971). Fungal sex hormones. The synthesis of (±)-7(t), 9(t)-trisporic acid B methyl ester. The stereochemistry at C9 of the trisporic acids. *Chemical Communications* 292–3.

ENDE, H. VAN DEN (1968). Relationship between sexuality and carotene synthesis in *Blakeslea trispora*. *Journal of Bacteriology* **96**, 1298–303.

ENDE, H. VAN DEN (1976). *Sexual interactions in plants*. London & New York: Academic Press.

ENDE, H. VAN DEN, WIECHMANN, A. H. C. A., REYNGOUD, D. J. & HENDRIKS, T. (1970). Hormonal interactions in *Mucor mucedo* and *Blakeslea trispora*. *Journal of Bacteriology* **101**, 423–8.

ENDE, H. VAN DEN, WERKMAN, B. A. & BRIEL, M. L. VAN DEN (1972). Trisporic acid synthesis in mated cultures of the fungus *Blakeslea trispora*. *Archiv für Mikrobiologie* **86**, 175–84.

ESLAVA, A. P., ALVAREZ, M. I. & DELBRÜCK, M. (1975). Meiosis in *Phycomyces*. *Proceedings of the National*

Academy of Sciences, U.S.A. **72**, 4076–80.

GOODAY, G. W. (1968). Hormonal control of sexual reproduction in *Mucor mucedo*. *New Phytologist* **67**, 815–21.

GOODAY, G. W. (1973). Differentiation in the Mucorales. *Symposium Society of General Microbiology* **23**, 269–94.

GOODAY, G. W. (1975). Chemotaxis and chemotropism in fungi and algae. In *Primitive sensory and communication systems* Edited by M. J. Carlile, pp. 155–204, London & New York: Academic Press.

GOODAY, G. W., FAWCETT, P., GREEN, D. & SHAW, J. (1973). The formation of fungal sporopollenin in the zygospore wall of *Mucor mucedo*: a role for the sexual carotenogenesis in the Mucorales. *Journal of General Microbiology* **74**, 233–9.

GREEN, E. (1927). The life history of *Zygorhynchus moelleri* Vuill., *Annals of Botany* **41**, 419–35.

HAWKER, L. E. (1957). *The physiology of reproduction in fungi*. Cambridge University Press.

HEISENBERG, M. & CERDÁ-OLMEDO, E. (1968). Segregation of heterokaryons in the asexual cycle of Phycomyces. *Molecular and General Genetics* **102**, 187–95.

HEPDEN, P. M. & HAWKER, L. E. (1961). A volatile substance controlling early stages of zygospore formation in *Rhizopus sexualis*. *Journal of General Microbiology* **24**, 155–64.

HESSELTINE, C. W. (1961). Carotenoids in the fungi Mucorales. *U.S. Department of Agriculture, Technical Bulletin* no. 1245, 33 pp.

ISOE, S., HAYASE, Y. & SAKAN, T. (1971). Sexual hormones of the Mucorales. The synthesis of methyl trisporate B and C. *Tetrahedron Letters*, 3691–4.

KNIEP, H. (1929). Vererbungserscheinungen bei Pilzen. *Bibliographia Genetica* **5**, 371–415.

KRAFCZYK, H. (1935). Die Bildung und Keimung der Zygosporen von *Pilobolus crystallinus* und sein heterokaryotisch Myzel. *Beitrage zur Biologie der Pflanzen* **23**, 349–96.

LING-YOUNG, H. (1930). Etude biologique des phénoménes de la sexualité chez les Mucorinées. *Revue de Botanique* **495**, 144–58.

MESLAND, D. A. M., HUISMAN, J. G. & ENDE, H. VAN DEN. (1974). Volatile sexual hormones in *Mucor mucedo*. *Journal of General Microbiology* **80**, 111–17.

NIEUWENHUIS, M. & ENDE, H. VAN DEN (1975). Sex specificity of hormone synthesis in *Mucor mucedo*. *Archives of Microbiology* **102**, 167–9.

OOTAKI, T. (1973). A new method for heterokaryon formation in *Phycomyces*. *Molecular and General Genetics* **121**, 49–56.

ORBAN, G. (1919). Untersuchungen über die Sexualität von *Phycomyces nitens*. *Beinhefte Botanischer Zentralblatt* **36**, 1–59.

PLEMPEL, M. (1960). Die zygotropische Reaktion bei Mucorineen. I. *Planta* **55**, 254–8.

PLEMPEL, M. (1962). Die zygotropische Reaktion bei Mucorineen. III. *Planta* **58**, 509–20.

PLEMPEL, M. (1963). Die chemischen Grundlagen der Sexualreaktion bei Zygomyceten. *Planta* **59**, 492–508.

PLEMPEL, M. & BRAUNITZER, G. (1958). Die Isolierung der Mucorineen-Sexualstoffe. I. *Zeitschrift für Naturforschung* **13b**, 302–5.

PLEMPEL, M. & DAWID, W. (1961). Die zygotropische Reaktion bei Mucorineen. II. *Planta* **56**, 438–46.

RESCHKE, T. (1969). Die Gamone aus *Blakeslea trispora*. Zur Struktur der Sexualstoffe aus Mucoraceae. I. *Tetrahedron Letters* **39**, 3435–9.

SATINA, S. & BLAKESLEE, A. F. (1930). Imperfect sexual reactions in homothallic and heterothallic Mucors. *Botanical Gazette* **90**, 299–311.

SCHIPPER, M. A. A. (1971). Induction of zygospore production in *Mucor saximontensis*, and agamic strain of *Zygorhynchus moelleri*. *Transactions of the British Mycological Society* **56**, 157–9.

SUTTER, R. P. (1970). Trisporic acid synthesis in *Blakeslea trispora*. *Science* **168**, 1590–2.

SUTTER, R. P. (1975). Mutations affecting sexual development in *Phycomyces blakesleeanus*. *Proceedings of the National Academy of Sciences, U.S.A.* **72**, 127–30.

SUTTER, R. P. (1976). Regulation of the first stage of sexual development in *Phycomyces blakesleeanus* and in other mucoreous fungi. In *Eukaryotic microbes as model developmental systems*. Edited by D. H. O'Day and P. A. Horgen. New York: Marcel Dekker Inc.

SUTTER, R. P. & RAFELSON, M. E. (1968). Separation of β-factor from stimulated β-carotene synthesis in mated cultures of *Blakeslea trispora*. *Journal of Bacteriology* **94**, 426–32.

SUTTER, R. P., CAPAGE, D. A., HARRISON, T. L. & KEEN, W. A. (1973). Trisporic acid biosynthesis in separate *plus* and *minus* cultures of *Blakeslea trispora*: identification by Mucor assay of two mating-type specific components. *Journal of Bacteriology* **114**, 1074–82.

SUTTER, R. P., HARRISON, T. L. & GALASKO, G. (1974). Trisporic acid biosynthesis in *Blakeslea trispora* via mating-type-specific precursors. *Journal of Biological Chemistry* **249**, 2282–4.

THOMAS, D. M. & GOODWIN, T. W. (1967). Studies on carotenogenesis in *Blakeslea trispora*. I. General observations on synthesis in mated and unmated strains. *Phytochemistry* **6**, 355–60.

THOMAS, D. M., HARRIS, R. C., KIRK, J. T. O. & GOODWIN, T. W. (1967). Studies on carotenogenesis in *Blakeslea trispora*. II. The mode of action of trisporic acid. *Phytochemistry* **6**, 361–6.

WEBER, G. F. & WOLF, F. A. (1927). Heterothallism in *Blakeslea trispora*. *Mycologia* **19**, 302–7.

WENGER, C. J. & LILLY, V. G. (1966). The effects of light on carotenogenesis, growth and sporulation of *Syzigites megalocarpus*. *Mycologia* **58**, 671–80.

WERKMAN, B. A. (1976). Localization and partial characterization of a sex-specific enzyme in homothallic and heterothallic Mucorales. *Archives of Microbiology* **109**, 209–13.

WERKMAN, B. A. & ENDE, H. VAN DEN (1973). Trisporic acid synthesis in *Blakeslea trispora*. Interaction between *plus* and *minus* mating types. *Archiv für Mikrobiologie* **90**, 365–74.

WERKMAN, B. A. & ENDE, H. VAN DEN (1974). Trisporic acid synthesis in homothallic and heterothallic Mucorales. *Journal of General Microbiology* **82**, 273–8.

WHITE, J. D. & SUNG, W. L. (1974). Alkylation of Hageman's ester. Preparation of an intermediate for trisporic acid synthesis. *Journal of Organic Chemistry* **39**, 2323–8.

WURTZ, T. & JOCKUSCH, H. (1975). Sexual differentiation in *Mucor*: trisporic acid response mutants and mutants blocked in zygophore development. *Developmental Biology* **43**, 213–20.

ZYCHA, H., SIEPMANN, R. & LINNEMAN, G. (1969). *Mucorales*. Berlin: Verlag von J. Cramer.

CHAPTER 14

Dikaryon Formation in Higher Basidiomycetes

LORNA A. CASSELTON

14.1 Introduction

In the typical life cycle of most higher basidiomycete fungi (the Hymenomycetes and Gasteromycetes) two functionally different mycelial states can be distinguished, the *homokaryon* and the *dikaryon*.

The *homokaryon* is the primary mycelium which develops on germination of a single haploid sexual spore. The term homokaryon derives from the fact that only a single nuclear type is present. In some species, but not all, the homokaryon is a *monokaryon*, a name often used synonymously with homokaryon but implying that each cell of the mycelium contains just a single nucleus. The homokaryon is capable of indefinite vegetative growth and is usually sexually sterile, but may produce abundant numbers of asexual spores.

A mycelium which contains nuclei of different genotypes is termed a *heterokaryon*. Heterokaryons generally arise as a result of anastomosis between homokaryotic hyphae of different origin. The *dikaryon* or secondary mycelium of the basidiomycete fungi is a specialized type of heterokaryon in which two genetically different haploid nuclei are associated in pairs in each cell. The dikaryotic state is the normal prerequisite for sexual reproduction, but like the homokaryon, the dikaryon is capable of indefinite vegetative growth, and it is the predominant mycelial state in nature. The two nuclei in each cell remain discrete during somatic cell divisions and it is only after the formation of the fruit-body that they eventually fuse in specialized reproductive cells, the *basidia*, to give diploid nuclei which almost immediately undergo meiosis. In general, four basidiospores develop on each basidium and each derives one of the four haploid nuclei produced at meiosis.

14.2 Morphological Observations

Morphological differences

The homokaryon and the dikaryon are characteristically different in morphology. The clamp connection is considered to be the diagnostic feature which distinguishes dikaryotic hyphae from homokaryotic hyphae (Fig 14.1a,b). Clamps are only present in the dikaryon and always occur at a septum because of their involvement in cell division. Although there are species in which the dikaryon does not have clamp connections (*e.g. Coprinus congregatus*) these must be considered atypical.

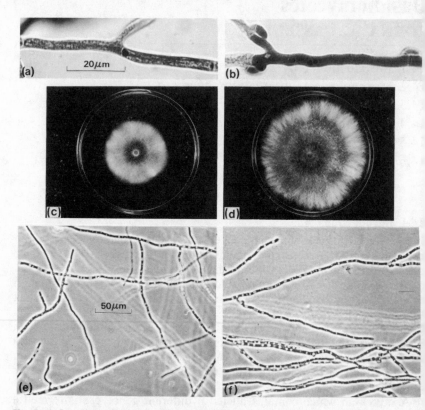

Fig. 14.1 Some characteristic differences between homokaryotic and dikaryotic mycelia. *Polyporus ciliatus* (a) Uninucleate homokaryotic cell with unclamped septa (b) Binucleate dikaryotic cell with clamp connections (Courtesy of K. Esser and U. Stahl). *Coprinus cinereus* Equal age colonies of (c) a homokaryon and (d) a dikaryon. (e) Wide-angled branching of homokaryotic hyphae (f) Acute-angled branching of dikaryotic hyphae.

Apart from clamp connections, there may be other characteristic differences. In *Coprinus cinereus*, for example, the dikaryotic hyphae are thicker than those of the homokaryon, approximately 7μm in diameter compared with 4μm; the angle of branching is very different, wide-angled in the homokaryon and acute-angled in the dikaryon (Fig. 14.1e,f) and radial growth of the dikaryon is much faster than that of the homokaryon with a corresponding reduction in hyphal density (Fig. 14.1c, d). In this species also the homokaryon produces uninucleate asexual spores (oidia) on aerial branches whereas the dikaryon produces subsurface binucleate chlamydospores.

Dikaryon formation—the morphogenetic sequence

Several events can be distinguished during dikaryon formation as illustrated in Fig. 14.2. The initial step involves fusion between genetically different homokaryotic hyphae. The fusion cells thus contain two different nuclei. There is an exchange of nuclei which is followed by a very interesting phenomenon first described in *Coprinus cinereus** by Buller (1931). Nuclei derived from one homokaryon migrate throughout the established hyphae of the other homokaryon. This is a reciprocal process so that both homokaryons become dikaryotized.

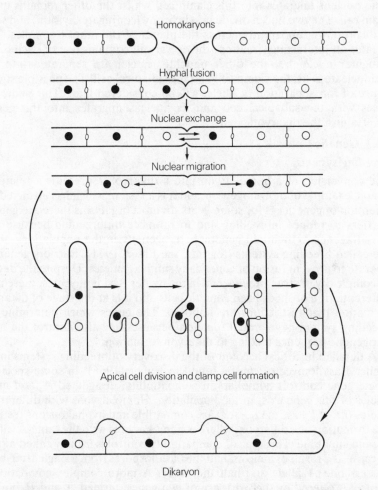

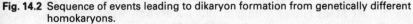

Fig. 14.2 Sequence of events leading to dikaryon formation from genetically different homokaryons.

---

* Classified by Buller as *Coprinus lagopus* and referred to by this name in all the literature cited.

As a result of nuclear migration, the apical cells of the mycelium contain two nuclei, one derived from each of the original homokaryons. These two nuclei become closely associated and constitute the dikaryotic pair, and it is from these cells that the dikaryotic mycelium develops. Nuclear migration therefore enables the dikaryon to escape from the surrounding mycelial mass into uncolonized substrate. Once the dikaryotic condition has been established, the regular distribution of the two nuclei in each cell is maintained by a complex cell division involving formation of the clamp connections.

Prior to cell division, a clamp cell develops on the side of the apical cell. One nucleus migrates into this clamp cell whilst the other remains in the main cell. The two nuclei divide in synchrony (conjugate division) and septa are laid down, apparently across the planes of the mitotic spindles. This creates an interesting situation; the apical cell contains one of each of the daughter nuclei but the other daughter nuclei are separated into two uninucleate cells, the clamp cell and the subapical cell. The possible significance of this separation of nuclei will be considered later. The clamp cell fuses with the subapical cell and its nucleus migrates into this cell to re-establish the dikaryotic pair.

14.3 Genetic Controls

Breeding systems

The conversion of a homokaryon into a dikaryon represents a relatively simple example of morphogenesis and it is a system which has attracted the attention of geneticists for some years. In most organisms there are genetic barriers to reduce inbreeding and to promote outbreeding because it is advantageous to maintain genetic heterogeneity in the population—these are called breeding systems (for refs. see Esser, 1971). Barriers generally operate to prevent fusion of genetically similar gametes. Despite the degree of complexity of structure reached in the higher Basidiomycetes, there is no differentiation of specialized gametic cells and it is at the stage of dikaryon formation that restrictions are imposed. The genes which determine the breeding system have a dual function because they also control the morphogenetic sequence leading to dikaryon formation.

A detailed historical account of the discovery of breeding systems in the higher Basidiomycetes can be found in Raper (1966). In some species a single gene controls homokaryon compatibility, designated A, and many alleles of this gene exist in the population. Homokaryons with different A alleles (*i.e.* $A1 + A2$; $A1 + A3$) are compatible and hyphal fusions result in the formation of a dikaryon, whereas homokaryons with the same A alleles are incompatible. This type of compatibility control is usually called *bipolar* because as a result of meiosis, half the basidiospores from a single fruit-body possess one A allele and half the other. A more complex compatibility control is offered by the existence of two genes, termed A and B, both of which have many alleles. Compatibility occurs only when alleles of both A and B are different (*i.e.* $A1B1 + A1B2$). If one or other allele is common, hyphal fusion is possible but not dikaryon formation. A and B are located in different chromosomes and four combinations of the A and B alleles are derived in the basidiospores from a single fruit body in equal frequency (*e.g.*

A1B1, A1B2, A2B1, A2B2). Species having the two gene system are therefore termed *tetrapolar*.

Providing a large number of *A* and *B* alleles exist, the chances of incompatible fusions arising in nature are small. In the tetrapolar species *Schizophyllum commune*, Raper, Krongelb & Baxter (1958) identified 96 different *A* alleles and 56 different *B* alleles from a world-wide sample of 114 homokaryons. In *Coprinus cinereus* 36 *A* alleles and 32 *B* alleles have been characterized (Day, 1963a).

It is more appropriate to refer to *A* and *B* as incompatibility factors rather than genes because it is now known from studies with *Schizophyllum commune* that they are not single genes but each composed of two closely linked genes designated α and β. α and β are separable by recombination, an event first clearly demonstrated by Papazian (1954) for the *A* factor and confirmed for both factors by the more extensive studies of Raper (Raper, Baxter & Ellingboe, 1960; Raper, Baxter & Middleton, 1958).

The large numbers of *A* and *B* allelic specificities found in nature are generated in an extremely economical way. The α and β loci of each factor have different alleles and the specificity of each factor is determined by its particular combination of α and β. Every combination has a unique specificity. In a cross involving the *A* factors $A\alpha1\beta1$ and $A\alpha2\beta2$, recombination will generate the new $\alpha\beta$ combinations $A\alpha1\beta2$ and $A\alpha2\beta1$. These have a different *A* specificity from the original $\alpha\beta$ combinations. Given the fact that in *Schizophyllum* at least 9 *A*α alleles and 32 *A*β alles exist (Raper *et al.*, 1960) a total of 288 possible *A* factors can be generated. Similarly, for the *B* factor, 9 *B*α and 9 *B*β alleles have been identified giving a possible 81 *B* factors (Parag & Koltin, 1971). This is a minimum estimate since more α and β alleles may exist than have so far been identified (see Koltin, Stamberg & Lemke, 1972).

It is assumed that this two gene structure of the *A* and *B* factors is true of all tetrapolar forms, though investigations have been few. The presence of α and β genes for the *A* factor of *Coprinus cinereus* (Day, 1963a) and *Collybia velutipes* (Takemaru, 1961) have been confirmed and α and β genes for the *B* factor of *Pleurotus ostreatus* (Terakawa, 1957, 1960) and *Collybia velutipes* (Takemaru, 1961).

From the point of view of an effective breeding system it is important that the α and β loci are closely linked in order to limit compatibility between the basidiospore progeny of the same fruit-body (sibs). In the absence of recombination between the α and β loci, the probability of compatible fusions between sib-homokaryons is 25%. Recombination between α and β, by generating different *A* and *B* factors, acts to increase sib compatibility. In *Schizophyllum commune* the frequency of recombination between the two genes of both factors is variable, depending on the general genetic background of the homokaryons and ranges from 1 to 17% for the *A* genes (Raper *et al.*, 1960) and 0.1 to 8% for the *B* genes (Koltin, Raper & Simchen, 1967).

It is interesting to note that there is no evidence for a two gene structure to the *A* factor in bipolar species even though the number of different *A* factors may be numerous, *e.g.* 33 in *Fomes cajanderi* (Neuhauser & Gilbertson, 1971) and 20 in *Polyporus palustris* (Flexer, 1969).

In a minority of species, some 10–15% (Whitehouse, 1949; Quintanilha & Pinto-Lopes, 1950) dikaryon formation occurs in the absence of hyphal fusions between different homokaryons. In at least some of these, apparent self-fertility results from the fact that each basidiospore contains two non-identical nuclei. This is exemplified by *Agaricus bisporus* (Raper, Raper & Miller, 1972). Basically, *A. bisporus* appears to be a bipolar species but only two basidiospores develop on each basidium, instead of the usual four, and each contains two nuclei which have different *A* factors. The requirement for dikaryon formation is thus satisfied at the time of spore germination and the homokaryotic phase is effectively by-passed.

The incompatibility factors are considered to be regulatory genes which control the action of the many other genes which must be involved in the morphogenetic sequence outlined in Fig. 14.2 (Raper, 1966). In bipolar species all these events are controlled by a single factor (Flexer, 1969) but in tetrapolar forms the *A* and *B* factors control different parts of the sequence. It is for this reason that the tetrapolar species are of particular interest. The two fungi which have been studied in most detail are *Schizophyllum commune* and *Coprinus cinereus*. This is because they are widely used in genetical studies and many biochemical and morphological mutants exist which have facilitated the study of dikaryon morphogenesis. Although the account which follows mainly relates to these two fungi, the conclusions to be drawn will undoubtedly provide a basis for understanding the events in dikaryon formation in all tetrapolar forms.

Heterokaryons—separation of the A and B sequences

Dikaryon formation is only possible when both *A* and *B* factor alleles are different, but hyphal fusions between incompatible homokaryons can give rise to heterokaryotic mycelia.

Three incompatible associations are possible depending on which factors are common: common $AB(A=B=)$, common $B(A \neq B=)$ and common A $(A=B \neq)$. The symbols in parenthesis were introduced by Raper as a useful way to indicate the factor which is common = and the factor which is different $\neq$. The fully compatible dikaryotic association is represented as $A \neq B \neq$.

When only one of the factors is common, part of the morphogenetic sequence leading to dikaryon formation takes place and this gives each of the two types of heterokaryon a distinctive hyphal morphology and makes it possible to determine which parts of the morphogenetic sequence are controlled by the different factors (Quintanilha, 1933; Papazian, 1950; Fulton, 1950; Raper & San Antonio, 1954; Swiezynski & Day, 1960*a*, *b*; Snider & Raper, 1965; Raper, 1966).

THE COMMON *B* HETEROKARYON $(A \neq B=)$ This arises only at the site of fusions between homokaryons with similar *B* factors showing that nuclear migration is blocked when the *B* alleles are the same. Many of the apical hyphal cells contain paired nuclei and at cell division a clamp cell is formed and the two nuclei undergo conjugate division. The normal sequence shown in Fig. 14.2 continues as far as septum formation which separates one of each of the daughter nuclei in the subapical cell and the clamp cell. However, the clamp cell fails to fuse with the subapical cell and its nucleus is trapped, leaving the subapical cell uninucleate. The unfused clamp cells are called pseudo or false clamps. Since it is only the failure of clamp cell fusion which prevents dikaryon formation, it is evident that one role of the clamp is to provide a mechanism to exclude the continued dikaryotic association of nuclei with common B alleles.

COMMON A HETEROKARYON $(A = B \neq)$ Nuclear exchange and nuclear migration follow hyphal fusions between homokaryons with similar A alleles. Active nuclear migration continues in the heterokaryotic mycelium so that nuclear distribution is highly irregular. Some cells are anucleate, others have one, two or more nuclei (Raper & Raper, 1966). Because the A factors are common there is no nuclear pairing or conjugate division and no clamp cells are formed. In both *Coprinus cinereus* and *Schizophyllum commune* the $A = B \neq$ heterokaryon has a distinctive type of growth. This is particularly marked in *S. commune* and was termed 'flat' by Papazian (1950). Growth is sparse with little aerial mycelium, the cells are distorted in shape and abundant short hyphal branches develop. This aberrant morphology can be correlated with certain biochemical changes to be described later.

COMMON AB HETEROKARYON $(A = B =)$ When both the A and B factors are common, no morphological changes can be detected as would be expected (Swiezynski & Day, 1960a; Middleton, 1964).

The partial sequence operating in common factor heterokaryons is controlled by the factor which is different and the sequence blocked is controlled by the factor which is common. This is summarized in Table 14.1. It follows that *the A factor controls conjugate nuclear division and clamp cell formation. The B factor controls nuclear migration, firstly during the establishment of the dikaryotic condition and secondly by clamp cell fusion.*

A and B factor mutations

Because they can display only part of the morphogenetic sequence leading to dikaryon formation, heterokaryons are generally infertile. This has provided a technique for obtaining mutations in both the A and B factors. Normal fruit-bodies produced by $A = B \neq$ heterokaryons of *Coprinus cinereus* (Day, 1963b) and *Schizophyllum commune* (Raper, Boyd & Raper, 1965) arose because of mutation in one of the A factors common to both nuclei. Mutation allowed the formation of a normal dikaryon and half the basidiospore progeny derived from the fruit-bodies carried the mutated factor. Fruiting of $A \neq B =$ heterokaryons may also result because of mutation in one of the common B factors. Mutated B factors in *S. commune* have been obtained (Parag, 1962; Koltin, 1968) but attempts to select similar mutations in *C. cinereus* have failed until very recently (R. W. Haylock & L. A. Casselton, unpublished results). In *C. cinereus* as in a number of other tetrapolar species (see Raper, 1966) $A \neq B =$ heterokaryons can occasionally fruit normally and it would appear that different B factors are not essential for sexual reproduction. In the absence of a selective technique, mutated factors are difficult to identify.

The mutated A and B factors obtained by fruiting heterokaryons do not have a new specificity. Mutation in all cases leads to self compatibility, in other words the complete breakdown of compatibility control. The mutated factor is compatible with itself, the factor from which it was derived and all

Table 14.1 Characteristics of the four types of heterokaryon formed in tetrapolar species.

Heterokaryon genetic constitution	Morphogenetic Sequence		Hyphal Morphology nuclear distribution
	Operating	Blocked	
$A \neq B \neq$ dikaryon ($A1B1 + A2B2$)	nuclear migration conjugate nuclear division clamp cell formation clamp cell fusion	—	
$A = B \neq$ common A ($A1B1 + A1B2$)	nuclear migration	conjugate nuclear division clamp cell formation	
$A \neq B =$ common B ($A1B1 + A2B1$)	conjugate nuclear division clamp cell formation	nuclear migration clamp cell fusion	
$A = B =$ common AB ($A1B1 + A1B1$)	—	nuclear migration conjugate nuclear division clamp cell formation	

* ○ $A1B1$ nucleus; ● $A2B2$ nucleus; ⊜ $A1B2$ nucleus; ⊜ $A2B1$ nucleus

other factors. These *primary* mutations, as they are called, offer no clue as to the origin of the large numbers of different α and β alleles of each factor which are found in nature. Self-incompatibility can be restored by a *secondary* mutation in the B factor but despite the large number of secondary mutations obtained (Raper *et al.*, 1965; Raper & Raudaskoski, 1968; Raper & Raper, 1973) none has ever generated a new wild type factor. Raper & Raudaskoski (1968) suggested that new alleles may originate from existing alleles by a series of mutational steps but all secondary mutations in the B factor cause loss of function rather than a change in function. This loss of function may be slight, the doubly mutant factor interacting normally with all wild type B factors except the one from which it was derived. In this latter interaction the mutant nucleus can migrate through cells of the wild type homokaryon but cannot promote the reciprocal nuclear migration. At the other extreme there is total sterility, a doubly mutated B factor which cannot interact with any wild type B factor to promote either nuclear migration or clamp cell fusion.

Primary A and B factor mutations are particularly interesting. A homokaryon carrying a primary A factor mutation resembles in every respect a $A \neq B =$ heterokaryon. The hyphae have binucleate apical cells in which conjugate nuclear division takes place and false clamp connections are formed. The primary A mutation is therefore equivalent to having two different A factors present and the morphogenetic sequence controlled by the A factor is turned on. Similarly, a homokaryon with a primary B factor mutation is identical in its aberrant morphology to a $A = B \neq$ heterokaryon. The B sequence is turned on and there is active nuclear migration leading to an irregular distribution of nuclei.

The availability of primary mutations in both the A and B factors of *Schizophyllum commune* made it possible to construct a homokaryon in which both mutated factors were present together in the same haploid nucleus. A fertile mycelium developed which resembled in all respects a normal dikaryon with binucleate cells and true clamp connections (Raper *et al.*, 1965).

Mutants are potentially far more useful than heterokaryons. Heterokaryons may be unstable and the distribution of the different nuclei is irregular. This means that only some of the cells of the mycelium are truly heterokaryotic. Every cell of a mutant exhibits A or B functions, depending on which factor is mutated, and therefore presents a completely homogeneous system for cytological and biochemical studies. What is only a transitory phenomenon in a compatible $A \neq B \neq$ interaction is a continuous process in primary mutants and can be studied in detail.

Ultrastructural studies

Little more can be learned about the morphogenetic sequence controlled by the A factor, but considerably more is known about nuclear migration controlled by the B factor. This was initiated by the ultrastructural studies of Giesy & Day (1965) which established the type of biochemical investigations which might ensue.

Nuclear migration at the onset of dikaryon formation involves extensive movement of nuclei through established homokaryotic hyphae. That this is possible is interesting because the septa between each cell have a very complex pore structure (Girbardt, 1961) termed the dolipore by Moore & McAlear (1962). This is illustrated in Figure 14.3a. The septum is swollen into a ring around the actual pore and the whole is surrounded by a multilayered dome-shaped membrane known as the parenthesome. Although the parenthesome is perforated, the pore often contains a plug of material which suggests that there is considerable restriction on movement of contents between cells. Certainly the size of the pore is too small to allow passage of nuclei.

In order that nuclei can migrate, the dolipore is dissolved away to leave a simple septum (Fig. 14.3b) through which nuclei can pass easily. Nuclear migration occurs after compatible hyphal fusions and in $A = B \neq$ heterokaryons. In both these mycelia some simple septa are found (Giesy & Day, 1965; Jersild, Mishkin & Niederpruem, 1967). No simple septa are found in the $A \neq B =$ heterokaryon where nuclear migration is blocked (Giesy & Day, 1965). That simple septa are degraded rather than stages in septal synthesis was confirmed by examining a primary B mutant homokaryon which mimics the $A = B \neq$ heterokaryon. Koltin & Flexer (1969) found that for some fifty hours after basidiospore germination, nuclear distribution and septal structure were normal in the B mutant but there was then a rapid change in nuclear distribution due to migration and, concomitant with this, a general dissolution of the septa.

Mayfield (1974) used a distinctive morphological mutant (puff) of Schizophyllum commune, in which the normal dikaryotic hyphae produced after migration of a wild type nucleus are particularly obvious. He was able to locate precisely the region of septal disruption and to observe nuclei passing through simple septa (Fig. 14.3b). His observations indicated that septal dissolution begins with the parenthesome and pore swelling on one side of the septum followed by erosion of the septum, and that the enzymes involved may be located in multivesicular bodies and vesicles found associated with dissolving septa. Marchant & Wessels (1974) also observed vesicles associated with septal dissolution in a B mutant homokaryon. In Coprinus cinereus we have found that multivesicular bodies are very commonly associated with septa in homokaryotic hyphae (even when dissolution is not occurring) and that they are actually invaginations of the plasmalemma surrounding the septum (Fig. 14.3c, d). When the invading nucleus promotes dissolution, the plasmalemma appears to separate from the septum, the small vesicles are released into the space created and into direct contact with the septum (Fig. 14.3e) (L. A. Casselton and J. B. Kirkham, unpublished results). Nuclear migration occurs very rapidly (1.5–2.5 mm/h in S. commune) (Snider & Raper, 1958). Thus one might suppose that the enzymes involved in septal dissolution are already present in a form ready for activation in these multivesicular bodies seen at the homokaryotic septa.

Nuclear motility

Nothing definite is known about the forces behind nuclear motility during migration. Mayfield (1974) noted that migration is accompanied by a flow of other cytoplasmic organelles leaving behind highly vacuolated hyphae. In contrast, Niederpruem (1969) made direct observations of nuclear movement in B mutants by phase contrast microscopy and saw no protoplasmic

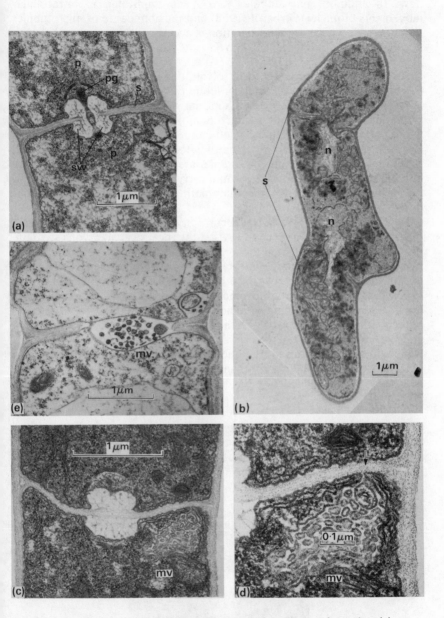

Fig. 14.3 Septal structure and nuclear migration during dikaryon formation. (a) dolipore septum of a homokaryon. s, septum; sw, septal swelling; p, parenthesome; pg, plug. (b) nuclear migration which initiates dikaryon formation after the conversion of dolipore septa to simple septa. s, septum; n, nucleus. (c) multivesicular body (mv) attached to a homokaryotic septum. (d) inset of (c) to show the point of attachment of the multivesicular body to the plasmalemma (arrow). (e) multivesicular body (mv) associated with a disrupted septum. (a), (c), (d) and (e) *Coprinus cinereus* (b) *Schizophyllum commune* by courtesy of J. E. Mayfield.

flow. In addition, microtubules have been implicated in intracellular movements of nuclei (Girbardt, 1968) and the abundance of microtubules found in cells where nuclear migration is occurring suggests that they may also be involved in intercellular movements (Raudaskoski, 1972; Raudaskoski & Koltin, 1973).

An experimental approach to resolving this problem was attempted by Casselton & Condit (1972) using a mutant of *Coprinus cinereus* which had a cytoplasmically inherited mitochondrial defect. By allowing dikaryon formation between this mutant and a homokaryon with normal mitochondria it was possible to determine whether nuclear exchange between the two homokaryons also included exchange of mitochondria as claimed by Watrud & Ellingboe (1973). Since the septa are disrupted to allow nuclear migration, any cytoplasmic flow would carry the mitochondria with it, and the resulting dikaryons would contain both mutant and normal mitochondria.

Reciprocal exchange of nuclei occurred and both homokaryons became dikaryotized. As can be seen in Fig. 14.4 the reciprocal dikaryons are quite distinct. This is because the mitochondrial component is entirely determined by the recipient homokaryon in each case. The vigorous dikaryon has only normal mitochondria whereas the other has only mutant mitochondria. This experiment cannot rule out movement of organelles already present within the recipient hyphae, it only demonstrates that there is no transfer of mitochondria between the original homokaryons. This in itself is significant and the experiment is open to an alternative interpretation. In organisms which produce specialized gametic cells, organelles are predominantly inherited from the maternal parent and the contribution of the male gamete is in some way excluded (see Sager, 1972). It is particularly interesting,

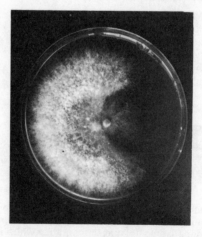

Fig. 14.4 Reciprocally formed dikaryons of *Coprinus cinereus* derived by mating a homokaryon with normal mitochondria and a mutant with defective mitochondria. Two discrete dikaryons have been formed because there has been no exchange of mitochondria. The fluffy fast growing dikaryon (left) has normal mitochondria whereas the sparse slow growing dikaryon (right) has only defective mitochondria.

therefore, that the mitochondria of the nuclear donor in dikaryon formation can be excluded even though the sexual interaction is between purely vegetative hyphae.

B factor functions defined

If one considers that the major function of the B factor in controlling nuclear migration is to promote dissolution of the dolipore septa then the other function attributed to B factor control, *vis*. clamp cell fusion during apical cell division in the dikaryon, is seen to be very similar. Casselton & Lewis (1966) pointed out that both functions involve localized enzymic degradation of cell wall or septum and associated membranes and it is possible that the same enzymes are involved.

Once the dikaryon is established, the switch is made from septal dissolution to clamp cell fusion. In the dikaryon the integrity of the septa is essential in order to maintain the binucleate cell condition and it was suggested by Moore (1965) that this could be reinforced if the dikaryotic septa were structurally different from homokaryotic septa. This idea was substantiated by studies with $A = B \neq$ diploid homokaryons of *Coprinus cinereus*.

$A = B \neq$ diploid homokaryons

Vegetative diploidy is not a normal feature of the life cycle but the technique for obtaining diploid homokaryons is very simple. An $A = B \neq$ heterokaryon constituted from two mutants having gene mutations causing different auxotrophic growth requirements produces uninucleate haploid oidia which are unable to germinate without special growth supplements. Rare fusion of non-identical nuclei gives rise to oidia with heterozygous diploid nuclei and these can be selected because they germinate on unsupplemented medium to give diploid homokaryons (Casselton, 1965).

$A = B \neq$ diploid homokaryons provide a different approach to studying the control of B *factor functions* because, although two different B factors are present, the diploid is neither equivalent to the heterokaryon from which it is derived, nor to the primary B mutant which behaves as if it has two different B factors. Full compatibility control is maintained in the diploid (unlike the mutant) and this has made it possible to detect the situations which promote septal dissolution or clamp cell fusion and the likely situation which triggers the switch in B function from septal dissolution to clamp cell fusion.

In the $A = B \neq$ heterokaryon simple septa are present because nuclear migration occurs. The $A = B \neq$ diploid homokaryon derived from this heterokaryon is identical in genotype, but unlike the heterokaryon, only dolipore septa are present (Casselton, Lewis & Marchant, 1971). Septal dissolution does not occur in the diploid and nuclear migration is blocked.

The only way in which the $A = B \neq$ heterokaryon and $A = B \neq$ diploid homokaryon can differ is in the intercellular distribution of the two different B factors. This is represented in Fig. 14.5, and is compared with the situation found in the dikaryon. In the $A = B \neq$ heterokaryon the two different B factors are in different haploid nuclei and are not always present in the same cell. Since septal dissolution is associated with nuclear migration, the migrating nucleus must pass from a cell having two different B factors to one having only one. This differential situation cannot arise in the diploid

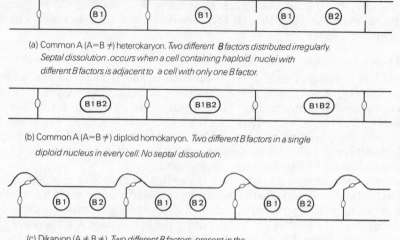

(a) Common A (A=B ≠) heterokaryon. *Two different B factors distributed irregularly.*
Septal dissolution .occurs when a cell containing haploid nuclei with
different B factors is adjacent to a cell with only one B factor.

(b) Common A (A=B ≠) diploid homokaryon. *Two different B factors in a single*
diploid nucleus in every cell. No septal dissolution.

(c) Dikaryon (A ≠ B ≠). *Two different B factors present in the*
two haploid nuclei in every cell. No septal dissolution.

Fig. 14.5 *B* factor distribution in cells of (a) *A = B ≠* heterokaryon, (b) *A = B ≠* diploid homokaryon and (c) *A ≠ B ≠* dikaryon.

homokaryon because the two different *B* factors are present together in the same nucleus and therefore in every cell. It does not arise in the dikaryon either where two different *B* factors are also present in every cell.

It can be concluded that *septal dissolution only occurs when adjacent cells contain nuclei with different B factors*—this is the situation which activates *B* function. An exactly similar situation obtains for clamp cell fusion. Casselton & Lewis (1966) tested for dikaryon formation between a number of differently constituted *A = B ≠* diploid homokaryons in combination with each other and with haploid homokaryons. Their findings in terms of the different clamp cell situations studied are summarized in Fig. 14.6. It can be seen that *clamp cell fusion only occurs when the clamp cell and the subapical cell have nuclei with different B factors.*

The switch in *B* factor function from septal dissolution to clamp cell fusion appears to be made when every cell contains two different *B* factors. Unlike the dikaryon, the *A = B ≠* diploid homokaryon cannot produce clamp cells because the *A* factors are the same, but the switch has been made because the septa can no longer be converted to simple septa. *A = B ≠* diploid nuclei can migrate through cells of a haploid homokaryon and can therefore induce septal dissolution, but haploid nuclei cannot migrate through cells of the diploid homokaryon (Casselton *et al.*, 1971). The diploid septa are apparently no longer susceptible to enzymic dissolution and the only way this could be effected is by a change in structure. If this interpretation is correct, the change in septal structure must be part of the *B* morphogenetic sequence because it occurs when the *A* factors are the same.

Mayfield (1974) noticed an increased deposition of new wall material following nuclear migration during dikaryon formation in *Schizophyllum commune*, and Casselton (1975) observed an unusual degree of wall thickening in older cells of *A = B ≠* diploid homokaryons of *Coprinus cinereus*.

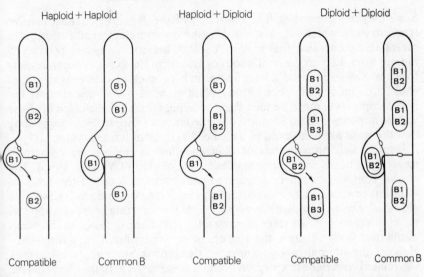

Fig. 14.6 Compatible and common *B* reactions between haploid and *A* = *B* ≠ diploid nuclei in clamp cell fusion.

Deposition of new wall material may be an important initial step in dikaryon formation acting to protect newly formed septa from enzymic dissolution.

14.4 Biochemical Correlations

Total protein spectra

The idea that the *A* and *B* factors are regulatory genes (Raper, 1966) means that they control the activity of many structural genes responsible for the proteins which bring about the morphogenetic sequence leading to dikaryon formation. Attempts to identify some of the proteins of the regulated component of the system were made by Raper & Esser (1961) and Wang & Raper (1969) using immunological and electrophoretic techniques. By comparing the protein spectra of homokaryons, a dikaryon and primary *A* factor and *B* factor mutants it was possible to identify protein bands which were present or absent when either the *A* or the *B* morphogenetic sequence was in operation.

Interesting as this type of analysis is, it provides little information as to the function of the proteins implicated in morphogenesis. The analysis is further complicated by the fact that some of the apparent differences in protein spectra were not due to the appearance of new proteins, but due to changes in position of enzyme proteins present in all mycelia. Isozyme patterns of various enzymes with no obvious role in morphogenesis (*e.g.* NADH-dehydrogenase activities) were surprisingly different in the mycelia compared indicating that a marked change in basic metabolism accompanies morphogenesis (Wang & Raper, 1970).

Role of lytic enzymes

A more direct approach to identifying specific enzymes involved in morphogenesis derives from the work of Wessels (1965) on cell wall structure in

Schizophyllum commune. The operation of the B sequence must involve lytic enzymes which effect septal dissolution and clamp cell fusion. It is reasonable to assume that in the $A = B \neq$ heterokaryon and primary B mutant homokaryon the continuous operation of the B sequence in respect to septal dissolution would be reflected in high levels of such enzymes. This could account for the aberrant morphology of $A = B \neq$ mycelia. At the microscopic level it can be seen that the hyphal walls are distorted in shape and protoplasmic extrusions are not uncommon (Raper, 1966) suggesting that the walls are weakened by enhanced activity of lytic enzymes.

Hyphal wall preparations of *Schizophyllum commune* are composed largely of glucans with only a small amount of chitin (Wessels, 1965). The glucan component can be separated into two fractions on the basis of solubility in alkali, an alkali-resistant glucan, *R-glucan*, and an alkali soluble glucan, *S-glucan*. The R-glucan component is interesting because it appears to act as a carbohydrate reserve in the dikaryon. The expansion of the pileus of the fruit-body occurs in the absence of external nutrients at a time when glucose, the normal carbon source in the growth medium, is exhausted. Correlated with pileus expansion is an increase in the activity of an enzyme (or enzymes) which Wessels has called *R-glucanase*, because it hydrolyses R-glucan, and a decrease in the amount of R-glucan in the cell wall. Removal of R-glucan from the cell wall can be measured as a change in the ratio of S-glucan:R-glucan (S/R glucan ratio) in total wall preparations. As R-glucan is removed the ratio obviously increases.

R-glucan degradation can be brought about in the absence of fruit-body development simply by removal of glucose from the growth medium and concomitant with this is the appearance of R-glucanase activity (Wessels, 1966). It is evident that R-glucanase activity in the dikaryon is induced by the removal of glucose, a characteristic of an enzyme subject to catabolite repression.

The implication of R-glucanase in dikaryon morphogenesis as well as fruit-body morphogenesis came from a study of the activity of this enzyme in homokaryons, heterokaryons and A and B factor mutants (Wessels & Niederpreum, 1967; Wessels, 1969). In homokaryons R-glucanase activity is always low and is unaffected by removal of glucose from the growth medium. In the $A = B \neq$ heterokaryon and primary B mutant homokaryon where the B sequence is continuously in operation to cause septal dissolution R-glucanase activity is abnormally high even in the presence of glucose. It is some ten times higher than that in the homokaryon and this is correlated with a two to three-fold increase in the S/R glucan ratio in the hyphal walls.

Wessels & Niederpruem (1967) have interpreted this data to mean that in the homokaryon R-glucanase is permanently repressed (genic repression) and is derepressed when different B factors are present following compatible hyphal fusions. This leads to septal dissolution and allows nuclear migration to establish the dikaryotic condition. Once the dikaryon is formed, the regulation of R-glucanase activity changes from genic repression to catabolite repression.

In the $A = B \neq$ heterokaryon and primary B mutant R-glucanase is permanently derepressed. Apart from septal dissolution, the removal of R-glucan from the walls would act to weaken them and account for much of the observed abnormality in hyphal morphology.

In $A \neq B$ = heterokaryons and primary A mutant mycelia in which the B morphogenetic sequence is not operating, R-glucanase activity is low and is not induced in the absence of glucose. In these mycelia, R-glucanase is apparently permanently repressed. Wessels & Niederpruem (1967) suggest that during the normal morphogenetic sequence it is the different A factors which act to repress the catabolite insensitive R-glucanase. However, the fact that the catabolite repressible enzyme is not synthesized in $A \neq B$ = mycelia shows that synthesis of both enzymes is under control of the B factor. The suggested regulation of R-glucanase is summarized in Table 14.2.

It has not been established whether the catabolite insensitive enzyme which is derepressed when septal dissolution occurs is the same as the catabolite repressible enzyme found in the dikaryon but the different mode of regulation would suggest that they are different gene products. There is some evidence to suggest that both enzymes can be synthesized by $A = B \neq$ mycelia. A single gene mutation (MII), quite distinct from the incompatibility factors, restores normal homokaryotic morphology to a homokaryon with a primary B factor mutation (Raper & Raper, 1966). As a result of this mutation an inactive form of the catabolite insensitive R-glucanase is present (Wessels & Koltin, 1972). This is unable to bring about septal dissolution but in the absence of glucose it can be seen that a functional catabolite repressible enzyme can still be synthesized (Wessels, 1969).

Direct confirmation of R-glucanase activity in septal dissolution was obtained by Janszen & Wessels (1970) using an *in vitro* technique in which isolated cell walls were variously treated with chitinase and R-glucanase and then examined in the electron microscope. A combination of both enzymes effected complete dissolution of the septa but left the hyphal walls intact. This experiment provides a beautiful illustration of the way in which developmental processes can be controlled by differences in structure—in this case glucan composition. The septa only contain the R-glucan because they are completely dissolved by the enzyme treatment. The wall, however, contains S-glucan and despite the fact that R-glucan was completely removed by the treatment they retained their integrity. Although Janszen and Wessels used enzyme extracted from the dikaryon for this *in vitro* experiment, the catabolite insensitive enzyme must be the enzyme which effects septal dissolution *in vivo* because dissolution occurs in the presence of glucose.

It is not known whether R-glucanase is also involved in clamp cell fusion, but the fact that $A = B \neq$ mycelia which have elevated R-glucanase activity also have a greatly increased incidence of hyphal fusions, suggests that it is.

Resistance to R-glucanase activity

Homokaryotic septa are disrupted when there is high R-glucanase activity in the mycelium but there is no evidence that this occurs in the dikaryon when R-glucanase activity increases at the time of fruit-body development and removes R-glucan from the cell walls. From genetic studies with *Coprinus cinereus* it was concluded that dikaryotic septa are probably structurally different from homokaryotic septa and are resistant to enzymic degradation (Casselton *et al.*, 1971). This was confirmed to some extent by comparing the degree of septal dissolution effected by *in vitro* treatment of homokaryotic

Table 14.2 Regulation of catabolite sensitive and catabolite insensitive R-glucanases by the incompatibility factors in *Schizophyllum commune* (Based on Wessels & Niederpruem, 1967 and Wessels, 1969).

Mycelium	Morphogenetic sequence in operation	R-glucanase activity + glucose	R-glucanase activity − glucose	Regulation of R-glucanases
Homokaryon $A = B \neq$ heterokaryon Ax Bmut.	none B sequence	low high	low high	Both enzymes repressed Both enzymes derepressed by different B factors (or B factor mutation)
$A \neq B =$ heterokaryon Amut Bx $A \neq B \neq$ dikaryon	A sequence A and B sequences	low low	low high	Both enzymes repressed by similar B factors Catabolite insensitive enzyme repressed by different A factors Catabolite sensitive enzyme derepressed by different B factors

and dikaryotic wall fragments of *Schizophyllum commune* (Wessels & Marchant, 1974). The number of dikaryotic septa dissolved was significantly less than the number of homokaryotic septa.

Changes in R-glucanase activity and susceptibility to enzymic degradation are obviously important during morphogenesis in *Schizophyllum commune*. In the dikaryon it is not only the septa that are resistant. The cell walls of the fruit-body·are more resistant than those of the vegetative mycelium and can, therefore, retain a normal composition when R-glucan is being hydrolysed in the vegetative mycelium (Wessels, 1966).

Impairment of energy conservation

There is another feature of the metabolism of $A = B \neq$ mycelia which may be related to R-glucanase activity. Raper & San Antonio (1954) were able to attribute the sparse growth of the $A = B \neq$ heterokaryon to inefficient utilization of glucose from the growth medium. More recently Hoffman & Raper (1971) examined this respiratory deficiency in more detail in the primary B mutant homokaryon. Mitochondria isolated from the B mutant produce the same amount of ATP as mitochondria from a normal homokaryon but it would appear that *in vivo*, the energy yielding process is partially uncoupled from energy conserving processes and the efficiency of glucose utilization is only 9% of that of a normal homokaryon (Hoffman & Raper, 1974).

The genetic situation ($A = B \neq$) which impairs respiratory efficiency is identical to that which derepresses R-glucanase and both events are reversed by the presence of different A factors. The two could well be related. Inefficient use of glucose from the growth medium may be the reason that storage glucan is utilized with the necessary derepression of R-glucanase and consequent dissolution of septa. These are of course transitory phenomena in the normal morphogenetic sequence leading to dikaryon formation.

There is another phenomenon which should not be overlooked. In many species it has been observed that some homokaryons can produce fruit-bodies even though this cannot lead to sexual reproduction because only a single haploid nuclear type is present. Homokaryotic (or monokaryotic) fruiting has been shown to the determined by a single gene in *Polyporus ciliatus*. Interestingly, the activity of this gene is completely suppressed by different B factors in a $A = B \neq$ heterokaryon (Esser & Stahl, 1975; Stahl & Esser, 1976). Normal growth of the dikaryon requires a higher respiratory activity than that of the homokaryon (Casselton & Kirkham, 1975) and a mitochondrial defect completely inhibits dikaryotic fruiting (Casselton & Condit, 1972). An impaired respiratory activity in $A = B \neq$ mycelia could, therefore, explain suppression of monokaryotic fruiting.

14.5 Concluding Remarks

The formation of the dikaryon involves many structural and metabolic changes demanding a precisely regulated sequence of gene activities. Once established, the exactly balanced 1:1 ratio of the two component nuclei means that each cell of the dikaryon has the equivalent of a diploid genome. However, the retention of two haploid nuclei, despite the complexity of the cell division required to maintain their exact distribution, offers a degree of flexibility that no diploid mycelium could achieve. The dikaryon is the predominant vegetative mycelium in nature. Under certain conditions, selection may favour a particular nuclear type and the dikaryotic association may break down. At the same time, new dikaryotic associations can constantly arise as a result of hyphal fusions between dikaryotic and homokaryotic hyphae.

A single gene mutation can eliminate the dikaryotic phase. This was discovered in *Schizophyllum commune* by Koltin and Raper (1968). The mutation causes a precocious fusion in the vegetative mycelium of the two nuclei which would normally comprise the dikaryotic pair. The resulting mycelium is a diploid monokaryon which is sexually fertile but exhibits none of the morphological characteristics of the dikaryon. The replacement of the dikaryon by a diploid monokaryon or homokaryon has occurred in nature in *Armellariella mellea* (Korhonen & Hintikka, 1974). It is significant that in this genus, the vegetative mycelium exhibits an unusual degree of differentiation into rhizomorphs, and cells of the rhizomorphs may contain many nuclei (Hintikka, 1973). In this situation it would be impossible to maintain the 1:1 ratio of the component genomes without diploidization. In the majority of species, where this secondary differentiation does not occur, it is clear that selection has operated to retain the advantage of the flexible dikaryotic condition.

References

BULLER, A. H. R. (1931). *Researches on fungi*. Volume IV. London: Longmans, Green & Co.

CASSELTON, L. A. (1965). The production and behaviour of diploids of *Coprinus lagopus*. *Genetical Research* 6, 190–208.

CASSELTON, L. A. (1975). Control of nuclear migration in Basidiomycetes. *Proceedings of the First Intersectional Congress of International Association of Microbiological Societies, Japan* Vol. I, 301–9.

CASSELTON, L. A. & CONDIT, A. (1972). A mitochondrial mutant of *Coprinus lagopus*. *Journal of General Microbiology* 72, 521–7.

CASSELTON, L. A. & KIRKHAM, J. B. (1975). Growth and ultrastructural studies on the mitochondrial mutant of *Coprinus lagopus*. *Archives of Microbiology* 106, 215–20.

CASSELTON, L. A. & LEWIS, D. (1966). Compatibility and stability of diploids in *Coprinus lagopus*. *Genetical Research* 8, 61–72.

CASSELTON, L. A., LEWIS, D. & MARCHANT, R. (1971). Septal structure and mating behaviour of common *A* diploid strains of *Coprinus lagopus*. *Journal of General Microbiology* 66, 273–8.

DAY, P. R. (1963a). The structure of the *A* mating-type factor in *Coprinus lagopus*: wild alleles. *Genetical Research* 4, 323–5.

DAY, P. R. (1963b). Mutations affecting the *A* mating-type locus in *Coprinus lagopus*. *Genetical Research* 4, 55–65.

ESSER, K. (1971). Breeding systems and their significance for genetic recombination. *Molecular and General Genetics* 110, 86–100.

ESSER, K. & STAHL, U. (1975). A genetic correlation between dikaryotic and monokaryotic fruiting in Basidiomycetes. *Proceedings of the First Intersectional Congress of International Association of Microbiological Societies, Japan* Vol. I, 294–300.

FLEXER, A. S. (1969). Bipolar incompatibility in the hymenomycete *Polyporus palustris*. *American Journal of Botany* 56, 410–17.

FULTON, I. W. (1950). Unilateral nuclear migration and the interactions of haploid mycelia in the fungus *Cyathus stercoreus*. *Proceedings of the National Academy of Sciences, U.S.A.* 36, 306–12.

GIESY, R. M. & DAY, P. R. (1965). The septal pores of *Coprinus lagopus* (Fr.) *sensu* Buller in relation to nuclear migration. *American Journal of Botany* 52, 287–94.

GIRBARDT, M. (1961). Licht-und Elektronmikroskopische Untersuchungen an *Polystictus versicolor* (L.) II. Die Fienstruktur von Grundplasma und Mitochondrien. *Archiv für Mikrobiologie* **39**, 351–9.

GIRBARDT, M. (1968). Ultrastructure and dynamics of the moving nucleus. *Symposia of the Society for Experimental Biology* **12**, 249–59.

HINTIKKA, V. (1973). A note on the polarity of *Armellariella mellea*. *Karstenia* **13**, 32–9.

HOFFMAN, R. M. & RAPER, J. R. (1971). Genetic restriction of energy conservation in *Schizophyllum*, *Science* **171**, 418–9.

HOFFMAN, R. M. & RAPER, J. R. (1974). Genetic impairment of energy conservation in development of *Schizophyllum*. Efficient mitochondria in energy-starved cells. *Journal of General Microbiology* **82**, 67–75.

JANSZEN, F. H. A. & WESSELS, J. G. H. (1970). Enzymic dissolution of hyphal septa in a Basidiomycete. *Antonie van Leeuwenhoek* **36**, 255–7.

JERSILD, R., MISHKIN, S. & NIEDERPRUEM, D. J. (1967). Origin and ultrastructure of complex septa in *Schizophyllum commune* development. *Archiv für Mikrobiologie* **57**, 20–32.

KOLTIN, Y. (1968). The genetic structure of the incompatibility factors of *Schizophyllum commune*. Comparative studies of primary mutations in the *B* factor. *Molecular and General Genetics* **102**, 196–203.

KOLTIN, Y. & FLEXER, A. S. (1969). Alteration of nuclear migration in *B*-mutant strains of *Schigophyllum commune*. *Journal of Cell Science* **4**, 739–49.

KOLTIN, Y. & RAPER, J. R. (1968). Dikaryosis: genetic determination in *Schizophyllum*. *Science* **160**, 85–6.

KOLTIN, Y., RAPER, J. R. & SIMCHEN, G. (1967). The genetic structure of the incompatibility factors of *Schizophyllum commune*. The *B* factor. *Proceedings of the National Academy of Sciences of the United States of America* **57**, 55–62.

KOLTIN, Y., STAMBERG, J. & LEMKE, P. A. (1972). Genetic structure and evolution of the incompatibility factors in higher fungi. *Bacteriological Reviews* **36**, 156–71.

KORHONEN, K. & HINTIKKA, V. (1974). Cytological evidence for somatic diploidization in dikaryotic cells of Armillariella mellea. *Archives of Microbiology* **95**, 187–92.

MARCHANT, R. & WESSELS, J. G. H. (1974). An ultrastructural study of septal dissolution in *Schizophyllum commune*. *Archives of Microbiology* **96**, 175–82.

MAYFIELD, J. E. (1974). Septal involvement in nuclear migration in *Schizophyllum commune*. *Archives of Microbiology* **95**, 115–24.

MIDDLETON, R. B. (1964). Evidence of common *AB* heterokaryosis in *Schizophyllum commune*. *American Journal of Botany* **51**, 379–87.

MOORE, R. T. (1965). The ultrastructure of fungal cells. In *The fungi*, Vol. I. p. 95, edited by G. C. Ainsworth and A. S. Sussman. New York: Academic Press.

MOORE, R. T. & MCALEAR, J. H. (1962). Fine structure of Mycota. 7. Observations on septa of Ascomycetes and Basidiomycetes. *American Journal of Botany* **49**, 86–94.

NEUHAUSER, K. S. & GILBERTSON, R. L. (1971). Some aspects of bipolar heterothallism in *Fomes cajanderi*. *Mycologia* **63**, 722–35.

NIEDERPRUEM, D. J. (1969). Direct studies of nuclear movements in *Schizophyllum commune*. *Archiv für Mikrobiologie* **64**, 387–95.

PAPAZIAN, H. (1950). Physiology of the incompatibility factors in *Schizophyllum commune*. *Botanical Gazette* **112**, 143–63.

PAPAZIAN, H. P. (1954). Exchange of incompatibility factors between the nuclei of a dikaryon. *Science* **119**, 691–3.

PARAG, Y. (1962). Mutations in the *B* incompatibility factor of *Schizophyllum commune*. *Proceedings of the National Academy of Science of the United States of America* **48**, 743–50.

PARAG, Y. & KOLTIN, Y. (1971). The structure of the incompatibility factors of *Schizophyllum commune*: constitution of the three classes of *B* factors. *Molecular and General Genetics* **112**, 43–8.

QUINTANILHA, A. (1933). Le problème de la sexualité chez les champignons. *Boletim da Sociedade Broteriana* **8**, 1–99.

QUINTANILHA, A. & PINTO-LOPES, J. (1950). Aperçu sur l'état actuel de nos connaisances concernant la 'conduite sexuelle' des espèces d'Hymenomycetes. *Boletim da Sociedade Broteriana* **24**, 115–290.

RAPER, C. A. & RAPER, J. R. (1966). Mutations modifying sexual morphogenesis in *Schizophyllum*. *Genetics* **54**, 1151–68.

RAPER, C. A. & RAPER, J. R. (1973). Mutational analysis of a regulatory gene for morphogenesis in *Schizophyllum*. *Pro-*

ceedings of the National Academy of Sciences, U.S.A. **70**, 1427–31.

RAPER, C. A., RAPER, J. R. & MILLER, R. E. (1972). Genetic analysis of the life cycle of *Agaricus bisporus. Mycologia* **64**, 1088–1117.

RAPER, J. R. (1966). *Genetics of sexuality in higher fungi.* New York: The Ronald Press Co.

RAPER, J. R. & ESSER, K. (1961). Antigenic differences due to the incompatibility factors in *Schizophyllum commune. Zeitschrift für Vererbungslehre* **92**, 439–44.

RAPER, J. R. & RAUDASKOSKI, M. (1968). Secondary mutations at the *Bβ* incompatability locus of *Schizophyllum. Heredity* **23**, 109–17.

RAPER, J. R. & SAN ANTONIO, J. P. (1954). Heterokaryotic mutagenesis in Hymenomycetes. I. Heterokaryosis in *Schizophyllum commune. American Journal of Botany* **41**, 69–86.

RAPER, J. R., BAXTER, M. G. & ELLINGBOE, A. H. (1960). The genetic structure of the incompatibility factors of *Schizophyllum commune*: the *A* factor. *Proceedings of the National Academy of Sciences, U.S.A.* **46**, 833–42.

RAPER, J. R., BAXTER, M. G. & MIDDLETON, R. B. (1958). The genetic structure of the incompatibility loci in *Schizophyllum. Proceedings of the National Academy of Sciences of the United States of America* **53**, 889–900.

RAPER, J. R., BOYD, D. H. & RAPER, C. A. (1965). Primary and secondary mutations at the incompatibility loci in *Schizophyllum. Proceedings of the National Academy of Sciences of the United States of America* **53**, 1324–32.

RAPER, J. R., KRONGELB, G. S. & BAXTER, M. G. (1958). The number and distribution of incompatibility factors in *Schizophyllum commune. American Naturalist* **92**, 221–32.

RAUDASKOSKI, M. (1972). Occurrence of microtubules in the hyphae of *Schizophyllum commune* during intercellular nuclear migration. *Archiv für Mikrobiologie* **86**, 91–100.

RAUDASKOSKI, M. & KOLTIN, Y. (1973). Ultrastructural aspects of a mutant of *Schizophyllum commune* with continuous nuclear migration. *Journal of Bacteriology* **116**, 981–8.

SAGER, R. (1972). Cytoplasmic genes and organelles. New York and London: Academic Press Inc.

SNIDER, P. J. & RAPER, J. R. (1958). Nuclear migration in the Basidiomycete

Schizophyllum commune. American Journal of Botany **45**, 538–46.

SNIDER, P. J. & RAPER, J. R. (1965). Nuclear ratios and complementation in common-*A* heterokaryons of *Schizophyllum commune. American Journal of Botany* **52**, 547–52.

STAHL, U. & ESSER, K. (1976). Genetics of fruit body production in higher Basidiomycetes. I. Monokaryotic fruiting and its correlation with dikaryotic fruiting in *Polyporus ciliatus. Molecular and General Genetics* **148**, 183–97.

SWIEZYNSKI, K. M. & DAY, P. R. (1960a). Heterokaryon formation in *Coprinus lagopus. Genetical Research* **1**, 114–28.

SWIEZYNSKI, K. M. & DAY, P. R. (1960b). Migration of nuclei in *Coprinus lagopus. Genetical Research* **1**, 129–39.

TAKEMARU, T. (1961). Genetic studies on fungi. X. The mating system in Hymenomycetes and its general mechanism. *Biological Journal Okayama University* **7**, 133–211.

TERAKAWA, H. (1957). The nuclear behaviour and the morphogenesis in *Pleurotus ostreatus. Scientific Papers of the College of General Education University of Tokyo* **7**, 61–88.

TERAKAWA, H. (1960). The incompatibility factors in *Pleurotus ostreatus. Scientific Papers of the College of General Education, University of Tokyo* **10**, 65–71.

WANG, C. S. & RAPER, J. R. (1969). Protein specificity and sexual morphogenesis in *Schizophyllum commune. Journal of Bacteriology* **99**, 291–7.

WANG, C. S. & RAPER, J. R. (1970). Isozyme patterns and sexual morphogenesis in *Schizophyllum. Proceedings of the National Academy of Sciences, U.S.A.* **66**, 882–9.

WATRUD, L. S. & ELLINGBOE, A. H. (1973). Use of cobalt as a mitochondrial vital stain to study cytoplasmic exchange in matings of the Basidiomycete *Schizophyllum commune. Journal of Bacteriology* **115**, 1151–8.

WESSELS, J. G. H. (1965). Morphogenesis and biochemical processes in *Schizophyllum commune* Fr. *Wentia* **13**, 1–113.

WESSELS, J. G. H. (1966). Control of cell wall glucan degradation during development in *Schizophyllum commune. Antonie van Leeuwenhoek* **32**, 341–55.

WESSELS, J. G. H. (1969). Biochemistry of sexual morphogenesis in *Schizophyllum commune*: effect of mutations affecting the incompatibility system on cell wall metabolism. *Journal of Bacteriology* **98**, 697–704.

WESSELS, J. G. H. & KOLTIN, Y. (1972). R-glucanase activity and susceptibility of hyphal walls to degradation in mutants of *Schizophyllum* with disrupted nuclear migration. *Journal of General Microbiology* **71**, 471–5.

WESSELS, J. G. H. & MARCHANT, R. (1974). Enzymic degradation of septa in hyphal wall preparations from a monokaryon and a dikaryon of *Schizophyllym commune*. *Journal of General Microbiology* **83**, 359–68.

WESSELS, J. G. H. & NIEDERPRUEM, D. J. (1967). Role of a cell wall glucan-degrading enzyme in mating of *Schizophyllum commune*. *Journal of Bacteriology* **94**, 1594–602.

WHITEHOUSE, H. L. K. (1949). Multiple allelomorph heterothallism in the fungi. *New Phytologist* **48**, 212–44.

CHAPTER 15

Hymenial Cytodifferentiation in Basidiomycetes

W. J. SUNDBERG

15.1 Introduction

Although genetically similar, hyphal branches do not necessarily exhibit the same features as their parent (Smith, 1966). Nowhere is this more obvious than in the basidiocarp of Basidiomycetes where structural complexity results from the formation of various associations and assumption of numerous morphological variations by a single morphogenetic unit—the filamentous fungal hypha.

Aspects of sexual morphogenesis in Basidiomycetes have interested many investigators (see Taber, 1966). Current research includes examination and elucidation of environmental (*e.g.*, Lu, 1974), internal physiological (*e.g.*, Cox & Niederpruem, 1975) and enzymatic (*e.g.*, Gooday, 1975) systems controlling basidiocarp growth and differentiation. Further, ultrastructural and ultrahistochemical studies (*e.g.*, McLaughlin, 1974; Gooday, 1975) are adding to our understanding of basidiocarp morphogenesis at the cellular-subcellular level.

Because a recent review on gross morphological and physiological aspects of morphogenesis exists (Taber, 1966), the information below on hymenial morphogenesis, basidial cytodifferentiation and sporogenesis is presented as a complement to that work.

15.2 General Hymenial Structure and Composition

The hymenium or fertile layer of the basidiocarp is a palisade of terminal segments of hyphae and their branches. Structural alteration resulting from cytodifferentiation of the hymenial elements often obscures its hyphal nature (Smith, 1966). The hymenium is comprised of sexual cells, the basidia (Figs. 15.1, 15.2a), and additionally, it may also contain various sterile,

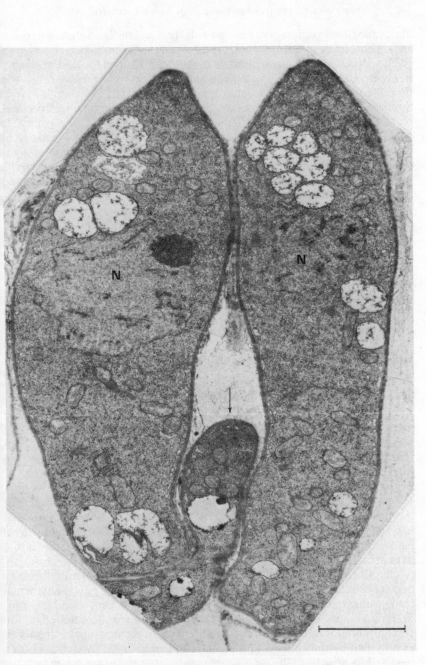

Fig. 15.1 Young basidia of *Mycena citricolor*. Note the lateral proliferation of a new basidium from the sub-basidial segment (arrow). Nuclei (N) of the large basidia are in the pachytene stage (division I) of meiosis. Scattered, mostly apical and basal vacuoles are evident. Scale line equals 2 μm. (W. J. Sundberg, unpublished results).

morphologically differentiated elements (*e.g.*, *cystidia*, *basidioles*, *setae*, *etc.*). Thus, its exact composition varies among different taxa.

The structure and, to a limited extent, the histochemical differences between sterile hymenial elements at the light microscope level are described and discussed in detail elsewhere (Lentz, 1954; Donk, 1964; Smith, 1966; Singer, 1975). Although some information on ultrastructural features of such cells exists (Clémencon, 1972*a*, *b*, 1975; Thielke, 1972; McLaughlin, 1973, 1974; Gull & Newsam, 1975), there is a paucity of developmental studies. Because they are hyphal in origin, detailed fine-structural and ultrahistochemical studies on ontogeny of the various sterile hymenial elements would be not only interesting, but would most likely add to our understanding of possible alternative mechanisms of cytodifferentiation available to the terminal fungal hyphal segment.

As the site of karyogamy, meiosis, and spore formation and dispersal, the basidium is the focal point of sexual morphogenesis in Basidiomycetes. Its morphology is utilized in systematics and has been employed in elucidation of phylogenetic relationships (Rogers, 1934). Basidial ontogeny, nuclear events and spore development have been subjects of extensive light microscope studies (Buller, 1909, 1922; Corner, 1948; Olive, 1953, 1965). Recent studies, many with an ultrastructural or more rarely an ultrahistochemical emphasis, have further elucidated events in basidial development. Some aspects of homobasidial cytodifferentiation are reviewed below in light of these works.

15.3 Basidial Origin and Proliferation

In *Poria latemarginata* (Dur. & Mont.) Cke., basidia develop as lateral outgrowths (Fig. 15.2i) from subhymenial hyphal segments; up to five are produced per segment (Setliff, MacDonald & Patton, 1972; Setliff, Hoch & Patton, 1974). However, in most homobasidiomycete species, the basidium (*e.g.*, Fig. 15.2a) arises via differentiation of the terminal cell of a hypha or hyphal branch (Harper, 1902; Corner, 1950, 1953; Donk, 1964; Smith, 1966). With few exceptions, its development concludes the apical growth of the branch which bears it (Corner, 1950, 1953; Donk, 1964). Continued basidial production and hymenial expansion occur as a result of lateral (sub-basidial) or, less frequently, percurrent proliferation of the fertile hyphal branches and their subsequent apical differentiation.

The amount of elongation occurring in the newly proliferated branch prior to terminal basidial differentiation distinguishes two general hymenial forms—the thickening and non-thickening types. When abundant, proliferation sometimes exerts a characteristic morphogenetic effect on configuration of the hymenium (Donk, 1964). In the non-thickening form, exemplified by mushrooms and beautifully diagrammed by Buller (1922), new basidia arise from the same subhymenial layer and at approximately the same height as the mature and collapsed ones (Donk, 1964). Elongation of sub-basidial segments is minimal. During proliferation, basidia in many species of the Aphyllophorales often develop at increasingly higher levels than their predecessors (Corner, 1950; Donk, 1964). The subtending cell elongates and penetrates outward between the older elements prior to apical basidial differentiation resulting in a thickening hymenium (Donk, 1964).

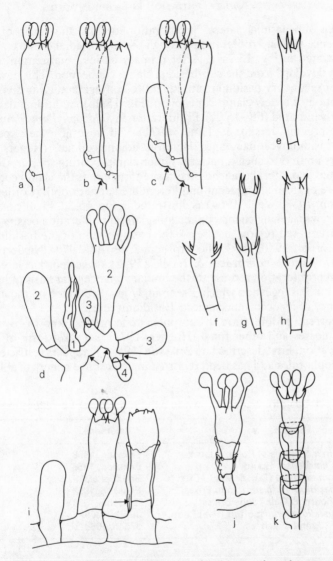

Fig. 15.2 Some methods of basidial proliferation. All diagramatic and not to scale. (a)–(c) Lateral proliferation through the sub-basidial clamp connection (arrows). (Adapted from Niederpruem *et al.*, 1971). (d) Lateral proliferation through both sub-basidial clamp connections (arrows) and other sub-basidial areas in the Corticiaceae. Arabic numerals indicate the sequence of basidial formation. (Redrawn from Eriksson, 1958). (e) Basidium of *Hymenochaete arida*. (f)–(h) Presumptive apical basidial proliferation in *Hymenochaete arida*. (i) Primary basidial development and percurrent proliferation in *Poria latemarginata*. Left: young primary basidium. Middle: mature primary basidium. Right: percurrent basidium surrounded by sheath-like primary basidial wall. A portion of the original basal septum is evident (arrow). (Adapted from Setliff *et al.*, 1974). (j) Percurrent proliferation in *Repetobasidium mirificum*. Note the collar-like remnants around newest basidium. (Redrawn from Eriksson, 1958). (k) Diagramatic representation of percurrent proliferation in *Repetobasidium mirificum*. (Redrawn from Hughes, 1971).

When sub-basidial lateral proliferation occurs, the locus of branch emergence varies. Proliferation through the sub-basidial clamp connection is illustrated in Fig. 15.2a–c. After primary basidial maturation, a lateral branch develops from the sub-basidial clamp connection. This new branch forms a secondary basidium subsequent to conjugate nuclear division and generation of a new clamp connection (Fig. 15.2b); additional basidia are similarly initiated (Fig. 15.2c). Proliferation in *Schizophyllum commune* Fr. (Niederpruem, Jersild & Lane, 1971) and several other species of Homobasidiomycetidae and Heterobasidiomycetidae (Rogers, 1936) appears limited to such a pattern. Other clamp-bearing species are not so restricted and lateral branches may issue from either the clamp connection as above or from an adjacent undifferentiated area or both (Corner, 1950; Eriksson, 1958; Wells, 1964) as illustrated in Fig. 15.2d. Finally, in species (and strains) lacking clamp connections, lateral poliferation occurs from an undifferentiated region just below the basal septum of the basidium (Fig. 15.1; Corner, 1950, 1953; Niederpruem & Jersild, 1972). Niederpruem *et al.* (1971) and Niederpruem & Jersild (1972) suggested that lateral sub-basidial proliferation represents the fundamental cellular basis of hymenial expansion, a suggestion which is seemingly justified based on the apparent frequency of its occurrence among Basidiomycetes.

Percurrent proliferation, a morphogenetic process known to occur in algae, mosses and other fungi (Hughes, 1971) is a rarer form of basidial regeneration. First described by Eriksson (1958), it appears restricted to the Aphyllophorales and has been reported in at least five genera (Table 15.1).

Table 15.1 Basidiomycetes known to exhibit percurrent proliferation

Species	Reference
Corticium incrustans v. Höhn. & Litsch.	Eriksson, 1958
Galzinia pedicellata Bourd.	Eriksson, 1958
Poria latemarginata (Dur. & Mont.) Cke.	Setliff *et al.*, 1972
Repetobasidium mirificum John Erikss.	Eriksson, 1958
R. vile (Bourd. & Galz.) John Erikss.	Eriksson, 1958
Subulicystidium longisporum (Pat.) Parm.	Jülich, 1969
S. nikau (Cunn.) Jülich	Jülich, 1969

Percurrent proliferation can be detected with the light microscope by the persistence of one or more collar-like sheaths around the newly formed cells (Fig. 15.2j). However, the exact nature of this ontogenetic process can only be revealed at higher magnification levels. In the one ultrastructural study on percurrent basidia performed to date, Setliff *et al.* (1972) determined that proliferation is initiated at the basal septum of the basidium after discharge of its basidiospores. The sub-basidial segment elongates upward, growing within the now evacuated basidium (Fig. 15.3a) and forcing remnants of at least part of the original basal septum to one side (Fig. 15.2i). Discrepancies at the light microscope level regarding the amount of involvement of the basal septum in the proliferation process (Hughes, 1971) indicate that our understanding of this process is far from clear. Eventually, as illustrated in Fig. 15.2i and Fig. 15.3a, the proliferating segment emerges through the

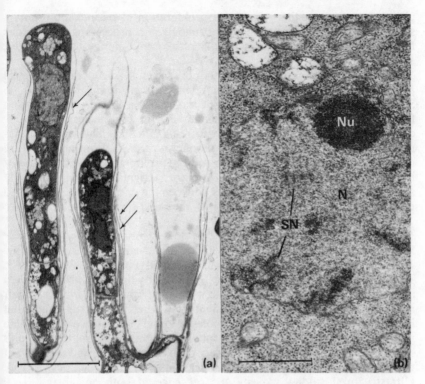

Fig. 15.3 (a) Young percurrent basidia of *Poria latemarginata* surrounded by the walls (arrows) of preceding ones. Note the dikaryotic nuclei and scattered apical and basal vacuoles in the central basidium. Scale line equals 5 μm. (E. C. Setliff, unpublished results) (b) Pachytene (division I) nucleus (N) in a basidium of *Mycena citricolor*. Synaptonemal complexes (SN) and the nucleolus (Nu) are conspicuous. Scale line equals 1 μm. (W. J. Sundberg, unpublished results).

apical opening resulting from dissolution of the original basidial wall and differentiates. Unlike *Repetobasidium vile* (Bourd. & Galz.) John Erikss. and *R. mirificum* John Erikss. (Ericksson, 1958), in which formation of the basal septum delimiting the new basidium occurs 1–2 μm above the site of the original one (Fig. 15.2j–k), new basal septa develop near the vicinity of the previous ones in *Poria latemarginata* (Setliff *et al.*, 1972).

Up to 11 percurrent proliferations per linear series were counted in *Poria latemarginata* while as few as one were described or illustrated or both in species studied by Eriksson (1958). The upper limit on the number of basidial proliferations which can be produced in a single percurrent linear series and the factors which control it are not known. However, Setliff *et al.* (1972) suggested that dessication may be a limiting factor in cultural conditions.

In *Corticium incrustans* v. Höhn. & Litsch., Eriksson (1958) noted percurrent replacement of an old basidium by a hypha, and occasional growth of basidial branches or hyphae laterally out through the walls of empty basidia occurs in *Poria latemarginata* (Setliff *et al.*, 1972). These phenomena suggest that percurrent proliferation of basidia represents a special form of a more general morphogenetic process in filamentous fungi—development of intrahyphal hyphae (see Calonge, 1968; States, 1975).

A presumptive third mechanism of basidial regeneration—apical proliferation—sometimes occurs in *Hymenochaete arida* Karst. (W. J. Sundberg, unpublished results). Appearance of the basidia suggests that following formation of primary sterigmata (Fig. 15.2e), the basidium proliferates apically (Fig. 15.2f) ultimately forming new secondary apical sterigmata (Fig. 15.2g–h). Tertiary proliferation was rare. The continuous nature of the walls of the original basidium and the proliferation and the more or less upright, turgid aspect of the primary sterigmata suggest probable structural and cytoplasmic continuity between the primary and proliferated elements—a feature not associated with the other forms of proliferation.

15.4 Basidial Growth and Enlargement

Basidia, which may range from cylindrical and almost hyphal-like (Fig. 15.2e; Corner, 1948) to more or less club-shaped (Fig. 15.2a; Wager, 1893; Thielke, 1969), presumably enlarge (Fig. 15.1) by tip growth because they are modified terminal cells of hyphae or hyphal branches. However, their mode of growth is not well documented. Corner (1948) suggested that increasing vacuolar size was responsible for initial basidial enlargement. He reported that dependant on the morphology of the basidium, the vacuoles subsequently disappear almost entirely or become smaller and localized at the basal region of the basidium as it becomes 'charged' with cytoplasm. Ultrastructural studies indicate the presence of both apical and basal vacuoles at this stage (Wells, 1965; Thielke, 1967, 1969).

The basidia continue to enlarge while karyogamy and subsequent nuclear events occur within (Wager, 1892, 1893; Thielke, 1969). Concomitant with this increase, existing vacuoles disappear almost completely in some species (Wager, 1892; Wells, 1965), or become aggregated at the base of the basidium (Wakayama, 1930) or increase in size (Thielke, 1967, 1969) in others. Thus, it appears that if vacuoles are involved in early basidial cytodifferentiation, their function may not be uniform.

Often basidia do not reach maximum volume until after completion of meiosis (Wakayama, 1930; Ehrlich & McDonough, 1949; Wells, 1965; Thielke, 1967; Chang & Chu, 1969; Chang & Ling, 1970). Basidia sometimes protrude beyond the level of the immature ones at the cessation of enlargement (Corner, 1948; Ehrlich & McDonough, 1949; Thielke, 1967). Corner (1948) proposed that basidial protrusion results from the development of basal vacuoles. However, in *Schizophyllum commune* (Wells, 1965) and *Poria latemarginata* (Hoch & Setliff, 1976), basal vacuolation occurs later as sporulation is initiated.

Cytoplasmic microtubules have been reported parallel to the long axis in the basidium during several pre-sporulation developmental stages (McLaughlin, 1970, 1971; Sundberg, 1971; Gull, 1975). Their orientation at these and subsequent stages suggest that they are of general importance in basidial cytodifferentiation (McLaughlin, 1973). Although Spooner (1975) noted the association of microtubules with elongation in some animal cells, they have not been so implicated in pre-sporulation basidia. Instead, they were suggested as organelles involved in nuclear movement, spindle orientation and as possible cytoskeletal elements (McLaughlin, 1971, 1973).

15.5 A Summary of Basidial Nuclear Behaviour and Meiosis

Detailed reviews on nuclear behaviour and meiosis in the basidium are available (Olive, 1953, 1965), including an exhaustive study by Wells (1977) correlating the recent ultrastructural and experimental approaches with previous work on these processes. Therefore, these events are only summarized below (see Wells, 1977 for details, variations and references) using *Schizophyllum commune* (Sundberg, 1971) and illustrative examples from *Mycena citricolor* (Berk. & Curt.) Sacc. (W. J. Sundberg, unpublished results) and *Poria latemarginata* (Setliff *et al.*, 1974) to acquaint the reader with internal changes characteristic of these stages in basidial cytodifferentiation.

In the young basidium, a pair of nuclei—the dikaryotic nuclei—lies near the mid-region of the cell (Fig. 15.3a). Karyogamy occurs with nucleolar fusion being somewhat delayed; the resulting diploid nucleus remains in a near central position and early (pre-metaphase I) stages shortly ensue. During the zygotene stage of prophase I, lateral elements of the synaptinemal complexes appear signalling the onset of chromosomal synapsis. Subsequently, more pronounced condensation of chromatin and formation of the central element of the synaptonemal complexes occur. The presence of completely formed, persistent, tripartite synaptinemal complexes (Figs. 15.1 and 15.3b), which are associated with genetic crossover, characterize the pachytene stage at the electron microscope level. In the diplotene and diakinesis stages, the meiotic nucleus migrates to near the basidial apex. Apparently the synaptonemal complexes break down at this point, but the ultrastructural aspects of the disruption process are poorly documented for Basidiomycetes. Likewise, the nucleolus disappears prior to the next stage. A transversely oriented spindle composed of numerous microtubules and associated with electron-opaque spindle pole bodies develops at or just prior to metaphase. During late metaphase I, the nuclear envelope becomes discontinuous and vesiculate or at least disrupted at the poles. Chromosomal disjunction and polar anaphase I movement are not synchronous. During anaphase I, the spindle appears composed of a core of parallel, continuous microtubules surrounded by chromosomes irregularly distributed along its length (Fig. 15.4a). Astral microtubules, although not abundant, are sometimes present in the cytoplasm. Once the chromosomes reach the spindle pole zones, telophase I and the associated nuclear envelope reformation occur. The two resulting nuclei undergo division II synchronously to produce four daughter nuclei. Second division seems similar to division I in its mechanics, appearance and orientation (Fig. 15.4b). However, it is more rapid and occurs on spindles with reduced numbers of microtubules.

15.6 Post-Meiotic Basidial Changes

Prior to sporulation, post-meiotic homobasidial cytodifferentiation often results in some characteristic structural and histochemical changes. Although they sometimes remain in the upper third of the basidium (Sass, 1929), in most species the nuclei descend to the basidial mid-region prior to sporogenesis (Wager, 1892, 1893; Harper, 1902; Sass, 1929; Wakayama,

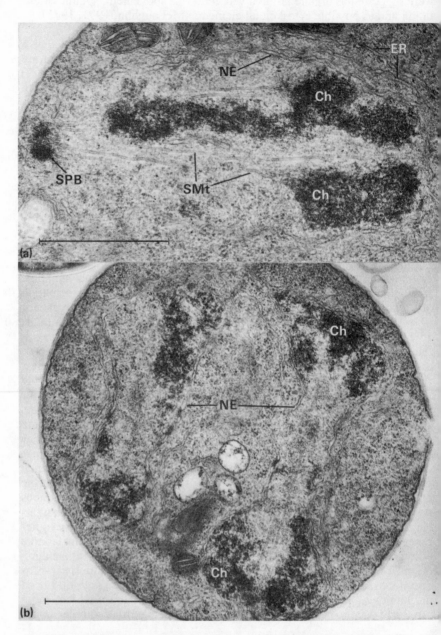

Fig. 15.4 Basidial cross sections illustrating meiosis in *Poria latemarginata*. Scale lines equal 1 μm. (Setliff *et al.*, 1974). (a) Anaphase I. (b) Late anaphase II. Ch, chromatin; ER, endoplasmic reticulum; NE, nuclear envelope, SMt, spindle microtubule; SPB, spindle pole body.

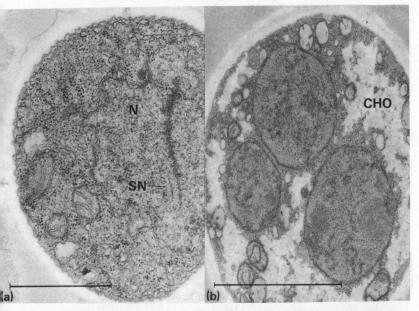

Fig. 15.5 Cross sections of basidia of *Schizophyllum commune* showing the change in cytoplasmic carbohydrate content. Scale line equals 1 μm. (Sundberg, 1971). (a) At the pachytene (division I) stage of meiosis, cytoplasmic carbohydrate is scant. A synaptonemal complex (SN) is present in the nucleus (N). (b) After completion of meiosis, cytoplasmic carbohydrate (CHO) deposits become extensive. Three of the four daughter nuclei are present.

1930; Smith, 1934; Ehrlich & McDonough, 1949; Sequeira, 1954; Wells, 1965). Rapid synthesis and extensive accumulation of glycogen was observed in post-meiotic basidia of *Schizophyllum commune* (Fig. 15.5; Wells, 1965; Volz, Heintz, Jersild & Niederpruem, 1968; Niederpruem & Wessels, 1969). Glycogen production was not strictly correlated with nuclear events (Sundberg, 1971), but, in general, its development could be used as a post-meiotic indicator. Alternatively, lipids were formed or present during maturation in several other species (Manocha, 1965; Thielke, 1967, 1968; McLaughlin, 1970; Hoch & Setliff, 1976).

15.7 Basidiosporogenesis

Initiation of sterigmata, the first step in basidiosporogenesis, occurs on completion of division I of meiosis in *Collybia maculata* (A. & S. ex Fr.) Quel. (Huffman, 1968) and toward the end of division II in *Agaricus bisporus* (Lange) Imbach (Evans, 1959). In other species, they originate subsequent to division II, usually during or soon after basipetal nuclear migration (Wager, 1892, 1893; Harper, 1902; Buller, 1909; Sass, 1929; Wakayama, 1930; Smith, 1934; Ehrlich & McDonough, 1949; Lu & Brodie, 1964; Wells, 1965; Lu & Raju, 1970; Raju & Lu, 1970; Hoch & Setliff, 1976). Evans (1959) postulated that formation of sterigmata was not synchronized with meiosis, and like Corner (1948), suggested attainment of maximum basidial size as the initiation stimulus.

Sterigmata appear to result from renewed tip growth at restricted loci in the basidial apex (McLaughlin, 1973). Seeing the need for localized softening of the basidial wall, Corner (1948) hypothesized the existence of specialized wall regions which he termed 'sterigmatal patches'. Further, he observed the development of a 'hyaline cap' at the basidial apex prior to formation of the sterigmata. Corner (1948) suggested that the cap represented a rearrangement of cytoplasmic contents and might be wall material for the future sterigmata. Ultrastructural studies have yet to support these observations and hypotheses. Noting the presence of microtubules extending from the basidium into the sterigmata, McLaughlin (1973) proposed that microtubules may perform a cytoskeletal function in formation of the asymmetric sterigmata. However, the mechanisms controlling the number and position of sterigmatal loci are poorly understood.

Sterigmata develop as blunt protuberances (Corner, 1948) that become tapered (Corner, 1948; Wells, 1965; McLaughlin, 1973; Hoch & Setliff 1976) and curved (Corner, 1948; McLaughlin, 1977, Fig. 3; Hoch & Setliff 1976, Fig. 8) as they enlarge. The presence of variable numbers of membrane-bound vesicles (Fig. 15.6a), which are 40–95 nm in diameter and appear to fuse with the plasmalemma, and the lack of ribosomes and other organelles at the sterigma tip in *Coprinus cinereus* (Schaeff. ex Fr.) S. F. Gray and *Boletus rubinellus* Peck (McLaughlin, 1973), and *Poria latemarginata* (Hoch & Setliff, 1976) suggest that sterigma growth is similar to that reported by Marchant, Peat & Banbury (1967) and Grove & Bracker (1970).

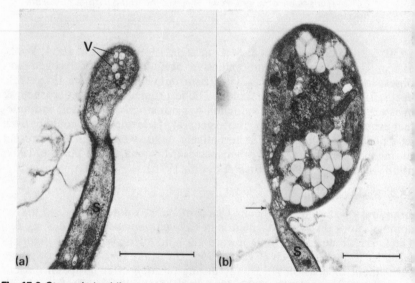

Fig. 15.6 Stages in basidiospore development in *Poria latemarginata*. Scale lines equal 1 μm. (Hoch & Setliff, 1976.) (a) Section through a sterigma (S) with expanding asymmetric basidiospore initial at apex. Note the numerous cytoplasmic Golgi vesicles (V). (b) Longitudinal section through a nearly mature basidiospore on the sterigma (S). Numerous lipid droplets are evident in the spore, and the cytoplasm of the sterigma and spore are still continuous. The hilar appendix region (arrow) is not well differentiated.

for the basidiomycete hyphal apex (McLaughlin, 1973). The vesicles appear to be derived from primitive unstacked Golgi cisternae in the basidium (McLaughlin, 1973; Hoch & Setliff, 1976). Light microscopic (periodic acid–Schiff's reaction) and ultrastructural (silver protein technique) cytochemistry elegantly demonstrated that in *Coprinus cinereus* (McLaughlin, 1972, 1974) they contain carbohydrates as do the Golgi cisternae. These vesicles most likely transport wall materials and may contain enzymes for wall synthesis (McLaughlin, 1973) which are exuded from the cell on vesicle–plasmalemma fusion during sterigma formation and subsequent basidio-spore development.

Basidiospore initiation is signalled by asymmetric sterigma tip expansion (Corner, 1948; Wells, 1965; Sundberg, 1971; McLaughlin, 1973, 1977; Hugueney, 1975; Hoch & Setliff, 1976). It occurs only after the sterigmata reach their maximum size (Corner, 1948; Ehrlich & McDonough, 1949) which seems reasonable if a mechanism homologous with hyphal tip growth is involved in expansion of the young spore (McLaughlin, 1973, 1977). What controls the shift in cytodifferentiation from the narrow tapered growth form of sterigmata to the enlarging one characteristic of the young basidiospore remains to be determined.

In a detailed ultrastructural study, McLaughlin (1977) recognized four ontogenetic stages in early basidiospore development in *Coprinus cinereus*—the inception, asymmetric growth, equal enlargement, and elongation stages. He correlated these stages with changes in the mechanism and orientation of growth in the spore, and like Corner (1948), suggested that different degrees of wall setting influenced the generation of spore shape. Tip growth, similar to that responsible for hyphal extension, occurs during these first two stages. Subsequently, the enlargement pattern suggests a shift to another morphogenetic mechanism with more diffuse intercalary or band growth (McLaughlin, 1977).

During the inception stage, the tip of the sterigma enlarged to form a more or less spherical spore initial and growth was more or less vertical (along a line parallel to the longitudinal axis of the sterigma). Internally, an electron-opaque structure—the hilar appendix body—appeared near the plasma membrane and adjacent to the future site of the hilar appendix. A similar structure was reported in *Coprinus comatus* (Mull. ex Fr.) S. F. Gray and *C. congregatus* (Bull. ex St. Amans) Fr. (Hugueney, 1975) but was not noted in other genera (Wells, 1965; Hoch & Setliff, 1976). Although its morphology changed in later stages (McLaughlin, 1977), the hilar appendix body persisted until the basidiospores were mature and subsequently disappeared (Hugueney, 1975). It was seemingly associated with formation of the hilar appendix and may be involved with the discharge mechanism (McLaughlin, 1977).

Asymmetric abaxial growth ensued and the hilar appendix developed (see Fig. 15.6b for location) during the second stage (McLaughlin, 1972). A change in the orientation of growth from vertical to oblique to the longitudinal axis of the sterigma resulted from the asymmetric aspect of growth and the greater increase in length than width (McLaughlin, 1977). A corresponding oblique stage was also noted in *Schizophyllum commune* (Wells, 1965) and *Poria latemarginata* (Hoch & Setliff, 1976). A band of subsurface vesicles, presumably exocytotic, sometimes occurred along the area of growth (McLaughlin, 1973). The hilar appendix consisted of a projecting, structurally modified, multilayered, wall thickening which became less apparent as maturation ensued and the remainder of the wall thickened. In agreement with Hugueney (1975), McLaughlin (1977) suggested that one layer—an electron-transparent zone—appears to be the site of droplet formation prior to basidiospore discharge described by Buller (1922) and others.

Spherical enlargement of the basidiospores, reported earlier by Wells (1965) in *Schizophyllum commune*, occurred during the equal enlargement stage (McLaughlin, 1977). A cylindrical layer of peripheral rough endoplasmic reticulum with an asymmetric distribution of ribosomes (smooth on the outer side, rough on the inner one) lined the lower part of the basidiospore and was oriented about the axis of growth (McLaughlin, 1977). Since it disappeared at the conclusion of this stage, it is most likely somehow involved in the spherical enlargement process. McLaughlin (1977) suggested that perhaps it functions in enzyme formation for wall synthesis or softening required during expansion.

In the elongation stage, orientation of growth shifts again—from oblique to the longitudinal axis of the sterigma back to vertical (McLaughlin, 1977). Elongation is apparently retarded until an enclosing wall layer present during the spherical growth stage ruptures apically. The basidiospores then elongate considerably with little or no increase in width, attain their near-mature spore shape (Fig. 15.6b) and their walls become considerably thickened.

Corner (1948) proposed that a piston-like pressure exerted by development of a basal vacuole was responsible for both basidiospore initiation and cytoplasmic movement into the differentiating spores. The presence (Thielke, 1968; Niederpruem & Wessels, 1969) and progressive elaboration of basal vacuoles during basidiosporogenesis (Wager, 1892; Corner, 1948; Manocha, 1965; Wells, 1965; McLaughlin, 1970; Hoch & Setliff, 1976) seemingly agree with Corner's hypothesis.

Alternatively, microtubules have been implicated (Rogers, 1973; McLaughlin, 1973, 1977; Gull, 1975; Hoch & Setliff, 1976). Microtubules extend from deep in the basidium up into the sterigmata and eventually into the developing basidiospores (McLaughlin, 1973, 1977; Hoch & Setliff, 1976). They may direct movement (McLaughlin, 1973) or provide a 'track' along which movement occurs (Hoch & Setliff, 1976) as Golgi vesicles required for wall formation and other organelles pass through the sterigmata into the basidiospores.

Finally, microfilaments may be involved in some species (Gull, 1975), but available evidence is fragmentary.

Involvement of microtubules in basidial nuclear movement has also been suggested due to their presence in the sterigmata (Sundberg, 1971; McLaughlin, 1973; Hoch & Setliff, 1976) and in the vicinity of the nuclei during basipetal post-meiotic migration (Sundberg, 1971) and migration into the basidiospores (Rogers, 1973; E. C. Setliff, unpublished results). However, as previously noted for somatic nuclei (Wilson & Aist, 1967; Girbardt, 1968), Wilson, Moller & Griffin (1967) reported that in *Fomes annosus* (Fr.) Cke. the spindle pole bodies were involved in nuclear migration and appeared to pull the nucleus through the upper basidium and sterigmata into the enlarging basidiospores. Spindle pole bodies have been found at the apices of migrating basidial nuclei during sporogenesis in other species using both light (Smith, 1934) and electron (Rogers, 1973; E. C. Setliff, unpublished results) microscopy. In somatic hyphae, Girbardt (1968) observed that nuclear movement was associated with microtubulues attached to the spindle pole body and only occurred when they were present. Contrary to these observations, in *Phanerochaete chrysosporium* Burds., serial sections through a nucleus entering a sterigma just behind its spingle pole body failed to uncover connections between the spindle pole body and microtubules (E. C. Setliff, unpublished results). Since the position of the spindle pole body suggests involvement in nuclear movement during sprorogenesis, clarification of its role and the motive force merit further study.

After entering the spore, the nucleus undergoes a mitotic division in many species of Homobasidiomycetidae (Sass, 1929; Wakayama, 1930; Smith, 1934; Ehrlich & McDonough, 1949; Olive, 1953; Sequeira, 1954; Evans, 1959; Lu & Brodie, 1964; Wells, 1965; Wilson *et al.*, 1967; Duncan & Galbraith, 1972; Hoch & Setliff, 1976); mature spores of many species are thus binucleate.

Other subsequent aspects of cytodifferentiation include separation of the basidiospore cytoplasm from that of the sterigma and development of mature wall structure. An electron-translucent zone separates the cytoplasm of the sterigma and basidiospore (Wells, 1965; Oláh & Reisinger, 1974; Hoch & Setliff, 1976). It appears that this zone forms prior to deposition of material continuous with the remainder of the developing spore wall (Hoch & Setliff, 1976). The mature or near mature basidiospore wall in *Schizophyllum commune* is seemingly continuous in the hilar region (Sundberg, 1971). In contrast, this area is histochemically differentiated and plug-like in *Coprinus comatus* (Oláh & Reisinger, 1974). Wall layers surrounding the remainder of the spore are not produced in a centripetal fashion during early growth as are subsequent increases in wall thickness (McLaughlin, 1977). Variations in mature basidiospore wall structure and their development are beyond the scope of this treatment; for details, the reader should consult Melendez-Howell (1967), Perreau-Bertrand (1967), Pegler & Young (1971), and Kühner (1973).

15.8 References

BULLER, A. H. R. (1909). *Researches on fungi. Vol. I.* London: Longmans, Green, and Co.

BULLER, A. H. R. (1922). *Researches on fungi. Vol. II.* London: Longmans, Green, and Co.

CALONGE, F. D. (1968). Origin and development of intrahyphal hyphae in *Sclerotina fructigena. Mycologia* **60**, 932–42.

CHANG, S. T. & CHU, S. S. (1969). Nuclear behaviour in the basidium of *Volvariella volvacea. Cytologia* **34**, 293–9.

CHANG, S. T. & LING, K. Y. (1970). Nuclear behavior in the basidiomycete, *Volvariella volvacea. American Journal of Botany* **57**, 165–71.

CLÉMENCON, H. (1972*a*). Die exkretorischen zystiden von *Baeospora myosura* (Agricales). *Zeitschrift für Pilzkunde* **38**, 55–71.

CLÉMENCON, H. (1972*b*). Die Phaeozystiden von *Fayodia deusta* (Agricales). *Zeitschrift für Pilzkunde* **38**, 73–87.

CLÉMENCON, H. (1975). Ultrastructure of hymenial cells in two boletes. *Beihefte zur Nova Hedwigia* **51**, 93–8.

CORNER, E. J. H. (1948). Studies in the basidium. I. The ampoule effect, with a note on nomenclature. *New Phytologist* **47**, 22–51.

CORNER, E. J. H. (1950). A monograph of *Clavaria* and allied genera. *Annals of Botany Memoirs* **1**, 1–740.

CORNER, E. J. H. (1953). The construction of polypores. I. Introduction: *Polyporus sulphureus*, *P. squamosus*, *P. betulinus* and *Polystictus microcyclus*. *Phytomorphology* **3**, 152–67.

COX, R. J. & NIEDERPRUEM, D. J. (1975). Differentiation in *Coprinus lagopus*. III. Expansion of excised fruit-bodies. *Archives of Microbiology* **105**, 257–60.

DONK, M. A. (1964). A conspectus of the families of Aphyllophorales. *Persoonia* **3**, 199–324.

DUNCAN, E. G. & GALBRAITH, M. H. (1972). Post-meiotic events in the Homobasidiomycetidae. *Transactions of the British Mycological Society* **58**, 387–92.

EHRLICH, H. G. & MCDONOUGH, E. S. (1949). The nuclear history in the basidia and basidiospores of *Schizophyllum commune* Fries. *American Journal of Botany* **36**, 360–3.

ERIKSSON, J. (1958). Studies in the Heterobasidiomycetes and Homobasidiomycetes–Aphyllophorales of the Muddus National Park in North Sweden. *Symbolae Botanicae Upsalienses* **16**, 1–172.

EVANS, H. J. (1959). Nuclear behaviour in the cultivated mushroom. *Chromosoma* **10**, 115–35.

GIRBARDT, M. (1968). Ultrastructure and dynamics of the moving nucleus. In *Aspects of cell motility*. pp. 249–59. Edited by P. L. Miller. London: Cambridge University Press.

GOODAY, G. W. (1975). The control of differentiation in fruit-bodies of *Coprinus cinereus*. *Reports of the Tottori Mycological Institute* **12**, 151–60.

GROVE, S. N. & BRACKER, C. E. (1970). Protoplasmic organization of hyphal tips among fungi: vesicles and Spitzenkörper. *Journal of Bacteriology* **104**, 989–1009.

GULL, K. (1975). Cytoplasmic microfilament organization in two basidiomycete fungi. *Journal of Ultrastructure Research* **50**, 226–32.

GULL, K. & NEWSAM, R. J. (1975). Ultrastructural organization of cystidia in the Basidiomycete *Agrocybe praecox*. *Journal of General Microbiology* **91**, 74–8.

HARPER, R. A. (1902). Binucleate cells in certain Hymenomycetes. *Botanical Gazette* **33**, 1–25.

HOCH, H. C. & SETLIFF, E. C. (1976). Sterigma and basidiospore development in *Poria latemarginata*. *Memoirs of the New York Botanical Garden* **28**, 98–104.

HUFFMAN, D. M. (1968). Meiotic behavior in the mushroom *Collybia maculata* var. *scorzonerea*. *Mycologia* **60**, 451–6.

HUGHES, S. J. (1971). Percurrent proliferations in fungi, algae, and mosses. *Canadian Journal of Botany* **49**, 215–31.

HUGUENEY, R. (1975). Morphologie, ultrastructure et developpement de l'apicule des spores de quelques Coprinacees: etude particuliere du punctum lacrymans. *Bulletin Mensuel de la Société Linnéene de Lyon* **44**, 249–56.

JÜLICH, W. (1969). Über die Gattungen *Piloderma* gen. nov. und *Subulicystidium* Parm. *Berichte der Deutschen Botanischen Gesellschaft* **81**, 414–21.

KÜHNER, R. (1973). Architecture de la paroi sporique des hyménomycètes et de ses differenciations. *Persoonia* **7**, 217–48.

LENTZ, P. L. (1954). Modified hyphae of Hymenomycetes. *Botanical Review (Lancaster)* **20**, 135–99.

LU, B. C. (1974). Meiosis in *Coprinus*. V. The role of light on basidiocarp initiation, mitosis, and hymenium differentiation in *Coprinus lagopus*. *Canadian Journal of Botany* **52**, 299–305.

LU, B. C. & BRODIE, H. J. (1964). Preliminary observations of meiosis in the fungus *Cyathus*. *Canadian Journal of Botany* **42**, 307–10.

LU, B. C. & RAJU, N. B. (1970). Meiosis in *Coprinus*. II. Chromosome pairing and the lampbrush diplotene stage of meiotic prophase. *Chromosoma* **29**, 305–16.

MANOCHA, M. S. (1965). Fine structure of the *Agaricus* carpophore. *Canadian Journal of Botany* **43**, 1329–33.

MARCHANT, R., PEAT, A. & BANBURY, G. H. (1967). The ultrastructural basis of hyphal growth. *The New Phytologist* **66**, 623–9.

McLAUGHLIN, D. J. (1970). Some aspects of hymenial fine structure in the mushroom *Boletus rubinellus*. *American Journal of Botany* **57**, 745 (Abstr.)

McLAUGHLIN, D. J. (1971). Centrosomes and microtubules during meiosis in the mushroom *Boletus rubinellus*. *Journal of Cell Biology* **50**, 737–45.

McLAUGHLIN, D. J. (1972). Golgi apparatus in the postmeiotic basidium of *Coprinus lagopus*. *Journal of Bacteriology* **110**, 739–42.

McLAUGHLIN, D. J. (1973). Ultrastructure of sterigmata growth and basidiospore formation in *Coprinus* and *Boletus*. *Canadian Journal of Botany* **51**, 145–50.

McLAUGHLIN, D. J. (1974). Ultrastructural localization of carbohydrate in the hymenium and subhymenium of *Cop-*

rinus. Evidence for the function of the Golgi apparatus. *Protoplasma* **82**, 341–64.

McLAUGHLIN, D. J. (1977). Basidiospore initiation and early development in *Coprinus cinereus. American Journal of Botany* (In press).

MELENDEZ-HOWELL, M. M. Recherches sur le pore germinatif des basidiospores. *Annales des Sciences Naturelles. Botanique et Biologie Vegetale Series 12* **8**, 487–638.

NIEDERPRUEM, D. J. & JERSILD, R. A. (1972). Cellular aspects of morphogenesis in the mushroom *Schizophyllum commune. Critical Reviews in Microbiology* **1**, 545–76.

NIEDERPRUEM, D. J. & WESSELS, J. G. H. (1969). Cytodifferentiation and morphogenesis in *Schizophyllum commune. Bacteriological Reviews* **33**, 505–35.

NIEDERPRUEM, D. J., JERSILD, R. A. & LANE, P. L. (1971). Direct microscopic studies of clamp connection formation in growing hyphae of *Schizophyllum commune.* I. The dikaryon *Archiv für Mikrobiologie* **78**, 268–80.

OLÁH, G. M. & REISINGER, O. (1974). L'ontogénie des téguments de la paroi sporale en relation avec le stérigmate et la gouttelette hilaire chez quelques Agarics mélanosporés. *Comptes rendus des seances de l'Academie des Sciences, Series D* **278**, 2755–8.

OLIVE, L. S. (1953). The structure and behaviour of fungus nuclei. *Botanical Review (Lancaster)* **19**, 439–586.

OLIVE, L. S. (1965). Nuclear behaviour during meiosis. In *The fungi.* Vol. I, pp. 143–161, edited by G. C. Ainsworth and A. S. Sussman. New York: Academic Press.

PEGLER, D. N. & YOUNG, T. W. K. (1971). Basidiospore morphology in the Agaricales. *Beihefte zur Nova Hedwigia* **35**, 1–210.

PERREAU-BERTRAND, J. (1967). Recherches sur la différenciation et la structure de la paroi sporale chez les Homobasidiomycètes a spores ornées. *Annales des Sciences Naturelles. Botanique et Biologie Vegetale. Series 12* **8**, 639–749.

RAJU, N. B. & LU, B. C. (1970). Meiosis in *Coprinus.* III. Timing of meiotic events in *C. lagopus* (sensu Buller). *Canadian Journal of Botany* **48**, 2183–6.

ROGERS, D. P. (1934). The basidium. *University of Iowa Studies in Natural History* **16**, 160–83.

ROGERS, D. P. (1936). Basidial prolifera-

tion through clamp-formation in a new *Sebacina. Mycologia* **28**, 347–62.

ROGERS, M. A. (1973). Ultrastructure of meiosis in *Coprinus stercorarius. American Journal of Botany.* **60** (Suppl.), 21–2. (Abstr.)

SASS, J. E. (1929). The cytological basis for homothallism and heterothallism in the Agaricaceae. *American Journal of Botany* **16**, 663–701.

SEQUEIRA, L. (1954). Nuclear phenomena in the basidia and basidiospores of *Omphalina flavida. Mycologia* **46**, 470–83.

SETLIFF, E. C., HOCH, H. C. & PATTON, R. F. (1974). Studies on nuclear division in basidia of *Poria latermarginata. Canadian Journal of Botany* **52**, 2323–33.

SETLIFF, E. C., MacDONALD, W. L. & PATTON, R. F. (1972). Fine structure of the septal pore apparatus in *Polyporus tomentosus, Poria latemarginata,* and *Rhizoctonia solani. Canadian Journal of Botany* **50**, 2559–63.

SINGER, R. (1975). *The Agaricales in modern taxonomy.* 3rd ed. Vaduz: Cramer.

SMITH, A. H. (1934). Investigations of two-spored forms in the genus *Mycena. Mycologia* **26**, 305–31.

SMITH, A. H. (1966). The hyphal structure of the basidiocarp. In *The fungi.* Vol. II, pp. 151–77, edited by G. C. Ainsworth and A. S. Sussman. New York: Academic Press.

SPOONER, B. S. (1975). Microfilaments, microtubules, and extracellular materials in morphogenesis. *Bioscience* **25**, 440–51.

STATES, J. S. (1975). Intrahyphal hyphae in the Basidiomycete, *Gloeophyllum (Lenzites) saepiarium. Mycologia* **67**, 417–20.

SUNDBERG, W. J. (1971). A study of basidial ontogeny and meiosis in *Schizophyllum commune* utilizing light and electron microscopy. *Ph.D. Dissertation. University of California, Davis.*

TABER, A. W. (1966). Morphogenesis in Basidiomycetes. In *The fungi.* Vol. II, pp. 387–412, edited by C. G. Ainsworth and S. A. Sussman. New York: Academic Press.

THIELKE, C. (1967). Die Feinstruktur der Basidien des Kulturchampignons. *Archiv für Mikrobiologie* **59**, 405–7.

THIELKE, C. (1968). Membransysteme in meiotischen Basidien. *Berichte der Deutschen Botanischen Gesellschaft* **81**, 183–6.

THIELKE, C. (1969). Die Substruktur der Zellen im Fruchtkörper von *Psalliota bispora*. *Mushroom Science* **7**, 23–30.

THIELKE, C. (1972). Zisterneaggregation bei hohern Pilzen. *Protoplasma* **75**, 335–9.

VOLZ, P. A., HEINTZ, C., JERSILD, R. & NIEDERPRUEM, D. J. (1968). Synaptinemal complexes in *Schizophyllum commune*. *Journal of Bacteriology* **95**, 1476–7.

WAGER, H. (1892). On the nuclei of the Hymenomycetes. *Annals of Botany (London)* **6**, 146–8.

WAGER, H. (1893). On the nuclear divisions in Hymenomycetes. *Annals of Botany (London)* **7**, 489–514.

WAKAYAMA, K. (1930). Contributions to the cytology of fungi. I. Chromosome number in Agaricaceae. *Cytologia* **1**, 369–88.

WELLS, K. (1964). The basidia of *Exidia nucleata*. I. Ultrastructure. *Mycologia* **56**. 327–41.

WELLS, K. (1965). Ultrastructural features of developing and mature basidia and basidiospores of *Schizophyllum commune*. *Mycologia* **57**, 236–61.

WELLS, K. (1977). Mitotic and meiotic divisions in the Basidiomycotina. In *Mechanism and control of cell division*, (in press). Edited by T. L. Rost and E. M. Gifford. Stroudsburg, Pennyslvania: Dowden, Hutchison & Ross.

WILSON, C. L. & AIST, J. R. (1967). Motility of fungal nuclei. *Phytopathology* **57**, 769–71.

WILSON, C. L., MOLLER, J. C. & GRIFFIN, B. R. (1967). Nuclear behavior in the basidium of *Fomes annosus*. *American Journal of Botany* **54**, 1186–8.

CHAPTER 16

Sexual Morphogenesis in the Ascomycetes

G. TURIAN

16.1 Introduction

Sexual reproduction in the fungi is controlled by two systems. The first is responsible for the sex or mating types which operate at the population level and regulate the balance between inbreeding and outbreeding. The second is concerned with the morphogenetic sequence leading through the gametes or their equivalents and their fertilization to the meiospores, the ascospores in Ascomycetes. The mating types determine the potential ability of different kinds of nuclei, or gametes which carry them, to fuse and undergo fertilization. Inbreeding is insured by self-fertilization or homomixis (= homothallism) while outbreeding is promoted by heteromixis (= heterothallism) including dimictic and diaphoromictic systems (see Table 8 in Burnett, 1975). In Ascomycetes, the heterothallic systems are of the dimictic type only in which the matings are controlled by two types of complementary nuclei (bipolar control), each containing one of the two alleles (A/a) existing at the same locus. There are, however, some species (*Podospora* genus) termed homodimictic (Burnett, 1975) where the outbreeding system has had some type of inbreeding device secondarily imposed upon it (heterogenic incompatibility, according to Esser & Kuenen, 1965).

As mentioned above, morphogenesis of reproductive structures is a precise process which is typically under the control of factors other than those which determine sexual compatibility *i.e.* genes, hormones and environment. Such morphogenesis can, however, be completely lacking in many Ascomycetes in which exchange of nuclei must occur by hyphal anastomosis or somatogamy. In its fullest expression, sexual morphogenesis leads to the differentiation of female and male organs, rarely on different mycelia (dioecism as known in Laboulbeniales), but usually on the same mycelium (in parallel with homomictic, dimictic or homodimictic situations).

In its most complex form, the female organ or archicarp includes a functional or gametic zone consisting of one cell or a chain of cells, the ascogonium; this generally coiled portion may be supported on a few basal cells and surmounted by a trichogyne, a cell that receives the male protoplasm and allows it to pass into the ascogonium (plasmogamy). The male organ may be a gametangium or antheridium as in certain Plectomycetes (*Penicillium wortmanni*) and Discomycetes (*Ascobolus magnificus, Pyronema confluens*) or a detached gametic cell or spermatium in Pyrenomycetes (*Neurospora crassa, Podospora anserina*) and several Discomycetes (*A. carbonarius, Sclerotinia spp.*). In the apogamous species (*Eurotium herbariorum, P. vermiculatum, A. equinus, etc.*) the male nuclei are not transferred to the ascogonium.

Following plasmogamy, karyogamy is usually delayed, and the dikaryotic condition initiated is prolonged for several cell generations. The special hyphae containing coupled haploid nuclei of both male and female origin are termed the ascogenous hyphae. These hyphae are generally broader than the associated vegetative hyphae and have thinner walls, and dense, easily stained contents. They are first multinucleate and then become septated into binucleate cells. The dikaryons they contain divide synchronously. The terminal cell in which nuclear fusion or karyogamy is to occur becomes crozier-shaped. Prior to diploidization, the two nuclei divide in the same plane in such a way as to leave one of each daughter nucleus in the tip of the cell which is then cut off by the formation of two septa to produce a binucleate ascus mother cell. Meiosis occurs immediately after the formation of the single diploid nucleus. It is normally followed by a single mitotic division producing 8 nuclei around which ascosporogenesis then occurs. As expressed by Smith & Berry (1974), the mechanism by which the nuclei and therefore the ascospores remain precisely ordered during meiosis and the subsequent mitosis in *Sordaria* and *Neurospora* presents a fascinating problem. It is related to spindle orientation, as skilfully shown by Srb's group (Pincheira & Srb, 1969) using morphological mutants such as *peak* of *N. crassa*.

The whole sexual process occurs in a protective structure known as the ascocarp. The initiation of this resistant, pseudoparenchymatous structure (external plectenchyma, internal prosenchyma) may occur from somatic, monokaryotic hyphae in a preformed stroma, before the onset of archicarp genesis (Ascoloculares); it may also be issued from vegetative hyphae induced by plasmogamy which proliferate from the stalk cells of the archicarp (*Pyrenoma* sp.). Such proliferation may also spontaneously follow archicarp formation but stop at a so-called protoperithecium stage as exemplified by mycelia of either mating type in heterothallic *Neurospora* species (*N. sitophila, N. crassa.*) The simplest type of ascocarp fully encloses the ascogenous system and refers to the cleistothecium of Aspergillales.

Immersed within a stroma (*Nectria–Hypocrea* organizational level) or free on the substrate hyphae (*Sordaria–Neurospora*), the flask-shaped perithecium of Pyrenomycetes opens at maturity by means of an ostiole. The completely open apothecium of Discomycetes bears the asci and separating paraphyses on its exposed surface.

16.2 Morphological Differentiation

The central and most constant initial sexual structure is the ascogonium. In Plectomycetes (Gymnoascales, Eurotiales–Aspergillales, *etc.*) it arises as a single, highly stainable, hyphal branch which coils in various ways. In Aspergilli (Eurotiales), no side antheridium has been demonstrated (Raper & Fennell, 1965). In the Penicillia (Eurotiales) of the *luteum* series, the sexual initials form coiled, paired branches, the antheridial one usually being thinner than the ascogonial. A regressive series has been described leading to hyphal functional substitutes (Gäumann, 1964; Turian, 1966*a*). Sexual organs may also develop side by side on the same hyphae or from different hyphae. After formation, protective hyphae arise from the base of the ascogonium and then surround and enmesh the ascogenous hyphae and asci. Such loose ascocarpic structures are superseded in Eurotiales by a distinct peridium formed of pseudoparenchyma (Butler, 1966).

In the Plectomycetes, the ascogenous hyphae ramify throughout the central ascocarpic prosenchyma and produce globose asci at all levels within the fructifications. In the Pyrenomycetes and Discomycetes, they typically grow out radially to form a plate or hollow disc in the lower part of the ascocarp. Asci are cylindrical or clavate and produced as a palisade (hymenium).

The two major lines of development in the Pyrenomycetes show not only a difference in origin of the protective hyphae adjacent to the asci, but they are most readily separated on the structure of the ascus wall. Roughly, the Ascoloculares have double-walled asci and are referred to as Bitunicatae while the Ascohymeniales possess single-walled asci and are therefore termed Unitunicatae (von Arx & Müller, 1975). In the bitunicate orders Dothideales and Pleosporales, the ascogonium arises below the surface of the vegetatively produced stromatic ascocarp and the bitunicate asci are formed in loculi (preformed in Pleosporales). In the important genera *Pleospora*, *Leptosphaeria*, and *Ophiobolus* (*Gäumannomyces*) the stromatic part of the uniloculate fructification is poorly developed; the ascogonia of *Ophiobolus* can therefore be easily observed when they appear among the tangles of *in vitro* grown mycelium as highly basophilic hyphal coils (Gindrat & Turian, 1967).

The Unitunicatae corresponding to the Ascohymeniales reach the highest form of development with the perithecium or the apothecium. Along the ontogenesis of such fruiting bodies, the developing (fertilized or not) ascogonium first forms a prosporophyte of ascogenous hyphae immersed in a nutrient-rich prosenchymatous trophophyte (Chadefaud, 1960; Letrouit-Galinou, 1973), constituting a centrum well protected by the external plectenchymatous wall. The maturing asci finally constitute the sporophyte while the separating paraphyses are of gametophytic origin (the ascogonial stalk cells).

A prosenchymatous centrum has been observed in one of the more primitive members of the series, the genus *Chaetomium*, in which the ascogonium is typically coiled (Whiteside, 1961). The trophophytic phase is decreasing at the benefit of the developing ascogenous apparatus. Its cells are multinucleate (syncytial character) and densely filled with lipid globules as detected by optical microscopy in young perithecia of *Gnomonia leptostyla* (Fayret, 1975) or by electron microscopy in the fructifications of *Neurospora* species (Hohl & Streit, 1975; Turian, 1975; Fig. 16.1).

Among the non-stromatic Pyrenomycetes, primary and secondary homothallic species are most easily studied as only one strain is required for

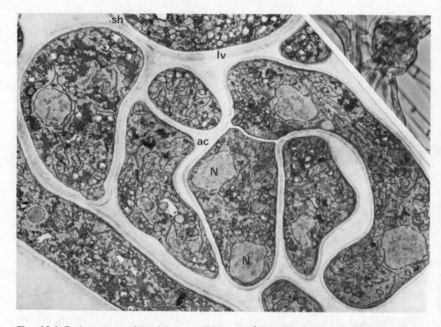

Fig. 16.1 Early protoperithecial stage (80 h at 25°C) of *Neurospora tetrasperma* with sheath hyphae (sh) surrounding central, fertile elements of the ascogonial coil (ac). Electron microscopy of a transverse thin section (×8000) fixed in glutaraldehyde-permanganate: ascogonium with dikaryotic cells (N + N); richness in lipid vesicles (lv) especially in the sheath elements. Insert, in vivo at the optical microscope (×800) (with N. Oulevey, 1971).

the whole sexual morphogenesis. In *Sordaria fimicola*, the initiation of the perithecium starts with a hook- or coil-shaped lateral outgrowth from a relatively old hypha, having septal pores plugged (Greis, 1936). Soon a new septum delimits the ascogonial coil from the hyphal compartment. This coil then becomes enveloped by neighbouring hyphae which begin to turn brownish and finally dark brown (protoperithecium stage). The perithecia become mature (ascogenesis) and start to discharge ascospores after about 5 days culture (Lindenmayer & Schoen, 1967).

In *Neurospora tetrasperma*, the first morphological sign of the sexual initial has been traced to a lateral bud on a septate hypha which extends to form a hook and a tightly coiled ascogonium (Viswanath-Reddy & Turian, 1975). This last, fertile structure with dense protoplasmic content contrasts with the sterile loose coils of narrow hyphae described as *lassos* or pseudoascogonia (Gindrat & Turian, 1967) appearing in unfavourable growth conditions (high temperature, *etc.*). The developing steps through the transitional protoperithecial stage to the fertile perithecium have been cytologically investigated by Colson (1934). In the strains of heterothallic *Neurospora* grown separately, morphogenesis stops at the protoperithecial stage. The emerging trichogynes are receptive to fertilizing microconidia acting as spermatia (Backus, 1939).

This morphogenetic sequence is still poorly documented by means of electron microscopy, especially in its initial stages. The coiled ascogonia inside developing protoperithecia of *Neurospora tetrasperma* are dense, often binucleate and surrounded by enveloping hyphae filled with lipid vesicles (Fig. 16.1). It is at the next stage of the ascogenous hyphae that dikaryons can be detected in *N. tetrasperma* (Turian, 1975) and *N. lineolata* (Hohl & Streit, 1975). Scanning electron microscopy complemented by a freeze-fracture technique has allowed the viewing of external and internal features from the protoperithecial stage to the ascospore-liberating perithecium of *N. crassa* (Harris, Howe & Roth, 1975).

In a few Discomycetes such as *Pyronema domesticum* (Moore-Landecker, 1963), the ascogonium differentiates as a massive hook-like structure while in others such as *Peziza quelepidotia*, it forms the more usual coiled structure, as beautifully illustrated by scanning electron microscopy (O'Donnell & Hooper, 1974).

In several species of *Chaetomium*, a cell of the ascogonial coil is fertilized by a slender antheridium (Whiteside, 1957). In *Venturia inaequalis*, both ascogonium and antheridium are still well differentiated while in *Glomerella*, the sexual organs are often indistinguishable from one another when they arise from adjacent hyphae (McGahen & Wheeler, 1951). Many monoecious Discomycetes among which *Pyronema confluens* (Moore-Landecker, 1963) and *Ascobolus magnificus* (Dodge, 1920) also differentiate an antheridium on the same hyphal branch as the ascogonium. It is the apical trichogynous cell of this last organ which receives and channels the male nucleus to the female cell; In *A. magnificus*, the central cell of the septate ascognium becomes the initial of the ascogenous hyphae (Dodge, 1920). The trichogyne can, however, fail to unite with a fertilizing agent in a few isolates of *P. domesticum* in which the ascogonia directly produced ascogenous hyphae (Moore-Landecker, 1975). No functional antheridium has been found near the coiled ascogonium in the Plectomycetous *Ophiostoma* (Turian, 1966a).

In several Pyrenomycetes and Discomycetes, the male structure is a phialidic organ, basipetally and sequentially budding uninucleate cells called microconidia, or, more specifically, spermatia. These male cells are produced in Pseudosphaeriales (*Mycosphaerella*) and the Sphaeriales (*Bombardia, Podospora,* occasionally *Sordaria*) as the only conidia and in a minor proportion compared to the mainly asexual macroconidia in both *Neurospora crassa* and *N. sitophila* (Turian & Bianchi, 1972). Among Discomycetes, microconidial formation (also produced with macroconidia) has been especially studied in the Sclerotiniaceae (Drayton's group, see Turian, 1966a).

In the Ascobolaceae, the sexual reaction of *Ascobolus stercorarius* results from the interaction between the directionally-oriented trichogyne of the ascogonium and a single arthrosporic oidium (Bistis' studies, 1956–7, see Turian, 1969).

16.3 Genetic Control

Homothallic forms

In homothallic forms, the sequence of sexual development is constitutive in contrast to heterothallic forms where it seems that all mating type genes function as switches to turn on morphogenesis when two compatible nuclei are associated (Fincham & Day, 1971).

The wild types of the best-studied homothallic Pyrenomycetes, *Glomerella cingulata*, *Sordaria macrospora* and *S. fimicola*, initiate their sequential morphogenesis as soon as environmental factors and internal conditions become permissive. It is only in spontaneous or induced variants obtained from nature or from laboratory cultures that the sequence is completely, totally (sterile forms) or partially (stage-blocked forms) interrupted. The 'loss mutations' first described in *G. cingulata* (Wheeler & McGahen, 1952) convert this basically homomictic species into potentially outbreeding, bisexual forms. Such self-sterile mutants behave like *A* and *a* or 'male' and 'female' strains. They were shown to recover full fertility through crosses, thus implicating complementary action of wild-type alleles (Wheeler, 1956). From the crossed perithecia, self-sterile homothallic forms were recovered through genetic recombination between the mutants. Substances liberated in the filtrates of wild-type cultures could alleviate a satisfactory proportion of the self-sterility as shown in the B2 mutant blocked at the nuclear fusion level in the ascus (Wheeler, 1954). Such diffusable agents are presumably the products of switched-on 'morphogenetic genes'.

From these cases of loss-mutations and alleviating effects following crosses it is a small step to the regular outbreeding situation of heterothallic species. This was shown with the two exceedingly closely linked arginine mutants of *Sordaria fimicola*, *a-3* and *St-59*; their asci and ascospores aborted or were non-viable when they were grown alone while they complemented each other and produced fertile asci when grown together (El-Ani & Olive, 1962). Only the two parental genotypes segregated out of 504 asci analysed. In this case, the homothallic *S. fimicola* resembled a dimicitic fungus in its segregation of two kinds of 'compatible' partners (Burnett, 1975). The occurrence of morphological mutants newly found to be cross-fertile in all conditions, the protoperithecial mutant *th* and the abortive ascospore mutant *m* of *S. fimicola* has been re-emphasized by El-Ani & Olive (1975) as illustrating a method by which typical balanced heterothallism may originate from homothallism in nature by means of non-overlapping, complementing mutations within one locus or two tightly linked loci controlling an essential phase in the sexual process. However, in an attempt to find complexity at the mating type (*mt*) locus of *Neurospora crassa*, Newmeyer, Howe & Galeazzi (1973) found no change in crossing ability among the 5019 recombinants tested, representing 235 000 viable ascospores. They concluded that if subunits exist, they are not more than 0.002 units apart. Moreover, no separation of the crossing-compatibility function and the heterokaryon-incompatibility was found. Taken together with the heterokaryon-incompatibility obtained when Metzenberg & Ahlgren (1973) introgressed the *N. tetrasperma* mating-type alleles into *N. crassa* (*Neurospora* strains of opposite mating-type are heterokaryon-incompatible during vegetative growth), Newmeyer's results suggest that mating type in *Neurospora* is either a single multifunctional locus, or else a complex region which produces a negligible frequency of viable recombinants.

In *Sordaria macrospora*, the numerous genetic blocks described by Esser & Straub (1956, 1958) as interrupting the morphogenetic sequence from the ascogonium to the mature perithecium have been repeatedly described and well-illustrated (see all treatises of fungal genetics cited, also Turian, 1969).

In *Neurospora*, several mutations interfere with protoperithecial morphogenesis (Westergaard & Hirsch, 1954). Among such protoperithecialess mutants, the tyrosinase mutants *ty-1* and *ty-2* present the additional interest that their induced tyrosinase activity by D-amino acids is not paralleled by an alleviation of their morphogenetic defect (Horowitz, Fling, McLeod & Sueoka, 1960). The new *ff-1* mutation of *N. crassa* specifically caused the prevention of the formation of protoperithecia and had no effect on vegetative morphology or nutritional requirements (Tan Sai Tee & Coy Choke, 1970). Interestingly, most of the seven to eight different female sterile mutants which failed to produce functional protoperithecia were shown to be able to function as female parents in heterokaryons (Mylyk & Threlkeld, 1974). The additional fact that vegetative morphology of these mutants differed from those of the wild type suggests that protoperithecial development is under an elaborate genetic control. Mutations to male sterility, obtained by ultraviolet light irradiation of macroconidia, have also been mapped in *N. crassa* (Weijer & Vigfusson, 1972; Vigfusson & Weijer, 1972).

In *Podospora anserina*, a genetic block which is effective only with the addition of mutations in two modifier genes (*mod-1* and *mod-2*) suppressed both the incompatibility process in strain confrontation ('barrage') and the formation of protoperithecia (Boucherie & Bernet, 1974). Interestingly, the block operated at the ribosomal level and prevented the translation of the messenger of a proteolytic enzyme (see sex protein, p. 328).

Perithecial colour mutants are known in many Pyrenomycetes which have been used to demonstrate that the perithecial wall is of maternal origin (see introduction by Howe & Benson (1974) to their new, orange *per -1* mutant).

Dimictic or heterothallic forms

In dimictic or heterothallic forms, a haploid individual can never breed with more than 50% of the whole population but it can always mate with half the progeny of the same zygote (Burnett, 1975). Such mating systems thus favour the opportunity for outcrossing in preventing the selfing of a single haploid. They are functioning through a unifactorial, bipolar expression of one of the two alleles A/a at the same locus as best illustrated by the classical, heterothallic *Neurospora crassa* and *N. sitophila* (Whitehouse, 1949).

Bipolar mating types are also known not only in other Pyrenomycetes such as *Sordaria heterothallis* (Fields & Maniotis, 1963) and Discomycetes such as many Ascobolaceae (*Ascobolus immersus, etc.*) but also in Plectomycetes, *Aspergillus heterothallicus* being however an exception among self-fertile *Aspergillus* species (Raper & Fennell, 1965).

Those phenomena described as bipolar mating systems correspond to *homogenic* incompatibility when they are contrasted with the *heterogenic* incompatibility defined by Esser (1966) as the inhibition of zygote formation between partners of the same species due to the heterogeneity (pluripolarity) of their incompatibility loci. Such a situation has been analysed in different strains of *Podospora anserina* and appears to extend to other fungal species (Esser & Blaich, 1973).

In many Ascomycetes, inbreeding has been secondarily and indirectly imposed on outbreeding as a consequence of morphological or cytological changes (Burnett, 1975). The best known examples of such homodimictic or constitutive homothallic systems are the 4-spored *Neurospora tetrasperma* and *Podospora anserina*. In both of these species two haploid nuclei of opposite mating type are distributed in each of the ascospores by distinctive cytological devices (well-illustrated in all treatises of fungal genetics cited).

Mating systems can be impaired through a mutational block resulting in a loss of recombination. Thus the potentially dimicitic *Fusarium* (*Hypomyces*) *solani* f. *cucurbitae* (mt^+ and mt^-) remains asexual in nature due to a combination of morphological blocks and the spatial distribution of the responsible mutant genes (Hansen & Snyder, 1946). This species has four morphological types determined at the two loci M and C which lead in laboratory cultures to the development of protoperithecia with trichogynes (M) and of conidia (C), respectively. Eight types, therefore, can occur and, as shown by Schippers & Snyder (1967), (see Burnett, 1975, p. 170), differences of gene frequency at the M, C or mt loci could reduce fertility in *Hypomyces*.

Helminthosporium (*Cochliobolus*) is also dimictic but two kinds of mutations can impair its efficiency (Nelson & Kline, 1964). In *Cochliobolus sativus*, extreme impairment of mating systems which suppresses meiosis and fertilization results in an amictic condition while otherwise keeping all the morphological attributes of sexuality. In the same species, sexual morphogenesis was recently shown to be prevented in progeny from one cross by a pre-crozier block (controlled by 3 genes), a post-crozier block and a preascospore block. White and red sterile strains developed from sectors in black monospore colonies (Hosford, Solangi & Kiesling, 1975). Biallelic heterothallism is also complicated in *Nectria cosmariospora* where protoperithecium production has been shown by Tayel & Hastie (1975) to be influenced by 2 more genes (G/g and P/p). No protoperithecia are formed in cultures with the allele p. The allele g determines white mycelium (with G the mycelium has a green pigmentation) and is epistatic to the P gene. In *Podospora arizonensis*, early sexual morphogenesis is normal and the asci develop from binucleate penultimate cells of the ascogenous hyphae; however, these nuclei do not fuse and directly undergo two successive mitotic divisions to give eight haploid ascospores; for unknown reasons, only four of these blacken normally (Mainwaring & Wilson, 1968).

The final processes of sexual morphogenesis, the shape of the asci and the size and shape of their ascospores have been reviewed elsewhere (see Srb, Nasrallah & Basl, 1973 for *Neurospora*; Turian, 1976 for Ascomycetes in general).

16.4 Nutritional–Environmental Controls

An abundance of assimilable carbohydrate has often been considered to be the most general condition favouring sexual morphogenesis in the Aspergillales (see Raper & Fennell, 1965). However, the concentration of nitrogen has been shown to be more important than that of glucose in determining the production of cleistothecia in *Penicillium vermiculatum* (Das Gupta & Nandi, 1957). In this species (and *P. wortmanni*), yellow fruit-body formation is best on low sugar medium or on polysaccharides as C source (Basu & Bhattacharyya, 1962) but is prevented by acetate which rather favours green conidiation and a diffusable, red mycelial pigmentation (G. Turian and J.-P. Haemmerli, unpublished results). There is also no perithecial production when *Neurospora* spp. (Turian, Seydoux & Volkmann, 1962) or *Leptosphaeria typhae* (Viala & Vidal, 1972) are grown on acetate media. In *Chaetomium globosum*, the formation of fructifications is stimulated by exogenous phosphoglycerate (Buston & Khan, 1956) and in *Sordaria destruens* by phosphoric esters of glucose and fructose (Hawker, 1957).

The enhanced formation of perithecia by *Sordaria macrospora* in the presence of an obstacle can be imitated by an inducing substance isolated from cornmeal (Bahn & Hock, 1973) while L-arginine has now been found to be an improved inducing substitute to the plant extract (Bahn & Hock, 1974). More generally, the carbon/nitrogen ratio is a controlling factor for sexual morphogenesis as clearly shown in *Neurospora* (Westergaard & Mitchell, 1947). Thus, too high nitrate in a 2% sucrose medium delays perithecial formation in *N. crassa* (Hirsch, 1954) and exhaustion of this ion in *Pyronema confluens* is even a prerequisite for initiation of apothecial formation (Robinson, 1926). The controlling role of the carbon/nitrogen ratio on the production of perithecia has also been emphasized for *Hypomyces solani* (Hix & Baker, 1964) and for *Sordaria finicola* (Hall, 1971).

A sufficient supply of thiamin is required for complete sexual morphogenesis for species of *Chaetomium* (Hawker, 1942; Lilly & Barnett, 1949) and *Ophiostoma* (Robbins & Ma, 1942 and others, see Turian, 1966a), and of biotin for species of *Sordaria* (Barnett & Lilly, 1947). The level of biotin is also important in relationship to lower levels of sucrose in *Neurospora* (Girbardt, 1952).

As for mineral elements, calcium exerts a positive effect (counteracting acidity) on perithecial production by *Chaetomium globosum* (Basu, 1952, see Hawker, 1957). Boron nearly doubled the number of perithecial initials produced by *Sordaria* grown in a liquid synthetic (sucrose) medium (Turian, 1955). Zinc seems to be required for normal formation of perithecia in *Neurospora* (Turian, 1966b). Manganese has recently been shown to control cleistothecial formation in *Aspergillus* (Zonneveld, 1975).

It is difficult to ascribe a stage-specific effect to most of the nutrients. Thus, it is simply by sequential priority that the traces of biotin required by *Sordaria fimicola* (Lilly & Barnett, 1951) first permit the normal appearance of the sexual initials before eventually impairing by deficiency the differentiation of asci and ascospores.

It is not only the quality but also the concentration of nutrients in the medium which are important in perithecial induction. This is well exemplified by *Neurospora tetrasperma* in which perithecial formation was inhibited in the second of two sequential crosses at different locations on the same mycelium (Calhoun & Howe, 1972). This effect was explained by a previous lowering of the concentration of nutrients in the medium in the second-cross location, owing to prior demand upon those nutrients by the developing perithecia in the first-cross location.

Most important among the physical agents controlling sexual morphogenesis is temperature. There is no general rule as to its differential effect on asexual and sexual sporogenetic stages. Thus, the cleistocarps of *Eurotium herbariorum* and the perithecia of *Ceratostomella fimbriata* are produced most readily on a suitable medium at a relatively high temperature (25–30°C) while the conidial (*Aspergillus* and *Graphium*) stages can develop at lower temperatures (see Hawker, 1957). In *Neurospora crassa*, the macroconidia are still readily formed at 37–40°C (Ojha & Turian, 1968) while fertile perithecia do not appear much above 25°C. At 30°C, only protoperithecia are formed and at 35°C even ascogonia are absent (Hirsch, 1954). This differential effect of temperature on particular stages in reproduction can be exploited for transfer experiments such as those of sterile, 35° cultures to 25°C after fertilization which then produce mature perithecia (Hirsch, 1954). The same thermosensitivity of morphogenesis occurs in *N. tetrasperma* in which synchronous formation of ascogonia can be induced in shift-down cultures from 37° to 25°C (Viswanath-Reddy & Turian, 1972). In other cases, such as that of *Pleospora herbarum*, perithecia are ultraviolet light-induced at 24–27°C but they do not mature unless the temperature is then reduced to 5–10°C (Leach, 1971).

The optimum initial pH for reproduction may also differ from that for vegetative growth of the same species, and the optimum value may change with the type and stage of reproduction. Thus, several Ascomycetes (*e.g.*, *Eurotium herbariarum*, *Chaetomium globosum*) form their fruit-bodies over a wide range of pH but differentiate viable ascospores only over a narrower range (Lockwood, 1937; Hawker, 1966).

The effects of aeration on morphogenesis may be due to one or more of a number of factors among which the concentration of oxygen is most important. It was shown early on that apothecial formation in *Ascobolus* has a more exacting requirement than mycelial growth (Green, 1930). Perithecia of *Neurospora sitophila* fail to form at 0.5% of O_2 (Denny, 1933); however, ascogonia can still differentiate in such conditions while protoperithecia require at least 1% O_2 and a trace of CO_2 to form (Viswanath-Reddy & Turian, 1975). CO_2 had previously been shown to stimulate sporocarp formation in *Chaetomium globosum* (Buston, Moss & Tyrrell, 1966).

The photocontrol of ascocarp formation has been less studied than that of asexual reproduction (Marsh, Taylor & Bassler, 1959; Carlile, 1965; Leach, 1971). Pioneering such photomorphogenetic studies, Robinson showed in 1926 that initiation of the sexual organs in *Pyronema* occurred only after 6 h exposure to blue light. Sexual initials of another Discomycete, *Sclerotinia sclerotiorum*, will however form in the dark, but light is necessary for the pileate disks (Purdy, 1956).

The formation of perithecia can require UV irradiation or, more generally, visible light (see ref. in Inoue & Furuya, 1970). From action spectra for perithecial formation in *Pleospora herbarum*, and other fungi, Leach & Trione (1966) suggested that the photoreceptor in the control of sexual and asexual reproduction was the same (P_{310}). In several species, alternate treatment with light and darkness promoted perithecial formation. Thus, in *Gelasinospora reticulospora*, light had to be given after a dark period to induce fruit-body formation (Inoue & Furuya, 1970). Blue or near-ultraviolet were the promotive light qualities (Inoue & Furuya, 1974). In *Leptosphaerulina trifolii*, mature asci only developed after a few days of incubation in darkness following monochromatic ultraviolet (240–360 nm) induction of perithecia (Leach, 1972).

Little is known about the differentiation of photosensitive hyphae during morphogenesis. Inoue & Furuya (1975) observed micromorphological changes in the hyphae of apically growing mycelium after the photoinduction of perithecia in *Gelasinospora reticulospora*. The perithecium initiation in other Ascomycetes such as *Sordaria fimicola* is not sensitive to light although the perithecia may attain a larger size under alternating, light–dark regime (Surapipith & Lindenmayer, 1969) which also affects the ascospore discharge from mature perithecia (Ingold, 1962; Walkey & Harvey, 1967).

16.5 Hormonal–Metabolic Controls

Hormonal regulation or activation of the sexual process has been reported for a number of fungal species, mainly Phycomycetes (see Gooday, 1974 and also Chapter 13), but only for a few Ascomycetes (see Machlis, 1972). It has only been shown that the trichogyne can alter its direction of growth towards microconidia-spermatia in *Neurospora sitophila* (Backus, 1939) or towards oidia-spermatia in *Ascobolus stercorarius* (Bistis, 1957) in a manner resembling a chemotactic response (see Turian, 1969).

An oestrogenic metabolite (F-2 or zearalenone, an undecenyl-resorcylic acid lactone) isolated from *Fusarium graminearum* has exerted a regulatory control over sexual reproduction in many of the Ascomycetes screened (from *Cochliobolus* to *Neurospora* and *Sordaria* spp., Nelson, Mirocha, Huisingh & Tijerina-Menchaca, 1968). That this compound is serving primarily as an hormonal-like regulator rather than a nutrient source has been inferred from its activity at low doses ($0.01–1.0\ \mu g$) and its stimulating effects limited to the early stages of perithecial formation in *Cochliobolus carbonum*. Positive responses resulting from exogenous applications of sterols in the presence of zinc have been interpreted as evidence that these compounds regulate the biosynthetic pathways participating in that sexual morphogenesis (Tijerina-Menchaca & Nelson, 1969). F-2 did not change the sterol profiles from treated crosses of *C. carbonum* but significantly increased RNA synthesis (Nelson, 1971). It is metabolized and it has been suggested that the gaseous products evolved, such as ethylene, could be the sexual inducers or regulators. Ethylene (with mercaptan and methyl-disulfide) has indeed been detected in crosses of *C. carbonum* treated with methionine, a compound stimulating reproduction in that species (Nelson, 1971). The fact that the prevention of perithecial development by sterol synthesis inhibitors could be partially alleviated by methionine suggests that sterol metabolism *per se* may not be essential for reproduction in *Cochliobolus* (Nelson, 1971). The situation may be similar in *Sordaria fimicola* where sterols have also been implicated in the control of perithecial formation and in which inhibition of that morphogenesis by hypocholesteremic drugs could not be reversed by either cholesterol or oleic acid alone (Elliot, 1969).

When provided exogenously in excess, metabolites such as acetate can switch the metabolism toward certain pathways, in particular the glyoxylate cycle, which are unable to sustain certain morphogenetic states. In *Neurospora*, acetate provided as single C source does not support development beyond the ascogonial stage; when provided to sucrose media, it counteracts the melanic protoperithecial morphogenesis while enhancing the carotenoidic macroconidiogenesis (Turian *et al.*, 1962). Energy relationships are certainly involved in the well-known beneficial, catabolic role of polysaccharides such as cellulose provided as single C source on ascocarp formation (*i.e.* in *Chaetomium* spp. see Hawker, 1957; Turian, 1966*a*). That anabolic processes are involved in parallel in sexual morphogenesis is suggested by the Mn-regulated synthesis of α-glucan, a reserve material, in the cell walls of developing cleistothecia in *Aspergillus nidulans* (Zonneveld 1975).

The use of selective metabolic inhibitors has provided a preliminary insight into the necessary pathways and their sequential functioning during sexual morphogenesis. In *Sordaria fimicola*, two inhibitors of terminal cytochrome respiration, cyanide and sodium azide, efficiently blocked perithecial initiation. Interestingly, these and all perithecial inhibitors were found to be good branching inhibitors, the converse not being true (Linden-mayer & Schoen, 1967). In *Neurospora tetrasperma*, while glycolytic inhibitors such as iodoacetate or deoxyglucose selectively interfered with the ascogonial differentiation, citric acid cycle inhibitors such as fluoro-acetate prevented the further development of the ascogonia into protoperithecia. This suggests that glycolysis is initially needed for the differentiation of ascogonia (early respiratory quotient > 1) while oxidative metabolism is required for the conversion of the ascogonial coils into protoperithecia (Viswanath-Reddy & Turian, 1975). By contrast, differenti-ation of male structures of *Neurospora*, e.g. microconidia-spermatia, was highly stimulated in the presence of the glycolytic inhibitor, iodoacetate (Rossier, Oulevey & Turian, 1973), with parallel enhancement of the activity of the enzymes at the beginning of the pentose phosphate cycle (Rossier & Turian, 1976). This metabolic control has been tentatively portrayed (Fig. 16.2) with reference to the most complete (4 stages) and studied morphogenetic sequence of *N. crassa*.

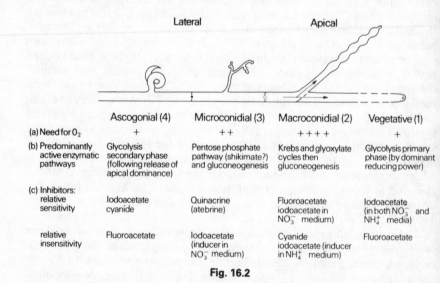

	Ascogonial (4)	Microconidial (3)	Macroconidial (2)	Vegetative (1)
(a) Need for O_2	+	+ +	+ + + +	+
(b) Predominantly active enzymatic pathways	Glycolysis secondary phase (following release of apical dominance)	Pentose phosphate pathway (shikimate?) and gluconeogenesis	Krebs and glyoxylate cycles then gluconeogenesis	Glycolysis primary phase (by dominant reducing power)
(c) Inhibitors: relative sensitivity	Iodoacetate cyanide	Quinacrine (atebrine)	Fluoroacetate iodoacetate in NO_3^- medium)	Iodoacetate (in both NO_3^- and NH_4^+ media)
relative insensitivity	Fluoroacetate	Iodoacetate (inducer in NO_3^- medium)	Cyanide iodoacetate (inducer in NH_4^+ medium)	Fluoroacetate

Fig. 16.2

Metabolism must also be oriented toward efficient nucleic acid synthesis. This could be presumed from the fact that ascogonia have a highly basophilic content reflected in the higher RNA/DNA ratios measured in protoperithecia-bearing mycelia compared to uniformly conidiated ones (Turian, 1966a). As could be expected, the perithecial initials of *Sordaria fimicola* were not formed in the presence of 5-fluorouracil, an inhibitor of RNA synthesis (Lindenmayer & Schoen, 1967). However, the inhibitors of RNA and protein synthesis, actinomycin D and cycloheximide, were inactive in the *Sordaria* system (the second blocked only the maturation stage) while both could efficiently prevent ascogonial induction at the 37 to 25°C step-down in *Neurospora tetrasperma* (Viswanath-Reddy & Turian, 1972).

Attempts have been made to correlate the development of the ascocarps with the synthesis of secondary metabolites such as pigments. In the case of *Pyronema confluens*, carotenoid-less mutants bearing fertile apothecia could be produced by treatment with ultraviolet light; two, apparently independent, photoreceptors are therefore involved in normal photomorphogenesis of *Pyronema* apothecia (Carlile & Friend, 1956). Melanin production and its enzymes, tyrosinase or laccase, have been associated with perithecium development in *Neurospora crassa* (Hirsch, 1954), *Podospora anserina* (Bu'Lock, 1961), *Glomerella cingulata* and *Hypomyces solani* (see Turian, 1966a). Here too, no necessary relationship occurs between dark pigment production and perithecial formation as shown (a) *nutritionally*, when *P. anserina* was grown on a glucose or a fructose medium to produce melanin but no mature perithecia, whereas perithecia developed abundantly with poor pigmentation on a sucrose medium (Esser, 1956, see Turian, 1966a); also in *Sordaria fimicola* when the arginine inducer (see p. 323) was omitted, and sterile cultures strongly pigmented (Bahn & Hock, 1974); and (b) *genetically*, when female sterile mutants of *N. crassa* were induced to produce tyrosinase in the presence of aromatic amino acids which had no alleviating effects on perithecial production (Horowitz *et al.*, 1960). A closer association has, however, been observed in *H. solani* where the period of cold required for the perithecial induction could be replaced by treatment of the vegetative mycelium with tyrosinase (Wilson & Baker, 1969). Phenoloxidase activity has also been shown to be necessary for normal fruit-body formation in *P. anserina* (Esser, 1968). From the ecological point of view, the melanism of most perithecial (rather than apothecial) structures could have some type of adaptive usefulness (protection against radiations, increased resistance to exogenous wall-lytic activities, *etc.*). As for the male, microconidial structures, which are melanic compared to the carotenoid-containing, asexual macroconidia in *Neurospora*, their melanization could be accounted for by the sharp increase in tyrosinase activity measured during their induction (Rossier & Turian, 1976).

16.6 Physiogenetic Interactions

In *Aspergilli*, hypothetical cytoplasmic elements have been assumed to superimpose their effects on those of nuclear factors in the control of perithecial frequency. Such a condition of cytoplasmic duality between asexual and sexual reproduction has been revealed by the loss of ability to produce perithecia in *Aspergillus nidulans* subcultured exclusively by hyphal tip transfer or conidia and its restoration from an ascospore subculture (Jinks, 1966).

In *Neurospora crassa*, the phenotypic range of protoperithecial formation also results from interaction of the genotype with a particular cellular state of the same or a similar system (Fitzgerald, 1963). It has further been shown by Itoh & Morishita (1971) that protoperithecial formation in the culture of heteroplasmonic hyphae obtained by making anastomosis between two strains was more rigorous than either one of the parental strains. This has led to the consideration that a cytoplasmic factor, possibly mitochondrial, is responsible for protoperithecial formation.

Relatively little is still known about the chemical basis of the bipolar compatibility versus incompatibility of heterothallic Ascomycetes. A complementary mechanism involving A and a mating gene products acting as the 'key-lock system' has been envisaged for zygote formation (Esser, 1966); this assumption implies that in the incompatibility combination, both partners form gene-like products unable to complement. The results of different authors (Zickler, 1952, with *Bombardia*; Kuwana, 1958, with *Neurospora*; and Bistis & Raper, 1963, with *Ascobolus*) do not contradict such a mechanism. Those of Itoh (1956) who induced protoperithecia to develop into perithecia by filtrates from cultures of the opposite mating type or from mated cultures support the complementary stimulant assumption; as could be expected, and contrarily to previous claims of Moreau's group (Moreau & Moruzi, 1931), the induced perithecia were sterile. Similar results have been obtained with lipid extracts from single and mated cultures of *N. crassa* and, surprisingly, a few ascospores were isolated from the resulting perithecia which segregated in an aberrant manner with respect to the mating-type alleles (Islam & Weijer, 1972). Unfortunately, the derived suggestion that physical contact between compatible mating partners is not essential (the sex hormones having been preliminarily characterized as volatile, unsaturated hydrocarbons, Islam, 1973) has never been accompanied by a conclusive demonstration that the induction has led to the isolation of ascospores carrying genetic markers from the single parent. More convincing are recent data demonstrating artificial induction, by a non-diffusible sex factor extracted from mated mycelia, of perithecial development culminating in the production of mature ascospores which carry genetic markers from the single parent only (Vigfusson & Cano, 1974). From preliminary enzymatic digestion tests, this sex inducer, termed erogen B according to Machlis' terminology (1972) appears to be a protein, a fact in line with a similar characterization of the yeast mating factors as being proteinaceous (see Van den Ende & Stegwee, 1971).

The sex protein of *Neurospora* is possibly related to the protein detected as an actinomycin- and cycloheximide-sensitive band in gel separated extracts from *N. tetrasperma* induced to sexual morphogenesis following the step-down of sterile cultures from 37 to 25°C (Viswanath-Reddy & Turian, 1972). An apparently homologous protein has also been described as phase specific in the heterothallic *N. crassa* where it was found to be absent in vegetative material but appeared and rapidly increased only after fertilization (Nasrallah & Srb, 1973). This co-called perithecial protein has electrophoretic variants, presumably with homologous roles, not only in *N. tetrasperma* but also in the true homothallic *N. terricola* (Srb *et al.*, 1973). It will be of high interest to know whether at fertilization the synthesis of the protein is induced or whether it is released from some previously bound form already present in the perithecial initials. The possible homology between the sex-perithecial protein and the proteolytic enzyme postulated by Boucherie & Bernet (1974) to be an early protein in protoperithecial differentiation of *Podospora* will also be worth further investigation.

16.7 References

ARX, J. A. VON & MÜLLER, E. (1975). A re-evaluation of the bitunicate Ascomycetes with keys to families and genera. *Studies in Mycology No. 9. Centraalbureau voor Schimmelcultures, Baarn.*

BACKUS, M. P. (1939). The mechanics of conidial fertilization in *Neurospora sitophila. Bulletin Torrey Botanical Club* **66**, 63–79.

BAHN, M. & HOCK, B. (1973). Morphogenese von *Sordaria macrospora*: Die Induktion der Perithezienbildung. *Berichte der Deutschen Botanischen Gesellschaft* **86**, 309–11.

BAHN, M. & HOCK, B. (1974). Morphogenese von *Sordaria macrospora*: Induktion der Perithezienbildung durch Arginin. *Berichte der Deutschen Botanischen Gesellschaft* **87**, 433–42.

BARNETT, H. L. & LILLY, V. G. (1947). The effects of biotin upon the formation and development of perithecia, asci and ascospores by *Sordaria fimicola* Ces. and de Not. *American Journal of Botany* **34**, 196–204.

BASU, S. N. & BHATTACHARYYA, J. P. (1962). Studies on the growth and sporulation of some species of *Penicillium. Journal of General Microbiology* **27**, 61–73.

BISTIS, G. N. (1957). Sexuality in *Ascobolus stercorarius*. II. Preliminary experiments on various aspects of the sexual process. *American Journal of Botany* **44**, 436–43.

BISTIS, G. N. & RAPER, J. R. (1963). Heterothallism and sexuality in *Ascobolus stercorarius. American Journal of Botany* **50**, 880–91.

BOUCHERIE, H. & BERNET, J. (1974). Protoplasmic incompatibility and female organ formation in *Podospora anserina*: properties of mutations abolishing both processes. *Molecular and General Genetics* **135**, 163–74.

BU'LOCK, J. D. (1961). Intermediary metabolism and antibiotic synthesis. *Advances in Applied Microbiology* **3**, 239–42.

BURNETT, J. H. (1975). *Mycogenetics*. London, New York: John Wiley & Sons.

BUSTON, H. W. & KHAN, A. H. (1956). The influence of certain micro-organisms on the formation of perithecia by *Chaetomium globosum. Journal of General Microbiology* **14**, 655–60.

BUSTON, H. W., MOSS, M. O. & TYRRELL, D. (1966). The influence of carbon dioxide on growth and sporulation of *Chaetomium globosum. Transactions of the British Mycological Society* **49**, 387–96.

BUTLER, G. M. (1966). Vegetative structures. In *The fungi*, Vol. II, pp. 83–112, edited by G. S. Ainsworth and A. S. Sussman, New York, London: Academic Press.

CALHOUN, F. & HOWE, H. B. JR. (1972). Nutrient-dependent inhibition of perithecial development due to sequential crosses on the same mycelium of *Neurospora tetrasperma. Planta, Berlin* **108**, 289–302.

CARLILE, M. J. (1965). The photobiology of fungi. *Annual Review of Plant Physiology* **16**, 175–202.

CARLILE, M. J. & FRIEND, J. (1956). Carotenoids and reproduction in *Pyronema confluens. Nature, London* **178**, 369–70.

CHADEFAUD, M. (1960). *Traité de botanique systématique*. Paris: Masson & Cie.

COLSON, B. (1934). The cytology and morphology of *Neurospora tetrasperma. Annals of Botany, London*, **48**, 211–24.

DASGUPTA, A. & NANDI, P. N. (1957). Role of nitrogen concentration on production of perithecia in *Penicillium vermiculatum* Dang. *Nature, London* **179**, 429–30.

DENNY, F. E. (1933). Oxygen requirements of *Neurospora sitophila* for formation of perithecia and growth of mycelium. *Contributions Boyce Thompson Institute* **5**, 95–102.

DODGE, B. O. (1920). The life history of *Ascobolus magnificus. Mycology* **12**, 115–34.

EL-ANI, A. S. & OLIVE, L. S. (1962). The induction of balanced heterothallism in *Sordaria fimicola. Proceedings of the National Academy of Sciences, U.S.A.* **48**, 17–9.

EL-ANI, A. S. & OLIVE, L. S. (1975). Genetics of *Sordaria fimicola*. IX. Linkage group II. *American Journal of Botany* **62**, 166–71.

ELLIOT, G. C. (1969). Effects of inhibitors of sterol synthesis on growth of *Sordaria* and *Phytophthora. Journal of General Microbiology* **56**, 331–43.

ENDE, H. VAN DEN & STEGWEE, D. (1971). Physiology of sex in Mucorales. *The Botanical Review* **37**, 22–36.

ESSER, K. (1966). Incompatibility. In *The fungi*, Vol. II, pp. 661–76, edited by G. C. Ainsworth and A. S. Sussman, New York, London: Academic Press.

ESSER, K. (1968). Phenol oxidases and morphogenesis in *Podospora anserina*. *Genetics* **60**, 281–8.

ESSER, K. & BLAICH, R. (1973). Heterogenic incompatibility in plants and animals. *Advances in Genetics* **17**, 107–52.

ESSER, K. & KUENEN, R. (1965). *Genetik der Pilze*. Berlin: Springer-Verlag.

ESSER, K. & STRAUB, J. (1956). Fertilität im Heterokaryon aus zwei sterilen Mutanten von *Sordaria macrospora* Auersw. *Zeitschrift für Induktive Abstammungs-Vererbungslehre* **87**, 625–6.

ESSER, K. & STRAUB, J. (1958). Genetische Untersuchungen an *Sordaria macrospora* Auersw. Kompensation und Induktion bei Genbedingten Entwicklungsdefekten. *Zeitschrift für Vererbungslehre* **89**, 729–46.

FAYRET, J. (1975). *Thèse Etat, Faculté des Sciences, Université Sabatier, Toulouse.*

FIELDS, W. G. & MANIOTIS, J. (1963). Some cultural and genetic aspects of a new heterothallic *Sordaria*. *American Journal of Botany* **50**, 80–5.

FINCHAM, J. R. S. & DAY, P. R. (1971). *Fungal genetics*. Oxford and Edinburgh: Blackwell Scientific Publications.

FITZGERALD, P. H. (1963). Genetic and epigenetic factors controlling female sterility in *Neurospora crassa*. *Heredity* **18**, 47–62.

GÄUMANN, E. (1964). Die Pilze. Basel: Birkhäuser Verlag.

GINDRAT, D. & TURIAN, G. (1967). Ascogone et anomalie d'enroulement hyphal chez *Gaeumannomyces graminis* (Sacc.) von Arx et Olivier et *Neurospora crassa* Shear et Dodge. *Journal of General and Applied Microbiology* **13**, 381–9.

GIRBARDT, M. (1952). Zur Frage der Fertilität bei *Neurospora crassa* und *Neurospora tetrasperma* in Abhängigkeit von Aussenfaktoren. *Flora* **139**, 477–525.

GOODAY, G. W. (1974). Fungal sex hormones. *Annual Review of Biochemistry* **43**, 35–49.

GREEN, E. (1930). Observations on certain Ascobolaceae. *Transactions of the British Mycological Society* **35**, 321–2.

GREIS, H. (1936). Entwicklungsgeschichte von *Sordaria fimicola* (Rob.). *Botanisches Archiv* **38**, 113–51.

HALL, R. (1971). Effect of carbon-nitrogen ratios on production of perithecia by *Sordaria fimicola*. *Canadian Journal of Microbiology* **17**, 132–4.

HANSEN, H. N. & SNYDER, W. C. (1946). Inheritance of sex in fungi. *Proceedings of the National Academy of Sciences, U.S.A.* **32**, 272–3.

HARRIS, J. L., HOWE, H. B. JR. & ROTH, I. L. (1975). Scanning electron microscopy of surface and internal features of developing perithecia of *Neurospora crassa*. *Journal of Bacteriology* **122**, 1239–46.

HAWKER, L. E. (1942). Effect of vitamin B_1 on concentration of glucose optimal for certain fungi. *Annals of Botany (N.S.), London* **6**, 631–6.

HAWKER, L. E. (1957). *The physiology of reproduction in fungi*. London and New York: Cambridge University Press.

HAWKER, L. E. (1966). Environmental influences on reproduction. In *The fungi*. Vol. II, pp. 435–469, edited by G. C. Ainsworth and A. S. Sussman, New York, London: Academic Press.

HIRSCH, H. M. (1954). Environmental factors influencing the differentiation of protoperithecia and their relation to tyrosinase and melanin formation in *Neurospora crassa*. *Physiologia Plantarum* **7**, 72–97.

HIX, S. M. & BAKER, R. (1964). Physiology of sexual reproduction in *Hypomyces solani* f. *cucurbitae*. I. Influence of carbon and nitrogen. *Phytopathology* **54**, 584–6.

HOHL, H. R. & STREIT, W. (1975). Ultrastructure of ascus, ascospore and ascocarp in *Neurospora lineolata*. *Mycologia* **67**, 367–81.

HOROWITZ, N. H., FLING, M., MCLEOD, H. L. & SUEOKA, N. (1960). Genetic determination and enzymatic induction of tyrosinase in *Neurospora*. *Journal of Molecular Biology* **2**, 96–104.

HOSFORD, R. M. JR., SOLANGI, G. R. M. & KIESLING, R. L. (1975). Inheritance in *Cochliobolus sativus*. *Phytopathology* **65**, 699–703.

HOWE, H. B. JR. & BENSON, E. W. (1974). A perithecial color mutant of *Neurospora crassa*. *Molecular and General Genetics* **131**, 79–83.

INGOLD, C. T. (1962). The reaction of fungi to light and the problem of photoperception. *Symposium of the Society for Experimental Biology* **16**, 154–69.

INOUE, Y. & FURUYA, M. (1970). Perithecial formation in *Gelasinospora reticulispora*. I. Effects of light at two different growth states. *Development, Growth and Differentiation* **12**, 141–50.

INQUE, Y. & FURUYA, M. (1974). Perithecial formation in *Gelasinospora reticulispora*. II. Promotive effects of near-ultraviolet and blue light after dark incubation. *Plant and Cell Physiology* **15**, 195–204.

INOUE, Y. & FURUYA, M. (1975). Perithecial formation in *Gelasinospora reticulispora*. IV. Action spectra for the photoinduction. *Plant Physiology* **55**, 1098–101.

ISLAM, M. S. (1973). Sex-hormones in *Neurospora crassa*. Further studies on its biological properties. *Mycopathologia et Mycologia Applicata* **51**, 87–97.

ISLAM, M. & WEIJER, J. (1972). Development of fertile fruit bodies (perithecia) in the single strain culture (Em A) of *Neurospora crassa*. *Folia Microbiologica* **17**, 316–9.

ITOH, T. (1956). Fruit-body formation in red bread mould, *Neurospora crassa*. Effect of culture filtrate on perithecial formation. *The Botanical Magazine, Tokyo* **69**, 369–72.

ITOH, T. & MORISHITA, K. (1971). Effect of cytoplasm on protoperithecium formation in *Neurospora crassa*. *Japanese Journal of Genetics* **46**, 7–15.

JINKS, J. L. (1966). Extranuclear inheritance. In *The fungi*. Vol. II, pp. 619–660, edited by G. C. Ainsworth and A. S. Sussman, New York, London: Academic Press.

KUWANA, H. (1958). Melanization in the mycelium due to the interaction of two strains of *Neurospora crassa*. *Botanical Magazine, Tokyo* **71**, 270–4.

LEACH, C. M. (1971). Regulation of perithecium development and maturation in *Pleospora herbarum* by light and temperature. *Transactions of the British Mycological Society* **57**, 295–315.

LEACH, C. M. (1972). An action spectrum for light-induced sexual reproduction in the Ascomycete fungus *Leptosphaerulina trifolii*. *Mycologia* **64**, 475–90.

LEACH, C. M. & TRIONE, E. J. (1966). Action spectra for light-induced sporulation of the fungi *Pleospora herbarum* and *Alternaria dauci*. *Photochemistry and Photobiology* **5**, 621–30.

LETROUIT-GALINOU, M. A. (1973). Sexual reproduction. In *The lichens*. pp. 59–90, edited by V. Ahmadjian and M. E. Hale, New York, London: Academic Press.

LILLY, V. G. & BARNETT, H. L. (1949). The influence of concentrations of nutrients, thiamin and biotin upon growth and formation of perithecia and ascospores by *Chaetomium convolutum*. *Mycologia* **41**, 186–96.

LILLY, V. G. & BARNETT, H. L. (1951). *Physiology of the fungi*. New York: McGraw-Hill.

LINDENMAYER, A. & SCHOEN, H. F. (1967). Selective effects of purine and

pyrimidine analogues and of respiratory inhibitors on perithecial development and branching in *Sordaria*. *Plant Physiology* **42**, 1059–70.

LOCKWOOD, L. B. (1937). Hydrogen ion concentration and ascus formation. *Mycologia* **29**, 289–90.

MACHLIS, L. (1972). The coming of age of sex hormones in plants. *Mycologia* **64**, 235–47.

MAINWARING, H. R. & WILSON, I. M. (1968). The life cycle and cytology of an apomictic *Podospora*. *Transactions of the British Mycological Society* **51**, 663–77.

MARSH, P. B., TAYLOR, E. E. & BASSLER, L. M. (1959). A guide to the literature on certain effects of light on fungi: reproduction, morphology, pigmentation, and phototropic phenomena. *Plant Disease Reporter. Supplement* **261**, 215–312.

McGAHEN, J. W. & WHEELER, H. E. (1951). Genetics of *Glomerella*. IX. Perithecial development and plasmogamy. *American Journal of Botany* **38**, 610–7.

METZENBERG, R. L. & AHLGREN, S. K. (1973). Behavior of *Neurospora tetrasperma* mating type genes introgressed into *N. crassa*. *Canadian Journal of Genetics and Cytology* **15**, 571–6.

MOORE-LANDECKER, E. (1963). The ontogeny of the apothecia of *Pyronema domesticum*. *American Journal of Botany* **50**, 37–44.

MOORE-LANDECKER, E. (1975). A new pattern of reproduction in *Pyronema domesticum*. *Mycologia* **67**, 1119–27.

MOREAU, F. & MORUZI, C. (1931). Recherches expérimentales sur la formation des périthèces chez les *Neurospora*. *Comptes Rendus de l'Académie des Sciences, Paris* **192**, 1476–8.

MYLYK, O. M. & THRELKELD, S. F. H. (1974). A genetic study of female sterility in *Neurospora crassa*. *Genetical Research, Cambridge* **24**, 91–102.

NASRALLAH, J. B. & SRB, A. (1973). Genetically related protein variant specifically associated with fruiting body maturation in *Neurospora*. *Proceedings of the National Academy of Sciences, U.S.A.* **70**, 1891–3.

NELSON, R. R. (1971). Hormonal involvement in sexual reproduction in the fungi, with special reference to F-2, a fungal oestrogen. In *Morphological and biochemical events in plant-parasite interaction*, pp. 181–205, edited by S. Akay and S. Ouchi, Tokyo.

NELSON, R. R. & KLINE, D. M. (1964). Evolution of sexuality and pathogenicity. IV. Effects of geographic origin and host association on the pathogenicity of isolates of *Helminthosporium* with similar conidial morphology. *Phytopathology* **54**, 1207–9.

NELSON, R. R., MIROCHA, C. J., HUISINGH, D. & TIJERINA-MENCHACA, A. (1968). Effects of F-2, an oestrogenic metabolite from *Fusarium*, on sexual reproduction of certain Ascomycetes. *Phytopathology* **58**, 1061–2.

NEWMEYER, D., HOWE, H. B. JR. & GALEAZZI, D. R. (1973). A search for complexity at the mating-type locus of *Neurospora crassa*. *Canadian Journal of Genetics and Cytology* **15**, 577–85.

O'DONNELL, K. L. & HOOPER, G. R. (1974). Scanning ultrastructural ontogeny of paragymnohymenial apothecia in the operculate discomycete *Peziza quelepidotia*. *Canadian Journal of Botany* **52**, 873–6.

OJHA, M. N. & TURIAN, G. (1968). Thermostimulation of conidiation and succinic oxidative metabolism of *Neurospora crassa*. *Archiv für Mikrobiologie* **63**, 232–41.

PINCHEIRA, G. & SRB, A. M. (1969). Cytology and genetics of two abnormal ascus mutants of *Neurospora*. *Canadian Journal of Genetics and Cytology* **11**, 281–6.

PURDY, L. H. (1956). Factors affecting apothecial formation by *Sclerotinia sclerotiorum*. *Phytopathology* **46**, 409–10.

RAPER, K. B. & FENNELL, D. I. (1965). *The genus Aspergillus*. Baltimore: Williams & Wilkins.

ROBBINS, W. J. & MA, R. (1942). Vitamin deficiencies of *Ceratostomella* and related fungi. *American Journal of Botany* **29**, 835–43.

ROBINSON, W. (1926). The conditions of growth and development of *Pyronema confluens* Tul. (*P. omphalodes* (Bull.) Fuckel). *Annals of Botany, London* **40**, 245–72.

ROSSIER, C. & TURIAN, G. (1976). Changes of some enzymatic activities in iodoacetate-treated microconidiating cultures of *Neurospora crassa*. *Archives of Microbiology* **108**, 317–19.

ROSSIER, C., OULEVEY, N. & TURIAN, G. (1973). Electron microscopy of selectively stimulated microconidiogenesis in wild type *Neurospora crassa*. *Archiv für Mikrobiologie* **91**, 345–53.

SMITH, J. E. & BERRY, D. R. (1974). *An introduction to biochemistry of fungal development*. London, New York, San Francisco: Academic Press.

SRB, A. M., NASRALLAH, J. B., BASL, M. (1973). Genetic control of the sexual reproductive apparatus of *Neurospora*.

Brookhaven Symposium in Biology **25**, 40–50.

SURAPIPITH, V. & LINDENMAYER, A. (1969). Thioguanine-dependent light sensitivity of perithecial initiation in *Sordaria fimicola*. *Journal of General Microbiology* **57**, 227–37.

TAN, SAI TEE & HO COY CHOKE (1970). A gene controlling the early development of the protoperithecium in *Neurospora crassa*. *Molecular and General Genetics* **107**, 158–61.

TAYEL, A. A. & HASTIE, A. C. (1975). Heterothallism and perithecium formation in *Nectria cosmariospora*. *Transactions of the British Mycological Society* **64**, 295–300.

TIJERINA-MENCHACA, A. & NELSON, R. R. (1969). The involvement of zinc and methylated compounds in sexual reproduction in *Cochliobolus carbonum*. *Phytopathology* **59**, 1053.

TURIAN, G. (1955). Recherches sur l'action de l'acide borique sur la fructification des *Sordaria*. *Phytopathologische Zeitschrift* **25**, 181–9.

TURIAN, G. (1966a). Morphogenesis in Ascomycetes. In *The fungi*, Vol. II, pp. 339–385, edited by G. C. Ainsworth and A. S. Sussman, New York, London: Academic Press.

TURIAN, G. (1966b). Quelques facteurs externes contrôlant la morphogenèse périthéciale et la prophyrie du *Neurospora tetrasperma*. *Revue Roumaine de Biologie, Série Botanique, Bucarest* **11**, 235–41.

TURIAN, G. (1969). *Différenciation fongique*. Paris: Masson & Cie.

TURIAN, G. (1975). Differentiation in *Allomyces* and *Neurospora*. *Transaction of the British Mycological Society* **64**, 367–80.

TURIAN, G. (1976). Spores in Ascomycetes. Their controlled differentiation. In *The fungus spore, form and function*, pp. 715–788, edited by D. J. Weber and W. M. Hess, New York: Wiley-Interscience.

TURIAN, G. & BIANCHI, D. E. (1972). Conidiation in *Neurospora*. *The Botanical Review* **38**, 119–54.

TURIAN, G., SEYDOUX, J. & VOLKMANN, D. (1962). Activité isocitratasique et type de sporulation chez *Neurospora tetrasperma* et chez *N. sitophila*, souche normale et mutant microconidien. *Pathologia and Microbiologia, Basel* **25**, 737–51.

VIALA, G. & VIDAL, G. (1972). Reproduction sexuée, croissance et métabolisme intermédiaire chez le *Leptosphaeria typhae*. *Physiologie Végétale* **10**, 481–94.

VIGFUSSON, N. V. & CANO, R. J. (1974). Artificial induction of the sexual cycle of *Neurospora crassa*. *Nature, London* **249**, 383–5.

VIGFUSSON, N. V. & WEIJER, J. (1972). Sexuality in *Neurospora crassa*. II. Genes affecting the sexual development cycle. *Genetical Research, Cambridge* **19**, 205–11.

VISWANATH-REDDY, M. & TURIAN, G. (1972). Temperature-induced synchronous differentiation of ascogonia in *Neurospora*. *Experientia* **28**, 99–100.

VISWANATH-REDDY, M. & TURIAN, G. (1975). Physiological changes during protoperithecial differentiation in *Neurospora tetrasperma*. *Physiologia Plantarum* **35**, 166–74.

WALKEY, D. G. A. & HARVEY, R. (1967). Spore discharge rhythms in Pyrenomycetes. III. Ascospore production and the quantitative and qualitative influence of light on spore discharge in *Sordaria macrospora*. *Transactions of the British Mycological Society* **50**, 241–9.

WEIJER, J. & VIGFUSSON, N. V. (1972). Sexuality in *Neurospora crassa*. I. Mutations to male sterility. *Genetical Research, Cambridge* **19**, 191–204.

WESTERGAARD, M. & HIRSCH, H. M. (1954). Environmental and genetic control of differentiation in *Neurospora*. *Proceedings Symposium Colston Research Society* **7**, 171–83.

WESTERGAARD, M. & MITCHELL, H. K. (1947). *Neurospora*. V. A. synthetic medium favouring sexual reproduction. *American Journal of Botany* **34**, 573–7.

WHEELER, H. E. (1954). Genetics and evolution of heterothallism in *Glomerella*. *Phytopathology* **44**, 342–5.

WHEELER, H. E. (1956). Linkage groups in *Glomerella*. *American Journal of Botany* **43**, 1–6.

WHEELER, H. E. & McGAHEN, J. W. (1952). Genetics of *Glomerella*. X. Genes affecting sexual reproduction. *American Journal of Botany* **39**, 110–9.

WHITEHOUSE, H. L. K. (1949). Heterothallism and sex in the fungi. *Biological Reviews* **24**, 411–47.

WHITESIDE, W. C. (1957). Perithecial initials of *Chaetomium*. *Mycologia* **49**, 420–5.

WHITESIDE, W. C. (1961). Morphological studies in the Chaetomiaceae. *Mycologia* **53**, 512–23.

WILSON, D. M. & BAKER, R. (1969). Physiology of production of perithecia and microconidia in *Hypomyces solani* f. sp. *cucurbitae*. *Transactions of the British Mycological Society* **53**, 229–36.

ZICKLER, H. (1952). Zur Entwicklungsgeschichte des Ascomyceten *Bombardia lunata* Zckl. *Archiv für Protistenkunde* **98**, 1–70.

ZONNEVELD, B. J. M. (1975). Sexual differentiation in *Aspergillus nidulans*. The requirement for manganese and its effect on α-1,3 glucan synthesis and degradation. *Archives of Microbiology* **105**, 101–4.

CHAPTER 17

Light-Induced Fungal Development

K. K. TAN

17.1 Introduction

Light affects many aspects of growth, development, reproduction and behaviour of fungi. Broadly speaking the effects of light on fungi can be classified (Page, 1965) as: (i) *non-morphogenetic effects*—in which light influences the rate or the direction of movement or growth of a structure or the synthesis of a compound; and (ii) *morphogenetic effects*—in which light induces or inhibits the formation of a structure. Non-morphogenetic responses may be oriented, the response bearing a spatial relationship to the source of illumination, *e.g.* phototaxis, phototropism, spore discharge; or non-oriented, either a stimulation or an inhibition of the rate of growth or the synthesis of a compound. Morphogenetic responses include germination, reproduction (asexual and sexual), and sclerotium formation.

It is important to realise that closely related species or different isolates of the same species may differ in their response to light. Jacob (1954), for instance, showed that *Pilobolus crystallinus* and *P. umbonatus* required light for sporangium formation whereas *P. longipes* and *P. sphaeroporus* had no such requirements. The physiological state of the fungus is an important consideration when considering a response to light. Both the quantity (dose) and the quality (wavelength) of light required must also be stressed. Thus in *Sphaerobolus stellatus*, light of wavelengths below 500 nm *i.e.* the blue region of the spectrum, is effective in promoting fruiting during the first 8 of the 14 days of the developmental period, and on the day of glebal discharge, light was not an influencing factor (Alasoadura, 1963). For the 4 or 5 days preceding glebal discharge, light of wavelengths above 550 nm, *i.e.* yellow-red, was found to be effective in the promoting of fruiting (Ingold & Nawaz, 1967).

It is also not unusual to find that light is both stimulatory and inhibitory. In *Botrytis cinerea* (Tan & Epton, 1973, 1974*a*), *Helminthosporium oryzae* (Honda, Sakamoto & Oda, 1968) and *Alternaria tomato* (Kumagai & Oda, 1969*b*), sporulation is promoted by near ultraviolet light (NUV), but inhibited by blue light. Light thus causes a complexity of responses in fungi.

The literature concerned with the phenomena of light responses is immense. An early comprehensive guide to the literature on light effects has been compiled by Marsh, Taylor & Bassler (1959). Reviews of light responses include Ingold (1962), Carlile (1965, 1970), Page (1965) and Leach (1971). References can also be found in several mycological books e.g. Hawker (1957), Cochrane (1958), Burnett (1968), Turian (1969) and Moore-Landecker (1972), and in review articles on fungal differentiation viz. Morton (1961), Cantino (1966), Hawker (1966), Taber (1966), Turian (1966), Trione & Leach (1969), Smith & Galbraith (1971) and Raper (1971). Phototropism has been extensively covered by Ingold (1962), Carlile (1965, 1970, 1975), Curry & Thimann (1961), Thimann (1967), Page (1968), and Bergman et al. (1969), and will not be discussed to any length in this chapter. Phototaxis has been dealt with in depth by Carlile (1975), and light-induced carotenoid synthesis by Batra (1971). Light-induced circadian rhythms and light-induced sclerotium formation are described elsewhere in this volume (Chapter 19).

This chapter is not intended to be a review of fungal light responses. The author feels that such a chapter would add little to the understanding of the phenomena of light effects and instead aspects which are fundamental to fungal photobiology will be discussed. Thus, action spectra and the identity of the photoreceptor(s) will be dealt with first. The primary and secondary consequences of light action will then be considered.

17.2 The Problem of Photoreceptor(s)

Only light which is absorbed by a molecule can be effective in producing a photochemical change in the molecule (Law of Photochemistry). It follows that if there is an ultraviolet (UV) photoresponse, then there must exist a receptor that can absorb light in the UV. If there is a blue response, then a receptor absorbing blue light must exist, and so on.

Fungal photoresponses are effected by wavelengths from the UV to the red end of the spectrum (Table 17.1). This may mean that there exist in fungi several photoreceptors, each absorbing in the specific regions or one receptor capable of absorbing the whole span of the spectrum. Discussions in subsequent sections should clarify this point.

Action spectra

The chemical identity of the photoreceptor involved in a photoresponse is usually deduced from an action spectrum of the response. An action spectrum is a plot of the relative effectiveness of different wavelengths in bringing about a response and should correspond to the absorption spectrum of the receptor pigment involved. For a theoretical consideration of action spectroscopy, Jagger (1967), Clayton (1970) or Shropshire (1972) should be consulted.

In action spectra studies, it is necessary to obtain dose–response curves for each of the wavelengths considered. The energy required to bring about a known level of response for each wavelength is then calculated and the reciprocal plotted against the wavelength—this is the action spectrum. For the level of response considered, it must be checked that the law of reciprocity holds, i.e. the response is only a function of the total dose (time × dose rate), otherwise, the action spectrum is not valid.

Table 17.1 Some fungal photoresponses and the effective wavelength
(Species of which detailed action spectra have been determined are shown by an asterisk.)

Ultraviolet	Near ultraviolet and blue	Yellow/red/far-red
Conidiation:	**Sporangium initiation:**	**Ascospore formation:**
*Alternaria chrysanthemia (Leach, 1964)	Phycomyces blakesleeanus (Bergman, 1972)	Saccharomyces carlsbergensis (Kelly & Gay, 1969)
A. dauci (Leach & Trione, 1966)		Leptosphaerina avenaria (Hogenson & Hosford, 1971)
A. kikuchiana (Ohmori & Nakajima, 1970a)	**Conidiation:**	
Helminthosporium oryzae (Honda, Sakamoto & Oda, 1968)	*Aspergillus ornatus (Stallings, 1970)	**Reversal of blue light inhibition of sporulation:**
H. teres (Onesirosan & Banttari, 1969)	*Penicillium isariiforme (Bennink, 1971)	*Alternaria solani (wild), A. solani (mutant) (Lukens, 1965)
Pyricularia oryzae (Ohmori & Nakajima, 1970b)	*Trichoderma viride (Gressel & Hartman, 1968; Kumagai & Oda, 1969a)	Botrytis cinerea (Tan, 1974c)
Stemphylium botryosum (Leach & Trione, 1966)		**Stimulation of spore discharge:**
*S. solani (Sproston, 1971)	**Circadian rhythm of conidiation:**	Sphaerobolus stellatus (Ingold, 1969)
Botrytis cinerea (Tan & Epton, 1973)	*Neurospora crassa (Sargent & Briggs, 1967)	Venturia inaequalis (Brook, 1969)
	Coremium formation:	**Inhibition of spore discharge:**
Reversal of blue light inhibition of sporulation:	*P. claviforme (Faraj Salman, 1971)	Ascobolus crenulatus (Ingold & Oso, 1969)
H. oryzae (Honda et al., 1968)	*P. isariiforme (Bennink, 1971)	
A. tomato (Kumagai & Oda, 1969b)		**Inhibition of spore germination:**
B. cinerea (Tan & Epton, 1974a)	**Perithecium formation:**	*Puccinia graminis (Calpouzos & Chang, 1971; Lucas et al., 1975)
	Ophiobolus graminis (Weste, 1970)	
Pycnidium formation:	*Nectria haematococca (Curtis, 1972)	P. recondita (Chang & Calpouzos, 1971)
*Ascochyta pisi (Leach & Trione, 1965)		
Septoria nodorum (Calpouzos & Lapis, 1970)	**Ascospore formation:**	
	Saccharomyces carlsbergensis S. cerevisiae (Kelly & Gay, 1969)	
Perithecium formation:		
*Pleospora herbarum (Leach & Trione, 1966)	**Fruit body initiation:**	
	*Favolus arcularius (Kitamoto, Suzuki & Furukawa,	

(Perkins & Gordon, 1959)

Sphaerobolus stellatus
(Alasaodura, 1963)

Positive phototropism:
Phycomyces blakesleeanus
(Curry & Gruen, 1959)
Pilobolus kleinii
(Page & Curry, 1966)
*Penicillium isariiforme
(Bennink, 1971)

Light growth response:
*Phycomyces blakesleeanus
(Delbrück & Shropshire, 1960)

Carotenoid synthesis:
*Neurospora crassa
(Zalokar, 1955)
*Fusarium aquaeductum (Rau, 1967a)

Inhibition of sporulation after the inductive phase:
Alternaria kikuchiana
(Ohmori & Nakajima, 1970a)
A. solani (Lukens, 1963)
A. tenuis (Aragaki, 1969)
A. tomato
(Aragaki, 1962;
Kumagai & Oda, 1968b)
H. oryzae (Honda et al., 1968)
*Stemphylium botryosum (Leach, 1968)
S. solani (Sproston, 1971)
Botrytis cinerea
(Tan & Epton, 1974a;
Tan, 1974a)

Inhibition of spore germination:
Puccinia graminis
(Calpouzos & Chang, 1971)
P. recondita
(Chang & Calpouzos, 1971)

(Leach, 1972)

Ascospore formation:
Leptosphaerulina spp.
(Leath, 1971)

Negative phototropism:
*Phycomyces blakesleeanus
(Curry & Gruen, 1959)

Melanogenesis:
Verticillium albo-atrum,
V. dahliae
(Brandt, 1964;
Gafoor & Heale, 1971)

Though action spectroscopy is useful, some limitations must be mentioned. Often, the action spectra obtained may not be distinctive enough to allow unequivocal identification of the receptor, or they may be distorted by screening pigments and hence not reflect the absorption spectrum of the receptor pigment. Notwithstanding these criticisms, action spectroscopy is about the only method available at present for the study of the identity of the photoreceptor, excluding direct spectrophotometry of the fungal material or extracts which requires more sophisticated instruments. In any case, direct spectrophotometry may be subjected to interference from pigments not concerned with the photoreception. A combination of the various methods should provide the necessary evidence for the identity of the receptor.

Near ultraviolet/blue photoresponses and the receptor involved

Near ultraviolet/blue photoresponses include sporulation (stimulation as well as inhibition), coremium, perithecium and basidocarp formation, phototropism, carotenoid synthesis and the inhibition of spore germination (Table 17.1). Light from the NUV, longer than 330 nm, to the blue region of the spectrum is effective. To avoid confusion with the reversible photoresponses effected by NUV and blue light described in a later section, these NUV/blue photoresponses will be regarded as 'blue' photoresponses.

All too often researchers tend to consider 'blue' photoresponses as being mediated by the same photoreceptor. However it seems to the present author that we may be dealing with two or more different categories of responses. The characteristic 'blue' photoresponses which most workers have referred to have action spectra which show no action beyond $c.520$ nm, a peak of maximum effectiveness around 450 nm, subsidiary peaks and shoulders at about 430 and 480 nm on either side of the major peak, and another peak of action in the NUV at $c.370$ nm. Action spectra of some of these 'blue' photoresponses are shown in Fig. 17.1.

Action spectra with a different profile from that described above have been obtained for *Penicillium isariiforme* (Bennink, 1971), *Alternaria tomato* (Kumagai & Oda, 1968) and *Stemphylium botryosum* (Leach, 1968). Studies of the effectiveness of different wavelengths in inhibiting germination of *Puccinia graminis* (Calpouzos & Chang, 1971) and *P. recondita* (Chang & Calpouzos, 1971) also showed peaks of action different from those obtained for the characteristic 'blue' photoresponses. Action spectra of these atypical 'blue' photoresponses are shown in Fig. 17.2. *Penicillium isariiforme* and *S. botryosum* show a maximum peak of action only at $c.480$ nm, whereas *A. tomato* and the two *Puccinia* species show a maximum at $c.410$–425 nm.

An interesting situation exists in the Basidiomycete *Favolus arcularius* where the action spectrum for photoinduction of pileus formation (Kitamoto, Horikoshi & Suzuki, 1974) shows the characteristic 'blue' profile, but that for primordium formation shows two additional peaks at $c.400$ and 510 nm (Kitamoto, Suzuki & Furukawa, 1972). We have here a species which possesses both the characteristic 'blue' photoreceptor and the atypical 'blue' photoreceptor.

The characteristic 'blue' action spectrum shows resemblance to the absorption spectra of carotenes and flavins, and these two groups of compounds have been suggested to be the receptor involved. Evidence for

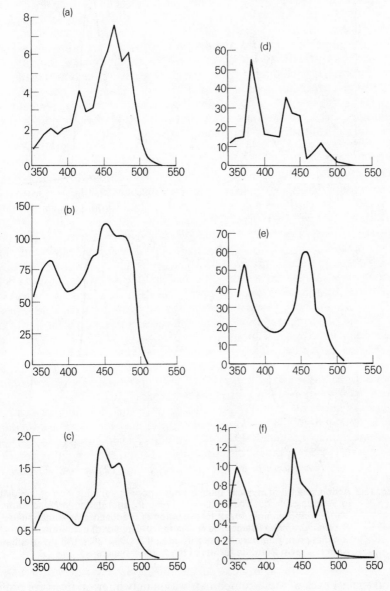

Fig. 17.1 Action spectra of some 'blue' photoresponses. (a) Inhibition of circadian rhythm of conidiation in *Neurospora crassa* (Sargent & Briggs, 1967); (b) carotenogenesis in *Fusarium aquaeductum* (Rau, 1967a); (c) positive phototropism in *Phycomyces blakesleeanus* (Curry & Gruen, 1959); (d) conidiation in *Trichoderma viride* (Kumagai & Oda, 1969a); (e) coremium formation in *Penicillium claviforme* (Faraj Salman, 1971); and (f) perithecium formation in *Nectria haematococca* (Curtis, 1972). (All the graphs were re-drawn from the results of the original papers).

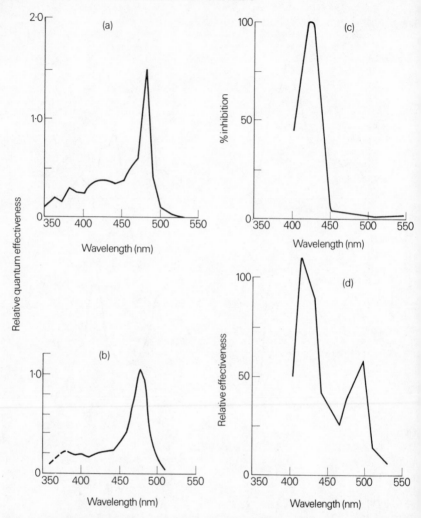

Fig. 17.2 Action spectra of atypical 'blue' photoresponses. (a) Inhibition of conidiation in *Stemphylium botryosum* (Leach, 1968); (b) sporulation in *Penicillium isariiforme* (Bennink, 1971); (c) inhibition of uredospore germination in *Puccinia graminis* (Calpouzos & Chang, 1971); and (d) inhibition of conidiation in *Alternaria tomato* (Kumagai & Oda, 1968). (All the graphs were re-drawn from the results of the original papers).

and against each of these compounds was mainly inferred from the profiles of the action spectra and from studies using inhibitors *e.g.* diphenylamine for proving carotenoid involvement or non-involvement, and atebrin, lyxoflavin and phenylacetic acid for flavoprotein. These inhibitors are non-specific, and conclusions drawn from such studies should be viewed with caution. Some of the arguments for and against carotenoid or flavoprotein as the receptor are summarized in Table 17.2.

It is only within the last two years that conclusive evidence has been published for flavin involvement in a species showing 'blue' photoresponses

Table 17.2 Carotenoid versus flavoprotein as photoreceptor

For	Against
Carotenoid	
1. Similar absorption spectrum 2. Diphenylamine affects *Trichoderma viride* (Kumagai & Oda, 1969*a*)	1. Albino mutant of *Pyronema confluens* still sensitive to light (Carlile & Friénd, 1956) 2. Carotenoid undetectable in *Collybia velutipes* (Aschan-Åberg, 1960) 3. Diphenylamine has no effect in *Pilobolus kleinii* (Page, 1956), *Aspergillus giganteus* (Trinci & Banbury, 1969) and *Favolus arcularius* (Kitamoto et al., 1972) 4. Inhibition by diphenylamine not reversed by carotene in *T. viride* (Gressel & Hartmann, 1968)
Flavoprotein	
1. Similar absorption spectrum 2. Inhibition of response overcome by added FMN in *Alternaria solani* (Lukens, 1963) 3. Lyxoflavin (an analog of riboflavin) has effect in *Pilobolus* (Page, 1962)	1. Phenylacetic acid and atebrin has no effect in *T. viride* (Kumagai & Oda, 1969*a*) 2. Inhibition by lyxoflavin not reversed by riboflavin in *T. viride* (Gressel & Hartmann, 1968) 3. Quinacrine, an antagonist for riboflavin, has no effect on *F. arcularius* (Kitamoto et al, 1972)

(Muñoz, Brody & Butler, 1974; Muñoz & Butler, 1975). Direct spectrophotometric measurements were made of mycelium of *Neurospora crassa* before and after irradiation with blue light. These showed a change in absorbance at 560 nm which clearly indicated the photoreduction of a b-type cytochrome. The photoreduction of a flavoprotein was also indicated. An action spectrum for the photoinduced absorbance change at 560 nm was determined and a profile similar to other 'blue' photoresponses was obtained. Flavin was suggested to be the photoreceptor involved as the photoreduction of cytochrome b and cytochrome c in soluble cell-free extracts could only be observed after the addition of flavin mononucleotide or flavin adenine dinucleotide. Furthermore, the blue light absorbing pigment was irreversibly bleached by long term irradiation. This resulted in a progressive loss of photoresponsiveness of cytochrome b—a result consistent with the photodestruction of flavins by prolonged irradiation. It seems quite conclusive that flavin is the receptor for this fungal system. Other fungal photoresponses with similar action spectra may also be mediated by the same photoreceptor. The long-standing controversy of whether carotenoid or flavoprotein is the photoreceptor has perhaps come to an end.

Ultraviolet photoresponses and the receptor involved

Here we are concerned with non-mutagenic and non-lethal photoresponses effected by UV radiation. These responses are mainly reproductive or processes associated with reproduction (see Table 17.1). The action spectra of some typical responses are shown in Fig. 17.3. Only wavelengths below 330 nm are effective and the peak of maximum effectiveness is around 280–290 nm. Subsidiary peaks on either side of this major peak are also

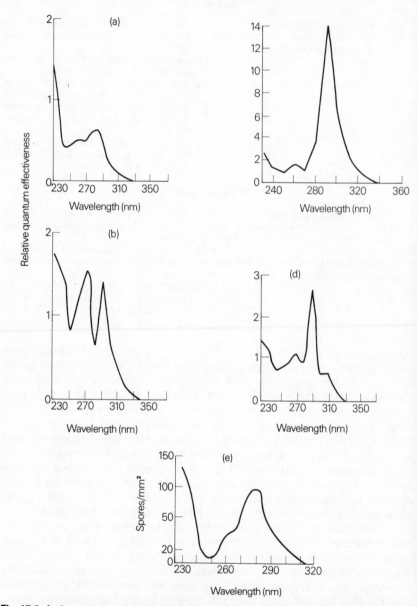

Fig. 17.3 Action spectra of some ultraviolet fungal photoresponses. (a) Conidiation in *Alternaria dauci* (from Leach & Trione, 1966), (b) *Stemphylium botryosum* (from Leach & Trione, 1966), and (e) *S. solani* (from Sproston, 1971); (c) pycnidium formation in *Ascochyta pisi* (from Leach & Trione, 1965); (d) perithecium formation in *Pleospora herbarum* (from Leach & Trione, 1966).

evident. The chemical identity of the receptor is still obscure. A compound P310 was suggested by Leach (1965), to be the photoreceptor, but to the present author's knowledge it has never been shown to be optically active. In subsequent studies, this compound was shown to be sporogenic and was even synthesized in the dark (Trione, Leach & Mutch, 1966). The studies dealing with P310 will be described later (p. 349).

Within the past decade, many physiological and biochemical studies have contributed to the understanding of some aspects of the UV receptor, especially in relation to its interaction with light of other wavelengths notably blue. This aspect will be considered later.

Red photoresponses and the receptor involved

While UV and 'blue' photoresponses have been firmly established for years, convincing reports of responses effected by the red end of the spectrum have not been available until fairly recently. Some of these responses are listed in Table 17.1. Action spectra of most of the responses have not been determined. An action spectrum for the reversal of the blue light inhibition of the terminal phase of sporulation in *Alternaria solani* (Lukens, 1965) shows a peak of activity at 600 nm for a wild species and peaks at 575 and 650–675 nm for a mutant. Semi-quinones which could be formed by flavin adenine dinucleotide or flavin mononucleotide when irradiated were suggested to be the receptor for the response effected by light of wavelengths 550–600 nm. The response in the 625–650 nm was suggested to be mediated by phytochrome, a well known plant photoreceptor. Various other workers also reported results indicative of phytochrome involvement. Brook (1969) showed that spore discharge in *Venturia inaequalis* could be promoted by far-red light and suppressed by subsequent red light irradiation. Calpouzos & Chang (1971) produced a crude action spectrum for the inhibition of spore germination in *Puccinia graminis* with peaks at 419 nm and 720 nm. This resembles that of the high irradiance responses of higher plants which some workers ascribe to phytochrome. More recently, Tan (1974c) showed red/far-red reversibility in the recovery from the blue-light inhibition of sporulation in *Botrytis cinerea* and Lucas, Kendrick & Given (1975) showed that simultaneous irradiation with ineffective red and inhibitory far-red light resulted in partial nullification of the inhibition of spore germination in *P. graminis* brought about by far-red alone. However until phytochrome is detected spectrophotometrically in fungi or isolated from fungi it cannot be said for certain that this is involved in fungal 'red' photoresponses. In the case of *B. cinerea*, the red/far-red response shows profound interaction with the UV and blue reversible receptor system.

Mycochrome—a reversible near ultraviolet–blue photoreceptor

PHYSIOLOGICAL EVIDENCE AND POSSIBLE MECHANISM OF PHOTORECEPTION. In three fungal species, *Helminosporium oryzae* (Honda *et al.*, 1968), *Alternaria tomato* (Kumagai & Oda, 1969b), and *Botrytis cinerea* (Tan & Epton, 1974a; Tan, 1974a) sporulation is controlled by a reversible NUV–blue photoreaction. Sporulation is promoted by NUV irradiation but is inhibited by blue light. Blue light given after NUV irradiation drastically reduced sporulation *i.e.* the inducing effect of near ultraviolet is partially negated. If this blue exposure is immediately followed by NUV irradiation, then sporulation is brought back to the original level. In a sequence of irradiation with alternating NUV and blue light, sporulation depended only on the quality of the last irradiation. Thus a sequence ending with NUV would result in the promotion of sporulation and one ending with blue, inhibition. Such results strongly suggest the existence of a photoreceptor which can be transformed repeatedly from a form that is active for sporogenesis to another that is inactive or even counter sporogenesis. Mycochrome was suggested to be such a reversible pigment system and was proposed to exist in two forms, the M_{NUV} for absorbing strongly in the NUV and the M_B form which absorbs strongly in the blue.

$$M_{NUV} \underset{B}{\overset{NUV}{\rightleftharpoons}} M_B$$

As a result of absorbing NUV light, the M_{NUV} form is converted to M_B form which causes the sporulation response. If blue light is applied immediately after the NUV irradiation, the M_B form is re-converted to the M_{NUV} form and sporulation does not ensue.

It is clear from further work that when blue light is applied, the sporulation process is arrested and developing conidiophores re-differentiate to form sterile mycelium (Honda, 1969; Kumagai & Oda, 1969b; Aragaki, Nishimoto & Hylin, 1973; Tan, 1974b). It appears that the M_{NUV} form is also active, but in this case for another developmental response, the formation of sterile mycelium.

Subsequent experiments by the author yielded results which showed profound interaction of this NUV–blue system with red and far-red light (Tan, 1975a, b). Far-red light also re-promoted the inhibition of sporulation caused by blue light. An intervening dark period between the blue light and NUV or far-red light reduced the 're-promotive power' of NUV or far-red light. Furthermore, as has been already briefly mentioned in the above section, the far-red effect could be reversed by red light irradiation. Blue light could also reverse the far-red effect. Such complex interactions led Tan (1975b) to propose a mechanism of photoreception in which two photoreceptor systems are interacting intimately.

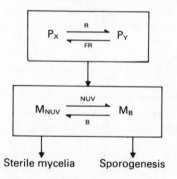

P_X and P_Y are the red and far-red absorbing forms of the red/far-red reversible photoreceptor, and M_{NUV} and M_B are the two forms of mycochrome.

An alternative photoreception system was also postulated (Tan, 1975b). A single photoreceptor was conceived and an intermediate(s) in the phototransformation of mycochrome was suggested to exist.

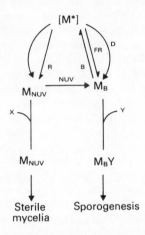

This postulation attempts to explain the results obtained and embodies the red/far-red response as well. For a detailed discussion, the reader is asked to consult the original papers.

ACTION SPECTRA AND THE IDENTITY OF MYCOCHROME   The identity of mycochrome is obscure. It is a pity that detailed action spectra studies have not been published for any of these very interesting reversible photoresponses. For the UV part of the reversible response, the author feels that the action would be no different from that of the UV responses already discussed (see p. 342). Statements that radiation of 310 nm is most effective for the induction of conidiophore formation in *Alternaria tomato* are contained in recent papers (Kumagai & Oda, 1973; Aragaki *et al.*, 1973). As for the blue light inhibition of response, 'crude' action spectra have been published for *A. tomato*. A broad peak of action between 340 and 540 nm was obtained by Aragaki (1962), and Kumagai & Oda (1968) found a major peak at about 410 nm and a minor peak at about 490 nm. It would be interesting to compare precise action spectra for the blue light inhibition in these three species showing photoreversibility with that for the blue light inhibition of sporulation in *A. solani* (Lukens, 1963), *Stemphylium solani* (Sproston, 1971), *S. botryosum* (Leach, 1968) and *Neurospora crassa* (Sargent & Briggs, 1967). Both *A. solani* and *S. solani* show a major peak of action at about 450 nm. The action spectra obtained for these two species however, are not precise enough to reveal any salient features. *N. crassa* shows characteristic flavin-type action and *S. botryosum* shows a peak of action at about 480 nm. An alternative to comparing action spectra would be to show reversibility of response inducible by NUV and blue light in these latter species, then we would know whether we are considering the same photoreceptor system for these two groups of responses or not.

SPECTROPHOTOMETRIC DETECTION OF MYCOCHROME   Mycochrome was recently detected spectrophotometrically in a particulate fraction of *Alternaria tomato* by Kumagai & Oda (1973). A light-minus-dark difference spectrum (irradiation at 310 nm) of a particulate fraction (the 140 000 g pellet of the supernatant from ground mycelium centrifuged at 5000 g) showed a dip at 310 nm with a shoulder at 340 nm and a peak at 400 nm. Prolonged irradiation caused the dip and peak to intensify and the dip shifted to about 300 nm. When the sample was irradiated with light of 410 nm after the NUV irradiation, the dip and the peak partially disappeared. These changes in absorbance could be reversibly repeated by alternating doses of blue and NUV light.

Further studies showed that the two forms of mycochrome are localised in different fractions of a two-layer sucrose density gradient (Kumagai, Yoshioka & Oda, 1976). M_{NUV}, the near ultraviolet absorbing form, is localised in the supernatant fraction. This fraction showed a decrease in absorbance in the NUV peak at c. 285 nm, when irradiated with 305 nm light, but reversibility of absorbance change was not observed. The optically active principle was precipitable by ammonium sulphate, but when the fraction was dialysed, the change in absorbance was lost. It was concluded that M_{NUV} is probably a low molecular weight material. M_B, the blue light absorbing form was found to be localized in the 70% (w/w) sucrose fraction as a particulate fraction. When this fraction was exposed to 305 nm light, the difference spectrum showed a dip in the NUV and a peak in the blue. This dip and peak intensified with the duration of irradiation. When the sample was irradiated with 400 nm light following the NUV exposure, both the dip and peak partially disappeared. These absorbance changes could be reversibly repeated by alternating exposures to NUV and blue. However, when the fraction was dialysed, though the absorbance changes as a result of NUV irradiation still occurred, reversibility by blue light was lost. The change in absorbance in the blue region which is much lower than for the sample not dialysed, was increased by the addition of the fraction containing M_{NUV}. Photoreversibility of absorbance changes was also restored. It appears therefore that M_{NUV} is either soluble in the cytoplasm or loosely bound to the particulate fraction and that M_B is tightly bound to the particulate fraction, probably to some membrane.

WIDESPREAD OCCURRENCE OF MYCOCHROME? Though only three species have been shown to exhibit the NUV–blue photoreversibility of sporulation, it is probable that more would be found if studies were extended to other species. There are indications that species whose response is induced by light and yet suppressed by continuous irradiation may show such reversibility. Four very probable species are *Alternaria solani* (Lukens, 1963) in which sporulation stimulated by white light was suppressed by blue light; *A. kikuchiana* (Ohmori & Nakajima, 1970a) and *Stemphylium botryosum* (Leach, 1967, 1968) in which sporulation was stimulated by NUV light and suppressed most effectively by blue light irradiation; and *S. solani* (Sproston, 1971) in which maximum inhibition of sporulation after induction by NUV occurred at 450 nm with pronounced inhibition from 350–500 nm. One needs only to go a step further to examine the effects of alternating NUV and blue irradiation, to see whether reversibility is present or not. The possibility of the widespread occurrence of such a NUV–blue reversible photoreceptor in fungi thus exists, operating in the control of various photo-processes like sporulation and reproduction. The production of P310, a compound absorbing maximally at 310 nm (see p. 349) appears to be under mycochrome control. Its synthesis is promoted by NUV light but suppressed by blue light (Tan & Epton, 1974b).

Blue–yellow reversible response

Another interesting case of antagonistic effects of different wavelengths is found in spore discharge of the Gasteromycete *Sphaerobolus stellatus*. This fungus requires light for fruiting. Development of the sporophore from initiation of fruit body to the time of discharge of glebal mass requires about 12 to 14 days. Alasoadura (1963) found that light of wavelengths below 500 nm was responsible for stimulation of fruiting during the first 8 days of the developmental period. On the day before glebal discharge, light had no effect, but for the 4 or 5 days preceding that, light above 550 nm *i.e.* yellow–red hastened the development (Ingold & Nawaz, 1967). Blue light, instead, had a retarding effect during this phase. The interaction of light of yellow/red and blue wavelengths on the later developmental stage was studied by Ingold (1969) and Ingold & Peach (1970). When cultures which have just started to discharge glebal mass were transferred from light to darkness, discharge occurred during the next 24 h at the same level as in cultures continuously illuminated, but thereafter it virtually ceased. However if such cultures after transfer to darkness were briefly interrupted by 1–2 h yellow or red light a day later, considerable discharge occurred subsequently, but if interrupted by blue light no discharge occurred (Ingold, 1969). When the yellow treatment was immediately followed by blue, no discharge occurred, but where this treatment was in reverse *i.e.* blue followed by yellow, subsequent discharge was abundant. The results thus show that the stimulatory effect of yellow light can be reversed by blue light. Repeated reversibility of response by alternating short exposure of blue and yellow light has also been demonstrated (Ingold & Peach, 1970). It is interesting to note that far-red was found to have the same effect as yellow light (Ingold & Nawaz, 1967). So this may be a case of antagonism between blue and far-red as well, a situation which is not very different from that found in *Botrytis cinerea* (Tan, 1975b). It is a pity that studies were not extended to NUV. Otherwise we would know whether the reversibility demonstrated for *S. stellatus* is similar to the mycochrome system or not.

A further example of antagonistic effects of light of different wavelengths has been shown (Ingold & Oso, 1969). In spore discharge of *Ascobolus crenulatus* blue light following a period of darkness causes immediate puffing, but yellow light has no effect. Simultaneous exposure to blue and yellow light is inhibitory to discharge. Repeated reversibility of response has not been shown, but studies extending into the UV end of the spectrum showed that 350 nm has the same action as blue (Oso, 1968).

17.3 The Consequence of Light Absorption

How light acts to bring about the observed response is virtually unknown. It is convenient to think in terms of the primary and secondary consequence of light action. Events that are detectable within minutes would be considered as primary and those that occur subsequently as secondary.

Primary consequence of light absorption

This would include the photochemical reactions that occur after light absorption. In the *Neurospora* system, where flavin was shown to be the receptor, photoreduction of cytochrome b is a primary step after light absorption. This may also be the mechanism for photoresponses of species showing characteristic flavin-type action spectra. As for mycochrome, how the NUV absorbing form is transformed to the blue absorbing form is still not known. It may involve a similar mechanism as for rhodopsin in animals or phytochrome in plants where isomerization or other intramolecular changes occur.

In *Phycomyces*, it has been shown that the level of adenosine $3' \cdot 5'$ cyclic monophosphate (cAMP) decreased within 1 min of irradiation (Cohen, 1974). Electrical signals have also been detected in this species within seconds of a light flash (Mogus & Wolken, 1974). These electrical signals were found to be related to the wavelength and intensity of the stimulus and the growth stage of the fungus. It seems likely that light may cause a change in membrane permeability and that this activates or inactivates the enzymes involved in cAMP metabolism thereby causing a change in concentration of cAMP. cAMP could of course act at various levels of the cellular metabolism *e.g.* substrate mobilization, gene expression and protein synthesis (see Bitensky & Gorman, 1973). It must be borne in mind that several different mechanisms may perhaps be involved depending on the type of response and we should not be too dogmatic about generalized mechanism of action.

Secondary consequence of light absorption

The author does not expect secondary events of light-induced processes to be any different from those inducible by other physical or chemical means. It is quite unlikely that different metabolic events would occur to effect the same observable response. Thus, if nucleic acids, proteins, carbohydrates and lipids are synthesized during reproduction, similar events would be expected to occur in light-induced reproduction.

P310   This is a compound or a group of compounds with absorption maxima at 310 nm and was first isolated by Leach (1965) from NUV-irradiated cultures of *Ascochyta pisi*, *Pleospora herbarum*, *Pyronema omphalodes*, *Alternaria chrysanthemi* and *Ophiobolus graminis*. P310 was absent or present in very low amounts in dark grown cultures, and it increased in relation to NUV dosage and with the time of induction. P310 was thought to be a photoreceptor initially but this was later considered improbable as it could be extracted from dark grown *A. pisi* on media which were able to support sporulation in the absence of light. A hormonal role was ascribed since it was found to be sporogenic for *A. pisi* and *P. herbarum* (Trione *et al.*, 1966) and for *Stemphylium solani* (Sproston, 1969). Trione *et al.* (1966) also extracted P310 from eight other fungi.

Studies by various workers within the last few years, however, cast doubts as to the significance of P310. Van den Ende & Cornelius (1970) extracted P310 from light-induced *Sclerotinia fructicola*, but found not correlation with sporulation. P310 was found in sporulating cultures of *Alternaria porri* and *A. tenuis* but not in *Choanephora cucurbitarum* and *Trichoderma viride*. Weste (1970) could not detect any P310 in her isolate of *Ophiobolus graminis* and *O. graminis* var. *avenae* which had been exposed to light. Neither could Stallings (1970) extract this compound from *Aspergillus ornatus*. Kuss (1970) working on *Stemphylium botryosum* showed that the absorbance in the region of 290–320 nm increased in cultures exposed to light. Production of P310 in continuously illuminated cultures correlated well with the sporulation curves for various nitrogen sources, and P310 continued to be synthesized in the dark after an initial exposure to light. However, P310 could not be extracted from cultures subjected to 24 h dark and then to continuous light, although sporulation was as profuse as in continuously illuminated cultures. The presence of extractable P310 was therefore found not to be required for sporulation.

Hite (1973) and Tan & Epton (1974*b*) were also able to extract P310 from NUV-induced cultures of *Botrytis cinerea*. P310 levels increased dramatically with time of irradiation. Possible relationship between P310 and sporulation was substantiated by the following results of Tan & Epton (1974*b*): (i) dark grown sporulating cultures contained P310; (ii) cultures irradiated with non-inductive blue light produced small amounts of the compound; and (iii) cultures irradiated with blue light immediately after photoinduction also showed reduction in the amount of P310. However, old cultures which were almost insensitive to NUV light yielded far too much P310 than could be accounted for by the amount of sporulation occurring. Sporogenicity of P310 could not be conclusively demonstrated as was also found by Hite (1973). The precise role of P310 is thus not clear. It appears to be associated with sporulation in some species, but not in others, and is produced as a result of irradiation in some species though it can also be synthesized in the dark.

Some physical and chemical characteristics of P310 have been obtained (Trione & Leach, 1969) and these are summarized in Table 17.3. The chemical structure has just been elucidated by N. Arpin of Universite Claude Bernard, Lyon, France (paper delivered at the physiology group

Table 17.3 Chemical and physical characteristics of P310

Physical properties
  M.W. 456(?)
  very soluble in water, dilute acids and alkali
  soluble in methanol, ethanol and pyridine
  insoluble in diethyl ether (35°C), petroleum ether (60°C), carbon tetrachloride (77°C), benzene (80°), ethylacetate (77°C) and P-dioxane (100°C).

Spectral characteristics
  λ_{max} 310 nm in water, dilute alkali, methanol, ethanol, pyridine and formamide
  λ_{max} 305 nm in acidic aqueous solutions, dimethylsulphoxide, dimethylformamide.
  Additional peaks at 277 nm and 282 nm in ethanol, methanol, methylpyrolidone and dimethylformamide

Stability
  stable at high temperature, even to autoclaving for 15 min at 121°C
  stable in acidic and alkaline solution

Chemical characteristics
  positive reactions for aldehyde, ketones, hydroxamic acids, amines, phenols and sugars
  presence of hydroxyl groups, carbonyl group, ethyl or ethoxy group; very few—CH, CH_2 or CH_3 groups.

---

meeting of the British Mycological Society at Warwick University, March 1976). P310 was found to be a methoxy-2, bis(hydroxymethyl)methylamino-3, hydroxy-5, hydroxymethyl-5, cyclohexene-2, one-1 with the following structure:

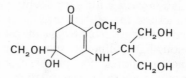

NUCLEOTIDES   Gettens & Shropshire (1963) reported at an American Society of Plant Physiologists meeting that the concentration of adenosine 5′ triphosphate (ATP) increased after sporangiophores of *Phycomyces* had been exposed to a saturating pulse of light. This may be of significance as the ATP would provide the energy for cell wall synthesis in the tropic response. It is regrettable that the work was not pursued further, and the link with cell wall synthesis was not established.

In light-induced circadian rhythm of conidiation in *Neurospora crassa*, levels of NADH, NADPH and NADP were lower in the conidiating area, while the NAD level was higher (Brody & Harris, 1973). However, the total pyridine nucleotide levels of the two areas were the same. This difference in pyridine nucleotide level would have far reaching effects on many areas of metabolism. Delmer & Brody (1975) showed that adenosine 5′ monophosphate (AMP) level, but not ATP or adenosine 5′ diphosphate, also oscillate with the circadian rhythm. The authors related this oscillation in AMP content to some circadian changes in mitochondria function. However the present author feels that this oscillation may be a reflection of cAMP metabolism. It would be of interest to look into this for *Neurospora*.

The level of cAMP in *Coprinus macrorhizus* increased during light-induced fruiting (Uno, Yamaguchi & Ishikawa, 1974). This increase is quite distant in time from the light treatment and is the result of enhanced activity of adenylate cyclase. Phosphodiesterase activity was also found to increase during fruiting but the peak was a day later. The changes in activity of these enzymes may be just a biochemical reflection of fruiting rather than an event specifically induced by light.

NUCLEIC ACID METABOLISM   The involvement of nucleic acids in photoinduced processes has been shown for several fungal species (Gressel & Galun, 1967; Valadon & Mummery, 1973; Tan, 1975c). Only in one reported case (Stallings, 1970) was nucleic acid not suggested to be specifically involved. 5-Fluorouracil inhibited conidiation in *Trichoderma viride* (Galum & Gressel, 1966) and *Botrytis cinerea* (Tan, 1975c), and carotenogenesis in *Verticillium agaricinum* (Valadon & Mummery, 1973). Incorporation of nucleic acid precursors, uridine and thymidine, into perchloric acid/ethanol insoluble material was inhibited in *B. cinerea*. An inhibitor of DNA synthesis, 5-fluorodeoxyuridine also inhibited conidiation and thymidine incorporation in *B. cinerea*, but carotenogenesis in *V. agaricinum* was only slightly affected though DNA and RNA levels were strongly affected. In *B. cinerea* application of these inhibitors after the photoinduction period was not as effective in inhibiting conidiation as application during the photoinduction period (Tan, 1975c). This suggests that most of the nucleic acids necessary for conidiation were synthesized during the photoinduction period. It would be interesting to see whether any 'new' species of RNA were synthesized as a result of light induction.

In the light-induced circadian rhythm of conidiation in *Neurospora crassa*, Martens & Sargent (1974) found that the RNA and DNA levels increased and decreased rhythmically. Incorporation of ^{3}H-uridine into nucleic acids was also rhythmic and was maximal just preceding the peaks of nucleic acid content and conidiation. Rhythms of RNA and DNA content were also observed in a non-conidiating 'fluffy' strain after irradiation, but the amplitude of oscillation was smaller.

PROTEINS AND ENZYMES   Much of the evidence for the requirements of protein synthesis in light-induced processes is from studies using inhibitors *e.g.* cycloheximide and puromycin. Conidiation in *Botrytis cinerea* (Tan, 1975c) and *Neurospora crassa* (Urey, 1971), and carotenogenesis in *Fusarium aquaeductum* (Rau, 1967b), *N. crassa* (Rau, Lindemann & Rau-Hund, 1968; Harding & Mitchell, 1968) and *Verticillium agaricinum* (Valadon & Mummery, 1973) are inhibited by cycloheximide and other inhibitors of protein synthesis. In *B. cinerea* (Tan, 1975c) and *V. agaricinum* (Valadon, Travis & Key, 1975) incorporation of radioactive leucine was also drastically reduced. Ribosomes extracted from light grown cultures of *V. agaricinum* were found to be more active in the incorporation of phenylalanine using poly(U) as the message, and the ribosome profile clearly shows an irradiance-dependent transformation of monoribosomes to polyribosomes (Valadon *et al.*, 1975).

It is also interesting to note that in *Fusarium aquaeductum* and *Neurospora crassa*, application of inhibitors progressively later resulted in more and more carotenoids being formed. The sequential synthesis of carotenogenic enzymes was suggested to occur for this photoresponse.

In *Aspergillus ornatus* (Stallings, 1970) and *Trichoderma viride* (Stavy, Stavy & Galun, 1970), enhanced protein synthesis was found not to be specifically associated with light-induction but rather with the actively growing peripheral mycelia which are most sensitive to light. Stavy *et al.* (1970) could not show an enhanced incorporation of phenylalanine, as a result of light, using either endogenous mRNA or poly(U). However, these results do not rule out the involvement of protein synthesis in these species as light. might preferentially cause the synthesis of some 'spore-specific' proteins.

The activity of some oxidative enzymes in several mould species was reported to be influenced by light (Chebotarev & Zemlyanukhin, 1973). Lactate dehydrogenase, malate dehydrogenase, succinate dehydrogenase, glutamate dehydrogenase, glucose oxidase and ascorbate oxidase in non-pigmented *Penicillium album* and *Fusarium oxysporum*, and in pigmented *P. glaucum* and *Trichoderma lignorum*, showed an increase in activity as a result of UV irradiation at 12 μJ/cm^2. This enhancement was more pronounced in the non-pigmented fungal species. Higher doses of UV (20–80 μJ/cm^2) decreased and even inactivated the various enzymes, different enzymes being sensitive to different UV dosage. Flavin enzymes were found to be more sensitive to the inactivating action. Visible light at 25 μW/cm^2 was found to decrease the activities of dehydrogenases, and wavelengths between 400 and 450 nm were found to be most effective in causing the decreased activity.

In *Polyporus hispidus* the synthesis of hispidin was stimulated by light (Nambudiri, Vance & Towers, 1973). The activity of phenylalanine ammonia lyase (PAL), an enzyme involved in the deamination of phenylalanine, and the conversion of cinnamate into *p*-coumarate, are enhanced by light. *p*-Coumaric acid hydroxylase, catalysing the conversion of *p*-coumarate into caffeate, was only detected in light-exposed cultures.

It has been mentioned previously that the enzymes involved in cAMP metabolism in *Coprinus macrorhizus* were stimulated during light-induced fruiting. Activities of DNA dependent RNA polymerases in *Verticillum agaricinum* were also increased following exposure to light (Valadon, 1974). In this species, polymerases I (nucleolar), II (non-nucleolar) and III (mitochondrial) of 4-day old cultures, and polymerases I and III of 7-day old cultures showed an increase in activity.

17.4 Substitutes for Light Requirement

Various compounds have been shown to substitute for the light requirement. Studies using such compounds may contribute to an understanding of some aspects of the mechanism of light action.

Conidiation in *Stemphylium solani*, a process normally requiring UV irradiation, could be induced by ergosterol and to a lesser extent by the 'carrier' solutions 5% dimethylsulphoxide (DMSO) and 2% ethanol used (Sproston & Setlow, 1968). It was suggested that free sterol was required for conidiation and this was released when DMSO or ethanol was added, or when the fungus was irradiated. In *Trichoderma viride*, Gressel, Strausbauch & Galun (1971) found that acetylcholine in the presence of eserine (an inhibitor of acetylcholine esterase) could induce conidiation, mimicking photoinduction. Acetylcholine alone, however, could not induce conidiation. The results suggest that the photoreceptor system may act through acetylcholine.

In another response, melanogenesis in *Verticillium albo-atrum* and *V. dahliae*, catechol was shown to reverse the NUV inhibition of synthesis (Brandt, 1965; Gafoor & Heale, 1971). NUV light was presumed to prevent synthesis or reduce the level of the melanin precursor, catechol.

The induction by light of carotenogenesis in *Fusarium aquaeductum* can be substituted by *p*-chloromercuribenzoate or *p*-hydroxymercuribenzoate (Rau, Feuser & Rau-Hund, 1967). Light and mercuribenzoate may act at the same site in the chain of regulatory mechanism, at SH groups, and it was suggested that the function of light is to photooxidize SH groups of a specific compound.

17.5 Concluding Remarks

It must be quite evident from this chapter that our knowledge of light-induced fungal photoresponses is still quite fragmentary. We know a little about the photoreceptors, a little about some of the metabolic events that occur following light absorption, and virtually nothing about the initial photochemical reactions. The literature is full of papers of a descriptive nature. Too many workers have been content with just describing the photoresponses while only a few are doing research aimed at understanding the basis of the observed responses. However, the trend may be changing. With more sophisticated biophysical and biochemical techniques and instrumentation available, researchers should have an added impetus to solving the many intriguing aspects of fungal photobiology. Then we may see the whole chain of events mapped out from the absorption of light to the manifestation of the response. Lastly the author would like to conclude by mentioning that light-induced processes have proved to be valuable systems for the study of development in eukaryotes. Examples are sporulation in *Dictyostelium discoideum* and *Physarum polycephalum*, and conidiation in *Neurospora crassa*, *Trichoderma viride* and *Botrytis cinerea*. In all these systems light is used to induce synchrony of the various developmental processes.

17.6 References

ALASOADURA, S. O. (1963). Fruiting in *Sphaerobolus* with special reference to light. *Annals of Botany* 27, 123–45.

ARAGAKI, M. (1962). Quality of radiation inhibitory to sporulation of *Alternaria tomato*. *Phytopathology* 52, 1227–8.

ARAGAKI, M. (1969). Inhibitory and stimulatory radiation effects on sporulation of *Alternaria tenuis*. *Phytopathology* 59, 1016.

ARAGAKI, M., NISHIMOTO, K. M. & HYLIN, J. W. (1973). Vegetative reversion of conidiophores in *Alternaria tomato*. *Mycologia* 65, 1205–10.

ASCHAN-ÅBERG, K. (1960). The production of fruit bodies in *Collybia velutipes*. III. Influence of quality of light. *Physiologia Plantarum* 13, 276–9.

BATRA, P. P. (1971). Mechanism of light-induced carotenoid synthesis in nonphotosynthetic plants. In *Photophysiology*, Vol. 6, pp. 47–76, edited by A. C. Giese. New York and London: Academic Press.

BENNINK, G. J. H. (1971). *Fotomorfogenese bij Penicillium isariiforme*, Ph.D. Thesis, Universiteit van Amsterdam, The Netherlands.

BERGMAN, K. (1972). Blue-light control of sporangiophore initiation in *Phycomyces*. *Planta* 107, 53–67.

BERGMAN, K., BURKE, P. V., CERDA-OLMEDO, E., DAVID, C. N., DELBRÜCK, M., FOSTER, K. W., GOODELL, E. W., HEISENBERG, M., MEISSNER, G., ZALOKAR, M. DENNISON, D. S. & SHROPSHIRE, W. JR. (1969). *Phycomyces*. *Bacteriological Reviews* 33, 99–157.

BITENSKY, M. W. & GORMAN, R. E. (1973). Cellular responses to cyclic AMP. *Progress in Biophysics and Molecular Biology* 26, 411–61.

BRANDT, W. H. (1964). Morphogenesis in *Verticillium*: effects of light and ultraviolet radiation on microsclerotia and melanin. *Canadian Journal of Botany* 42, 1017–23.

BRANDT, W. H. (1965). Morphogenesis in *Verticillium*: reversal of the near-UV effect by catechol. *Bioscience* 15, 669–70.

BRODY, S. & HARRIS, S. (1973). Circadian rhythms in *Neurospora*: spatial differences in pyridine nucleotide levels. *Science* 180, 498–500.

BROOK, P. J. (1969). Stimulation of ascospore release in *Venturia inaequalis* by far red light. *Nature, London* 222, 390–2.

BURNETT, J. H. (1968). *Fundamentals of mycology*. London: Edward Arnold (Publishers) Ltd.

CALPOUZOS, L. & CHANG, H-S. (1971). Fungus spore germination inhibited by blue and far red radiation. *Plant Physiology* 47, 729–30.

CALPOUZOS, L. & LAPIS, D. B. (1970). Effects of light on pycnidium formation, sporulation and tropism by *Septoria nodorum*. *Phytopathology* 60, 791–4.

CANTINO, E. C. (1966). Morphogenesis in aquatic fungi. In *The Fungi*, Vol. 2, pp. 283–337, edited by G. C. Ainsworth and A. S. Sussman. New York and London: Academic Press.

CARLILE, M. J. (1965). The photobiology of fungi. *Annual Review of Plant Physiology* 16, 175–202.

CARLILE, M. J. (1970). The photoresponses of fungi. In *Photobiology of microorganisms*, pp. 309–44, edited by P. Halldal. London and New York: Wiley-Interscience.

CARLILE, M. J. (1975). Taxes and tropisms: diversity, biological significance and evolution. In *Primitive sensory and communication systems*, pp. 1–28, edited by

M. J. Carlile. London, New York and San Francisco: Academic Press.

CARLILE, M. J. & FRIEND, J. (1956). Carotenoids and reproduction in *Pyronema confluens*. *Nature, London* **178**, 369–70.

CHANG, H-S. & CALPOUZOS, L. (1971). Germination of uredospores of *Puccinia recondita* inhibited by blue, red and far red light. *Phytopathology* **61**, 887–8.

CHEBOTAREV, L. N. & ZEMLYANUKHIN, A. A. (1973). Effect of visible light and ultraviolet rays on the activity of oxidative enzymes in molds. *Microbiology* **42**, 196–200.

CLAYTON, R. K. (1970). *Light and living matter: a guide to the study of photobiology*, **Vol. 1**, *The physical part*. New York: McGraw-Hill Book Company.

COCHRANE, V. W. (1958). *Physiology of fungi*. New York: John Wiley and Sons.

COHEN, R. J. (1974). Cyclic AMP levels in *Phycomyces* during a response to light. *Nature, London* **251**, 144–6.

CURRY, G. M. & GRUEN, H. E. (1959). Action spectra for the positive and negative phototropism of *Phycomyces* sporangiophores. *Proceedings of the National Academy of Science of the United States of America* **45**, 797–804.

CURRY, G. M. & THIMANN, K. V. (1961). Phototropism: the nature of the photoreceptor in higher and lower plants. In *Progress in photobiology*, pp. 127–134, edited by B. C. Christensen and B. Buchmann. Amsterdam: Elsevier Publishing Company.

CURTIS, C. R. (1972). Action spectrum of the photoinduced sexual stage in the fungus *Nectria haematococca* Berk. and Br. var. *cucurbitae* (Synder and Hansen) Dingley. *Plant Physiology* **49**, 235–9.

DELBRÜCK, M. & SHROPSHIRE, W. JR. (1960). Action and transmission spectra of *Phycomyces*. *Plant Physiology* **35**, 194–204.

DELMER, D. P. & BRODY, S. (1975). Circadian rhythms in *Neurospora crassa*: oscillation in the level of an adenine nucleotide. *Journal of Bacteriology* **121**, 548–53.

ENDE, G. VAN DEN & CORNELIUS, J. J. (1970). The induction of sporulation in *Sclerotinia fructicola* and some other fungi and the production of 'P310'. *Netherlands Journal of Plant Pathology* **76**, 183–91.

FARAJ SALMAN, A. G. (1971). Das Wirkungs spektrum der licht-abhangigen zonierung der Koremian von zwei mutanten von *Penicillium claviforme* Bainier. *Planta* **101**, 117–21.

GAFOOR, A. & HEALE, J. B. (1971). Near-U.V. irradiation and morphogenesis in *Verticillium*. *Microbios* **3**, 131–41.

GALUN, E. & GRESSEL, J. (1966). Morphogenesis in *Trichoderma*: suppression of photoinduction by 5-fluorouracil. *Science* **151**, 696–8.

GETTENS, R. H. & SHROPSHIRE, W. (1963). Light-induced biochemical changes in *Phycomyces*. *Plant Physiology* **38**, supplement p. iv.

GRESSEL, J. & GALUN, E. (1967). Morphogenesis in *Trichoderma*: photoinduction and RNA. *Developmental Biology* **15**, 575–98.

GRESSEL, J. B. & HARTMANN, K. M. (1968). Morphogenesis in *Trichoderma*: action spectrum of photoinduced sporulation. *Planta* **79**, 271–4.

GRESSEL, J., STRAUSBAUCH, L. & GALUN, E. (1971). Photomimetic effect of acetylcholine on morphogenesis in *Trichoderma*. *Nature, London* **232**, 648–9.

HARDING, R. W. & MITCHELL, H. K. (1968). The effect of cycloheximide on carotenoid biosynthesis in *Neurospora crassa*. *Archives of Biochemistry and Biophysics* **128**, 814–18.

HAWKER, L. E. (1957). *The physiology of reproduction in fungi*. London and New York: Cambridge University Press.

HAWKER, L. E. (1966). Environmental influences on reproduction. In *The fungi*, vol. 2, pp. 435–69, edited by G. C. Ainsworth and A. S. Sussman. New York and London: Academic Press.

HITE, R. E. (1973). Substances from *Botrytis cinerea* associated with sporulation and exposure to near-ultraviolet radiation. *Plant Disease Reporter* **57**, 760–4.

HOGENSON, R. O. & HOSFORD, R. M., JR. (1971). Sexual reproduction in *Leptosphaeria avenaria* f. sp. *triticea* induced by wavelengths of light greater than 560 mμ. *Mycologia* **63**, 958–63.

HONDA, Y. (1969). Studies on effects of light on the sporulation of *Helminthosporium oryzae*. *Bulletin of the Institute of Agricultural Research, Tohoku University* **21**, 63–132.

HONDA, Y., SAKAMOTO, M. & ODA, Y. (1968). Blue and near ultraviolet reversible photoreaction on the sporulation of *Helminthosporium oryzae*. *Plant and Cell Physiology, Tokyo* **9**, 603–7.

INGOLD, C. T. (1962). The reaction of fungi to light and the problem of photoperception. *Symposia of the Society for Experimental Biology* **16**, 154–69.

INGOLD, C. T. (1969). Effect of blue and yellow light during the later developmental stages of *Sphaerobolus*. *American Journal of Botany* **56**, 759–66.

INGOLD, C. T. & NAWAZ, M. (1967). Sporophore development in *Sphaerobolus*: effect of blue and red light. *Annals of Botany* **31**, 469–77.

INGOLD, C. T. & OSO, B. A. (1969). Light and spore discharge in *Ascobolus*. *Annals of Botany* **33**, 463–71.

INGOLD, C. T. & PEACH, J. (1970). Further observations on fruiting in *Sphaerobolus* in relation to light. *Transactions of the British Mycological Society* **54**, 211–20.

JACOB, F. (1954). Die Rolle des Lichtes in Entwicklungsgang der sporangientrager von *Pilobolus*-Arten. *Wissen-schaftliche Zeitschrift der Martin-Luther-Universitat, Halle-Wittenberg* **4**, 125–32.

JAGGER, J. (1967). *Introduction to research in ultraviolet photobiology*. Englewood Cliffs, New Jersey: Prentice-Hall Inc.

KELLY, M. S. & GAY, J. L. (1969). The action of visible radiation on the formation and properties of *Saccharomyces* ascospores. *Archiv für Mikrobiologie* **66**, 259–72.

KITAMOTO, Y., HORIKOSHI, T. & SUZUKI, A. (1974). An action spectrum for photinduction of pileus formation in a Basidiomycete, *Favolus arcularius*, *Planta* **119**, 81–4.

KITAMOTO, Y., SUZUKIA, A. & FURUKAWA, S. (1972). An action spectrum for light-induced primordium formation in a Basidiomycete, *Favolus arcularius* (Fr.) Ames. *Plant Physiology* **49**, 338–40.

KUMAGAI, T. & ODA, Y. (1968). Cited in Y. Oda (1970). Mycochrome system—a photoregulatory system participating in the morphogenesis of fungi. *Shizen* **25**, 56–64.

JUMAGAI, T. & ODA, Y. (1969*a*). An action spectrum for photoinduced sporulation in the fungus *Trichoderma viride*. *Plant and Cell Physiology, Tokyo* **10**, 387–92.

KUMAGAI, T. & ODA, Y. (1969*b*). Blue and near ultraviolet reversible photoreaction in conidial development of the fungus *Alternaria tomato*. *Development, Growth and Differentiation* **11**, 130–42.

KUMAGAI, T. & ODA, Y. (1973). Blue and near ultraviolet reversible photoreaction with intracellular particulate fraction of the fungus, *Alternaria tomato*. *Plant and Cell Physiology, Tokyo* **14**, 1107–12.

KUMAGAI, T., YOSHIOKA, N. & ODA, Y. (1976). Further studies on the blue and near ultraviolet reversible photoreaction with an intracellular particulate fraction of the fungus, *Alternaria tomato*. *Biochemica et Biophysica Acta* **421**, 133–40.

KUSS, F. R. (1970). *Light regulated morphogenesis of Stemphylium botryosum*, Ph.D. Thesis, Cornell University, U.S.A.

LEACH, C. M. (1964). The relationship of visible and ultraviolet light to sporulation of *Alternaria chrysanthemi*. *Transactions of the British Mycological Society* **47**, 153–8.

LEACH, C. M. (1965). Ultraviolet-absorbing substances associated with light-induced sporulation in fungi. *Canadian Journal of Botany* **43**, 185–200.

LEACH, C. M. (1967). Interaction of near-ultraviolet light and temperature on sporulation of the fungi *Alternaria, Cercosporella, Fusarium, Helminthosporium*, and *Stemphylium*. *Canadian Journal of Botany* **45**, 1999–2016.

LEACH, C. M. (1968). An action spectrum for light inhibition of the 'terminal phase' of photosporogenesis in the fungus *Stemphylium botryosum*. *Mycologia* **60**, 532–46.

LEACH, C. M. (1971). A practical guide to the effect of visible and ultraviolet light on fungi. In *Methods in microbiology*, vol. 4, pp. 609–64, edited by C. Booth. London and New York: Academic Press.

LEACH, C. M. (1972). An action spectrum for light-induced sexual reproduction in the ascomycete fungus *Leptosphaerulina trifolii*. *Mycologia* **64**, 475–90.

LEACH, C. M. & TRIONE, E. J. (1965). An action spectrum for light-induced sporulation in the fungus *Ascochyta pisi*. *Plant Physiology* **40**, 808–12.

LEACH, C. M. & TRIONE, E. J. (1966). Action spectra for light-induced sporulation of the fungi *Pleospora herbarum* and *Alternaria dauci*. *Photochemistry and Photobiology* **5**, 621–30.

LEATH, K. T. (1971). Quality of light required for sporulation by *Leptosphaerulina*. *Phytopathology* **61**, 70–2.

LUCAS, J. A., KENDRICK, R. E. & GIVAN, C. V. (1975). Photocontrol of fungal spore germination. *Plant Physiology* **56**, 847–9.

LUKENS, R. J. (1963). Photoinhibition of sporulation in *Alternaria solani*. *American Journal of Botany* **50**, 720–4.

LUKENS, R. J. (1965). Reversal by red light of blue light inhibition of sporulation in *Alternaria solani*. *Phytopathology* **55**, 1032.

MARSH, P. B., TAYLOR, E. E. & BASSLER, L. M. (1959). A guide to the literature on certain effects of light on fungi: reproduction, morphology, pigmentation and phototropic phenomena. *Plant Disease Reporter*, Supplement no. 261.

MARTENS, C. L. & SARGENT, M. L. (1974). Circadian rhythms of nucleic acid metabolism in *Neurospora crassa*. *Journal of Bacteriology* 117, 1210–15.

MOGUS, M. A. & WOLKEN, J. J. (1974). *Phycomyces*: electrical response to light stimuli. *Plant Physiology* 53, 512–13.

MOORE-LANDECKER, E. (1972). *Fundamentals of the fungi*. Englewood Cliffs, New Jersey: Prentice-Hall, Inc.

MORTON, A. G. (1961). The induction of sporulation in mould fungi. *Proceedings of the Royal Society of London B* 153, 548–69.

MUÑOZ, V., BRODY, S. & BUTLER, W. L. (1974). Photoreceptor pigment for blue light responses in *Neurospora crassa*. *Biochemical and Biophysical Research Communications* 58, 322–7.

MUÑOZ, V. & BUTLER, W. L. (1975). Photoreceptor pigment for blue light in *Neurospora crassa*. *Plant Physiology* 55, 421–6.

NAMBUDIRI, A. M. D., VANCE, C. P. & TOWERS, G. H. N. (1973). Effect of light on enzymes of phenylpropanoid metabolism and hispidin biosynthesis in *Polyporus hispidus*. *Biochemical Journal* 134, 891–7.

OHMORI, K. & NAKAJIMA, M. (1970a). Effect of light on sporulation of *Alternaria kikuchiana* Tanaka. *Annals of the Phytopathological Society of Japan* 36, 11–6.

OHMORI, K. & NAKAJIMA, M. (1970b). Effect of light on sporulation of *Pyricularia oryzae* Cavara. *Annals of the Phytopathological Society of Japan* 36, 319–24.

ONESIROSAN, P. T. & BANTTARI, E. E. (1969). The effect of light and temperature upon sporulation of *Helminthosporium teres* in culture. *Phytopathology* 59, 906–9.

OSO, B. A. (1968). *Spore discharge and ascus structure in Discomycetes*, Ph.D. Thesis, University of London.

PAGE, R. M. (1956). Studies on the development of asexual reproductive structure in *Pilobolus*. *Mycologia* 48, 206–24.

PAGE, R. M. (1962). Light and the asexual reproduction of *Pilobolus*. *Science* 138, 1238–45.

PAGE, R. M. (1965). The physical environment for fungal growth 3. Light. In *The fungi*, vol. 1, pp. 559–74. Edited by G. C.

Ainsworth and A. S. Sussman. London and New York: Academic Press.

PAGE, R. M. (1968). Phototropism in fungi. In *Photophysiology*, Vol. 3, pp. 65–90. Edited by A. C. Giese. New York and London: Academic Press.

PAGE, R. M. & CURRY, G. M. (1966). Studies on the phototropism of young sporangiophores of *Pilobolus kleinii*. *Photochemistry and Photobiology* 5, 31–40.

PERKINS, J. H. & GORDON, S. A. (1969). Morphogenesis in *Schizophyllum commune*. II. Effects of monochromatic light. *Plant Physiology* 44, 1712–16.

RAPER, J. R. (1971). Growth and reproduction of fungi. In *Plant physiology*, Vol. 4A, pp. 167–230, edited by F. C. Steward. New York and London: Academic Press.

RAU, W. (1967a). Uber die Lichtabhangige carotenoidsynthese. I. Das Wirkiengsspektrum von *Fusarium aquaeductum*. *Planta* 72, 14–28.

RAU, W. (1967b). Untersuchungen uber die Lichtabhangige carotinoidsynthese. II. Ersatz der Lichtinduktion durch Mercuribenzoat. *Planta* 74, 263–77.

RAU, W., FEUSER, B. & RAU-HUNG, A. (1967). Substitution of p-chloro- or p-hydroxymercuribenzoate for light in carotenoid synthesis by *Fusarium aquaeductum*. *Biochimica et Biophysica Acta* 136, 589–90.

RAU, W., LINDEMANN, I. & RAU-HUND, A. (1968). Untersuchungen uber die Lichtabhangige carotinoidsyntheses. III. Die Farbostoffbildung von *Neurospora crassa* in Submerskultur. 80, 309–16.

SARGENT, M. L. & BRIGGS, W. R. (1967). The effects of light on a circadian rhythm of conidiation in *Neurospora*. *Plant Physiology* 42, 1504–10.

SHROPSHIRE, W., JR. (1972). Action spectroscopy. In *Phytochrome*, pp. 161–181, edited by K. Mitrakos and W. Shropshire, Jr. London and New York: Academic Press.

SMITH, J. E. & GALBRAITH, J. C. (1971). Biochemical and physiological aspects of differentiation in the fungi. *Advances in Microbial Physiology* 5, 45–134.

SPROSTON, T. (1969). Cited in E. Trione and C. M. Leach (1969).

SPROSTON, T. (1971). An action spectrum for ultraviolet-induced sporulation in the fungus *Stemphylium solani* Weber. *Photochemistry and Photobiology* 14, 571–6.

SPROSTON, T. & SETLOW, R. B. (1968). Ergosterol and substitutes for the ultraviolet radiation requirement for conidia formation in *Stemphylium solani*. *Mycologia* **60**, 104–14.

STALLINGS, F. O. (1970). *The physiological control of light-induced conidiation in Aspergillus oranatus* Ph.D. Thesis, University of Wisconsin, U.S.A.

STAVY, R., STAVY, L. & GALUN, E. (1970). Protein synthesis in aged and young zones of *Trichoderma* colonies. *Biochimica et Biophysica Acta* **217**, 468–76.

TABER, W. A. (1966). Morphogenesis in Basidiomycetes. In *The fungi*, Vol. 2, pp. 387–412, edited by G. C. Ainsworth and A. S. Sussman. New York and London: Academic Press.

TAN, K. K. (1974a). Complete reversibility of sporulation by near ultraviolet and blue light in *Botrytis cinerea*. *Transactions of the British Mycological Society* **63**, 203–5.

TAN, K. K. (1974b). Blue-light inhibition of sporulation in *Botrytis cinerea*. *Journal of General Microbiology* **82**, 191–200.

TAN, K. K. (1974c). Red-far-red reversible photoreaction in the recovery from blue-light inhibition of sporulation in *Botrytis cinerea*. *Journal of General Microbiology* **82**, 201–2.

TAN, K. K. (1975a). Recovery from the blue-light inhibition of sporulation in *Botrytis cinerea*. *Transactions of the British Mycological Society* **64**, 223–8.

TAN, K. K. (1975b). Interaction of near-ultraviolet, blue, red, and far-red light in sporulation of *Botrytis cinerea*. *Transactions of the British Mycological Society* **64**, 215–22.

TAN, K. K. (1975c). Effect of inhibitors of RNA and protein synthesis on light-induced synchronous conidiation in *Botrytis cinerea*. *Proceedings of the Society for General Microbiology* **2**, 86.

TAN, K. K. & EPTON, H. A. S. (1973). Effect of light on the growth and sporulation of *Botrytis cinerea*. *Transactions of the British Mycological Society* **61**, 145–57.

TAN, K. K. & EPTON, H. A. S. (1974a). Further studies on light and sporulation in *Botrytis cinerea*. *Transactions of the British Mycological Society* **62**, 105–12.

TAN, K. K. & EPTON, H. A. S. (1974b). Ultraviolet-absorbing compounds associated with sporulation in *Botrytis cinerea*. *Transactions of the British Mycological Society* **63**, 157–67.

THIMANN, K. V. (1967). Phototropism. In *Comprehensive biochemistry*, Vol. 27, pp. 1–29, edited by M. Florkin and E. H. Stotz. Amsterdam: Elsevier Publishing Company.

TRINCI, A. P. J. & BANBURY, G. H. (1969). Effect of light on growth and carotenogenesis of the tall conidiophores of *Aspergillus giganteus*. *Transactions of the British Mycological Society* **52**, 73–86.

TRIONE, E. J. & LEACH, C. M. (1969). Light-induced sporulation and sporogenic substances in fungi. *Phytopathology* **59**, 1077–83.

TRIONE, E. J., LEACH, C. M. & MUTCH, J. J. (1966). Sporogenic substances isolated from fungi. *Nature, London* **212**, 153–64.

TURIAN, G. (1966). Morphogenesis in Ascomycetes. In *The fungi*, Vol. 2, pp. 339–85, edited by G. C. Ainsworth and A. S. Sussman. New York and London: Academic Press.

TURIAN, G. (1969). *Differenciation fongique*. Paris: Masson and Company.

UNO, I., YAMAGUCHI, M. & ISHIKAWA, T. (1974). The effect of light on fruiting body formation and adenosine $3':5'$-cyclic monophosphate metabolism in *Coprinus macrorhizus*. *Proceedings of the National Academy of Science of the United States of America* **71**, 479–83.

UREY, J. C. (1971). Enzyme patterns and protein synthesis during synchronous conidiation in *Neurospora crassa*. *Developmental Biology* **26**, 17–27.

VALADON, L. R. G. (1974). The effect of light on the composition of DNA-dependent RNA polymerases of *Verticillium agaricinum*. *Physiologia Plantarum* **32**, 233–9.

VALADON, L. R. G. & MUMMERY, R. S. (1973). Effect of certain nucleic acid and protein inhibitors on carotenogenesis in *Verticillium agaricinum*. *Physiologia Plantarum* **28**, 254–8.

VALADON, L. R. G., TRAVIS, R. L. & KEY, J. L. (1975). Light-induced activation of cytoplasmic protein synthesis in *Verticillium agaricinum*. *Physiologia Plantarum* **34**, 196–200.

WESTE, G. (1970). Factors affecting vegetative growth and the production of perithecia in culture by *Ophiobolus graminis* II. Variations in light and temperature. *Australian Journal of Botany* **18**, 11–28.

ZALOKAR, M. (1955). Biosynthesis of carotenoids in *Neurospora*. Action spectrum of photoactivation. *Archives of Biochemistry and Biophysics* **56**, 318–25.

CHAPTER 18

Temperature-Induced Fungal Development

J. G. ANDERSON

18.1 Introduction

There are strict temperature limits within which the metabolic processes of any fungal organism or indeed any other life form can take place. To this extent therefore it could be said that any physiological change or differentiation process in fungi is temperature dependent or temperature induced. However, apart from the general basic dependence of metabolism on a suitable temperature environment, specific instances can be recognized where temperature appears to play a special additional role as an inducer of developmental change in filamentous fungi. When considering such situations, however, it must be remembered that any change in temperature will simultaneously alter other physicochemical characteristics of the environment and that consequently the nature of induction processes which result from changes in temperature may well be quite complex. Thus the effects on cellular metabolism are likely to be less specific than for example those involved in 'light-induced' fungal development.

The diverse effects of temperature on different aspects of fungal physiology have been dealt with in various reviews (for refs. see Anderson & Smith, 1976). This review will be restricted to temperature effects on spore germination and hyphal morphogenesis in filamentous fungi. With some species, spore germination can be induced or activated by temperature treatment. With others there is evidence that the processes involved in the generation and maintenance of the filamentous hyphal form can be profoundly affected by temperature with the result that alternative morphological forms are produced.

Emphasis will be placed on studies which have been concerned with the possible physiological mechanisms involved in temperature induced developmental phenomena. Considerable information has appeared on these topics, not least, one suspects, because such systems have provided particularly useful and convenient models for studying developmental changes. Thus the functional alterations involved in the developmental change can be initiated by simple temperature manipulation and studies can then be conducted on the physiological and biochemical changes which follow the alteration of the thermal environment.

18.2 Activation of Spore Germination by Temperature

Fungal spores which show constitutive dormancy *i.e.* those which fail to develop even when provided with conditions under which vegetative development will proceed, require an activation treatment to commence germination. The nature of dormancy in fungal spores and spores of other organisms and the variety of physical and chemical factors which can serve as activators of germination have been extensively considered (Sussman, 1966*a*; Sussman & Halvorson, 1966; Sussman & Douthit, 1973). Also, the possible mechanisms whereby temperature or chemicals might activate germination have been discussed in detail by Sussman (1976).

Exposure to elevated temperatures is an effective activation treatment with some fungal spores while with others exposure to cold may be necessary. In some cases temperature fluctuations appear to be required while with others temperature interacts with certain chemicals. Examples of fungi which fall into each of these categories are given in Table 18.1. Cold induced

Table 18.1. Influence of temperature on the activation of spore germination in various fungi

Organism	Activation conditions	Reference
Neurospora tetrasperma	50–65° for 10–20 min ⎫	
Phycomyces blakesleeanus	50° for 3 min ⎬	see Sussman (1976)
Mucor miehi	45° for 5 h ⎭	
Byssochlamys fulva	75° for 5–10 min	
B. nivea	75° for 5 min	Yates, Seaman & Woodbine (1968)
Chaetomium globosum	37° for 1 h	Chapman & Fergus (1975)
	Cold treatment	
Tilletia asperifolioides	1–2° for months ⎫	Duran & Safeeulla (1968)
Urocystis fraseri	1–2° for months ⎭	
	Cold followed by heat treatment	
Puccinia graminis	Freezing then 40° for 10 min	McDonald & Strange (1976)
	Combined chemical and heat treatment	
Coprinus radiatus	Furfural and 45° for 4 h	Mills & Eilers (1973)
Chaetomium rectopilium	K acetate and 55° for 1 min	Fergus & Delwiche (1975)

germination is found particularly with basidiomycete spores and is little understood (Hess & Weber, 1976). Consequently, only heat-induced activation will be discussed here particularly with reference to studies with ascospores of *Neurospora tetrasperma* and the sporangiospores of *Phycomyces blakesleeanus*.

Heat activation of Neurospora ascospores

The activation of dormant ascospores of *Neurospora tetrasperma* by heat shock or chemical treatment has been extensively studied (Sussman, 1966a,b, 1976). The hypothesis developed by Sussman and co-workers to explain activation by heat or chemicals centres around the metabolism of trehalose. Dormant ascospores do not use trehalose even though it is present in considerable amount; instead they consume lipids. However, as soon as activation is accomplished trehalose is consumed along with lipids. It is considered that activation occurs because of the flood of glucose released by the utilization of trehalose.

Both trehalose and its hydrolase trehalase are present in considerable amounts prior to activation. Thus a key requirement for understanding activation has been to define the nature of the restraint upon trehalose catabolism in the dormant ascospore. Various mechanisms have been considered and the elegant studies by Hecker & Sussman (1973) appear to offer a probable explanation. Using immunofluorescent techniques, Hecker & Sussman (1973) discovered that trehalase is associated with the innermost ascospore wall. Trehalose however is located in the cytoplasm and thus would appear to be physically separated from its hydrolase. The association of trehalase with the ascospore wall also explains why trehalase is more stable to temperatures of 60° or 65° in intact ascospores than in extracts (Yu, Sussman & Wooley, 1967) since it has been found that ground ascospore walls protect trehalase against heating at the temperatures required to activate germination (Hecker & Sussman, 1973). From these results it has been concluded that activation whether by heat or chemical treatment probably involves an increase in the permeability of the ascospore plasma membrane allowing trehalose to diffuse to the vicinity of its hydrolase thereby providing the energy and intermediates for germination (Hecker & Sussman, 1973).

Heat activation of Phycomyces sporangiospores

Studies on metabolic changes following heat activation of *Phycomyces blakesleeanus* sporangiospores have recently been reviewed by Grove (1976). The process shows some similarities with *Neurospora* ascospores in that there is an increased glycolytic metabolism, a decrease in the reserve substance trehalose and the appearance of glucose initially in the cell contents and then later in the culture medium (Rudolph & Ochsen, 1969). Differences during and after heating have however been noted. In *Neurospora* heat activation does not increase trehalase activity whereas in *Phycomyces* there is a transient 10- to 15-fold increase in activity which can be repeated by repeating the heat shock, thus suggesting that in *Phycomyces*, heat treatment converts temporarily an inactive form of trehalase into an active form (Van Assche, Carlier & Dekeersmaeker, 1972). Whereas Sussman and co-workers relate activation to the release of glucose from trehalose utilisation this mechanism is not favoured by Rudolph and co-workers for *Phycomyces* sporangiospores. They point out that dormant spores do not germinate when supplied with external glucose to which they are permeable. Instead they suggest that increased glycolytic activity is more essential in spore activation than trehalose breakdown and that consequently heat might activate some step(s) in glycolysis (Rudolph & Ochsen, 1969; Rudolph & Furch, 1970).

While there is no direct implication of membrane involvement during heat activation in these biochemical studies, an ultrastructural study by Malhotra & Tewari (1973) has shown that important changes occur in the plasma membrane of *Phycomyces blakesleeanus* sporangiospores shortly after heat shock treatment. Using freeze fracture technique Malhotra & Tewari (1973) observed that the plasma membrane of *P. blakesleeanus* spores showed, in addition to the 5 to 10 nm particles which can be observed in the plasma membrane of various cells, two types of unusual 30 to 35 nm particles which they termed homogeneous particles and compound particles. Soon after heat shocking the compound particles broke up into sub-particles whereas the homogeneous particles aggregated and disappeared prior to germination. Because of the apparent effects of heat shock on these structures it was suggested that these membranous particles may be related to the controlled dormancy and germination of *Phycomyces* sporangiospores. It is interesting also that Tu & Malhotra (1973) have detected the presence of adenyl cyclase in the plasma membrane of these spores by histochemical techniques. The possibility that the initiation of germination in *P. blakesleeanus* spores may be regulated through adenyl cyclase, which may be activated upon heat-shocking, has been suggested by Malhotra & Tewari (1973).

Further evidence for membrane involvement in heat activation

Several other studies have been carried out which support the idea of membrane involvement in certain temperature-induced metabolic changes in fungal spores. Reversible changes in membrane structure and permeability are thought to be caused in rust urediospores, by cycles of extreme cold and subsequent warming (Maheshwari & Sussman, 1971). Also Mandels & Maguire (1972) have shown that endogenous respiration in spores of *Myrothecium verrucaria* can be stimulated by various treatments including heating at 50°. They suggest that such treatments may act by causing permeability changes in membranes which confine endogenous reserves.

It is relevant for comparative purposes to look briefly at evidence for membrane involvement in the heat activation of spores of the Myxomycete *Dictyostelium discoideum*. For additional information on the activation and germination of these spores consult Hohl (1976). According to a hypothesis developed by Cotter and co-workers heat shock and other activation treatments induces a partial helix-coil transition in a 'dormant' regulatory protein located in the inner mitochondrial membrane which is responsible for restricted oxidative phosphorylation. The induction leads to a metastable state of the protein which then decays to a relaxed state functional in spore swelling by allowing ATP production to commence (Cotter, 1973; Cotter & George, 1975). Ultrastructural evidence has also been provided which indicates that the site most sensitive to thermal-induced activation and the site most easily destroyed by supraoptimal heating in these spores may be the inner mitochondrial membrane (Cotter & George, 1975). It was observed that if the usual activation period of 30 min at 45° was extended or if higher temperatures were used the cristae of mitochondria were disrupted while other cellular structures showed no evidence of thermal damage.

There is therefore substantial evidence to show that the heat shock treatment required to activate germination in the spores of several fungi does cause significant changes to cellular membranes and that this may well be where the heat trigger mechanism is located. It is however more difficult to make generalizations concerning which of the membrane constituents might be involved in the heat mediated changes. On the basis of thermodynamic considerations Cotter (1973) and Sussman (1976) have concluded that it is at the level of the macromolecules that the candidates for heat mediated transitions should be sought. Of the various macromolecules which might mediate thermal effects Cotter (1973) considers that it is proteins and not lipids which are most likely to be involved. In particular, protein denaturation has been suggested to be the process responsible for the activation of bacterial and fungal spores (Sussman & Halvorson, 1966; Gould & Hurst, 1969).

Sussman (1976) speculates that the thermodynamic data do not exclude lipid involvement and considers that there are good reasons for invoking a key role for the lipid components of membranes. According to Sussman (1976) if phospholipids can diffuse rapidly in the plane of the membrane and if this property is combined with their ability to activate certain enzymes then it is possible that high temperatures might cause the movement of essential lipids to the site of enzymes requiring them for activity and thus initiate reactions that could lead to germination. Indeed, Malhotra & Tewari (1973) consider that the apparent movement and aggregation of particles in the plane of the plasma membrane of *Phycomyces* sporangiospores could be explained if the substrate in which the macromolecules are suspended were to become fluid during heat shocking. They suggest that this may be due to heat initiating a phase change in the lipid layer of the membrane converting its molecular core into a fluid form which is consistent with the fluid mosaic model proposed for functional biological membranes (Singer, 1973).

18.3 Effects of Temperature on the Isotropic Growth (Swelling) Phase of Germination

Once germination has been initiated by nutrients, temperature, or other factors depending on the spore type, normal germination will proceed in the presence of favourable environmental conditions. Disregarding the many differences that occur between spore types, the first stage of germination after activation generally involves the formation of an enlarged (usually spherical) cell which results from wall changes which probably involve the uniform isotropic depositon of cell wall polymers (see Smith, Gull, Anderson & Deans, 1976). Once the spore has achieved maximum or near maximum size the pattern of wall synthesis becomes polarized in that wall polymer deposition becomes limited to one or a few areas within the spore and these subsequently become the sites of incipient germ tube emergence (Bartnicki-Garcia, 1973).

The duration and extent of isotropic growth prior to germ tube emergence can vary depending on environmental conditions and in some fungi temperature can have a profound effect. In particular, with some species, elevated temperature in the presence of appropriate nutritional conditions can favour a prolongation of the spherical growth phase and in some cases can completely 'prevent' germ tube production. The spherical cell forms produced by such treatments range from quite complex cells which can develop into a multicellular structure to much simpler types which remain as non-dividing unicellular forms (Fig. 18.1).

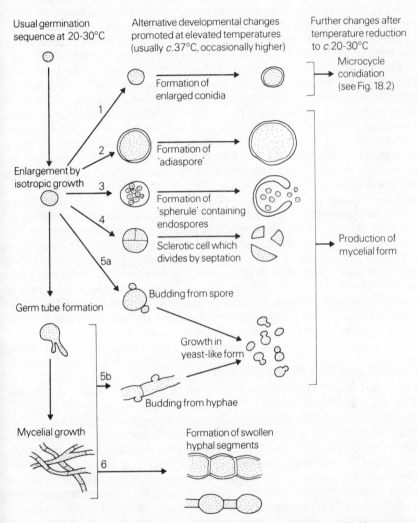

Fig. 18.1 Developmental changes in certain fungi which can be controlled by temperature treatment. Structural changes are shown diagrammatically and the variations in morphological detail to be found among the species listed are not included.
1. *Aspergillus niger, Penicillium urticae, Neurospora crassa*, 2. *Emmonsia crescens*, 3. *Coccidioides immitis*, 4. *Phialophora pedrosi, P. verrucosa, Cladosporium carrionii*, 5(ab). *Histoplasma capsulatum, Paracoccidioides brasiliensis, Blastomyces dermatitidis, Sporothrix schenckii*, 6. Various saprophytic and pathogenic fungi.

Spherule and sclerotic cell forms of pathogenic fungi

With a number of fungi the spherical cell form which develops is characteristic of an *in vivo* parasitic phase. Among these are the dematiaceous fungi which cause chromomycosis and which produce sclerotic cells *in vivo* and also a number of other pathogenic fungi in which the *in vivo* form is described as a 'spherule'. Generally, the spherule and sclerotic cell forms develop from spores although with some organisms development from swollen hyphal segments is also possible during *in vitro* cultivation. The importance of elevated temperature (37°) in the generation of these forms has been demonstrated in various studies (for refs. see Mariat, 1964).

The sclerotic cells produced by *Phialophora pedrosoi, P. verrucosa* and *Cladosporium carrionii* under the appropriate temperature and nutritional conditions are single, rounded and generally thick-walled cells, 6–12 μm or more in diameter which do not bud but multiply by producing septa. Fungi which can produce a 'spherule' form include *Coccidioides immitis, Emmonsia crescens* and *Rhinosporidium seeberi*. The most remarkable of these appears to be the spherule or adiaspore of *E. crescens* which *in vivo* can apparently enlarge from the initial spore diameter of 4 μm, becoming thick-walled and multinucleate, and reach up to 920 μm in diameter perhaps ranking these as the largest known single fungal element (Austwick, 1966). The spherules of *C. immitis* are 10–80 μm or more in diameter while those of *R. seeberi* are up to 300 μm in diameter and when mature the spherules of both these fungi contain a large number of endospores. Although the developmental processes responsible for the generation of these parasitic forms from original spores are likely to be quite complex, nevertheless, it is clear that the extensive spherical cell growth which occurs must involve considerable isotropic deposition of cell wall material. Additional details concerning the precise conditions for the induction of these forms together with a fuller description of the morphological changes involved have been given by Mariat (1964) and Anderson & Smith (1976).

Enlarged spore forms of saprophytic fungi

Temperature effects on the isotropic growth phase of various saprophytic fungi have also been reported. The use of a critical supraoptimal temperature has been shown to selectively allow the enlargement or spherical growth phase while inhibiting germ tube formation with the conidia of *Aspergillus niger* (Anderson & Smith, 1972), *Neurospora crassa* (Cortat & Turian, 1974) and *Penicillium urticae* (Sekiguchi, Gaucher & Costerton, 1975*a*). This treatment induces remarkable physiological changes since the enlarged cells produced by this treatment can display a rapid 'microcycle' condition (see chapter 11) under appropriate temperature and nutritional conditions (see Fig. 18.2).

With *Penicillium urticae* 'normal' germination at 28° involves spore enlargement from *c* 3 μm to *c* 4.5 μm followed by germ tube production whereas at 37° the spores enlarge to *c* 6.5 μm and germ tube formation does not occur. Ultrastructural studies have been carried out during the prolonged spore enlargement of these spores at 37° (Sekiguchi *et al.*, 1975*a*). The most striking changes observed were the disruption of the distinctive rodlet-patterned surface of the cell wall of the conidium, the apparent conversion of distinctive linear invaginations in the plasma membrane to round depressions and the initiation of nuclear division and new organelle synthesis. It was suggested that the spore wall changes which were observed might be the result of spore wall growth since the spores increased their size without any reduction of the overall thickness of the spore wall.

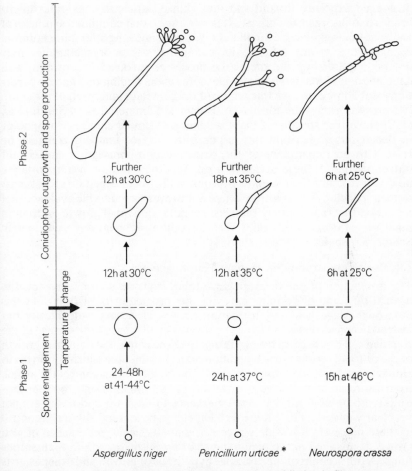

Fig. 18.2 The use of temperature to induce microcycle conidiation in filamentous fungi. *Phase 2 in *P. urticae* also involves nitrogen limitation.

With *Aspergillus niger*, incubation of the conidia at 44° or at slightly lower temperatures in the presence of added CO_2 (Anderson & Smith, 1972; Kuboye, Anderson & Smith, 1976) causes extensive conidial enlargement from c 3.5 µm to c 20 µm while inhibiting germ tube formation. The most dramatic ultrastructural changes observed during the formation of these large cells were the considerable increase in the number of nuclei and the production of an extensively thickened multilayered cell wall. This latter observation clearly showed that isotropic wall deposition continued during the prolonged spherical growth which occurred at elevated temperatures.

It is interesting that a somewhat similar phenomenon has been described with *Rhizopus stolonifer* sporangiospores in response to cold treatment. Ultrastructural studies on spores held at 0° for up to 35 days demonstrated that a progressive thickening of the complete inner spore wall occurred although in this case no accompanying excessive cell enlargement was reported (Matsumoto, Buckley, Sommer & Shalla, 1969). On ultrastructural evidence mitochondria were the organelles damaged first by the prolonged exposure to cold and they showed discontinuous outer membranes and disorganized cristae. It is interesting that continued synthesis of wall material was observed in this study to continue long after mitochondrial changes were evident. A further indication that respiratory metabolism may not be required for the metabolic processes involved in isotropic wall deposition in *R. stolonifer* sporangiospores is the finding that anoxic conditions also allow excessive thickening of the complete inner wall of the spore but prevent germ tube formation (Bussel & Sommer, 1973). Smith *et al.* (1976) have also speculated that the somewhat similar changes which occur in *Aspergillus niger* conidia in response to elevated temperatures may be related to an impairment of the oxidative phosphorylation capacity of mitochondria at these temperatures. Whether or not a fermentative metabolism is sufficient to allow isotropic wall growth in certain spores must remain speculative at this time. There is however considerable evidence that mitochondrial functions are required for polar or hyphal growth to occur in many fungi (Storck & Morrill, 1971; Zorzopulos, Jobbagy & Terenzi, 1973; Schulz, Kraepelin & Hinkelmann, 1974).

Effects of Temperature on Hyphal Morphogenesis

The production of one or more germ tubes from the spore represents the normal outcome of germination with the vast majority of fungi. At this outgrowth stage cell wall formation no longer occurs isotropically but instead becomes restricted to one or several small areas of the surface of the rounded cell where germ tubes will emerge. The continued polarized growth of the cell wall results in the formation of the hyphal tube which continues to elongate by the apical growth process. The factors which control the switch from isotropic to polarized growth during germination and the mechanism(s) responsible for germ tube emergence through the spore wall are not yet clearly understood. While much has still to be learned about the control of these processes it is widely accepted that a vesicular mechanism of wall growth is involved in the maintenance of apical tip growth in filamentous fungi. A full account of the vesicular hypothesis of growth and other aspects of apical tip growth in filamentous fungi is given in Chapter 4.

Effects of temperature on germ tubes and other outgrowths

While germ tube elongation and the formation of an extensive mycelium by operation of the apical growth process is the usual pattern of development in filamentous fungi, in some species temperature treatment can induce alternative developmental changes. Consider firstly the process of 'microcycle' conidiation which can be induced to occur from the spores of several fungi after these have undergone cellular enlargement by incubation at certain critical supraoptimal temperatures which prevent germ tube emergence (see previous section and Fig. 18.2). With *Neurospora crassa* incubation of conidia for 15 h at 46° followed by incubation at 25° leads to the production of 'germ tubes' from the swollen conidia which, instead of elongating, produce proconidia by a process of basifugal budding and consequently can be considered as simple conidiophores (Cortat & Turian, 1974). With *Penicillium urticae* microcycle conidiation has been achieved by incubation of conidia in a complete medium at 37° for 24 h followed by replacement to a nitrogen poor medium at 35°. At the lower temperature the 'germ tubes' which are produced from the enlarged conidia undergo highly abbreviated growth and then sporulate (Sekiguchi, Gaucher & Costerton, 1975b,c).

Although microcycle conidiation in *Neurospora crassa* and *Penicillium urticae* involves the production from the enlarged spores of structures similar to germ tubes which then sporulate this is not the case with *Aspergillus niger*. When *A. niger* conidia are maintained at 44°, or at slightly lower temperatures in the presence of added CO_2 for 24 h followed by incubation at 30° this leads to the direct production of one or more large conidiophores which are morphologically quite different structures from germ tubes (Anderson & Smith, 1971). Thus in this case the usual intervening stage of hyphal development prior to sporulation is completely by-passed.

From numerous physiological studies which have been carried out the generalization usually holds that sporulation in fungi occurs as a result of nutrient exhaustion or other growth limiting conditions (see Lovett, 1975 and Chapter 11). With *Penicillium urticae* the induction of microcycle conidiation involves an initial temperature check to development followed by exposure to nitrogen limitation. With *Aspergillus niger* and *Neurospora crassa* however nutrient limitation is not imposed and the microcycle conidiation which occurs in these organisms is in some way caused by physiological changes which take place during the period of restricted growth at supraoptimal temperatures. For *N. crassa*, Cortat & Turian (1974) have speculated that the heat treatment may inactivate the vesicular materials, aggregated in the tips of elongating hyphae, which are presumed to support glycolytic enzyme activities. This inhibition of glycolysis would allow the functioning of the oxidative and gluconeogenic metabolic pathways which Turian & Bianchi (1972) have suggested is required to sustain active conidiogenesis in *N. crassa*.

Other less dramatic alterations to the morphology of developing germ tubes in response to elevated temperatures have been reported in several fungi. Thus vesicles (swellings) can be induced in the germ tubes of germinating urediospores of several rust species in response to a variety of shock treatments including elevated temperatures (Hansen & Patton, 1975). Swellings and excessive branch formation in the germ tubes of *Aspergillus niger* have also been observed at supra-optimal incubation temperatures (Anderson & Smith, 1972). Also the *in vivo* temperature may be at least in part responsible for the distorted and bulbous hyphae which are frequently observed in lesions during the course of the somewhat rare infections by *Absidia*, *Mucor* and *Aspergillus* spp. (Rippon, Conway & Domes, 1965; Austwick, 1965). The formation, in certain pathogenic fungi, of swollen hyphal segments which can further differentiate into the more or less spherical 'spherule' and sclerotic cell forms has already been mentioned in the preceding section. It is reasonable to suggest that such morphological changes in germ tubes and hyphae may result from a partial inhibition of the apical growth process which is responsible for generating the usual tubular form of the hyphal filament.

Mould-yeast dimorphism

Many different environmental factors may be involved in the expression of fungal dimorphism but in a wide range of fungi including both pathogenic and essentially non-pathogenic forms temperature or temperature acting in conjunction with other environmental conditions can exert a powerful influence. Whatever the underlying physiological mechanism(s) of dimorphism, the importance of the localized nature of wall growth required for hyphal development as opposed to the isotropic wall growth required for yeast-like development is well recognized (Bartnicki–Garcia & Lippman, 1969; Bartnicki–Garcia, 1973). In the numerous examples of thermally induced or thermally influenced dimorphism in filamentous fungi the yeast-like phase is promoted at elevated temperatures whereas the filamentous form is favoured at lower temperatures.

In many cases the budding yeast-like cells can develop either directly from enlarged spores, thus representing an alternative temperature induced form of germination or they can be produced by budding from the hyphae. These dimorphic changes are therefore consistent with the general tendency (Fig. 18.1) of elevated temperatures to promote spherical cell forms in susceptible fungi by processes involving isotropic wall deposition.

There have been a large number of studies carried out on thermal dimorphism in various pathogenic fungi. For the purposes of this review emphasis will be placed on those studies which have been concerned with the possible physiological mechanisms whereby temperature might influence dimorphic changes. The general topic of fungal dimorphism is considered in Chapter 9 and additional details concerning thermally induced dimorphism have been discussed by Gilardi (1965), Romano (1966) and Anderson & Smith (1976).

Ultrastructural events accompanying the phase transitions which result from temperature changes have been clearly documented for a number of the pathogenic fungi. Comparison of events in *Paracoccidioides brasiliensis*, *Sporothrix schenckii*, *Blastomyces dermatitidis*, and *Histoplasma capsulatum* following the 37° to *c*. 25° change which induces the yeast-like to mould transformation suggests that each of these fungi undergoes this morphological transformation in a markedly similar sequence of events at the ultrastructural level (Garrison, Lane & Field, 1970). The yeast-like cells undergo phase conversion with the formation of a transitional cell which has a mixture of both yeast-like and hyphal characteristics. In all cases the formation of discrete intracytoplasmic membrane systems was observed to immediately precede the formation of the transitional cell. Other similarities were the presence of Woronin bodies at the septal area separating the parent yeast-like cell from the transitional cell, the origin of the cell wall of the transitional cell from the inner portion of the yeast-like cell and the appearance of intra-yeast and intra-hyphal hyphae.

Comparison of ultrastructural changes during the mould to yeast-like transformation in *Sporothrix schenckii* (Garrison, Boyd & Mariat, 1975) with that in *Histoplasma farciminosum* and *Phialophora dermatitidi.* (Oujezdsky, Grove & Szaniszlo, 1973) shows that this transformation occurs by a similar direct budding process of yeast cells from the hyphae by blastic action.

Thus the marked similarity in ultrastructural reorganization in these pathogenic dimorphic fungi might suggest that the phase transitions proceed under the control of very similar regulatory mechanisms. It has not been possible however to correlate directly the ultrastructural alterations with specific metabolic changes and as a result of the range of biochemical studie which have been carried out various regulatory mechanisms have been proposed.

Originally Nickerson & Edwards (1949) considered that the yeast-like to mould conversion in *Blastomyces dermatitidis* and *Paracoccidioides brasiliensis* resulted from the selective inhibition of cell division without simultaneous inhibition of growth and that thermal inactivation of enzymes might play an important role in the phase transitions of these dimorphic fungi. More recent studies on these organisms and on *Histoplasma* spp. have been mainly concerned with a comparison of the ultrastructural and chemical characteristics of the walls of the mycelial and yeast phases. With *P. brasiliensis* (Kanetsuna, Carbonnell, Azuma & Yamura, 1972), *B. dermatitidis* (Kanetsuna & Carbonell, 1971) and with *H. capsulatum* and *H. farciminosum* (San-Blas & Carbonell, 1974) it has been found that in all cases except *H. farciminosum* the wall of the yeast form contained mainly α-glucan and that in all cases the wall of the mycelial form contained primarily β-glucan. With *P. brasiliensis*, various other enzymatic and chemical differences have also been found between the two phases and a detailed structural model has been proposed to explain thermal dimorphism in this organism (Kanetsuna *et al.*, 1972). This model assumes that the spherical shape of the yeast-like form is due to the predominance of short rod-like fibres characteristic of the α-1,3-glucan and that the long narrow fibres of the β-1,3-glucan contribute to the mycelial form. However San-Blas, San-Blas & Cova (1976) have found that the yeast forms of strains of *P. brasiliensis* which they studied contained large amounts (about 30%) of the β-1,3-glucan thus casting some doubt on the assumed role of this polysaccharide in the formation of the mycelial form.

Several other studies have been carried out on *Histoplasma capsulatum* which have compared the physiological and chemical characteristics of the yeast-like and mould forms. In particular, special attention has been paid to the biochemical events associated with RNA metabolism during the phase transitions of this organism. It has been observed that just after the shift from 37° to 23° which induces the yeast to mould conversion, net RNA synthesis stops for some time (Cheung, Kobayashi, Schlessinger & Medoff, 1974). Further studies demonstrated that RNA polymerases isolated from the yeast cells showed differential responses to incubation at 23° and 37° and that the predominant species of RNA polymerase is rapidly but reversibly inactivated at elevated temperatures (Boguslawski, Schlessinger, Medoff & Kobayashi, 1974). This led to the suggestion that the conversion process might require some change in the RNA synthesizing components of the cell. Subsequent experiments (Boguslawski, Kobayashi, Schlessinger & Medoff, 1975) have revealed that in contrast to extracts from the yeast phase those from mycelia contained an inhibitor of RNA polymerase which they named 'histin'. There is as yet no information on the role played by histin in *H. capsulatum*; however, Boguslawski *et al.* (1975) have speculated that a temperature dependent critical level of histin in the cell could conceivably be a determinant in the phase transitions of this organism.

Although important physiological differences have been found between the yeast-like and mycelial forms of the various pathogenic dimorphic fungi, it must be recognized that the mechanisms which regulate the phase transitions are not yet well understood. When searching for physiological explanations, however, it may not prove too fruitful to seek for mechanisms which are solely temperature dependent. It must be remembered that dimorphic transformations occur in many other fungi in response to quite different environmental conditions. In studies on dimorphism in a number of *Mycotypha* (Mucorales) species elevated temperature was only one of a number of environmental factors which enhanced a budding yeast-like phase (Schulz *et al.*, 1974; Hall & Kolankaya, 1974). In these studies it was found that, in general, environmental conditions or metabolic inhibitors which depressed respiration stimulated the yeast phase. Schulz *et al.* (1974) speculated that the primary effects of environmental influences such as temperature may be to alter membrane structure and permeability and that regulation of metabolic activities at the cell membrane level plays a decisive role in the phenotypic determination of dimorphism in *Mycotypha*. Considering the numerous documented effects of temperature on cellular membranes which have already been referred to in this review one wonders if this generalization might also apply to other examples of thermally induced dimorphism and indeed to other environmentally determined forms of dimorphism in filamentous fungi.

18.5 Conclusions

In this review a variety of 'temperature induced' developmental phenomena have been discussed. Since emphasis has been placed on considering the possible physiological mechanisms involved it is pertinent at this stage to see what conclusions if any can be made. It is important to realise that with a number of the examples discussed the temperature effect is not unique in the sense that other environmental conditions can substitute for temperature. Thus for example certain chemicals can be used in place of heat shock to activate spore germination in *Neurospora* and *Phycomyces* (see Sussman, 1976). Dimorphic changes in many fungi do not require temperature alteration, and in those which do, temperature is frequently not the only factor involved. The transformation to yeast-like growth or the development of other more or less spherical cell forms often requires a complex interaction between temperature and other environmental factors particularly carbon dioxide and sulphydryl containing compounds. Hyphal abnormalities similar to those described here have been observed in numerous fungi in response to a variety of stress conditions other than supraoptimal temperatures. Finally, forms of 'microcycle sporogenesis' have been reported to occur in certain filamentous fungi again under conditions which did not involve temperature alteration. Thus temperature effects on differentiation processes must be viewed in this context and ideally any explanations of the mechanisms of action must leave room to accommodate explanations as to how other environmental conditions can lead to similar developmental changes.

When considering the temperature treatments themselves it would seem appropriate to distinguish between effects stemming from exposure to supra-maximal temperatures and effects which result from changes in temperature within the range over which growth can occur. In the former situation which corresponds to that used for the heat activation of certain fungal spores there is considerable evidence that the heat effect (or its chemical counterpart) acts at the level of the cell membranes. In addition to the studies already discussed which were specifically concerned with heat activation, various reports have appeared dealing with the effects of supra-maximal temperatures on the ultrastructure of filamentous fungi and these have invariably cited membrane damage as one of the most obvious cytolog-ical features (Anderson & Smith, 1977). While various membrane systems are affected, the mitochondrial membranes appear to be particularly vulner-able. Apparently this finding is not restricted to the filamentous fungi since Brock (1974) has concluded that for living organisms in general, the mitochondrial membrane is especially sensitive to heat and that mitochon-dria disappear when cells are heated a few degrees above their maximum temperature.

It is relevant also to consider briefly the thermophilic fungi which have evolved the ability to grow and differentiate at temperatures which are supra-maximal for the vast majority of fungal species. In a detailed discus-sion on the nature of fungal thermophilism, Crisan (1973) concluded that the ability of the thermophilic fungi to grow at high temperatures could best be explained in terms of the possession of a high degree of 'ultrastructural thermostability'. This hypothesis implied in particular that the thermophilic fungi must have exceptionally thermostable membrane systems. It is reason-able to conclude therefore that at around maximal and at supra-maximal temperatures changes in membrane structure and permeability can be expected to occur and that this may well be involved in the induction of any developmental change which follows a period of exposure to such temperatures.

Developmental changes induced in growing cells by less severe alterations of temperature may be less easy to understand since the effects on active metabolic processes are likely to be particularly complex. Many develop-mental changes in filamentous fungi can be influenced by alteration of the growth temperature. In this review however attention has been drawn to the ability of a wide variety of 'normally' filamentous fungi to grow in a more or less spherical cell form in response to incubation at elevated temperatures. The most intensively studied of these fungi have been those which can exhibit mould-yeast dimorphism. Various differences in chemical composi-tion and in enzymatic activities have been detected between the mould and yeast forms but it is not yet clear whether these are of causal significance in form alteration or whether they simply reflect some aspect of altered metabolism related to the change in growth temperature. It is predictable that many such temperature dependent differences in cell composition will occur. For example, it is now well established that both the quantity and composition of the cellular lipids of filamentous fungi and other organisms can be considerably influenced by growth temperature (Crisan, 1973).

The alternative to invoking a key role for differential synthesis of cell components in the determination of the spherical or filamentous form is the possibility that it is the localization of wall synthesis *per se* which is the dominant factor. During apical growth in filamentous fungi large numbers of membrane bounded vesicles accumulate in the hyphal apices and apparently contribute to growth by fusing with the plasma membrane. It may be that elevated temperatures or other appropriate environmental conditions might in some way influence the formation or the migration or function of the membranous vesicular materials. Considerable ultrastructural organization appears to be involved in the process of apical growth as evidenced by the complex gradient of cytological changes which have been observed in the tips of growing hyphae. Could it be that certain environmental conditions including elevated temperature might affect the establishment or maintenance of this complex ultrastructural organization thus acting by influencing organelle distribution with the consequence that isotropic rather than polarized wall growth occurs?

Clearly, much work remains to be done to determine the actual mechanisms involved in these deceptively 'simple' morphological changes in filamentous fungi. However, with the continuing improvements which are being made in the techniques for both ultrastructural and biochemical studies together with the more widespread use of the chemostat culture technique for the rigorous control of environmental parameters during dimorphic transformations significant results on these problems should be forthcoming in the forseeable future.

18.6 References

ANDERSON, J. G. & SMITH, J. E. (1971). The production of conidiophores and conidia by newly germinated conidia of *Aspergillus niger* (microcycle conidiation). *Journal of General Microbiology* **69**, 185–97.

ANDERSON, J. G. & SMITH, J. E. (1972). The effects of elevated temperatures on spore swelling and germination in *Aspergillus niger*. *Canadian Journal of Microbiology* **18**, 289–97.

ANDERSON, J. G. & SMITH, J. E. (1976). Effects of temperature on filamentous fungi. In *Inhibition and inactivation of vegetative microbes*. Symposium Society for Applied Bacteriology. pp. 191–218, edited by F. A. Skinner and W. G. Hugo. London and New York: Academic Press.

AUSTWICK, P. K. C. (1965). Pathogenicity. In *The genus aspergillus* pp. 82–126, edited by K. B. Raper and D. I. Fenell. Baltimore: Williams and Wilkins.

AUSTWICK, P. K. C. (1966). The role of spores in the allergies and mycoses of man and animals. In *The fungus spore*. pp. 321–37, edited by M. F. Madelin. London: Butterworths.

BARTNICKI-GARCIA, S. (1973). Fundamental aspects of hyphal morphogenesis. *Symposium Society for General Microbiology* **23**, 245–67.

BARTNICKI-GARCIA, S. & LIPPMAN, E. (1969). Fungal morphogenesis: cell wall construction in *Mucor rouxii*. *Science* **165**, 302–4.

BOGUSLAWSKI, G., SCHLESSINGER, D., MEDOFF, G. & KOBAYASHI, G. (1974). Ribonucleic acid polymerases of the yeast phase of *Histoplasma capsulatum*. *Journal of Bacteriology* **118**, 480–5.

BOGUSLAWSKI, G., KOBAYASHI, G. S., SCHLESSINGER, D. & MEDOFF, G. (1975). Characterisation of an inhibitor of ribonucleic acid polymerase from the myelial phase of *Histoplamsma capsulatum*. *Journal of Bacteriology* **122**, 532–7.

BROCK, T. D. (1974). *Biology of microorganisms*. Englewood Cliffs, New Jersey: Prentice-Hall.

BUSSEL, J. & SOMMER, N. F. (1973). Lomasome development in *Rhizopus stolonifer* sporangiospores during anaerobiosis. *Canadian Journal of Microbiology* **19**, 905–7.

CHAPMAN, E. S. & FERGUS, C. L. (1975). Germination of ascospores of *Chaetomium globosum*. *Mycologia* **67**, 1048. 10–48.

CHEUNG, S. C., KOBAYASHI, G. S., SCHLESSINGER, D. & MEDOFF, G. (1974). RNA metabolism during morphogenesis in *Histoplasma capsulatum*. *Journal of General Microbiology* **82**, 301–7.

CORTAT, M. & TURIAN, G. (1974). Conidiation of *Neurospora crassa* in submerged culture without mycelial phase. *Archiv für Mikrobiologie* **95**, 305–9.

COTTER, D. A. (1973). Spore germination in *Dictyostelium discoideum*. 1. The thermodynamics of reversible activation. *Journal of Theoretical Biology* **41**, 41–51.

COTTER, D. A. & GEORGE, R. P. (1975). Germination and mitochondrial damage in spores of *Dictyostelium discoideum* following supra-optimal heating. *Archives of Microbiology* **103**, 163–8.

CRISAN, E. V. (1973). Current concepts of thermophilism and the thermophilic fungi. *Mycologia* **65**, 1171–98.

DURAN, R. & SAFEEULLA, K. M. (1968). Aspects of teliospore germination in some North American smut fungi. *Mycologia* **60**, 231–43.

FERGUS, C. L. & DELWICHE, C. J. (1975). The effects of nutrients and temperature on germination and viability of the ascospores of *Chaetomium rectopilum*. *Mycologia* **67**, 722–32.

GARRISON, R. G., BOYD, K. S. & MARIAT, F. (1975). Ultrastructural studies of the mycelium-to yeast transformation of *Sporothrix schenckii*. *Journal of Bacteriology* **124**, 959–68.

GARRISON, R. G., LANE, J. W. & FIELD, M. F. (1970). Ultrastructural changes during the yeast like to mycelial phase conversion of *Blastomyces dermatitidis* and *Histoplasma capsulatum*. *Journal of Bacteriology* **101**, 628–35.

GILARDI, G. L. (1965). Nutrition of systemic and subcutaneous pathogenic fungi. *Bacteriological Reviews* **29**, 406–24.

GOULD, G. W. & HURST, A. (1969). *The bacterial spore*. New York and London: Academic Press.

GROVE, S. N. (1976). Form and function in zygomycete spores. In *The fungal spore: form and function*, pp. 559–89, edited by D. J. Weber and W. M. Hess. New York and London: John Wiley and Sons.

HALL, M. J. & KOLANKAYA, N. (1974). The physiology of mould-yeast dimorphism in the genus *Mycotypha* (Mucorales). *Journal of General Microbiology* **82**, 25–34.

HANSEN, E. M. & PATTON, R. F. (1975). Types of germination and differentiation of vesicles by basidiospores of *Cronartium ribicola*. *Phytopathology* **65**, 1061–71.

HECKER, L. I. & SUSSMAN, A. S. (1973). Localization of trehalase in the ascospores of *Neurospora*. *Journal of Bacteriology* **115**, 592–9.

HESS, W. M. & WEBER, D. J. (1976). Form and function of basidiomycete spores. In *The fungal spore: form and function*, pp. 645–713, edited by D. J. Weber and W. M. Hess. New York and London: John Wiley and Sons.

HOHL, H. R. (1976). Resistant structures in the Myxomycetes. Part 2: Acrasiomycetes. In *The fungal spore: form and function*, pp. 463–98, edited by D. J. Weber and W. M. Hess. New York and London: John Wiley and Sons.

KANETSUNA, F. & CARBONELL, L. M. (1971). Cell wall composition of the yeastlike and mycelial forms of *Blastomyces dermatitidis*. *Journal of Bacteriology* **106**, 946–8.

KANETSUNA, F., CARBONELL, L. M., AZUMA, I. & YAMURA, Y. (1972). Biochemical studies on the thermal dimorphism of *Paracoccidioides brasiliensis*. *Journal of Bacteriology* **110**, 208–18.

KUBOYE, A. O., ANDERSON, J. G. & SMITH, J. E. (1976). Control and autolysis of a spherical cell form of *Aspergillus niger*. *Transactions of the British Mycological Society* (In Press).

LOVETT, J. S. (1975). Growth and differentiation of the water mold *Blastocladiella emersonii*: cytodifferentiation and the role of ribonucleic acid and protein synthesis. Bacteriological Reviews **39**, 345–404.

MAHESHWARI, R. & SUSSMAN, A. S. (1971). The nature of cold-induced dormancy in urediospores of *Puccinia graminis tritici*. *Plant Physiology* **47**, 389–95.

MALHOTRA, S. K. & TEWARI, J. P. (1973). Molecular alterations in the plasma membrane of sporangiospores of *Phycomyces* related to germination. *Proceedings of the Royal Society of London, Series B* **184**, 207–16.

MANDELS, G. R. & MAGUIRE, A. (1972). Endogenous metabolism of fungal spores. Stimulation by physical and chemical means. *Plant Physiology* **50**, 425–31.

MARIAT, F. (1964). Saprophytic and parasitic morphology of pathogenic fungi. In *Microbial behaviour, 'In vivo and in vitro'*. pp. 85–111, edited by H. Smith and J. Taylor. London: Cambridge University Press.

MATSUMOTO, T. T., BUCKLEY, P. M., SOMMER, N. F. & SHALLA, T. A. (1969). Chilling-induced ultrastructural changes in *Rhizopus stolonifer* sporangiospores. *Phytopathology* **59**, 863–7.

MCDONALD, E. A. & STRANGE, R. N. (1976). Effects of temperature shocks, hydration and leaching on the subsequent germination of uredospores of *Puccinia striiformis*. *Transaction of the British Mycological Society* **66**, 555–7.

MILLS, G. L. & EILERS, F. I. (1973). Factors influencing the germination of basidiospores of *Coprinus radiatus*. *Journal of General Microbiology* **77**, 393–401.

NICKERSON, W. J. & EDWARDS, G. A. (1949). Studies on the physiological bases of morphogenesis in fungi. 1. The respiratory metabolism of dimorphic pathogenic fungi. *Journal of General Physiology* **33**, 41–55.

OUJEZDSKY, K. B., GROVE, S. N. & SZANISZLO, P. J. (1973). Morphological and structural changes during the yeast-to-mold conversion of *Phialophora dermatitidis*. *Journal of Bacteriology* **113**, 468–77.

RIPPON, J., CONWAY, T. P. & DOMES, A. L. (1965). Pathogenic potential of *Aspergillus* and *Penicillium* species. *Journal of Infectious Diseases* **115**, 27–32.

ROMANO, A. H. (1966). Dimorphism, In *The fungi*, Vol. 2. pp. 181–209, edited by G. C. Ainsworth, and A. S. Sussman. New York and London: Academic Press.

RUDOLPH, H. & FURCH, B. (1970). Untersuchungen zur Aktivierung von Sporenhomogenaten durch Warmebehandlung. *Archiv für Mikrobiologie* **72**, 175–81.

RUDOLPH, H. & OCHSEN, B. (1969). Trehalose—Umsatz wärmeaktivierter Sporen von *Phycomyces blakesleeanus*. *Archiv für Mikrobiologie* **65**, 163–71.

SAN-BLAS, G. & CARBONELL, L. M. (1974). Chemical and ultrastructural studies on the cell walls of the yeastlike and mycelial forms of *Histoplasma farciminosum*. *Journal of Bacteriology* **119**, 602–11.

SAN-BLAS, F., SAN-BLAS, G. & COVA, L. J. (1976). A morphological mutant of *Paracoccidioides brasiliensis* strain IVIC Pb9. Isolation and wall characterization. *Journal of General Microbiology* **93**, 209–18.

SCHULZ, B. E., KRAEPELIN, G. & HINKELMANN, W. (1974). Factors affecting dimorphism in *Mycotypha* (Mucorales): correlation with the fermentation/respiration equilibrium. *Journal of General Microbiology* **82**, 1–13.

SEKIGUCHI, J., GAUCHER, G. M. & COSTERTON, J. W. (1975a). Microcycle conidiation in *Penicillium urticae*: an ultrastructural investigation of spherical spore growth. *Canadian Journal of Microbiology* **21**, 2048–58.

SEKIGUCHI, J., GAUCHER, G. M. & COSTERTON, J. W. (1975b). Microcycle conidiation in *Penicillium urticae*: an ultrastructural investigation of conidial germination and outgrowth. *Canadian Journal of Microbiology* **21**, 2059–68.

SEKIGUCHI, J., GAUCHER, G. M. & COSTERTON, J. W. (1975c). Microcycle conidiation in *Penicillium urticae*: an ultrastructural investigation of con-idiogenesis. *Canadian Journal of Microbiology* **21**, 2069–83.

SINGER, S. J. (1973). Molecular organization of membranes. *Neurosciences Research Progress Bulletin* **11**, 9–12.

SMITH, J. E., GULL, K., ANDERSON, J. G. & DEANS, S. G. (1976). Organelle changes during fungal spore germination. In *The fungal spore: form and function*, pp. 301–52, edited by D. J. Weber and W. M. Hess. New York and London: John Wiley and Sons.

STORCK, R. & MORRILL, R. C. (1971). Respiratory-deficient, yeast-like mutants of *Mucor*. *Biochemical Genetics* **5**, 467–79.

SUSSMAN, A. S. (1966a). Dormancy and spore germination. In *The fungi*, Vol. 2, pp. 733–64, edited by G. C. Ainsworth and A. S. Sussman. New York and London: Academic Press.

SUSSMAN, A. S. (1966b). Types of dormancy as represented by conidia and ascospores of *Neurospora*. In *The fungus spore*. pp. 235–56, edited by M. F. Madelin. London: Butterworths.

SUSSMAN, A. S. (1976). Activators of fungal spore germination. In *The fungal spore: form and function*, pp. 101–39, edited by D. J. Weber and W. M. Hess. New York and London: John Wiley and Sons.

SUSSMAN, A. S. & DOUTHIT, H. A. (1973). Dormancy in microbial spores. *Annual Review of Plant Physiology* **24**, 311–52.

SUSSMAN, A. S. & HALVORSON, H. O. (1966). *Spores: their dormancy and germination*. New York and London: Harper and Row.

TU, J. C. & MALHOTRA, S. K. (1973). Histochemical localization of adenyl cyclase in the fungus *Phycomyces blakesleeanus*. *Journal of Histochemistry and Cytochemistry* **21**, 1041–6.

TURIAN, G. & BIANCHI, D. E. (1972). Conidiation in *Neurospora*. *Botanical Reviews* **38**, 119–54.

VAN ASSCHE, J. A., CARLIER, A. R. & DEKEERSMAEKER, H. H. (1972). Trehalase activity in dormant and activated spores of *Phycomyces blakesleeanus*. *Planta* **103**, 327–33.

YATES, A. R., SEAMAN, A. & WOODBINE, M. (1968). Ascospore germination in *Byssochlamys nivea*. *Canadian Journal of Microbiology* **14**, 319–25.

YU, S. A., SUSSMAN, A. S. & WOOLEY, S. (1967). Mechanisms of protection of trehalase against heat-inactivation in *Neurospora*. *Journal of Bacteriology* **94**, 1306–12.

ZORZOPULOS, J., JOBBAGY, A. J. & TERENZI, H. F. (1973). Effects of ethylenediaminetetraacetate and chloramphenicol on mitochondrial activity and morphogenesis in *Mucor rouxii*. *Journal of Bacteriology* **115**, 1198–204.

CHAPTER 19

Circadian Rhythms

G. LYSEK

19.1 Introduction

'The molecular basis of circadian rhythms has remained almost a complete mystery' (Feldman & Hoyle, 1974).

This statement, initiating a previously published paper, indicates that in fungi, as in other organisms, the knowledge of rhythms is still limited. Nevertheless, fungi are suitable organisms for experimental studies. In addition to their known advantages their rhythms exhibit a large variety with respect to periods, activities and underlying mechanisms. A unique feature of the fungi are rhythmic mutants, *i.e.* mutants, which exhibit spontaneous rhythms as a consequence of their mutation. From these mutants it is hoped to obtain information about the mechanism of rhythmicity and of circadian rhythms at the cellular and molecular level. The initial statement, however, demonstrates that these expectations have not yet been fulfilled, although fungi have contributed considerably to the actual knowledge of circadian rhythms.

19.2 The Types of Rhythms in Fungi

The rhythmic activities of filamentous fungi include all stages between vegetative mycelial growth and the discharge of mature spores. The following examples illustrate the variety: formation of rings (bands, zonations) by periodic alterations of mycelial growth pattern or sporulation; periodic spore discharge; alternating bioluminescence; oscillations in the evolution of CO_2, in the content of ATP or NADH, and in the synthesis of DNA; and mitotic rhythms. Glycolytic oscillations and rhythmic cAMP release, have not been observed in filamentous fungi.

The rhythms enumerated are either endogenous, *i.e.* created by the fungus itself, or are due to environmental cues. The periods depend mainly on the mode of generation: exogenously caused rhythms follow the alterations of the environment (light, temperature, humidity) and thus exhibit periods of exactly 24 h (diurnal rhythms), while spontaneous oscillations usually have periods which are not synchronized to exogenous factors.

In a few fungi, however, periods of about 24 h are found in constant darkness (DD) and at constant temperature. These rhythms represent the type most common in other organisms and are called *circadian*. In accordance to the usual terminology circadian rhythms are defined by a free running period of about 24 h (circa = about; dies = day) in constant darkness, and by a low Q_{10} of about 1, which means that the period does not alter with the temperature (Bünning, 1973). Circadian periods do not depend on the environmental cycles, but are entrained by light or temperature alterations, *i.e.* their phases are synchronized by the environmental cues.

These circadian rhythms are widespread in nature because of their ecological value. In fungi, however, noncircadian rhythms are dominant. Sweeney (1969) argues that fungal rhythms may represent the 'raw material' from which circadian oscillations have been selected during evolution. Thus, noncircadian fungal rhythms may help us to understand circadian rhythms and hence are included in this chapter.

19.3 Morphological Rhythms

Periodic changes of growth type and fructification are the most common type of fungal rhythms. During the development of the colony the density or the form of the mycelia and/or the number of reproductive organs vary periodically. The resulting concentric rings have been characterized as zonations, rings, or bands. In nature they are represented by the rings of conidia on *Monilia*-infected apples or by the fairy rings of mushrooms and toadstools. In cultures zonations were first noted by Werner (1898). To demonstrate their relation to the morphogenesis of the fungal colony, the term 'morphological rhythms' was introduced by Feldman (see Russel, 1972). Since other periodic activities in fungi may be traced to these morphological rhythms, they are discussed in more detail.

Rhythmic mutants

The investigation of morphological rhythms was stimulated by the isolation of mutants which grew rhythmically as a consequence of the mutation. Such strains are now known in some laboratory fungi and have been studied frequently, as demonstrated in Table 19.1.

The value of these strains depends on the creation by those mutants of a system consisting of an organism exhibiting rhythms (the mutant) and a non-rhythmic control (the wild strain). Since these mutants are mainly monogenic their differences are due to only one primary effect. In addition, in these strains the 'clock' itself is likely to be affected, while in non-rhythmic mutants of organisms exhibiting rhythms, any event between the 'master clock' and its manifestation may be affected. Therefore, these mutants facilitate the monitoring of the alterations joint to the expression of oscillations, and thus have contributed considerably to the actual understanding of fungal rhythms.

Non-circadian branching patterns

In the most elementary form the rhythmic growing colonies produce repeated zones of highly branched mycelia (sympodial or clock-type growth). This pattern is known from the *clock* mutants of *Ascobolus immersus* and *Neurospora crassa* (Table 19.1), but also from wild types as

Table 19.1 Mutants used in the study of fungal rhythms. (The key references give publications concerned with the rhythmicity and not the first citation of the mutants)

Fungus	Mutant	Type of rhythm	Key reference
Ascobolus immersus	clock	non-circadian branching rhythm	Berliner, Neurath & Yankovich (1965)
	wave (vague)	non-circadian formation of aerial hyphae	Yu-Sun (1964)
Neurospora crassa	acetate non-utilizing	non-circadian conidiation	Halaban & Feldman (1973)
	band	circadian formation of aerial hyphae circadian conidiation	Bianchi (1964) Sargent & Woodward (1969)
	clock	non-circadian branching rhythm	Sussman et al. (1964)
	frequency	alterations of circadian conidiation	Feldman & Hoyle (1973)
	patch (pro-, 21863)	diurnal formation of aerial hyphae	Pittendirgh, Bruce, Rosensweig & Rubin (1959)
	timex (=band/inv⁻)	diurnal conidiation	Sargent & Woodward (1969)
Neurospora tetrasperma	5–3M	diurnal conidiation	Nysterakis (1972)
Penicillium claviforme album		diurnal formation of coremia	Faraj-Salman (1970)
	olivicolor	diurnal formation of coremia	Faraj-Salman (1970)
Podospora anserina	circulosa	non-circadian formation of aerial hyphae	Nguyen Van (1967)
	undulata	non-circadian branching rhythm, clock-type	Nguyen Van (1967)
	zonata	non-circadian branching rhythm, clock-type	Lysek (1972)

Pilaira anomala (Fletcher, 1970). It was studied in *Neurospora crassa* by Sussman, Lowry & Durkee (1964), in *Ascobolus immersus* by Chevaugeon (1959), and in *Podospora anserina* by Nguyen van (1967). Lysek (1972, 1974) proposed the following mechanism for the morphogenesis of the clock-type mutant *zonata* of *Podospora anserina* (Fig. 19.1).

After inoculation, the hyphae grow radially on and into the solid medium. At the surface the hyphae have a declining rate of elongation since the internodes are increasingly reduced. Thus the number of branches increases in relation to the distance covered. The hyphae in the substrate, however, elongate uniformly at a constant rate. They outgrow the superficial mycelium and reach the surface distal to the superficial front and check its further radial elongation. The staled mycelia form a zonation by continued branching. The hyphae, which have emerged and become surface hyphae, now themselves elongate at a declining rate and finally are replaced by new

Fig. 19.1 Diagram of a cross section of a Petri dish showing the morphogenesis of the zonations in the mutant *zonata* of *Podospora anserina* (Adapted from Lysek, 1972, 1974).

substrate mycelia. As is seen in Fig. 19.1, this replacement of the superficial mycelia is repeated during the entire growth of the colony.

The differences between the surface and the substrate hyphae of this mutant depend on their sensitivity to oxygen. At normal oxygen tensions the hyphae grow at a declining rate forming increasing numbers of branches (sympodial growth). At low oxygen tensions, such as inside the medium, the hyphae elongate at a constant rate (see p. 383).

According to Fig. 19.1 the zonations are found merely at the surface and depend on a continuous growth rate of the substrate mycelium. Both features are reported from rhythmically growing fungi. It also results from the model that the periods are reset by inoculation as found by Berliner & Neurath (1965). The periods depend on the relation between the elongation of the superficial and the substrate hyphae and hence vary with respect to the experimental conditions and to internal factors. This is seen in the case of desynchronization of the bands by interrupting the lateral (tangential) contact of the hyphal branches. As is seen in Fig. 19.2, this interruption

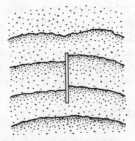

Fig. 19.2 Desynchronization of a clock-type growth rhythm by preventing hyphal contact (From Nguyen Van, 1967).

leads to differences in the size of the bands and thus to different periods in the adjacent but separated parts of a single zonation. This type of synchronization by hyphal contact (self synchronization) depends on the synchronization of the metabolism similar to oscillating yeast (Kraepelin & Franke 1973).

Effect of light and temperature cycles

From the model described in Fig. 19.1 it follows that any effect may cause growth bands, if it checks part of the mycelial front and permits the rest to advance sufficiently in front of the retarded hyphaes. Environmental cues of this kind are light and temperature. They affect the superficial hyphae more than those inside the medium, which are less or later exposed. Again the resulting retardation is associated with an enhanced consumption of oxygen, which is available only at the surface. Thus the check is mainly found in the superficial region.

The effects of light and temperature were studied by Sagromsky (1952). Using various fungi she found that the onset of illumination or the decline of temperature affected the hyphal tips. Light between 390 and 530 nm was active in producing bands. The reaction was very sensitive, an illumination of 6 lx following darkness or a temperature decline of 1°C were sufficient to produce a zonation in *Penicillium* sp. Her explanation that these temperature elicited rings were caused 'by different growth rates of the mycelia' agrees with the model given in Fig. 19.1.

It may be added that Molz (1907) had already proposed a similar model which was based on a role of the substrate hyphae. At the elevated temperatures during illumination, the gelatine (which he used instead of agar) melted and the hyphae submerged into the medium. At low temperatures (in the night) the gelatine solidified and the hyphae grew to the surface. Thus one ring per temperature or light cycle was formed.

Post-zonation development

With the action of light and/or temperature the zonations are not yet completed. Once a zonation is initiated there are checked and still elongating hyphae close together. The first hyphae have terminated their vegetative growth and may differentiate, forming pigments, aerial hyphae, branches, sporangiophores, or fruiting bodies. This post-zonation development alters the original pattern of resting and elongating hyphae: aerial mycelia may cover the line where the substrate or the unchecked hyphae replace the retarded hyphae. On the other hand, bands may become more distinct or even visible in the derivative form of the distribution of propagative organs. The resulting close relation between mycelial zonations and perithecial bands was shown by Lysek (1974) with *Podospora anserina* and other fungi (Fig. 19.3). In *Neurospora crassa* the bands of conidiophores, which are formed by aerial hyphae, are found at the distal regions of the rings.

The post-zonation development, the alteration of initiated bands by mycelial differentiation, indicates a possibility to trace all types of fungal morphological rhythms to a common concept. This would include the fairy rings of the Agaricales. Although their annual periodicity is obscured by the long intervals of the basidiocarp production, it is recognized by periodical recording (Ingold, 1974).

The diurnal periodicity of fruiting may also be involved in rhythmic spore discharge. According to Hodgkiss & Harvey (1971) in the Pyrenomycetes these rhythms are linked closely to the alterations of light and temperature, and thus may follow the diurnal fruiting with a delay due to the maturation of the spores. Simultaneously, the periodicity of atmospheric humidity may

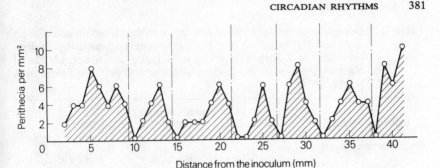

Fig. 19.3 Identity of mycelial and perithecial bands in *Pestalotia rhododendri*. ⧄ Number of perithecia; ——·—— edge of zonations. (From Lysek, 1974; Lysek & von Witsch, 1974*a*).

also be involved. This is interpreted as an ecological adaptation of phytopathogenic fungi, the germination of which is favoured by humidity.

Circadian rhythms

Rhythms of this type are rarely found in fungi. Some of the 24 h-oscillations have a 'circadian component' and mediate between circadian and non-circadian rhythms. Jerebzoff (1965) reported light dependent or endogenously elicited rhythms with periods of about 24 h, which could be manipulated by chemicals. Burnett, Lincoln & Carpenter (1969) described a clock-type rhythm in *Penicillium diversum* with periods of 24 h, which, however, was not affected by light at all and thus did not fit the criteria of circadian rhythms (see p. 377). An 'endogenous diurnal rhythm' was found by Uebelmesser (1954) and by Bruce, Weight & Pittendrigh (1960) with spore discharge of *Pilobolus sphaerosporus*. The observed rhythm had periods of about 24 h, was entrained by light or temperature cycles, continued in DD, and had a low Q_{10}. It was thus regarded as genuine circadian.

The best studied circadian rhythm in fungi, however, is exhibited by the mutant *band* (*bd*) of *Neurospora crassa*. It originates from the double mutant *timex*, which additionally is invertase lacking (*inv⁻*) (Table 19.1). The strain *band* shows a banding of conidiophores which is circadian according to the usual criteria: it is entrained by light-dark-cycles (LD), continues in permanent darkness (DD) and damps out in continuous light (LL) (Table 19.2). As in the case of the diurnal rhythms, blue light is active, and the photoreceptor is of the flavin type (Muñoz & Butler, 1975). Temperature does not markedly alter the period of conidiation, and the Q_{10} is near 1. It was shown by Sargent and Kaltenborn (1972) that on suitable media *band*-like circadian conidiation is induced by aeration in various wild strains of *Neurospora crassa*. This allows the conclusion by these authors that in this fungus 'physiological and developmental regulation has a circadian component'. Bünning (1973), however, doubts the genuine circadian character of the conidiation of *band*. Unfortunately, evidence from other fungi is still lacking.

One may ask whether these circadian oscillations may be correlated to other morphological rhythms. Since the conidiophores are produced by the

Table 19.2 Dependence on light and temperature of the circadian conidiation (induced by aeration) and the non-circadian branching rhythm (induced by L-sorbose) of the wild types 65-1A and 65-20a of *Neurospora crassa*. For abbreviations see text (From Feldman & Hoyle, 1974).

Light conditions	Period length (h)		
	20°C	25°C	30°C
Circadian conidiation			
DD	22.0 ± 0.5	21.0 ± 1.6	21.1 ± 2.6
LD 12 : 12	23.9 ± 0.3	24.0 ± 0.3	23.9 ± 0.4
LL	no rhythm	no rhythm	no rhythm
Non-circadian hyphal branching			
DD	93.3 ± 4.6	32.6 ± 6.6	21.6 ± 2.2
LD 12 : 12	85.8 ± 13.8	34.0 ± 8.7	18.1 ± 1.6
LL	97.3 ± 4.4	28.0 ± 2.7	18.6 ± 2.0

underlying mycelium their circadian formation depends on synchronous rhythms in mycelial growth or development. The correlation between both types of rhythms is confirmed also by the formation of mycelial rings in *band* when conidiation does not occur (Bianchi 1964; Sargent & Kaltenborn, 1972), and by the constant growth rate during circadian conidiation (Woodward & Sargent, 1973). The strict dependence on the nutrient supply and on aeration also allows linkage to mycelial rhythms (see 19.4). Both types differ by their sensitivity against environmental factors, as shown in Table 19.2. This, however, does not imply independent basic mechanisms, as argued by Feldman & Hoyle (1974). In terms of the established model, the circadian periods may be due to a constant ratio of the growth rates of the different mycelia of the colony, resulting in the formation of one band per day. The low Q_{10} may result from the fact that this ratio is not altered when temperature induced changes in elongation occur. This one may assume since the circadian conidiation is not affected by different growth rates on different carbon sources (Sargent & Kaltenborn, 1972).

19.4 Metabolic Alterations

The study of different periodic activities and of rhythmic mutants has also improved the investigation of the endogenous background. The available results give some insight into the metabolic alterations in fungal circadian rhythms.

Carbohydrate metabolism

First experiments about the alterations leading to fungal rhythms demonstrated an influence of non-metabolizable hexoses. L-sorbose caused rhythmic branching in *Neurospora crassa* (Sussman *et al.*, 1964) and in *Podospora anserina* D-galactose and L-sorbose induced clock-type zonations (Lysek & Esser, 1971). D-galactose is known to interfere with the turnover of the activated glucose (UDPG) and thus with the synthesis of polysaccharides while De Terra & Tatum (1961) found that L-sorbose caused inhibition of hyphal wall synthesis. Simultaneously, these hexoses caused an enhancement of carbohydrate catabolism. It was concluded that during clock-type growth catabolism is increased while the synthesis of hexose polymers is reduced (Fig. 19.4) (Lysek, 1971). The inhibitory delay of the increased catabolism prevented the rhythmic growth of the clock-type mutant *zonata* of *P. anserina*. An analysis of the phosphorylated intermediates confirmed these findings, but also demonstrated an involvement of the lipids (see p. 385).

Respiration

The increased catabolism of carbohydrates also effects respiration. A correlation between branching rhythms and respiration was found by Lysek (1972), who prevented zoned growth by keeping air off the agar surface or by inhibiting respiration. Crocken & Tatum (1968) observed an increased oxygen uptake by colonies of *Neurospora crassa* growing on L-sorbose, *i.e.* when exhibiting rhythms. Woodward & Sargent (1973) reported a periodic enhancement of CO_2 production in *Neurospora* synchronous to circadian conidiation. The dependence on aeration of the circadian conidiation also underlines the role of respiration.

These results indicate an enhanced respiratory activity coupled to fungal rhythms. Since carbohydrate turnover and respiration depend on substrate supply, the dependence on the available nutrients is a common feature of fungal rhythms (Jerebzoff, Jerebzoff-Quintin & Lambert, 1974; Sargent & Kaltenborn, 1972).

Energy turnover

The connecting link between accelerated carbohydrate catabolism and respiration is regarded as being the energy turnover. Inhibiting respiratory electron transport abolished the clock-type rhythm of the mutant *zonata*, while uncoupling of oxidative phosphorylation elicited mycelial bands (Lysek 1971). In *Neurospora crassa band* Delmer & Brody (1975) found circadian oscillations in the ATP dependent energy charge with the minimum occurring during conidiation. These were interpreted by the authors as due to a circadian uncoupling'. This uncoupling and the resulting lack of available ATP may be balanced by the increased turnover of glycolysis, tricarboxylic acid cycle, and respiration (Fig. 19.4), and thus may account for the altered nutrient consumption and the CO_2 production previously described. Thus the reduced hyphal elongation and dry weight production of rhythmically growing fungi result from a shift from synthetic to degradative metabolic pathways (Lysek & Esser, 1971).

Furthermore, energy supply and rhythmic activity are correlated with the (non-circadian) bioluminescence oscillations observed by Bühler & Bünning (1965) in *Omphalia flavida*. The correlation of (circadian) rhythms and energy supply is also found in other organisms; Wagner *et al.* (1975) characterized this linkage as one of the main features of the biological clock.

Synthetic reactions

The circadian conidiation and other forms of rhythmic fructification depend on the synchronous generation of precursors and of energy. The

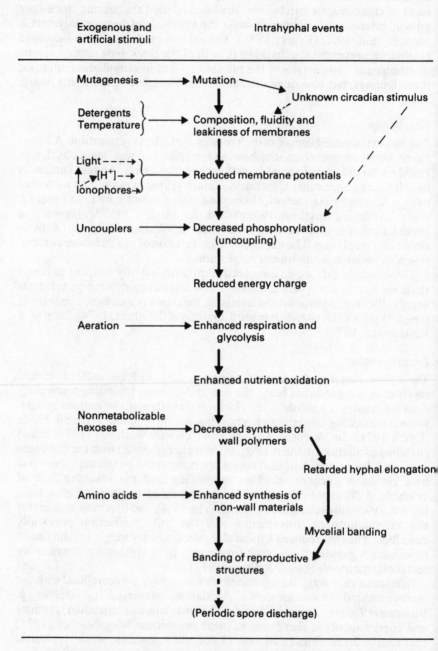

Exogenous and artificial stimuli | Intrahyphal events

Mutagenesis ⟶ Mutation

Unknown circadian stimulus

Detergents
Temperature } ⟶ Composition, fluidity and
leakiness of membranes

Light - - - →
↑
▼[H⁺] - →
Ionophores→ } ⟶ Reduced membrane potentials

Uncouplers ⟶ Decreased phosphorylation
(uncoupling)

Reduced energy charge

Aeration ⟶ Enhanced respiration and
glycolysis

Enhanced nutrient oxidation

Nonmetabolizable
hexoses ⟶ Decreased synthesis of
wall polymers

Retarded hyphal elongation

Amino acids ⟶ Enhanced synthesis of
non-wall materials

Mycelial banding

Banding of reproductive
structures

(Periodic spore discharge)

Fig. 19.4 Scheme summarizing the alterations leading to clock-type zonations and to banding of reproductive organs in the oscillating regions of a fungal colony (Adapted from Lysek, 1974; Lysek & von Witsch, 1974a).

'circadian uncoupling' thus may be synergistic to the ATP requirement of synthetic pathways. Therefore, the accelerated catabolism is involved in all types of post-zonation development. The difference between mere mycelial and other rhythms may be, whether the generated energy is wasted mainly by uncoupling as in *zonata* (a mutant not forming mature perithecia or conidia), or is used predominantly for aspects of differentiation. In the latter case the intermediates of the degradative pathways may also be used as metabolic precursors.

The involvement of anabolic metabolism and its intermediates in fungal rhythms has been found by several investigators. The pyrimidine nucleotides exhibit differences in the rings formed by the mutant *band*. In the conidiating area NADH, NADPH, and NADP show reduced concentrations, while NAD is enhanced as compared to the nonconidiating region of a newly formed band (Brody & Harris, 1973).

The effect of amino acids on fungal rhythms was studied by Jerebzoff (1965). In several fungi, morphological rhythms were induced, enhanced or altered by the application of amino acids. A specific role was ascribed to arginine, isoleucine and β-methyl-valeric acid. This latter substance exhibited a rhythm generating activity in *Aspergillus niger* (Jerebzoff & Hortala, 1974).

The synthesis of nucleic acids is also involved in circadian conidiation. Accordingly, it was found to be coupled to a synchronously alternating ^{3}H-uridine incorporation and to oscillating levels of RNA and DNA (Martens & Sargent, 1974). The key role of these substances in the conidiation of *Neurospora crassa* was demonstrated by Turian & Bianchi (1972). These periodic alterations in the nucleic acid content leads to the mitotic rhythms found by Valla (1974) in the hyphal tips of *Polyporus arcularius*.

Hyphal membranes

While the molecular mechanisms of the biological clock are still obscure, experimental evidence points to a possible involvement of the hyphal membranes. Membrane involvement in fungal rhythms has been inferred from the dependence of rhythms on external pH. Low pH values favour rhythmic growth and/or sporulation. The pH is thought to interfere with the proton potentials at the hyphal membranes (Fig. 19.4). The generation of rhythms by light presents further evidence for it is considered to act upon the hyphal membranes or via substances located in the membranes, like flavins or carotinoids. Alterations of the temperature may alter the fluidity of the membranes and thus induce zonations. The influence of K^+-Na^+-equilibria in the isoleucine elicited bands of *Aspergillus niger* also favour a role of membranes (Jerebzoff, Jerebzoff-Quintin & Nysterakis, 1970).

Direct evidence was found by an analysis of the phospholipids of the mutant *zonata* (Lysek, 1976). It revealed a markedly reduced synthesis of phospholipids and a correlated accumulation of phosphatidic acid indicating a block in the biosynthesis of phospholipids. The resulting damage of membrane structure is regarded as the primary cause of the rhythmic growth pattern.

A link between membrane disruption and rhythms has been demonstrated by Faraj-Salman (1971). He induced light dependent rhythms by application of alcohols. These and detergents elicited growth rhythms in the uniformly growing wild strain of *Podospora anserina* (Lysek & von Witsch 1974*b*). These substances partially solubilized the membrane lipids, which allowed increased fluxes of ions across the membranes and thus affected the transmembrane potentials—a condition known to occur at low pH values. These conclusions were supported by the effect of ionophorous antibiotics: nigericin, monensin and nonactin induced light dependent rhythms in the wild strain of *Podospora anserina* (Lysek & von Witsch, 1974*b*).

This offers a possible understanding of membrane activity. The increased fluxes of ions mediated by the ionophores balance the charges on both sides of the hyphal membranes and hence affect their potential. Among the activities dependent on these potential is phosphorylation, which is based on proton gradients (Mitchell, 1974). Thus the involvement of metabolic energy in fungal rhythms also includes a central role of the membranes. These relations are summarized in Fig. 19.4.

This scheme (Fig. 19.4) results mainly from experiments with the mutant *zonata* but may be extended to circadian rhythms. The described alterations of metabolism or metabolites fit into this scheme, suggesting an involvement of the hyphal membranes also in circadian conidiation. Although experimental evidence is still lacking, a role for hyphal membranes in the circadian oscillations of the mutant *band* has been postulated by Woodward & Sargent (1973), and by Halaban (1975). This corresponds to other organisms in which rhythms are assumed to be coupled to the biomembranes (Sweeney, 1969; Bünning, 1973). The site of the generation of the circadian alterations, however, is still obscure.

19.5 Conclusions

As shown, it is possible to summarize the present knowledge about fungal circadian and non-circadian rhythms. It is also seen that most of the experimental evidence is congruent to one another and to results from organisms other than filamentous fungi.

However, the total evidence is still poor. Beside mere descriptions, detailed studies of fungal rhythms have only been carried out with Ascomycetes. The two species studied exhaustively using rhythmic mutants, *Neurospora crassa* and *Podospora anserina*, belong to one family, *Sordariaceae*. Apart from these, only the *Aspergillaceae* have contributed more than sporadic evidence. From certain groups of fungi no rhythms have been reported, *e.g.* Oömycetales, Phragmobasidiomycetes; while with some others a few types have been studied, *e.g.* Zygomycetes (*Pilaira*), and Holobasidiomycetes (*Agrocybe aegerita, Marasmius oreades*). Thus, the actual knowledge derives from too few fungi.

The same holds true regarding rhythmic activities. Investigations have been made mainly with the morphological rhythms, using the advantages of the mutants—of which, however, not even the primary mutational effect is known. Other types like spore discharge, bioluminescence, or mitotic rhythms are described, while there is no evidence about the physiological background. Therefore, an increasing knowledge of the rhythms of different types and from an increasing number of fungal species has to be achieved to eliminate the gaps in our present understanding.

19.6 References

BERLINER, M. D. & NEURATH, P. W. (1965). The band forming rhythm of *Neurospora* mutants. *Journal of Cellular and Comparative Physiology* **65**, 183–93.

BERLINER, M. D., NEURATH, P. W. & YANKOVICH, B. M. (1965). The rhythms of *Ascobolus* clock mutants. *American Journal of Botany* **52**, 635–6.

BIANCHI, D. E. (1964). An endogenous circadian rhythm in *Neurospora*. *Journal of General Microbiology* **35**, 437–45.

BRODY, S. & HARRIS, S. (1973). Circadian rhythms in *Neurospora*: spatial differences in pyrimidine nucleotide levels. *Science* **180**, 498–500.

BRUCE, V. G., WEIGHT, F. & PITTENDRIGH, C. S. (1960). Resetting the sporulation rhythm in *Pilobolus* with short light flashes of high intensity. *Science* **131**, 728–30.

BUHLER, A. & BÜNNING, E. (1965). Periodisches Leuchten bei *Omphalia flavida*. *Archiv für Mikrobiologie* **52**, 80–2.

BÜNNING, E. (1973). *The physiological clock—circadian rhythms and biological chronometry*. 3rd Edition. Heidelberg: Springer-Verlag.

BURNETT, J. A., LINCOLN, R. G. & CARPENTER, B. H. (1969). Fungal endogenous rhythms expressed by spiral figures. *Science* **166**, 763–4.

CHEVAUGEON, J. (1959). Sur le déterminisme interne du rythme de croissance chez un mutant 'vague' de l'*Ascobolus immersus*. *Comptes Rendus de l'Académie des Sciences (Paris)* **248**, 1841–4.

CROCKEN, B. & TATUM, E. L. (1968). The effect of sorbose on metabolism and morphology of *Neurospora*. *Biochimica et Biophysica Acta* **156**, 1–8.

DELMER, D. P. & BRODY, S. (1975). Circadian rhythms in *Neurospora crassa*: oscillation in the level of adenine nucleotide. *Journal of Bacteriology* **121**, 548–53.

DE TERRA, N. & TATUM, E. L. (1961). Colonial growth of *Neurospora*. *Science* **134**, 1066–8.

FARAJ-SALMAN, A. G. (1970). Induktion einer endogenen Rhythmik der Koremienbildung durch Alkohol bei einer Mutante von *Penicillium claviforme* Bainer und einer Varietät davon. *Biochemie und Physiologie der Pflanzen* **161**, 42–9.

FARAJ-SALMAN, A. G. (1971). Zur Induktion einer endogenen Rhythmik bei Mutanten des Pilzes *Penicillium claviforme* Bainer. I. Wirkungsweise von Alkoholen. *Archiv für Protistenkunde* **113**, 306–13.

FELDMAN, J. F. & HOYLE, M. N. (1973). Isolation of circadian clock mutants of *Neurospora crassa*. *Genetics* **75**, 605–13.

FELDMAN, J. F. & HOYLE, M. N. (1974). A direct comparison between circadian and noncircadian rhythms in *Neurospora crassa*. *Plant Physiology* **53**, 928–30.

FLETCHER, H. J. (1970). Some effects of light on the vegetative growth of *Pilaira anomala*. *Journal of General Microbiology* **60**, 281–2.

HALABAN, R. (1975). Glucose transport-deficient mutant of *Neurospora crassa* with an unusual rhythmic growth pattern. *Journal of Bacteriology* **121**, 1056–63.

HALABAN, R. & FELDMAN, J. F. (1973). Circadian periodicity in acetate non-utilizing mutants. *Neurospora Newsletter* **20**, 20.

HODGKISS, I. J. & HARVEY, R. (1971). Effects of temperature on spore discharge rhythms in pyrenomycetes. *Transactions of the British Mycological Society* **56**, 225–34.

INGOLD, C. T. (1974). Growth and death of a fairy ring. *Bulletin of the British Mycological Society* **8**, 74–5.

JEREBZOFF, S. (1965). Manipulation of some oscillating systems in fungi by chemicals. In *Circadian clocks*, edited by J. Aschoff, pp. 183–189. Amsterdam: North Holland.

JEREBZOFF, S. & HORTALA, M.-T. (1974). Importance de la chaine carbonée de l'isoleucine dans le déclenchement du rythme endogène de zonation chez *Aspergillus niger*. *Comptes Rendus de l'Academie des Sciences (Paris)* **278**, 1479–82.

JEREBZOFF, S., JEREBZOFF–QUINTIN, S. & LAMBERT, E. (1974). *Aspergillus niger*: characteristics of endogenous medium and low frequency rhythms. *International Journal of Chronobiology* **2**, 131–44.

JEREBZOFF, S., JEREBZOFF-QUINTIN, S. & NYSTERAKIS, M. F. (1970). Rôles de l'équilibre K^+-Na^+ et de l'isoleucine dans le déclenchement du rythme interne de zonation chez *Aspergillus niger*. *Comptes Rendus de l'Academie des Sciences (Paris)* **271**, 1522–5.

KRAEPELIN, G. & FRANCKE, G. (1973). Self-synchronization in yeast and other fungi. *International Journal of Chronobiology* **1**, 163–72.

LYSEK, G. (1971). Rhythmic mycelial growth in *Podospora anserina*. III. Effect of metabolic inhibitors. *Archiv für Mikrobiologie* **78**, 330–40.

LYSEK, G. (1972). Rhythmic mycelial growth in *Podospora anserina*. VI. An attempt to elucidate the growth pattern of a clock mutant. *Archiv für Mikrobiologie* **87**, 129–37.

LYSEK, G. (1974). Zonierungen und Hexenringe, Morphologische Differenzierungen bei Pilzen. *Naturwissenschaftliche Rundschau* **27**, 449–55.

LYSEK, G. (1976). Alterations of the phospholipids in a rhythmically growing mutant of *Podospora anserina*. *Biochemie und Physiologie der Pflanzen* **169**, 207–12.

LYSEK, G. & ESSER, K. (1971). Rhythmic mycelial growth in *Podospora anserina* (Ascomyc.). II. Evidence for a correlation with carbohydrate metabolism. *Archiv für Mikrobiologie* **75**, 360–71.

LYSEK, G. & VON WITSCH, H. (1974a). Lichtabhängige Zonierungen bei Pilzen und ihre physiologischen Grundlagen. *Berichte der Deutschen Botanischen Gesellschaft* **87**, 207–13.

LYSEK, G. & VON WITSCH, H. (1974b). Rhythmisches Mycelwachstum bei *Podospora anserina*. VII. Der Einfluß oberflächenaktiver Substanzen und Antibiotika im Dunkeln und im Licht. *Archives of Microbiology* **97**, 227–37.

MARTENS, C. L. & SARGENT, M. L. (1974). Circadian rhythms of nucleic acid metabolism in *Neurospora crassa*. *Journal of Bacteriology* **117**, 1210–15.

MITCHELL, P. (1974). A chemiosmotic molecular mechanism for proton-translocating adenosine triphosphatases. *FEBS-Letters* **43**, 189–94.

MOLZ, E. (1907). Über die Bedingungen der Entstehung der durch *Sclerotinia fructigena* erzeugten 'Schwarzfäule' der Äpfel. *Zentralblatt für Bakteriologie, 2. Abteilung* **27**, 175–88.

MUÑOZ, V. & BUTLER, W. L. (1975). Photoreceptor pigment for blue light in *Neurospora crassa*. *Plant Physiology* **55**, 421–6.

NGUYEN VAN, H. (1967). *Étude de rythmes internes de croissance chez le Podospora anserina*. Thèses. Paris, Masson.

NYSTERAKIS, F. (1972). Importance de la souche et de la composition du milieu dans la photoinduction d'un rythme circadien de zonations chez *Neurospora*. *Comptes Rendus de l'Academie des Sciences (Paris)* **274**, 1667–70.

PITTENDRIGH, C. S., BRUCE, V. G., ROSENSWEIG. N S. & RUBIN, M. L. (1959). A biological clock in *Neurospora*. *Nature, London* **184**, 169–70.

RUSSELL, P. (1972). Report on the sixth *Neurospora* information conference. *Neurospora Newsletter* **19**, 3–5.

SAGROMSKY, H. (1952). Der Einfluß des Lichtes auf die rhythmische Konidienbildung von *Penicillium*. *Flora* **139**, 300–13.

SARGENT, M. L. & KALTENBORN, S. (1972). Effects of medium composition and carbon dioxide on circadian conidiation in *Neurospora*. *Plant Physiology* **50**, 171–5.

SARGENT, M. L. & WOODWARD, D. O. (1969). Genetic determinants of circadian rhythmicity in *Neurospora crassa*. *Journal of Bacteriology* **97**, 861–6.

SUSSMAN, A. S., LOWRY, R. J. & DURKEE, T. (1964). Morphology and genetics of a periodic colonial mutant of *Neurospora crassa*. *American Journal of Botany* **51**, 243–52.

SWEENEY, B. M. (1969). *Rhythmic phenomena in plants*. New York: Academic Press.

TURIAN, G. & BIANCHI, D. E. (1972). Conidiation in *Neurospora*. *The Botanical Review* **38**, 119–54.

UEBELMESSER, E. R. (1954). Über den endonomen Tagesrhythmus der Sporangienträgerbildung von *Pilobolus*. *Archiv für Mikrobiologie* **20**, 1–33.

VALLA, G. (1974). Rhythms of nuclear divisions and septation in growing mycelia of a basidiomycete mushroom, *Polyporus arcularius* (Batsch) ex Fr.: variation of the rhythm periods in relation to the temperature of cultivation. *Journal of Interdisciplinary Cycle Research* **5**, 223–30.

WAGNER, E., DEITZER, G. F., FISCHER, S., FROSH, S., KEMPF, O., STROEBELE, L. (1975). Endogenous oscillations in pathways of energy transduction as related to circadian rhythmicity and photoperiodic control. *Biosystems* **7**, 68–76.

WERNER, C. (1898). Die Bedingungen der Conidienbildung bei einigen Pilzen. *Inaugural-Dissertation* Basel/Frankfurt a.M.

WOODWARD, D. O. & SARGENT, M. L. (1973). Circadian rhythms in *Neurospora*. In *Behaviour of micro-organisms*; *Proceedings of the 10th International Congress for Microbiology*, edited by A. Pérez-Miravete, pp. 282–296. Mexico City.

YU-SUN, C. C. C. (1964). Biochemical and morphological mutants of *Ascobolus immersus*. *Genetics* **50**, 987–98.

CHAPTER 20

Cell Ageing and Autolysis

Z. FENCL

20.1 Introduction

In contrast to small molecules, macromolecules are subject to spontaneous degradation in time. The changes in the properties of macromolecules are reflected in the whole organism and represent one of the principal causes of natural ageing of living matter. To be sure, cells take care of resynthesizing subcellular units but some of the changes, *e.g.* in nuclear material and in cell walls, are irreversible and hence bear directly on the viability of the cell. Disorganization of cell structures, which is the final result of ageing and after which autolysis sets in, can be either accelerated or slowed down by physical conditions and by the number of available energy sources or building blocks required for anabolic processes. Likewise, the half-lives of various organelles can vary and may thus influence ageing. Maaløe & Kjeldgaard (1966) have shown that the time taken for chromosome replication is constant and therefore independent of the growth rate. It is also known that a lower rate of cell wall synthesis in yeasts growing at a high growth rate is reflected in an increase in the size of the cells (Vraná & Fencl, 1972). A less favourable surface-volume ratio can affect metabolism and hence ageing; in fact, it can bring about cell death (Mortimer & Johnson, 1959). In a similar way the thickness and composition of the cell wall might affect the transport of substances across the cell membrane and thus bring about a material imbalance of the cell. This is one of the factors affecting ageing as it triggers changes in the concentration of intermediates in cells which in turn is reflected by slower growth. Fencl, Řičica & Kodešová (1972), basing their deductions on the different periods required for reaching a steady state of enzymes in yeast cells following a change in the dilution rate in a chemostat, concluded that a steady growth rate alone does not necessarily mean a constant rate of synthesis of the various enzymes in cells. Balanced growth, as defined by Maaløe & Kjeldgaard (1966)—when all the processes in a cell proceed at an equal rate—is in fact an idealized state which may exist rather exceptionally, and the higher the complexity of the organism, the more difficult it is to attain.

It can thus be deduced that ageing is an irreversible process which is a function of time and which is affected by environmental conditions. The cell begins to age at the moment of its origin. This holds equally for microbial cells where the problem of ageing has been treated in the context of equality or inequality of the mother and the daughter cells. It has been shown that division of the genetic·material and of the wall is not random, but is subject to firm laws. Cole & Hahn (1962) showed for *Streptococcus pyogenes* that the original wall is retained in only one of the cells formed by division. However, this finding does not seem to be generally valid as it could not be confirmed in *Salmonella typhimurium* (May, 1963). The original strand of a nucleic acid remains intact in a chromosome for several generations in one of the divided cells which thus appears as a mother cell. On the basis of these findings and of the number of dead cells in a growing population, Maruyama & Hayashi (1966) suggested that in *Bacillus megaterium* the mother cells cannot produce a further generation after three divisions. On the other hand, Beran, Málek, Streiblová & Lieblova (1967) found in *Saccharomyces cerevisiae* as many as 16 bud scars which indicates that the cell has produced at least 16 generations. Mortimer & Johnson (1959) observed one cell budding 23 times and saw no cessation of budding.

A simple experimental technique making it possible to distinguish between mother and daughter cells of different relative age and a distribution of the population into the fraction of mother and daughter cells has provided a number of indications of inequality of cells in a population. These differences between mother and daughter cells increase with decreasing growth rate (Vraná, 1976). From this, one might conclude that the cultivation conditions alone are not decisive for the ageing process since cells with different physiological properties appear in the medium in which homogeneity is ensured both physically and chemically. These changes in the properties of the daughter and the mother cells may be accounted for only on the basis of different relative age, *i.e.* by the influence of time.

The problems of inequality of bacterial cells and their ageing have long been appreciated (Málek, 1959; Powell, 1959). In fungal cultures it has long been obvious (Reinhardt, 1892) that the filament is differentiated in its different parts both from the point of view of reproduction and of hyphal growth. Since a number of these problems of differentiation, including that of the reproductive cycle, are treated in other parts of this book attention will be focussed on some problems dealing with irreversible, time-dependent differentiation in fungi when cultivation conditions affect merely the amplitude of these changes.

From the point of view of cultivation conditions, fungal growth can be divided into surface and submerged growth. In submerged growth one may distinguish between mycelium growing diffusely and mycelium growing in pellets although there is a gradual transition between the two extremes (Galbraith & Smith, 1969). Diffuse growth resembles more that of unicellular micro-organisms (Fencl, Řičica, Munk & Novák, 1967) while growth in pellets bears some characteristics of surface growth.

20.2 Ageing of Hyphae

In spite of the difference brought about by the method of cultivation hyphae retain some phenomena in common, which can be indicated as symptoms of ageing.

A growing septate hypha can be separated into at least three zones, *viz.* (i) an apical zone with the Spitzenkörper, a central pore in the wall apex, and cytoplasmic vesicles participating in wall elongation through lysis and resynthesis of the wall; (ii) a subapical zone rich in plasma components, such as nuclei, mitochondria, ribosomes, and endoplasmic reticulum; and (iii) a vacuolization zone in which the size of vacuoles increases with distance from the apex, *i.e.* with the age of the compartments (see also Chapter 3).

As part of the hypha, the cell wall also undergoes an ageing process. Marchant & Smith (1968) and Strunk (1963) have indicated a structural difference between the wall of the terminal growth apex and that of the lower part of the hypha. Autoradiography of wall synthesis in *Aspergillus*, *Phytophthora*, *Neurospora* and *Schizophyllum* using labelled N-(acetyl-^{3}H)glucosamine (Katz & Rosenberger, 1971; Gooday, 1971) has shown that although wall elongation is limited to the apical part of the growing hypha, thickening of the walls may take place even in the parts more distant from the apex. Like the synthesis of hyphal walls, their autolysis is also affected by their age. Mitchell & Sabar (1966) found that maximum lysis of *Pythium*, exhibited by release of glucose from cells, occurs in six day-old hyphae. When preparing protoplasts from *Aspergillus niger* Musílková & Fencl (1968) showed that more protoplasts were obtained from the youngest parts of the hyphae (Table 20.1), since the apical part of the hyphae was lysed more rapidly than the older wall, which was more resistant to the effect of the snail gut enzymes. The observations on the compositional and structural inequality of the hyphal walls in fungi caused by ageing are in good agreement with the later findings of Gull & Trinci (1974), who studied ageing of walls in *Botrytis cinerea*, and of Chang & Trevithick (1974) who provided evidence that the release of enzymes into the medium takes place in the apical part of hyphae. One can thus deduce a relationship between the morphological and the physiological functions of hyphal parts of different age.

Table 20.1 Dependence of the amount of released protoplasts on the age of hyphae of *Aspergillus niger.* From Musílková & Fencl (1968)

Hyphal age in h	Number of protoplasts × 10⁵ related to 1 mg dry weight of mycelium		
	incubation in h		
	1	2	3
15	14.06	32.50	35.65
18	8.66	21.83	32.73
24	2.27	3.05	6.58
40	0.88	1.75	4.62

Differentiation between the young and the old parts of the hypha is enhanced by the septa which separate the hypha into compartments. Even if the number of septa per unit length of hypha and hence the size of the compartments is determined by cultivation conditions, the formation of the septa reflects a certain age. Septation also requires the formation of a certain number of nuclei (Clutterbuck, 1970; King & Alexander, 1969). By closing the septa the difference between the old and the new part of the hypha is intensified. Since a compartment does not grow further, the only possibility of growth is offered by branching when one may also observe the active participation of the septa (Trinci, 1974). According to our observations, branching is preceded by a shortening of the corresponding compartment which is lengthened again after the formation of a new terminal apex (Fencl & Machek, unpublished results). This points to the closing of the septum and an increase of the pressure within the compartment. There is also the possibility of the phenomenon of septum closing playing an important role in the survival of parts of hyphae during autolysis.

The problem of the relationship between hyphal age and the rate of synthesis of nuclear material and its division remains somewhat unclear. It is known that unicellular organisms possess a constant period for chromosome division e.g. in bacteria (Maaløe & Kjeldgaard, 1966) and in yeast (Meyenburg, 1969) and that this is not affected by the growth rate. This is in agreement with the findings of Rosenberger & Kessel (1967, 1968) who observed in hyphae of Aspergillus nidulans that the nucleus divides synchronously, i.e. at the same rate, and the time taken for mitosis is independent of the growth rate. The first two nuclei are localized in the apex of the hypha. However, these observations refer to very young hyphae attaining lengths between 10 and 15 μm. It is not known what happens to these nuclei when the hypha attains a length of several hundred μm. On the other hand, Nishi, Yanagita & Maruagama (1968) showed that in Aspergillus oryzae hyphae 100 μm long, the time required for DNA synthesis was different in young and in old parts of the hypha and that the drop in reproduction rate of the nuclear cycle is gradual along the length of a stationary cultivated hypha. In the old parts of a hypha the synthesis of nuclear material and nuclear division are about 1.5–2 times slower than in the first 10 μm of the hypha. In view of the existence of septa and the fixation of nuclei on the endoplasmic reticulum (Rosenberger & Kessel, 1968) it appears that older nuclei divide more slowly. Even if the slower division were caused by a longer time required for the formation of precursors without which the nucleus cannot begin to divide, while the division itself is independent of time, the differences in the time required for doubling of the number of nuclei in a hypha existing in a homogeneous medium clearly suggests that differentiation is affected by ageing.

Besides these time-dependent changes in the individual subcellular units, there exists a number of biochemical findings indicating the possibility of physiological changes in the fungal hypha during the course of ageing.

In his classical paper Zalokar (1959b) showed in Neurospora crassa that the rate of protein and nucleic acid formation does not decrease substantially in hyphal regions distant from the growing point. But using cytochemical methods, it was found that hyphal tips, i.e. the first 50–100 μm of the growing apex of the hypha, differed substantially from the rest of the young hypha. Hyphal tips were slightly richer in protein and significantly richer in protein-bound sulfhydryl groups and ribonucleic acids. From this he concluded that new protoplasm is formed throughout the hypha and transported to the tip by active cytoplasmic streaming.

On the other hand, Nishi et al. (1968) have reported unpublished work by Yanagita on submerged hyphae of Aspergillus in which he had failed to demonstrate protoplasmic streaming from the older to the apical parts.

These different findings may result from using different fungi as well as different cultivation techniques. It appears, however, that in surface cultures nutrients must flow from substrate to the aerial hyphae. Without this their growth could not be ensured. The two types of surface-grown hyphae contained different enzymes (Zalokar, 1959a). Similarly, Trinci (1974), studying hyphal growth on a membrane, assumed a continuous supply of the enzyme-containing vesicles including the wall building blocks from the older toward the apical parts of the hyphae.

Even if according to Zalokar (1959b) protein synthesis along a hypha grown under submerged conditions proceeds at an approximately equal rate, the proteins formed are not equivalent and different enzymes may be synthesized in different loci, depending on the age of the individual compartments of the hypha (Zalokar, 1959a, b; Yanagita, 1966; Yanagita & Namachi, 1967; Nagasaki, 1968a). Since the hypha exists in a homogeneous medium, the localization of the formation of these enzymes is a function of age. In the younger parts of hyphae alkaline phosphatase is synthesized in the nuclei while in the older parts it is found in the plasma. Protease is synthesized in *Aspergillus* hyphae only after they have reached a length of 40 μm. The tip of the hypha does not produce the enzyme. Yanagita & Namachi (1967) concluded from these findings that ageing depends on the distance from the apex and that this distance determines the physiological age of the hypha.

However, it is often difficult to distinguish whether the observed biochemical changes in hyphae are evoked by ageing or are only due to cultivation conditions. With the aid of autoradiography, changes in the rate of RNA and protein synthesis in *Aspergillus niger* hyphae have been studied during different phases of culture development (Fencl, 1970; Machek & Fencl, 1973). The growth curve shown in Fig. 20.1 can be separated into the first exponential phase, the lag phase, a second exponential phase and the stationary phase. The change in the growth rate between the first and the second exponential phase is in good agreement with the work of Trinci (1974) who used isolated hyphae to show that during the first hours after germination the growth rate of hyphae increases but later the hyphal tip grows at a constant rate. Because of branching, the growth of the entire mycelium appears to be exponential. During the lag phase septation and initiation of branching may be observed. At the same time, the rate of RNA synthesis began to decrease in the parts more distant from the tip (Fig. 20.2). During the first exponential phase, in agreement with Zalokar (1959b), we found no differences in the rate of RNA and protein synthesis along the length of the hypha (Fig. 20.2). In the second exponential phase when the fungus does not attain the growth rate observed during the first phase, the rate of synthesis of both components in the apical part did not change but decreased in the parts more distant from the tip although the culture was not limited by carbon source (Fig. 20.1). A shift-up to the starting medium brought about, during the first 15 min, a recovery of the rate of RNA and protein synthesis to that of the exponential level (Fig. 20.3). It seems the effect is connected more with synthesis of limiting factors than with the age of the hypha.

During the stationary phase some parts of the hypha irreversibly lose their ability to synthesize RNA and protein (Fencl, Machek & Novák, 1969) and they apparently begin to autolyse. In other parts of the hypha the rate of

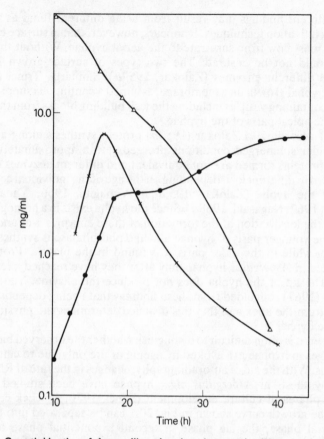

Fig. 20.1 Growth kinetics of *Aspergillus niger* in submerged culture. △—△, glucose concentration in the medium; ●—●, mycelial dry weight; ×—×, RNA content.

synthesis is decreased by 15–20% as compared to hyphae from the exponential phase (Fig. 20.2). After a shift-up the first hyphae to recover are the apical parts while the older parts retain their decreased rate of synthesis; this might indicate a lack of equivalence of hyphal compartments caused by age. On the other hand, on the assumption of protoplasmic streaming from the older to the younger parts of the hypha these results may be interpreted as being due to the transport of nutrients. Thus the apical part was not exposed to starvation as was the older part of the hypha and therefore it retained a higher regeneration ability. The quick recovery to the maximum rate of RNA synthesis and protein synthesis after a shift-up in the apical part of hyphae which had passed through the stationary phase of culture development indicates that this decrease in the rate of synthesis is not caused by time but by starvation.

Time plays a role in the compartments distant from the tip as these older parts of hyphae irreversibly lose their biosynthetic activity. The origin of this phenomenon may lie in changes caused by age or by greater starvation due to a continuous outflow of nutrients toward the apex. Then the localization

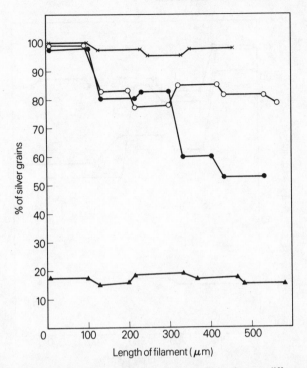

Fig. 20.2 Rate of RNA synthesis in a hypha of *Aspergillus niger* at different phases of population development. The rate of synthesis was followed autoradiographically. ×—×, hypha from the first exponential phase; phase; O—O, hypha from the lag phase; ●—●, hypha from the second exponential phase; ▲—▲, hypha from the stationary phase.

of the compartment in the hypha alone might relate to ageing. However, this hypothesis is not generally valid since in some older parts of a hypha there is the potential possibility of passing to a physiologically younger state if a compartment forms a new branch. Then this part of the hypha will then increase its rates of RNA and protein synthesis to those of the apical part (Machek & Fencl, 1973).

In spite of these open questions as to the extent in which some of the changes in the biochemical activity of the individual compartments are a function of nutrient limitation or of time, it is possible, on the basis of biochemical and morphological observations, to speak of ageing of individual parts of hyphae as a real fact.

20.3 Ageing of Culture

Practical experience from fermentation processes, especially from fungal production of antibiotics, led in the past to empirical relationships between the phase of development (culture age) and antibiotic production. Thus it was claimed that production of penicillin is associated with the stationary phase of growth. On changing cultivation conditions in later years it was possible to make the cells produce penicillin both during growth and the stationary phase. In most cases, the production of antibiotics, enzymes, *etc.*,

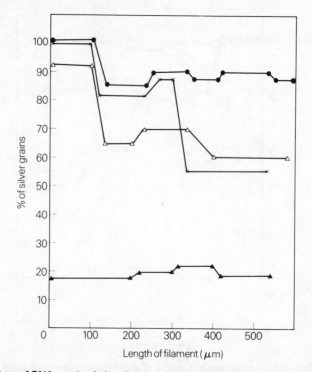

Fig. 20.3 Rate of RNA synthesis in a hypha of *Aspergillus niger* of the second exponential and the stationary phase after a shift-up. The rate of synthesis was followed by autoradiography. ×—×, hypha from the second exponential phase; ●—●, hypha from the second exponential phase after a shift-up; △—△, hypha from the stationary phase; ▲—▲, hypha from the stationary phase after a shift-up. The shift-up period was 15 min.

is not a function of age of hypha in a culture but simply a function of the cultivation conditions, including various substrate limitations. This can be further easily demonstrated with the production of glucose oxidase by *Aspergillus niger* (Fencl *et al.*, 1967).

In a batch cultivation, the production of this enzyme is associated with growth while at the onset of the stationary phase the degradation mechanism of the carbon substrate is shifted to the glycolytic pathway. The Q_{O_2} of the mycelium then decreases from 1100 μl O_2 to less than one-half while the RQ increases from less than 0.1 to 0.5 (Table 20.2). In a continuous three-stage system in which fresh medium was fed both to the first and to the third fermenter, conditions arose in the second stage that corresponded to the beginning of the stationary phase; however, the inflow of a fresh medium to the third stage where the 'oldest' mycelium was present subjected this mycelium to a continuous shift-up. In this way the biochemical properties of the mycelium changed to the level of the starting, exponentially growing culture. There is a certain analogy with the increased rate of RNA synthesis in hyphae after a shift-up (Fig. 20.3) with the difference that the recovery is practically complete.

Table 20.2 Changes in Q_{O_2} and RQ of *Aspergillus niger* in batch and three-stage continuous cultivation with the inflow of fresh medium into the first and third fermenter. From Fencl *et al.* (1967)

Fermentation method	Origin of the sample	Q_{O_2}	RQ
Batch	exponential phase	1100	0.08
	stationary phase	450	0.55
Continuous	1st stage	1150	0.08
	2nd stage	550	0.45
	3rd stage	1110	0.10

In a diffusely growing agitated culture of a fungus it is difficult to speak of its age in the sense that we can speak about the age of individual compartments in the hyphae. The shearing forces and other factors cause breaking of the tips or breakdown of hyphae and, moreover, they affect branching and thus the number of terminal apices. In this way, the ratio of physiologically older to younger compartments of hyphae is changed. In spite of this constant rejuvenation of hyphae it has not been possible to attain in a continuous culture the maximum growth rate of a batch culture (Novák & Fencl, 1973; S. J. Pirt, personal communication). This is probably associated with the fact that in a continuous culture the hyphae are septate and hence their growth rate corresponds rather to the second exponential growth phase (Fig. 20.1). One may thus speak about a culture age only from the point of view of young, nonseptate, and of older, septate, hyphae while transition from the second to the first phase is not possible. In a diffusely growing septate culture further differentiation is not due to ageing but rather reflects changes in the external cultivation conditions. This view is supported by the findings of Trinci (1974) who found during unlimited growth of isolated hyphae that their tips grow at a constant rate even after the mycelium has reached a length of several millimetres. Continuous branching then ensured an exponential growth of the mycelium as a whole.

In contrast to a diffusely growing culture, when the fungus grows in pellets there is a more substantial differentiation in the filaments caused by the transport of nutrients, including oxygen, from the peripheral parts to the inside of the pellets, and by an excretion of the metabolic products to the outside of the pellet. This differentiation may be influenced by self toxicity by excreted metabolites. Reining & Glasgow (1971) found that a mutant of *Neurospora crassa* secretes a mucopolysaccharide which inhibits its growth.

Yanagita (1966) showed in pellets of *Aspergillus niger* and in *Penicillium urticae* a structural and cytochemical differentiation. Only a thin layer in the peripheral region of the pellet is active in biosynthesis. Phosphate that had been incorporated into hyphae is not transported outward from the pellet into newly growing hyphae but remains mainly in its original place (Yanagita & Kogané, 1963). On the pellet surface there are RNA-rich hyphae while the internal older parts of hyphae possess a low level of RNA and hence a restricted possibility of protein synthesis. One of the results of ageing of the internal hyphae of some fungi is the formation of fructification organs and of conidia.

The gradient in nutrient transport along the hypha offers a comparison between growth in pellets and surface culture with the distinction that while differentiation in surface colonies occurs two-dimensionally, the cytological differentation of pellets may be regarded as a modification of that of the surface colony in three dimensions (Yanagita & Kogané, 1963). This comparison is not quite exact since in a surface culture, oxygen is transported against the direction of transport of other nutrients, *i.e.* there is a distinction between aerial and substrate mycelia. Differences in the respiratory systems in the two parts of the surface mycelium have already been mentioned (Zalokar, 1959*a*). For this reason, differentiation of a surface mycelium cannot be taken as two-dimensional, *i.e.* in the sense of a gradient between the old centre of the colony and the young periphery, but as three-dimensional, since the oldest hyphae are the first substrate hyphae at the area of contact with the medium while young hyphae grow into the medium at the same rate as that of the aerial mycelium (Trinci, 1969).

In spite of the fact that the colony receives nutrients from the entire contact area with the medium, Tanaka & Yanagita (1963) found structural differentiation of colonies caused by ageing. They recognize a peripheral part with a typical complement of nuclei and mitochondria, a productive zone with irregularly extended mitochondria, and a fruiting zone with vacuoles in the hyphae. From the margin to the centre of the colony, the hyphal walls thicken. The authors also proved that this morphological differentiation has a biochemical basis. Nagasaki (1968*b*) determined a time dependence between growth of substrate mycelium, aerial mycelium and the conidia-bearing apparatus. At the same time, he demonstrated that the entire course of ageing of a colony up to the maturation of spores, which takes about 10 days, is accompanied by changes in enzyme activity. Furthermore, the hyphae contain two isoenzymes for many of the enzyme functions, one operating in a neutral, the other in an acid medium. Young hyphae produce enzymes of the first type, older hyphae of the second type since during growth the medium is acidified. Whether these changes in enzymes are actually a symptom of ageing or are simply caused by altered conditions could not be established as there is a lack of experiments with shifts. Ng, Smith & Anderson (1972) showed that an alteration of the environment can prevent initiation of conidiophores although vegetative growth is not restricted. Likewise, it is known from surface production of citric acid by *Aspergillus niger* that, depending on cultivation conditions, a nine-day mycelium may or may not have conidia. From this it can be concluded that, besides physical environmental conditions, nutrient supply and disposal of intermediary metabolites will control the formation of reproductive organs even when a certain age of hypha is a prerequisite for the formation of conidia.

On the other hand, Fencl & Novák (1971) have reported previously unpublished results of Z. Fencl, E. Ujcová, L. Seichart and M. Musílková on the production of citric acid by a surface culture of *Aspergillus niger*, where it was found that the maximum specific rate of acid synthesis occurred in a four-day mycelium. Preformed mycelium of different ages was layered on a medium removed from different days of cultivation and the specific rate of citric acid synthesis measured in short-term experiments. It was shown that, irrespective of the prefermentation of the medium, the maximum production of citric acid was found in a four-day mycelium. The quality of the medium only modulated the level of this production. This might be used as evidence that a certain biochemical activity of the surface mycelium is a function of time rather than of cultivation conditions; in other words, that it is a consequence of ageing.

Another symptom of colony ageing is the change of growth rate in the course of growth. From the exponential growth phase it passes to a linear-radial growth. It does not appear that this slowing down is caused by lack of energy sources (Trinci, 1969) but rather that it is a function of colony area which contains more and more old, lysing compartments as time goes on. Colony growth stops partly as a consequence of formation of toxic products by the colony.

A submerged, diffusely growing fungal culture behaves rather as a culture of unicellular micro-organisms, *i.e.* its differentiation is determined by the cultivation conditions and it grows independently of time. On the other hand, in cultures growing submerged in pellets or on the surface, one can establish the role of time in biochemical and morphological differentiation, as well as their interrelationships, displayed by localization of biochemical activity in a certain morphologically distinct colony or pellet.

20.4 Autolysis

Autolysis is considered as the last stage of culture development even when it is observed in parts of hyphae during the stationary phase. Even in continuous cultivation lysed parts of hyphae can occur, and it is difficult to decide whether these represent fragments from dead parts of the mycelium growing on the walls or whether they are old hyphae which remain in the fermenter much longer than would correspond to the mean residence time. The main cause of lysis is apparently a material imbalance in fungal hyphae caused either by internal or external factors. Of the internal factors it could be a disturbance of the organelles or an accumulation of toxic metabolites. The external factors comprise physical, chemical and enzyme influences which disturb the intracellular structures of the cell wall, as well as a lack of nutrients, especially of sources of energy. This lack of energy source may be caused by exhaustion of the supply from the medium or by a specific deficiency in a compartment due to its particular localization in a hypha or colony. This may have to do with compartments with one septum closed while nutrients pass through the pore in the other septum toward the hyphal tip; it may also be caused by the specific microclimate about parts of hyphae generated in the colony which complicates or fully prevents metabolic exchange. For instance, the measured critical concentration of oxygen which is limiting for fungal growth is much higher in the medium than the true concentration of dissolved oxygen inside the pellets.

With isolated unbalanced hyphae, physiological age is subject to change in the direction tip to base (Nagasaki, 1968*b*), while in cultures, the physiological age of the individual compartments in the hypha is not strictly determined by the distance from the tip, *i.e.* by the relative age. Thus the occurrence of lysed parts of hyphae is rather random. Trinci & Righelato (1970) concluded from this randomness that the culture is cytologically heterogeneous at all stages of development and that the age of a culture cannot be used as an indicator of the sequence of morphological changes.

The fact that there exist simultaneously lysed and fully intact live compartments indicates that the pores in septa in autolysing hyphae must be closed and that the undamaged parts of hyphae apparently utilize low-molecular weight substances formed by lysis as substrates for their growth. Judging from the slow loss of nitrogen-containing substances in *Nectria galligena* (Lahoz Beltrá & Ballesteros, 1970) or in *Aspergillus* (Behr, 1930) the living parts of the filament take up these substrates more rapidly than they can be released into the medium. Growth of intrahyphal hyphae has been observed in an autolysed compartment (Lowry & Sussman, 1966; Trinci & Righelato, 1970). From this point of view, autolysis can be considered as a way to survive under starvation conditions.

Exhaustion of nutrients from the medium represents one of the most common reasons for a culture gradually reaching autolysis. The transition to autolysis is gradual and lack of nutrients is reflected in the hyphae during the pre-autolytic phase by increased differentiation.

With *Penicillium chrysogenum*, Righelato, Trinci, Pirt & Peat (1968) observed that a lowered amount of glucose caused increased vacuolization and breakdown of nucleic acids. Lahoz, Reyes, Beltrá & Garcia-Tapa (1967) observed in *Aspergillus terreus* a loss of amino acids from the pool before autolysis set in, while, concomitantly, there was an increase in mycelium hydrolases when preparing amylases from *A. niger*. We achieved a much higher production of the enzymes by starving the mycelium (Z. Fencl, K. Beran and M. Burger, unpublished results). Morton, Dickerson & England (1960) demonstrated protease activity in *P. griseofulvum* in the course of starvation of the fungus. Whether this activity was due to protease synthesis or due to relieving the effect of protease inhibitors, as is known to occur in yeast, is not yet clear.

Nagasaki (1968*b*) associated the transition of hyphal parts to the preautolytic phase not only with nutrient limitation but also with age. He noted that ribosomes were degraded in older cells and this paralleled the presence of a high RNase activity with an optimum at pH 4.5, *i.e.* an enzyme which is active when the culture is older and has generated a more acid medium. A decrease of RNA in resting cells or in cells which are beginning to autolyse appears to be common to all micro-organisms.

On the basis of experiments with continuous culture of *Penicillium chrysogenum*, Righelato *et al.* (1968) concluded that the age of the mycelium does not appear to control the ageing process but rather that ageing and lysis are determined by the amount of available energy. Transition to the preautolytic phase in *P. chrysogenum* occurs at the moment when the amount of glucose supplied to the culture approaches the maintenance energy which was calculated to be 0.022 g glucose/g mycelium dry weight/h. If the amount of available energy source drops below this level, autolysis sets in. The rate of degradation of the individual components of biomass depends on the cultivation conditions under which the culture has been grown, *i.e.* it is a function of culture history.

When the fungus is grown in excess nitrogen and growth is limited by the source of carbon, the protein component of the cell is degraded most rapidly

(Lahoz & Miralles, 1970; Lahoz, Ballesteros & Jimeno, 1974). Conversely, if excess carbon is present during cultivation, autolysis raises the amount of reducing substances; furthermore, while polysaccharides are normally lysed slowly, lipids are preferentially degraded (Lahoz *et al.*, 1967). Thus monosaccharides can always be detected in the lysing mycelium (Lahoz *et al.*, 1970).

Trinci & Righelato (1970) observed an initial decrease in the level of protein, RNA and DNA in starved cultures of *P. chrysogenum*. They also studied the resistance of the individual organelles during autolysis. During the early stages of starvation the granular structure of protoplasm was lost due to ribosomal lysis. Likewise, the nuclei became less electron-dense and lost their granular structure except near the nucleolus. Furthermore, it does not appear that membranes are affected since they tend to accumulate during autolysis. Mitochondria are also much more stable than ribosomes. The authors attribute this phenomenon to the stability of phospholipids.

Trinci & Righelato (1970) further confirmed that autolysis does not proceed synchronously in the entire filament but only in its individual compartments. In a lysed compartment the decomposition of organelles of the same type is synchronous, *i.e.* it is catalysed by cell free enzymes. From this, one can conclude that, in contrast with other organisms, lysosomes and autophagy do not play a substantial role in the autolysis of cytoplasm in fungi. Even if vesicles with lysosomal function have been discovered in fungi (Matile, 1966; Thornton, 1968), their function is rather associated with the lysis and resynthesis of the cell wall (Mahadevan & Mahodkar, 1970; Trinci, 1974).

It is generally assumed that micro-organisms produce enzymes which can hydrolyse the polymers of their walls and that they are essential for cell growth (Forsberg & Rogers, 1971). Mahadevan & Mahodkar (1970) found a positive correlation between the amount of lytic enzymes in the cell wall and apical growth and frequency of branching. Fungi are believed to transport these hydrolytic enzymes enclosed in vesicles towards the tips of hyphae where branching sets in. The composition of the enzyme content of the vesicles differs, depending on the composition of hyphal walls. Usually the presence of glucanases and proteases can be demonstrated. The vesicles may mediate the transport of other lytic enzymes, such as chitinases, pectinases, and amylases from the hyphae to the medium. However, hyphae contain their own protective substances through which they withstand lysis by their own enzymes. Wessels & Koltin (1972) demonstrated that only after removal of this barrier will endocellular glucanases and chitinases attack the walls. A similar protection exists apparently in the septa and spore walls, where the protective function is played by melanin or a heteropolysaccharide (Chu & Alexander, 1972).

Because of this protection the cell wall is only slowly autolysed (Trinci, 1974) although the autolytic enzymes are localized directly in hyphal walls. Polacheck & Rosenberger (1975) studied autolysis of isolated pure walls of *Aspergillus nidulans*. During wall incubation they observed successive release of glucose, galactose, *N*-acetylglucosamine, amino acids and soluble oligosaccharides. During the first hours autolysis proceeded linearly, but after about 3% of the wall material had been hydrolysed, the rate decreased and the remaining cell walls were resistant to the enzymes. The first to be autolysed are the hyphal tips which is not due to a specific localization of autolysins but to a decreased resistance of the young cell wall.

In contrast to the inside of hyphae, where autolysis affects compartments irregularly, the lysis of walls proceeds regularly from the tip to the older sections, thus reflecting the chemical composition of the walls.

Hyphal protoplasm undergoes lysis because of either a decreased amount or defects in composition of organelles, or else to a lack of maintenance energy. Vacuolization and disruption of organelles, including an increased hydrolase activity may be observed first of all in the older parts of hyphae but the physiological age and the age of the compartment need not coincide. For this reason there is no regularity in the lysis of hyphae from the base toward the tip and the lysed compartments are localized irregularly in different parts of the hypha.

20.5 General Conclusions

Changes in the wall composition along the hypha, the slowing down of nuclear synthesis, the activation of nucleases and decompositon of ribosomes in the older sections of hyphae, as well as the slowing down of biosynthetic reaction after a shift-up, indicate the inequivalence of the individual compartments of hyphae which can occur even in a homogeneous cultivation medium. In view of the fact that this inequivalence is associated with parts of hyphae of different age, ageing can be viewed as a fundamental cause of these changes. Both the autolysis and the increased biosynthetic activity of old compartments where branching has occurred show that the absolute age of the compartments is not identical with its physiological age. The physiological age is determined, among other things, by the localization of the compartments in a hypha. One of the irreversible steps in ageing is the formation of septa.

While in isolated hyphae the process of ageing can be followed with some ease, a submerged, diffusely growing, fungal culture, because of branching and breaking of hyphae, behaves as a bacterial culture and symptoms of ageing are difficult to discern. If it tends to the stationary phase and to autolysis this is due to starvation rather than ageing.

The spatial distribution of pellets or of surface cultures affects hyphal differentiation and is time-dependent. Particularly in surface cultures parts of the colony of a certain age can be associated with certain biochemical activity. Even if the role of time in the development of hyphae and fungal colonies is clearly demonstrable, the mechanisms which bear on the rate of the ageing process (with the exception of starvation) are unclear.

20.6 References

BEHR, G. (1930). Über Autolyse bei *Aspergillus niger. Archiv für Mikrobiologie* **1**, 418–44.

BERAN, K., MÁLEK, I., STREIBLOVÁ, E. & LIEBLOVÁ, J. (1967). The distribution on the relative age of cells in yeast populations. In *Microbial physiology and continuous culture* pp. 57–67, edited by E. O.

Powell, C. G. T. Evans, K. E. Strange and D. W. Tempest. Her Majesty's Stationary Office.

CHANG, P. L. J. & TREVITHICH, J. R. (1974). How important is secretion of exoenzymes through apical cell walls of fungi? *Archiv für Mikrobiologie* **101**, 281–99.

CHU, S. B. & ALEXANDER, M. (1972). Resistance and susceptibility of fungal spores to lysis. *Transactions of the British Mycological Society* **58**, 489–92.

CLUTTERBUCK, A. J. (1970). Synchronous nuclear division and septation in *Aspergillus nidulans. Journal of General Microbiology* **60**, 133–5.

COLE, R. M. & HAHN, J. Y. (1962). Cell wall replication in *Streptococcus pyogenes. Science* **135**, 722–4.

FENCL, Z. (1970). Comments on differentiation and product formation in molds. *Biotechnology and Bioengineering* **12**, 845–7.

FENCL, Z., MACEK, F. & NOVÁK, M. (1969). Kinetics of product formation in multistage continuous culture. In *Fermentation advances* pp. 301–322, edited by D. Perlman. New York: Academic Press.

FENCL, Z. & NOVÁK, M. (1971). Prediction of the product formation in continuous cultivation of micro-organisms. In *Recent advances in microbiology* pp. 427–433, edited by A. Perez-Miravete and D. Peláéz. *X. International Congress for Microbiology, Mexico.*

FENCL, Z., ŘIČICA, J. & KODEŠOVÁ, J. (1972). The use of the multi-stage chemostat for microbial product formation. *Journal of Applied Chemistry and Biotechnology* **22**, 405–16.

FENCL, Z., ŘIČICA, J., MUNK, V. & NOVÁK, M. (1967). Physiological changes in filamentous organisms as a function of growth rate. In *Microbial physiology and continuous culture* pp. 186–95, edited by E. O. Powell, C. G. T. Evans, R. E. Strange and D. W. Tempest. Her Majesty's Stationary Office.

FORSBERG, C. & ROGERS, H. J. (1971). Autolytic enzymes in growth of bacteria. *Nature, London* **229**, 273–3.

GALBRAITH, J. C. & SMITH, J. E. (1969). Filamentous growth of *Aspergillus niger* in submerged shake culture. *Transactions of the British Mycological Society* **52**, 234–46.

GOODAY, G. W. (1971). An autoradiographic study of hyphal growth of some fungi. *Journal of General Microbiology* **67**, 125–33.

GULL, K. & TRINCI, A. P. J. (1974). Detection of areas of wall differentiation in fungi using fluorescent staining. *Archiv für Mikrobiologie* **96**, 57–9.

KATZ, D. & ROSENBERGER, K. F. (1971). Hyphal wall synthesis in *Aspergillus nidulans:* effect of protein synthesis inhibition and osmotic shock on chitin

insertion and morphogenesis. *Journal of Bacteriology* **108**, 184–90.

KING, S. B. & ALEXANDER, L. J. (1969). Nuclear behaviour, septation and hyphal growth of *Alternaria solani. American Journal of Botany* **56**, 249–59.

LAHOZ, R., BALLESTEROS, A. M. & JIMENO, L. (1974). Influence of temperature on the autolytic phase of growth of *Aspergillus niger. Annals of Botany* **38**, 661–116.

LAHOZ, R., BELTRÁ, R. & BALLESTEROS, A. M. (1970). Biochemical changes in cultures of *Nectria galligena* during the autolytic phase of growth. *Annals of Botany* **34**, 205–11.

LAHOZ, R., & MIRALLES, M. (1970). Influence of the level of the carbon source on the autolysis of *Aspergillus niger. Journal of General Microbiology* **62**, 271–6.

LAHOZ, R., REYES, F., BELTRÁ, R. & GARCIA-TAPA, C. (1967). The autolysis of *Aspergillus terreus* in a physiologically acid medium. *Journal of General Microbiology* **49**, 259–65.

LOWRY, R. J. & SUSSMAN, A. S. (1966). Intra-hyphal hyphae in 'clock' mutants of *Neurospora. Mycologia* **58**, 541–9.

MAALØE, O. & KJELDGAARD, N. O. (1966). *Control of macromolecular synthesis.* New York: Benjamin.

MACHEK, F. & FENCL, Z. (1973). Differentiation of filamentous micro-organisms as a basis for understanding product formation. *Biotechnology and Bioengineering Symposium* **4**, 129–42.

MAHADEVAN, P. R. & MAHODKAR, U. R. (1970). Role of enzymes in growth and morphology of *Neurospora crassa. Journal of Bacteriology* **101**, 941–7.

MÁLEK, I. (1959). Discussion part of the symposium on the continuous cultivation of micro-organisms held in Praha in June 1958. *Folia Microbiologica* **4**, 390–408.

MARCHANT, R. & SMITH, G. D. (1968). A serological investigation of hyphal growth in *Fusarium culmorum. Archiv für Mikrobiologie* **63**, 85–94.

MARUYAMA, Y. H. & HAYASHI, K. (1966). Some aspects of cell age distribution function. *Journal Fermentation Technology, Japan* **44**, 227–32.

MATILE, P. (1966). Inositol deficiency resulting in death: an explanation of its occurrence in *Neurospora crassa. Science* **151**, 86–8.

MAY, J. W. (1963). The distribution of cell-wall label during growth and division of *Salmonella typhimurium. Experimental Cell Research* **31**, 217–21.

MEYENBURG, K. (1969). Katabolit-Repression und der Sprossungszyklus von *Saccharomyces cerevisiae*. Zürich, Dissertationsdruckerie, Leeman, A. G.

MITCHELL, R. & SABAR, N. (1966). Autolytic enzymes in fungal cell walls. *Journal of General Microbiology* 42, 39–42.

MORTIMER, R. K. & JOHNSON, J. R. (1959). Life span of individual yeast cells. *Nature, London* 183, 1951–2.

MORTON, A. G., DICKERSON, A. G. F. & ENGLAND, D. J. F. (1960). Changes in enzyme activity of fungi during nitrogen starvation. *Journal of Experimental Botany* 2, 116–24.

MUSÍLKOVÁ, M. & FENCL, Z. (1968). Some factors affecting the formation of protoplasts in *Aspergillus niger*. *Folia Microbiologica Czech.* 12, 235–9.

NAGASAKI, S. (1968a). Cytological and physiological studies on phosphatases in developing cultures of *Aspergillus niger*. *Journal of General and Applied Microbiology* 14, 263–77.

NAGASAKI, S. (1968a). Physiological aspects of various enzymes in relation to the culture age of *Aspergillus niger* mycelia. *Journal of General and Applied Microbiology* 14, 147–61.

NG, W. S., SMITH, J. E. & ANDERSON, Y. G. (1972). Changes in carbon catabolic pathways during synchronous development of conidiophores of *Aspergillus niger*. *Journal of General Microbiology* 71, 495–504.

NISHI, A., YANAGITA, T. & MARUAGAMA, Y. (1968). Cellular events occurring in growing hyphae of *Aspergillus oryzae* as studied by autoradiography. *Journal of General and Applied Microbiology* 14, 171–82.

NOVÁK, M. & FENCL, Z. (1973). Kinetic analysis of the relationship between batch and continuous cultivation of *Aspergillus niger*. *Biotechnology and Bioengineering Symposium* 4, 43–52.

POLACHECK, Y. & ROSENBERGER, R. F. (1975). Autolytic enzymes in hyphae of *Aspergillus nidulans*: their action on old and newly formed walls. *Journal of Bacteriology* 121, 332–7.

POWELL, E. O. (1959). Discussion part of the symposium on the continuous cultivation of micro-organisms held in Praha in June 1958. *Folia Microbiologica* 4, 190–408.

REINHARDT, M. O. (1892). Das Wachstum der Pilzhyphen. Ein Beitrag zur Kenntnis des Flächenwachstums vegetabilischer Zellmembranen. *Jahrbücher fur Wissenschaftliche Botanik* 23, 33–60.

REINING, J. L. & GLASGOW, J. E. (1971). Mucopolysaccharide which regulates growth in *Neurospora*. *Journal of Bacteriology* 106, 882–9.

RIGHELATO, R. C., TRINCI, A. P. J., PIRT, S. J. & PEAT, A. (1968). The influence of maintenance energy and growth rate on the metabolic activity, morphology and conidiation of *Penicillium chrysogenum*. *Journal of General Microbiology* 50, 399–412.

ROSENBERGER, R. F. & KESSEL, M. (1967). Synchrony of nuclear replication in individual hyphae of *Aspergillus nidulans*. *Journal of Bacteriology* 94, 1464–9.

ROSENBERGER, R. F. & KESSEL, M. (1968). Nonrandom sister chromatid segregation and nuclear migration in hyphae of *Aspergillus nidulans*. *Journal of Bacteriology* 96, 1208–13.

STRUNK, CH. (1963). Über die Substruktur der Hyphenspitzen von *Polystictus versicolor*. *Zeitschrift für Allgemeine Mikrobiologie* 3, 265–74.

TANAKA, K. & YANAGITA, T. (1963). Electron microscopy on ultrathin sections of *Aspergillus niger*. *Journal of General and Applied Microbiology* 9, 101–18.

THORNTON, R. M. (1968). The fine structure of *Phycomyces*. I. Autophagic vesicles. *Journal of Ultrastructural Research* 21, 269–79.

TRINCI, A. P. J. (1969). A kinetic study of the growth of *Aspergillus nidulans* and other fungi. *Journal of General Microbiology* 57, 11–24.

TRINCI, A. P. J. (1974). A study of the kinetics of hyphal extension and branch initiation of fungal mycelia. *Journal of General Microbiology* 81, 225–47.

TRINCI, A. P. J. & RIGHELATO, R. C. (1970). Changes in constituents and ultrastructure of hyphal compartments during autolysis of glucose starved *Penicillium chrysogenum*. *Journal of General Microbiology* 60, 239–49.

VRANÁ, D. (1976). Daughter cells as an important factor in determining the physiological state of yeast populations. *Biotechnology and Bioengineering* 18, 297–309.

VRANÁ, D. & FENCL, Z. (1972). Dependence of size of cells, their substance and the ribonucleic acid content on the dilution rate in a two-stage continuous cultivation of *Candida utilis*. In *Yeasts: models in science and technics* pp. 621–31, edited by A. Kocková-Kratochvílová and E. Minárik. Bratislava.

WESSELS, J. G. H. & KOLTIN, Y. (1972). R-glucanase activity and susceptibility of hyphal walls to degradation in mutants of *Schizophyllum* with disrupted nuclear migration. *Journal of General Microbiology* **71**, 471–15.

YANAGITA, T. (1966). Dynamic aspects of fungal growth with special reference to enzyme formation. *Journal of Fermentation Technology* **44**, 313–20.

YANAGITA, T. & KOGANÉ, F. (1963). Cytochemical and physiological differentiation of mold pellets. *Journal of General*

and Applied Microbiology **9**, 179–87.

YANAGITA, T. & NAMACHI, Y. (1967). Kinetic analysis of the region of protease formation in the hypha of *Aspergillus niger. Journal of General and Applied Microbiology* **13**, 227–35.

ZALOKAR, M. (1959a). Enzyme activity and cell differentation in *Neurospora. American Journal of Botany* **46**, 555–9.

ZALOKAR, M. (1959b). Growth and differentiation of *Neurospora* hyphae. *American Journal of Botany* **46**, 602–10.

CHAPTER 21

Cytoplasmic Inheritance and Senescence

G. TURNER

21.1 Introduction

The existence of cytoplasmic genes in higher plants was suggested as long ago as 1909 following observations on non-Mendelian inheritance of leaf variegation in *Mirabilis jalapa* (Correns, 1909) and *Pelargonium zonale* (Baur, 1909), and studies on similar behaviour in other higher plants were later reported (Rhoades, 1946). One of the first reports of cytoplasmic inheritance in fungi was that by Johnson (1946) who observed maternal inheritance of certain race characteristics in Rust Fungi, and detailed studies soon began on *Neurospora crassa* (Mitchell & Mitchell, 1952) and *Saccharomyces cerevisiae* (Ephrussi, 1953). Many studies were carried out in the 1950s into possible examples of cytoplasmic inheritance in a variety of fungi, though a revival of interest followed the establishment of the presence of DNA in mitochondria (Nass & Nass, 1963a,b; Nass, Nass & Afzelius, 1965) and gave a new direction to the studies. To this day, many of the reported examples of cytoplasmic inheritance (also referred to as 'extrachromosomal' and 'extranuclear') have not been ascribed to any particular genome, and it is possible that they reside in extranuclear, extramitochondrial genetic elements, though there is as yet no molecular evidence for such a genome in filamentous fungi. Certain reported examples of cytoplasmic inheritance may have been the result of chromosome abnormalities, since these have been known to give non-Mendelian behaviour during genetic tests (Jinks, 1963). Great care is thus needed in the interpretation of results, and although a number of criteria suitable for recognition of cytoplasmic mutants have been listed by various authors (Jinks, 1963; Esser & Kuenen, 1967) some of these criteria are peculiar to particular organisms and are not generally applicable. The possibility sometimes exists that a particular cytoplasmic condition may be the result of infection with an extracellular genetic element such as a fungal virus (Hollings & Stone, 1971) rather than a mutation in a cytoplasmic gene that is an essential part of the genetic make-up of the organism, and it is not always easy to eliminate this possibility.

Senescence is the term applied to loss of vigour and degeneration following continuous vegetative growth of an organism. Although there are many examples of this, not all are necessarily associated with cytoplasmic mutation, and only those which are will be discussed.

I will confine this chapter to dealing with a number of well-established examples of cytoplasmic inheritance in filamentous fungi. A number of books and reviews have given detailed treatment of all the various reports of cytoplasmic inheritance in fungi, as well as in other organisms (Jinks, 1963, 1966; Esser & Kuenen, 1967; Sager, 1972).

21.2 Cytoplasmic Mutants of *Neurospora crassa*

Characteristics of the mutants

Most of the known cytoplasmic mutants of *Neurospora crassa*, together with some of their properties, are listed in Table 21.1, though some not included have been lost (Gillie, 1970). It can be seen that these mutants grow slowly

Table 21.1 Cytoplasmic mutants of *Neurospora crassa* (After Bertrand & Pittenger, 1972*a*; Gillie, 1970.)

Group	Mutant	Phenotype	Mutagenic origin[a]	Criteria of extranuclearity[b]	Suppression by *f*	Suppression by *su-1*[c]
I	[*mi*-1] *poky*	Slow growth	SP	1,3,4	+	−
	[*SG*-1]	Slow growth	ACR	1,3	+	−
	[*SG*-3]	Slow growth	SP	1	+	−
	[*stp-B*1]	Slow growth	SP	1,3	+	−
	[*exn*-1]	Slow growth	NTG	1,3	+	−
	[*exn*-2]	Slow growth	NTG	1,3	+	−
	[*exn*-3]	Slow growth	SP	1,3	+	−
	[*exn*-4]	Slow-growth	NTG	1,3	+	−
II	[*mi*-3]	Intermediate growth	SP	1,3	−	+
III	[*stp*]	Stop/start growth	UV	2,3	−	−
	[*stp-A*]	Stop/start growth	SP	2,3	−	−
	[*stp-A*18]	Stop/start growth	SP	2,3	−	−
	[*stp-B*2]	Stop/start growth	SP	2,3	−	−
	[*stp-C*]	Stop/start growth	NTG	2,3	−	−
	[*abn*-1]	Stop/start growth	SP	2,3,4	nt[d]	nt
	[*abn*-2]	Stop/start growth	SP	2,3,4	nt	nt

[a]SP: spontaneous mutation; ACR: acriflavin-induced; NTG: nitrosoguanidine-induced; UV: ultraviolet light induced.

[b]1: the variant shows maternal inheritance; 2: the variant shows non-Mendelian segregations; 3: the variant shows transmission in heterokaryon and related tests; 4: the variant shows infective spreading through a culture.

[c]*f* and *su*-1 are nuclear suppressors of the slow growth phenotype

[d]nt: not tested

and/or irregularly, and all those listed have abnormal cytochrome spectra, suggesting some mitochondrial defect.

The extranuclearity of [*mi*–1] (originally called *poky*), the first-known cytoplasmic mutant of Neurospora, was established by observation of its inheritance in sexual crosses (Mitchell & Mitchell, 1952). In sexual crosses of *Neurospora crassa*, a single microconidium derived from one parent fertilizes the protoperithecium formed by the other parent of opposite mating type, and a perithecium is formed containing many asci, each of which contains eight ascospores. It was observed that [*mi*–1] was inherited only via the protoperithecial parent, this being termed 'maternal inheritance' (Fig. 21.1). Thus ascospores from any one perithecium were all

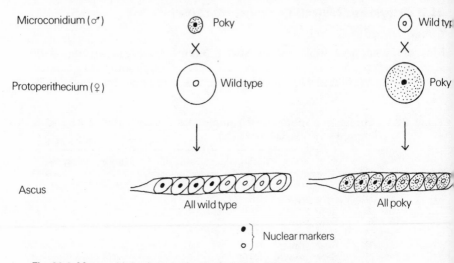

Fig. 21.1 Maternal inheritance of *poky* [*mi*-1] in *Neurospora crassa*.

[*mi*–1] if the protoperithecium was derived from the [*mi*–1] strain, while in a reciprocal cross where the protoperithecium was formed from the wild-type strain, all the ascospores were wild type. Known nuclear markers, however, segregated in the expected 1 : 1 Mendelian ratio within the perithecium and within individual asci, no difference being observed in reciprocal crosses. This criterion of maternal inheritance has been demonstrated in a number of the extranuclear mutants isolated (Table 21.1). Some of the extranuclear mutants, however, are female sterile, *i.e.*, unable to act as protoperithecial parent, and cannot be tested in this way. When used as the microconidial parent, it is observed that the mutation is not transmitted to the progeny, even though nuclear markers segregate normally. Another exception to the above criterion has been observed in the extranuclear mutant [*AC*-7] (Srb, 1963). This mutation is not transmitted to the progeny when crossed with the wild type, even if used as the protoperithecial parent. Furthermore, in reciprocal crosses between [*AC*-7] and [*SG*], it was observed that most of the perithecia contained [*SG*] ascospores, even when [*SG*] was the microconidial parent. This suggests that extranuclear genes of paternal origin are able to enter the perithecium.

Another criterion of extranuclearity is the ability of the character in question to be transmitted independently of known nuclear genes in heterokaryons (heterokaryon test, Jinks, 1963, Fig. 21.3). Pittenger (1956) demonstrated this following the formation of heterokaryons between [mi-1] and [mi-4] (now lost). Fairly stable heterokaryons containing a mixture of these two mutations were formed, and conidia could later be isolated where 'reassortment' had occurred, that is, each of the mutations emerging associated with a different nucleus from the one with which it entered the heterokaryon.

Finally, some of these characters exhibit 'infectivity' in that they tend to spread through the cytoplasm of a wild-type strain following hyphal anastomosis between strains. Once again, the infectivity is independent of nuclear migration. One of the problems of this last criterion is that it could be the result of infection by a fungal virus rather than a mutation in a fungal gene. Mitchell & Mitchell (1952) attempted to eliminate this possibility for [mi-1] by showing that the character was not transferred from mutant to wild type in mixed cultures where no heterokaryon formation occurred.

Complementation between mutants

On the basis of cytochrome spectra, growth behaviour, female fertility and reaction to two nuclear suppressors, f and su-1, Bertrand & Pittenger (1972a) divided the known cytoplasmic mutants into 3 groups (Table 21.1). The suppressor f was shown by Mitchell & Mitchell (1956) to be a nuclear suppressor restoring almost wild-type growth rate to [mi-1] without correcting the cytochrome defects. Conversely, f did not suppress [mi-3]. Suppressor su-1 was isolated by Bertrand (1971) as a suppressor of [mi-3] slow growth, again without restoring a wild-type cytochrome spectrum. Pittenger (1956) demonstrated complementation between extranuclear mutants [mi-1] and [mi-4] in stable heterokaryons resulting in restoration of near-normal growth without restoration of normal cytochrome spectrum, and similar tests were later carried out for complementation between mutants of the assigned groups (Bertrand & Pittenger, 1972b). It was found that mutants of groups I and II complemented those of group III, but that mutants within the same group did not complement each other, nor was complementation observed between [mi-3], the only group II mutant, and mutants of group I. On the basis of these results, it was suggested that the groups might represent 3 cistrons within the mitochondrial genome. However, it is important to note that no stable recombinants have yet been obtained from heterokaryons formed between complementing mutants.

Methods for isolation of extranuclear mutants

Study of extranuclear mutants naturally depends on their availability and ease of isolation. However, there is no easy way of obtaining a variety of such mutants in filamentous fungi. Gillie (1970) used redox dyes such as tetrazolium incorporated into the growth medium as indicators of respiratory deficiency (impaired ability to reduce the dye) and demonstrated that known extranuclear mutants could be distinguished from wild-type on such media. However, attempts to isolate new cytoplasmic mutants were unsuccessful. A cytoplasmic, mitochondrial mutant acu-10 has been isolated in the Basidiomycete Coprinus lagopus by selection for mutants unable to grow on acetate as sole carbon and energy source. This mutant grew slowly on glucose, and had altered cytochrome spectrum and mitochondrial morphology (Casselton & Condit, 1972; Casselton & Kirkham, 1975). Similar attempts to obtain respiratory-deficient mutants in *Aspergillus nidulans* produced only nuclear mutants (Houghton, 1970).

Since many of the extranuclear mutants of fungi show suppressiveness, Bertrand, McDougall & Pittenger (1968) attempted to select for mutants by growing *Neurospora* continuously in growth tubes, and looking for growth abnormalities. They succeeded in isolating cytoplasmic mutants with stop/start growth, [*stp*] (Table 21.1).

Many of the known extranuclear mutants of *Saccharomyces cerevisiae* (Thomas & Wilkie, 1968; Avner & Griffiths, 1973) and *Aspergillus nidulans* (see p. 412) are ones resistant to known inhibitors of mitochondrial function but, to date, no similar mutants have been reported for *Neurospora crassa*.

The effect of extranuclear mutation on the mitochondrion

The whole mycelia of [*mi*-1] and [*mi*-3] were examined with a hand spectroscope (Mitchell, Mitchell & Tissieres, 1953) and it was observed that the cytochrome spectra differed markedly from that of the wild type. Both mutants lacked cytochrome aa_3 and contained excess amounts of cytochrome c. [*mi*-3] still contained cytochrome b, though this was not visible in [*mi*-1]. It has since been shown that all the other mutants listed in Table 21.1 have abnormal cytochrome spectra resembling that of [*mi*-1] (Bertrand & Pittenger, 1972a). It was also noted that the spectrum of older mycelium of [*mi*-1] was much closer to the wild-type spectrum than that of young cultures. A similar observation was made with succinate oxidase activity, which was low in young cultures and higher in older cultures (Haskins, Tissieres, Mitchell & Mitchell, 1953). A parallel was drawn between these mutants and the respiratory-deficient, cytoplasmic mutants observed in yeast (Ephrussi, 1953), though in the latter organism, respiratory deficiency was complete, the mutant being unable to grow on non-fermentable substrates. Tissieres, Mitchell & Haskins (1953) further demonstrated that a large proportion of the respiration of [*mi*-1] was cyanide insensitive, and more detailed studies have since been carried out on this pathway. It is sensitive to salicyl hydroxamate, an inhibitor of cyanide-insensitive respiration in higher plants (Schonbaum, Bonner, Storey & Bahr, 1971) and forms only a very small proportion of the total respiration in young, wild-type mycelium, increasing somewhat in older cultures. However, in both extranuclearly and nuclearly-inherited mitochondrial mutants with cytochrome abnormalities, it becomes the major electron transport pathway (Colvin, Sauer & Munkres, 1973; Edwards & Kwiecinski, 1973) and seems to involve site I oxidative phosphorylation but not sites II and III (von Jagow, Weiss & Klingenberg, 1973). The branchpoint with the cyanide-sensitive pathway is in the region of ubiquinone, thus bypassing the cytochromes. Although it is tempting to suggest that this may represent a way of overcoming lesions in the 'normal' electron transport pathway, further studies are needed to assess the physiological importance of this alternative pathway during the growth of the mutants.

Following the establishment of the general presence of DNA in mitochondria (Nass *et al.*, 1965) Reich & Luck (1966) demonstrated the maternal inheritance of mitochondrial DNA in *Neurospora*. Around the same time, the technique of microinjection was used to demonstrate the mitochondrial nature of the mutation [abn-1] (Diacumakos, Garnjobst & Tatum, 1965). They microinjected mitochondria, isolated and purified from the mutant, into wild-type hyphal compartments, and produced the mutant condition in the recipient strains. Taken together, the above results tend to support the hypothesis that at least some of the extranuclear mutants result from mutations in the mitochondrial DNA of the organism. This DNA has been isolated and characterised (Agsteribbe, Kroon & van Bruggen, 1972; Clayton & Brambl, 1972) as a 20 μm circumference circular molecule with a molecular weight of 40×10^6 daltons.

More recently, detailed examination of the electron transport systems and cytochromes of [mi-1] and [mi-3] has been carried out, confirming and extending the results of Mitchell and colleagues (Lambowitz, Slayman, Slayman & Bonner, 1972; Lambowitz & Bonner, 1974; von Jagow et al., 1973). It seems likely that the multiple lesions caused by the [mi-1] mutation may be the result of defective mitochondrial ribosomes (Rifkin & Luck, 1971) leading to a lowered rate of mitochondrial protein synthesis. Young cultures of [mi-1] are deficient in the small mitochondrial ribosome subunit, though this is not so marked in older cultures. However, no such defect was observed in [abn-1] or [mi-3], supporting the proposal that these are genetically different from [mi-1] (Bertrand & Pittenger, 1972b). It is pertinent to note that in mammalian cells (Aloni & Attardi, 1971) and in yeast (Reijnders, Kleisen, Grivell & Borst, 1972) it has been demonstrated that mitochondrial ribosomal RNA is coded for by mitochondrial DNA.

Accumulation of large amounts of cytochrome c may be the result of incomplete binding of the cytochrome to the mitochondria, perhaps through lack of a mitochondrially-synthesized component. It has been shown that two species of cytochrome c are present in Neurospora crassa, differing by one amino acid (Scott & Mitchell, 1969). It appeared that cytochrome c_{II} was made first, then converted to c_I, the bound form. c_{II} accumulated in young [mi-1], while it was only a minor component in wild type and old [mi-1].

The cause of mitochondrial abnormalities in other mutants has not yet been elucidated, though vesicles containing 33S RNA have been observed within the mitochondria and cytoplasm of [abn-1]. Although described as 'virus-like particles', infectivity has not been demonstrated, and it was suggested that these particles may be, at least in part, products of mitochondrial DNA (Küntzel et al., 1973).

21.3 Cytoplasmic Mutants of Aspergillus

Morphological variants of Aspergillus nidulans

Reports of cytoplasmic mutants in Aspergillus nidulans were numerous during the 1950s and early 1960s (Jinks, 1954; Jinks, 1963), but none has been amenable to biochemical studies. Furthermore, it seems possible that at least some of these may now be explained by chromosome abnormalities or non-Mendelian behaviour of nuclear genes. For example, aneuploidy may have been responsible for some of the 'minute' variants (Faulkner & Arlett, 1964) reported as examples of cytoplasmic mutants (Upshall, 1971; Käfer & Upshall, 1973; A. Upshall, personal communication). These mutants were mostly somatically unstable, showing persistent segregation to yield wild-type. Similar somatic segregation was also observed in the 'red' mutant (Grindle, 1964) also held to be cytoplasmic. Unfortunately, these original mutants are now no longer available, and cannot be re-examined.

Mitochondrial mutants of Aspergillus nidulans

Subsequent to the growing interest in mitochondrial genetics, Rowlands and Turner (1973) reported the isolation of a number of stable mutants resistant to the mitochondrial ATPase inhibitor oligomycin, one of which appeared to be extranuclear. This mutant, (*oliA1*), which is also slow growing on drug-free media and has an altered cytochrome spectrum (Fig. 21.2) was

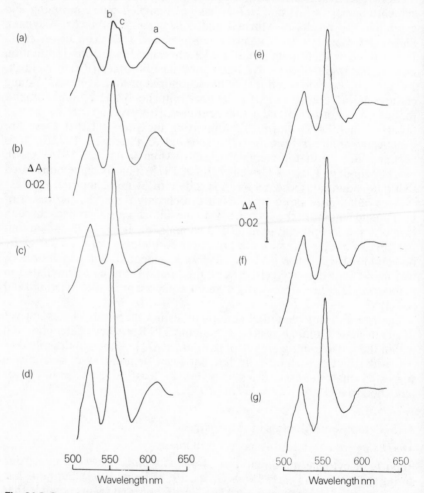

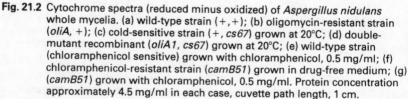

Fig. 21.2 Cytochrome spectra (reduced minus oxidized) of *Aspergillus nidulans* whole mycelia. (a) wild-type strain (+,+); (b) oligomycin-resistant strain (*oliA*, +); (c) cold-sensitive strain (+, *cs67*) grown at 20°C; (d) double-mutant recombinant (*oliA1*, *cs67*) grown at 20°C; (e) wild-type strain (chloramphenicol sensitive) grown with chloramphenicol, 0.5 mg/ml; (f) chloramphenicol-resistant strain (*camB51*) grown in drug-free medium; (g) (*camB51*) grown with chloramphenicol, 0.5 mg/ml. Protein concentration approximately 4.5 mg/ml in each case, cuvette path length, 1 cm.

crossed with the wild type and the perithecia were examined. It was found that in all cases, whether selfed or hybrid, the perithecia contained ascospores which were either all oligomycin resistant or all sensitive. Known nuclear markers, on the other hand, segregated in the usual Mendelian pattern in hybrid perithecia.

In heterokaryon tests (Fig. 21.3), reassortment of oligomycin resistance/sensitivity with nuclear markers was detected. In addition, it was found that heteroplasmons containing mixtures of mutant and wild type could not be maintained as such for long, in contrast to heteroplasmons of *Neurospora crassa*. Instead, segregation rapidly occurred to give homoplasmic sectors when heteroplasmons were allowed to grow out from a small

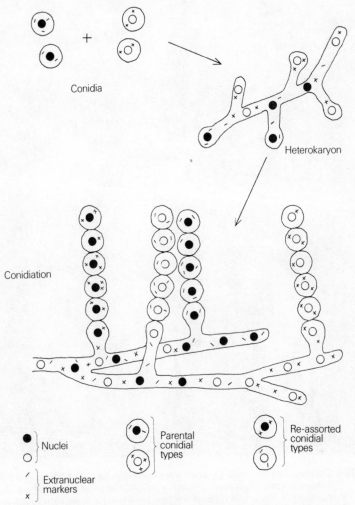

Fig. 21.3 The heterokaryon test in *Aspergillus nidulans*. Conidia of mutant and wild-type strains are mixed and allowed to grow together to form a heterokaryon. Conidia taken from the heterokaryon are tested for 'reassortment', *i.e.*, reassociation of nuclear/extranuclear markers. No recombination is observed between nuclear markers.

inoculum of mycelium (Fig. 21.4), even though the mycelium remained heterokaryotic. This clearly demonstrated the independent behaviour of the oligomycin resistance/sensitivity characters from the nuclear markers.

Shortly after isolation of the oligomycin-resistant mutant, Waldron & Roberts (1973) isolated a cold-sensitive mutant (*cs67*) which grew normally at 37°C but slower than the wild type at 20°C, exhibiting cytochrome

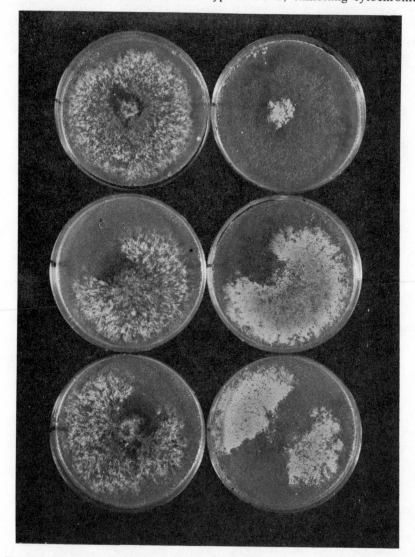

Fig. 21.4 Sectoring of rutamycin resistance/sensitivity, *i.e.*, (*oliA1*)/(+), in 8 day old heterokaryons of *Aspergillus nidulans*. Each horizontal pair of plates consists of the original heterokaryon (left) and its replica on rutamycin-containing medium (right) obtained by replicating with damp velveteen. The sectors are independent of nuclear markers, since yellow and white conidiating areas can be seen to be well-mixed all over the heterokaryons (left-hand column) (from Rowlands & Turner, 1973, with permission).

abnormalities at the lower temperature (Fig. 21.2). Genetically, this mutant behaved essentially like (*oliA1*) in the tests described above, and was deemed to be one of the same class.

Gunatilleke, Scazzocchio & Arst (1975) reported the isolation of two extranuclear chloramphenicol-resistant mutants at a single locus, (*camA*), which again appeared to be of the same class as the previous two mutants.

Recombination of extranuclear markers in heterokaryons

In view of the genetic similarities between these mutants, attempts were made to demonstrate recombination between them. To bring about contact of the extranuclear elements, heterokaryons were constructed between strains carrying the various mutations and conidia isolated before sectoring had occurred (Rowlands & Turner, 1974*b*, 1975). Conidia were then plated and colonies scored for the possible phenotypes. It was found that recombination was possible between all three extranuclear loci yielding stable double mutants and wild type, though reciprocal recombinants did not occur with equal frequency. Furthermore, since the degree of mixing of cytoplasms can vary from one heterokaryon to another, a parameter was needed for standardization of recombination frequency. Instead of expressing recombinants as a percentage of the total conidia tested, they may be expressed as a percentage of the reassorted types (*i.e.*, those conidia in which the combination of nucleus/extranuclear marker is different from the parental one, Fig. 21.3). This then makes some allowance for the degree of cytoplasmic mixing. The results showed equal recombination between all three markers, suggesting either that the loci are equidistant on a circular genome, or that the recombination figures obtained represent those for unlinked genes. While the latter alternative is more plausible, it does seem that the mitochondria of *Aspergillus nidulans* contain covalently closed circular DNA with a circumference of approximately 10 μm, molecular weight 20×10^6 daltons (López-Pérez & Turner, 1975). Although the mitochondrial genome of *Aspergillus* does not seem to be large, it is not known how many rounds of recombination occur when mitochondria come into contact, thus 'unlinked' mitochondrial genes cannot be directly compared with unlinked nuclear genes.

Recombination of extranuclear markers has also been observed during sexual crosses (Mason & Turner, 1975) but since in heterokaryon-compatible strains of *Aspergillus nidulans* heterokaryons are formed and cytoplasmic mixing can occur before onset of the sexual phase, it is possible that extranuclear recombination occurs in the heterokaryon only.

Maternal inheritance

Aspergillus nidulans is homothallic, thus formation of perithecia does not require opposite mating types as in *Neurospora crassa*. Nevertheless, it does seem possible that differentiated male and female organs are involved in formation of perithecia, each contributing one nucleus to form the dikaryotic ascogenous hyphae. To support this idea, Apirion (1963) studied mutants of *Aspergillus* in which the ascospore and perithecial wall colour was blue or colourless rather than the usual red. He observed that ascospores in hybrid perithecia obtained from crosses of mutant with wild type were always of one colour only, even though the genes for colour determination were nuclear and segregated 1 : 1 in the ascospores of hybrid perithecia. In other words, the colour of the ascospores within a perithecium was not determined by the genotype of the ascospores themselves. He suggested that the genotype of the protoperithecium (female organ) determined the ascospore colour and perithecial wall colour. Before such mutants may be used for demonstration of maternal inheritance in *Aspergillus*, one further problem remains. Heterokaryon compatible strains are normally used for genetic studies in this organism, and the sexual cycle is preceded by heterokaryon formation. Unfortunately, heterokaryon formation gives an opportunity for extranuclear markers associated with one nucleus to become reassorted with the nucleus from the other strain (*cf.* heterokaryon test) so blurring the origin of the extranuclear markers. Handley (1975) overcame a similar problem in *A. amstelodami* by using heterokaryon incompatible strains in which the sexual cycle could take place without permitting prior heterokaryon formation. Such strains have now been employed in *A. nidulans*, and the results indicate maternal inheritance of the extranuclear markers.

Mitochondrial alterations in the extranuclear mutants

Unlike some of the nuclear oligomycin-resistant mutants of *Aspergillus*, (*oliA 1*) does not show any cross resistance to venturicidin (Rowlands & Turner, 1974*a*) but does exhibit an increase in sensitivity to triethyltin, both drugs being inhibitors of the mitochondrial ATPase. All oligomycin-resistant mutants so far isolated are cross resistant to the closely related inhibitor rutamycin. *In vitro* studies with (*oliA 1*) mitochondria have shown a five-fold increase in the resistance level of the ATPase to oligomycin, and this probably accounts for the *in vivo* resistance (Rowlands & Turner, 1976). Similar results have been reported for extranuclear oligomycin-resistant mutants of *Saccharomyces cerevisiae* (Griffiths & Houghton, 1974) and there is clear evidence that part of the ATPase complex is synthesized on the mitochondrial ribosomes in yeast (Tzagoloff & Meagher, 1972) and *Neurospora crassa* (Jackl & Sebald, 1975).

In addition to the oligomycin resistance, (*oliA 1*) grows rather more slowly than the wild type on drug-free media and has an abnormal cytochrome spectrum in which cytochrome c is approximately 1.5 times the wild-type level (Fig. 21.2). In all the genetic studies performed on this mutant, this slow growth character has never been separated from the oligomycin resistance, and appears to be a result of the same mutation. It is possible that an alteration in a single inner membrane polypeptide is responsible for all the effects observed.

The cold-sensitive mutant cannot be distinguished from the wild type at 37°C, but growth at 20°C results in an abnormal cytochrome spectrum in which cytochrome c is approximately twice the wild-type level, while there is partial loss of cytochromes aa$_3$ and b (Fig. 21.2; Rowlands & Turner, 1974*a*; Turner & Rowlands, 1976). A combination of these alterations is seen in the double mutant recombinant (*oliA* 1, *cs* 67). As in *Neurospora*, it has also been observed that all mutants with cytochrome alterations develop high levels of cyanide-insensitive respiration, though in the case of *Aspergillus*, it does not appear that the mutants actually use this cyanide-insensitive pathway to provide energy for growth. The link between cold sensitivity and the multiple mitochondrial lesions is as yet unknown, though it is conceivable that a defect in the mitochondrial ribosomes may be responsible.

The chloramphenicol-resistant mutants (*camA112*) and (*camA141*) isolated by Gunatilleke *et al.* (1975) show no unusual features on drug-free medium, but a recently isolated extranuclear chloramphenicol-resistant mutant (*camB51*) which maps close to the (*camA*) locus has quite a different phenotype (Lazarus & Turner, 1976). On drug-free medium, it grows more slowly than either the wild type or (*oliA1*), and has an abnormal cytochrome spectrum resembling that of the wild-type strain grown on chloramphenicol and that of (*cs67*) grown at 20°C (Fig. 21.2). When transferred to medium containing chloramphenicol, it grows as well as it does on drug-free medium and may be described as 'chloramphenicol indifferent'. The gene involved in these mutations is as yet unknown, though it does seem possible that the mutations are affecting the mitochondrial ribosomes, the site of action of chloramphenicol.

These results provide circumstantial evidence that the extranuclear mutations are located in the mitochondrial genome of *Aspergillus nidulans*, though experiments such as microinjection of mitochondria have not been attempted. Although, in yeast, location of certain extranuclear markers on the mitochondrial DNA has been clearly demonstrated by observing the simultaneous loss of markers and DNA in petite mutants (Thomas & Wilkie, 1968; Nagley & Linnane, 1972), such tests are not possible in obligate aerobes.

Nuclear/extranuclear interactions

Combinations of (*oliA1*) with a variety of nuclear oligomycin-resistant mutants have been made (Rowlands & Turner, 1976) leading to enhanced *in vivo* resistance to oligomycin in the double mutants. In one case, O^R15, where the nuclear mutation alone resulted in a fivefold increase in the *in vitro* resistance of the mitochondrial ATPase to oligomycin, the combined strain $O^R15/(oliA1)$ possesses an ATPase approximately 20 times more resistant to oligomycin than that of the wild type. Two other double mutants have regained the wild-type growth rate and cytochrome spectrum. In all these cases, it is apparent that the nuclear gene product is interacting with that coded by the extranuclear gene at the level of the mitochondrial inner membrane.

Both nuclear and extranuclear suppressors of (*cs67*) have been isolated by Waring & Scazzocchio (1976) which partially or fully restore wild-type growth rate at 20°C.

Vegetative death in Aspergillus glaucus

Jinks (1956, 1959) reported a senescence phenomenon in *Aspergillus glaucus* which in many ways resembled that of *Podospora anserina* (see p. 420). The senescent condition, which occurred after 5 months to 2 years of vegetative growth, led to the appearance of gaps in growing colonies, resulting from swelling and lysis of the hyphae. In the early stages of senescence, or vegetative death (*vgd*) as it was called, the non-senescent state could be recovered by cutting and transfer of hyphal tips from the healthier areas of the colony, but this became more difficult with time. Most hyphal transfers rapidly gave rise to the senescent state. An intense brown pigment was produced in the area of lysis, and it became impossible to maintain the senescent strains by further transfer. A loss of conidial viability and ability to produce cleistothecia was also observed during the onset of senescence. As in *Podospora*, the condition proved to be infective and suppressive in heterokaryons formed with wild-type strains.

No biochemical studies were carried out on the original *vgd* strains, and they are no longer available, but the observations made were compatible with the mutation of a cytoplasmic gene.

'Ragged' mutant of Aspergillus amstelodami

In order to study the effect of vegetative incompatibility on cytoplasmic infection, Caten (1972) used a mutant of *Aspergillus amstelodami* which in some ways resembled the vegetative death condition of *A. glaucus*. *A. amstelodami* is a member of the *A. glaucus* species group, but is not the same strain as was used to obtain *vgd*. Although this recent isolate has low conidial viability, sectors of lysing hyphae, and was shown to be a suppressive, cytoplasmically-inherited condition, further detailed studies (Handley, 1975) have revealed differences from *vgd*, and the name 'ragged' (*rgd*) has been adopted to avoid confusion. Although *vgd* could not be maintained by repeated subculture and eventually died (Jinks, 1959) *rgd* is not truly senescent. Once the condition has appeared, it may be maintained indefinitely by hyphal transfer or plating of conidia, even though their viability is low. No pigment is produced in lysing areas, and the ability to produce cleistothecia is retained. In long term linear growth experiments, stop–start growth was observed, resembling that seen in the cytoplasmic stopper mutants of *Neurospora crassa*, and the wild type can be recovered from *rgd* by repeated plating of conidia chosen each time from the largest colonies. It was suggested that a mixture of normal and lethal extranuclear elements might be present in *rgd*, the lysis of hyphae occurring when the mycelium became homoplasmic for the mutant element.

In heterokaryons made between *rgd* and wild-type, the *rgd* condition rapidly spreads throughout the heterokaryon, though this infection can be effectively prevented by heterokaryon incompatibility (Caten, 1972; Handley, 1975). Like *Aspergillus nidulans*, *A. amstelodami* is homothallic, but by use of cleistothecial colour mutants and incompatible strains, maternal inheritance of *rgd* has been clearly demonstrated.

An altered cytochrome spectrum has been observed in mitochondria isolated from (*rgd*), and it was suggested that mutant mitochondria may be responsible for the condition (Handley & Caten, 1973).

21.4 Barrage Phenomenon, Senescence and Mitochondrial Mutants of Podospora

Barrage phenomenon

Podospora anserina is an Ascomycete with a life cycle resembling that of *Neurospora crassa*, though lacking asexual reproduction via conidia. *S* and *s* represent alleles at one of the nuclear loci responsible for protoplasmic incompatibility (Bernet, 1965) and if strains containing opposite alleles are allowed to grow together, a 'barrage' is formed in the area of contact. This results from hyphal fusion followed by death of the mycelium in this area of contact (Rizet, 1952). Despite this somatic incompatibility, sexual crosses may be performed between S and s strains, but instead of the expected $1:1$ Mendelian segregation of these alleles, it was observed that s was replaced by a novel phenotype designated s^s (Fig. 21.5a). This did not give a barrage with either s or S, and numerous experiments on this system (Beisson-Schecroun, 1962) have suggested the involvement of a cytoplasmic factor(s).

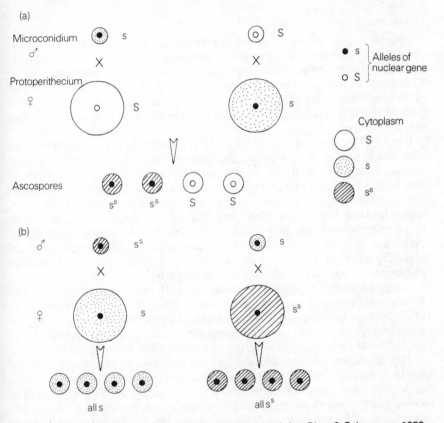

Fig. 21.5 Inheritance of barrage in *Podospora anserina* (after Rizet & Schecroun, 1959; Beisson-Schecroun, 1962).

In the s strain, it seems that a nuclear allele s together with a cytoplasmic factor (s) are responsible for the observed phenotype, *i.e.*, incompatibility with S strain. (s^s) represents the cytoplasmic state of the s strain in which (s) has been lost, and thus the incompatibility is no longer expressed. During sexual crosses between $(s) \times (s^s)$ strains, a maternal inheritance pattern is observed, suggesting the involvement of cytoplasmic genes (Fig. 21.5b). In addition, the (s) state is infectious, and following anastomosis between (s) and (s^s), (s^s) cytoplasm reverts to (s). This 'infection' is not accompanied by nuclear migration. By production of protoplasts (Belcour, 1976) containing small volumes of cytoplasm, it has been possible to obtain (s^s) from (s), presumably as a result of the chance loss of the (s) cytoplasmic factor.

In crosses between S and s, it seems that the S allele eliminates the (s) factor from the zygote, resulting in a $1:1$ ratio of S/s^s. The nature of the cytoplasmic factor (s) is unknown, and there is no evidence to connect it with the mitochondrion. It is now believed that (s) is not a DNA gene, but possibly a protein molecule able to induce and regulate its own synthesis by switching on nuclear gene s (Beisson-Schecroun, 1962; L. Belcour, personal communication).

Senescence

Prolonged vegetative propagation of *Podospora anserina* leads to degeneracy of the mycelium, accompanied by swelling and bursting of hyphal tips, cessation of growth and eventually death of the growth region (Rizet, 1953). The length of the juvenile or non-senescent state is affected by a variety of factors, including a nuclear gene and the temperature at which the organism is maintained, but a cytoplasmic mutation seems to be responsible for the senescence itself.

Rizet (1957) carried out crosses via spermatization between different combinations of juvenile and senescent strains. If the juvenile strain is used as the protoperithecial ($♀$) parent and fertilized with conidia from a senescent strain, juvenile offspring are always obtained from the acospores formed. On the other hand, if the senescent strain provides the protoperithecium, some perithecia give rise to senescent strains and some to juvenile strains, the proportion depending on the strains used rather than the degree of senescence of the parent. No segregation of juvenile/senescence was observed within individual asci. Thus the phenomenon appears to be inherited maternally.

Senescence also appears to be infectious. If compatible senescent and juvenile strains are allowed to grow together and anastomose, the migration of senescence into the juvenile mycelium may be observed even though no nuclear migration takes place (Marcou & Schecroun, 1959) suggesting the migration of a cytoplasmic determinant responsible for senescence.

The development of senescence has been studied by Marcou (1961) and Smith & Rubinstein (1973a, b), and it was observed that the median length of growth before senescence is apparent is strain dependent. The onset of senescence seems to be triggered by a single event, namely the appearance or mutation of a cytoplasmic determinant responsible for senescence. Once present, multiplication of the determinant occurs, the rate of this (transformation rate) again being strain dependent, until sufficient are present within the mycelium to give the visible morphological changes associated with senescence. Smith & Rubinstein (1973a) showed that treatment of the mycelium with cyanide or cycloheximide significantly extended the median length of growth, though a number of inhibitors of mitochondrial function, including chloramphenicol, antimycin A, oligomycin and dinitrophenol did not. There is thus at present no clear evidence to implicate mitochondria in the senescence process, though this possibility cannot be excluded. A deficiency of cytochrome aa₃ has been observed recently in the spectra of ageing mycelium (L. Belcour, personal communication).

It is also interesting to note that the characteristic growth parameters associated with senescence (*e.g.*, median length of growth, transformation rate) of the various strains are maternally inherited. In addition, a nuclear gene (the plus mating type allele of race S) is also able to influence the timing of senescence (Smith & Rubinstein, 1973b). Holliday (1969) has suggested that accumulating errors of protein synthesis (Orgel, 1963) may account for senescence as observed in *Aspergillus glaucus* (vegetative death) and *Podospora*, rather than a cytoplasmic genetic determinant. This would imply that nuclear genes alone controlled the timing of senescence, which does not seem to be the case.

Senescence resembling that seen in *Podospora anserina* has also been observed in *Pestalozzia annulata* (Chevaugeon & Digbeu, 1960). Its cytoplasmic nature has been demonstrated by observation of infective transmission independent of the nuclei following anastamosis with normal strains.

Mitochondrial mutants

Using protoplasts of *Podospora anserina* as starting material, Belcour (1975) has succeeded in isolating a number of stable cytoplasmic mutants identified by their growth characteristics. These mutants showed a slow germination rate, fewer aerial filaments and slightly diminished female fertility. All showed maternal inheritance patterns, and when cytoplasms of wild type and mutant were allowed to mix in compatible strains, the wild type seemed to be dominant. Preliminary analysis of the cytochrome spectra suggested abnormalities which implies that the mutations are mitochondrial in nature.

A chloramphenicol-resistant mutant (*cap 1-r*) has now been isolated, which shows maternal inheritance in crosses carried out by spermatization, and as with the other extranuclear mutants, is at a selective disadvantage relative to the wild-type strain if crosses are made by confrontation in compatible strains (Belcour & Begel, 1976). When this mutant is grown on drug-free medium, although the growth rate of the mycelium is similar to that of the wild-type strain, germination of the ascospores is impaired as in the other mutants. Examination of the cytochrome spectrum revealed a marked loss of cytochrome aa_3 and 2–3 fold increase in cytochrome c concentration, and an increase in the level of cyanide-insensitive respiration was also observed in whole mycelium. It was suggested that the mutation conferred resistance to chloramphenicol at the level of the mitochondrial ribosome, but at the same time impaired the normal functioning of the ribosome. This would lead to a reduction in the rate of mitochondrial protein synthesis even on drug-free medium, and give rise to the abnormal cytochrome spectrum. This mutant has many similarities to the 'chloramphenicol indifferent' mutant of Lazarus & Turner (1976).

Extranuclear crosses have been performed between the various mitochondrial mutants, and wild-type recombinants obtained in some cases, but no double mutants. This is not surprising in view of the selective disadvantage of all the mutants against wild-type mitochondria.

21.5 Concluding Remarks

While it is quite clear that many instances of cytoplasmic inheritance in filamentous fungi are the result of mitochondrial mutations, most probably located on the mitochondrial DNA, there still remain some puzzling questions about the nature and location of other well-studied phenomena such as barrage and senescence in *Podospora anserina*. *Saccharomyces cerevisiae* contains more than one species of extrachromosomal DNA (Clark-Walker,

1972; Griffiths, Lancashire & Zanders, 1975) and it would be interesting to examine the DNA species of *Podospora* now that mitochondrial mutants have been isolated in this organism.

Finally, the characteristic behaviour of extranuclear genes during the life cycle of the organism, namely the phenomena of maternal or uniparental inheritance and somatic segregation of heteroplasmons, may have evolutionary significance. Sager (1972) has suggested that this may be a means of limiting extranuclear recombination and maintaining clonal uniformity. In addition, protection against deleterious, infective cytoplasmic conditions and fungal viruses may be provided by the presence of nuclear genes controlling heterokaryon incompatibility (Caten, 1972).

21.6 References

AGSTERIBBE. E., KROON, A. M. & VAN BRUGGEN, E. F. J. (1972). Circular DNA from mitochondria of *Neurospora crassa*. *Biochimica et Biophysica Acta* **269**, 299–303.

ALONI, Y. & ATTARDI, S. (1971). Expression of the mitochondrial genome in HeLa cells. IV. Titration of mitochondrial genes for 16s, 12s and 4s RNA. *Journal of Molecular Biology* **55**, 271–6.

APIRION, D. (1963). Formal and physiological genetics of ascospore colour in *Aspergillus nidulans*. *Genetical Research* **4**, 276–83.

AVNER, P. R. & GRIFFITHS, D. E. (1973). Studies on energy-linked reactions. Genetic analysis of oligomycin-resistant mutants of *Saccharomyces cerevisiae*. *European Journal of Biochemistry* **32**, 312–21.

BAUR, E. (1909). Das Wesen und die Erblichkeitsverhältnisse der 'Varietates albomarginatae hort.' von *Perlargonium zonale*. *Zeitschrift für Vererbungslehre* **1**, 330–51.

BEISSON-SCHECROUN, J. (1962). Incompatibilité cellulaire et interactions nucléoplasmiques dans les phénomènes de 'barrage' chez le *Podospora anserina*. *Annales de Génétique* **4**, 3–50.

BELCOUR, L. (1975). Cytoplasmic mutations isolated from protoplasts of *Podospora anserina*. *Genetical Research* **25**, 155–61.

BELCOUR, L. (1976). Loss of a cytoplasmic determinant through formation of protoplasts in *Podospora*. *Neurospora Newsletter* 23.

BELCOUR, L. & BEGEL, O. (1977). Mitochondrial genes in *Podospora anserina*: recombination and linkage. Submitted for publication.

BERNET, J. (1965). Mode d'action des gènes de "barrage' et relation entre l'incompatibilité cellulaire et l'incompatibilité sexuelle chez *Podospora anserina*. *Annales des Sciences Naturelles, 12e Série* (*Botanique*) **6**, 611–768.

BERTRAND, H. (1971). A nuclear suppressor of the cytochrome defects of the *mi-3* extranuclear mutant of *Neurospora crassa*. *Canadian Journal of Genetics and Cytology* **13**, 626. (Abstract.)

BERTRAND, H., McDOUGALL, K. J. & PITTENGER, T. H. (1968). Somatic cell variation during uninterrupted growth of *Neurospora crassa* in continuous growth tubes. *Journal of General Microbiology* **50**, 337–50.

BERTRAND, H. & PITTENGER, T. H. (1972*a*). Isolation and classification of extranuclear mutants of *Neurospora crassa*. *Genetics* **71**, 521–33.

BERTRAND, H. & PITTENGER, T. H. (1972*b*). Complementation among cytoplasmic mutants of *Neurospora crassa*. *Molecular and General Genetics* **117**, 82–90.

CASSELTON, L. A. & CONDIT, A. (1972). A mitochondrial mutant of *Coprinus lagopus*. *Journal of General Microbiology* **72**, 521–7.

CASSELTON, L. A. & KIRKHAM, J. B. (1975). Growth and ultrastructural studies on the mitochondrial mutant of *Coprinus lagopus*. *Archives of Microbiology* **106**, 215–20.

CATEN, C. E. (1972). Vegetative incompatibility and cytoplasmic infection in fungi. *Journal of General Microbiology* **72**, 221–9.

CHEVAUGEON, J. R. & DIGBEU, S. (1960). Un second facteur cytoplasmique infectant chez le *Pestalozzia annulata*. *Comptes Rendus de l'Academie des Sciences* **251**, 3043–5.

CLARK-WALKER, G. D. (1972). Isolation of circular DNA from a mitochondrial fraction from yeast. *Proceedings of the National Academy of Sciences of the United States of America* **69**, 388–92.

CLAYTON, D. A. & BRAMBL, R. M. (1972). Detection of circular DNA from mitochondria of *Neurospora crassa*. *Biochemical and Biophysical Research Communications* **46**, 1477–82.

COLVIN, H. J., SAUER, B. L. & MUNKRES, K. D. (1973). Respiration of wild type and extrachromosomal mutants of *Neurospora crassa. Journal of Bacteriology* **116**, 1314–21.

CORRENS, C. (1909). Vererbungsversuche mit blass (gelb)grünen und buntglättrigen Sippen bei *Mirabilis jalapa, Urtica pilulifera* und *Lunaria annua. Zeitschrift für Verebungslehre* **1**, 291–329.

DIACUMAKOS, E. G., GARNJOBST, L. & TATUM, E. L. (1965). A cytoplasmic character in *Neurospora crassa*. The role of nuclei and mitochondria. *Journal of Cell Biology* **26**, 427–43.

EDWARDS, D. L. & KWIECINSKI, F. (1973). Altered mitochondrial respiration in a chromosomal mutant of *Neurospora crassa. Journal of Bacteriology* **116**, 610–18.

EPHRUSSI, B. (1953). *Nucleo-cytoplasmic relations in micro-organisms*. Oxford: Clarendon Press.

ESSER, K. & KUENEN, R. (1967). *Genetics of fungi*. Berlin and Heidelberg: Springer-Verlag.

FAULKNER, B. M. & ARLETT, C. F. (1964). The 'minute' cytoplasmic variant of *Aspergillus nidulans. Heredity* **19**, 63–73.

GILLIE, O. J. (1970). Methods for the study of nuclear and cytoplasmic variation in respiratory activity of *Neurospora crassa*, and the discovery of three new genes. *Journal of General Microbiology* **61**, 379–95.

GRIFFITHS, D. E. & HOUGHTON, R. L. (1974). Studies on energy-linked reactions: modified mitochondrial ATP-ase of oligomycin-resistant mutants of *Saccharomyces cerevisiae. European Journal of Biochemistry* **46**, 157–67.

GRIFFITHS, D. E., LANCASHIRE, W. E. & ZANDERS, E. D. (1975). Evidence for an extra-chromosomal element involved in mitochondrial function: a mitochondrial episome? *FEBS Letters* **53**, 126–30.

GRINDLE, M. (1964). Nucleo-cytoplasmic interactions in the 'red' cytoplasmic variant of *Aspergillus nidulans. Heredity* **19**, 75–95.

GUNATILLEKE, I. A. U. N., SCAZZOCCHIO, C. & ARST, H. N. Jr. (1975). Cytoplasmic and nuclear mutations to chloramphenicol resistance in *Aspergillus nidulans. Molecular and General Genetics* **137**, 269–76.

HANDLEY, L. (1975). *Cytoplasmic inheritance in Aspergillus amstelodami*. University of Birmingham: Ph.D thesis.

HANDLEY, L. & CATEN, C. E. (1973). Vegetative death: a mitochondrial mutation in *Aspergillus amstelodami. Heredity* **31**, 136. (Abstract.)

HASKINS, F. A., TISSIERES, A., MITCHELL, H. K. & MITCHELL, M. B. (1953). Cytochromes and the succinic acid oxidase system of poky strains of *Neurospora. Journal of Biological Chemistry* **200**, 819–26.

HOLLIDAY, R. (1969). Errors in protein synthesis and clonal senescence in fungi. *Nature, London* **221**, 1224–8.

HOLLINGS, M. & STONE, O. M. (1971). Viruses that infect fungi. *Annual Review of Phytopathology* **9**, 93–118.

HOUGHTON, J. A. (1970). A new class of slow-growing non-perithecial mutants of *Aspergilllus nidulans. Genetical Research* **16**, 285–92.

JACKL, G. & SEBALD, W. (1975). Identification of two products of mitochondrial protein synthesis associated with mitochondrial adenosine triphosphatase from *Neurospora crassa. European Journal of Biochemistry* **54**, 97–106.

VON JAGOW, G., WEISS, H. & KLINGENBERG, M. (1973). Comparison of the respiratory chain of *Neurospora crassa* wild-type and the *mi*-mutants *mi*-1 and *mi*-3. *European Journal of Biochemistry* **33**, 140–57.

JINKS, J. L. (1954). Somatic selection in fungi. *Nature, London* **174**, 409–10.

JINKS, J. L. (1956). Naturally occurring cytoplasmic changes in fungi. *Comptes Rendus des Travaux du Laboratoire Carlsberg (Série Physiologique)* **26**, 183–203.

JINKS, J. L. (1959). Lethal suppressive cytoplasms in aged clones of *Aspergillus glaucus. Journal of General Microbiology*, **21**, 397–409.

JINKS, J. L. (1963). Cytoplasmic inheritance in fungi. In *Methodology in basic genetics*, pp. 325–54, Edited by W. J. Burnette. San Francisco: Holden-Day.

JINKS, J. L. (1966). *Extrachromosomal inheritance*. Englewood Cliffs, New Jersey: Prentice-Hall.

JOHNSON, T. (1946). Variation and the inheritance of certain characters in Rust Fungi. *Cold Spring Harbor Symposia on Quantitative Biology* **11**, 85–93.

KÄFER, E. & UPSHALL, A. (1973). The phenotypes of the eight disomics and trisomics of *Aspergillus nidulans. Journal of Heredity* **64**, 35–8.

KÜNTZEL, H., BARATH, Z., ALI, I., KIND, J. & ALTHAUS, H.-H. (1973). Virus-like particles in an extranuclear mutant of *Neurospora crassa. Proceedings of the National Academy of Sciences of the United States of America* **70**, 1574–8.

LAMBOWITZ, A. M. & BONNER, W. D. Jr. (1974). The mitochondrial b-cytochromes of the wild type and *poky* strains of *Neurospora crassa. Journal of Biological Chemistry* **249**, 2886–90.

LAMBOWITZ, A. M., SLAYMAN, C. W., SLAYMAN, C. L. & BONNER, W. D. Jr. (1972). The electron transport components of wild type and *poky* strains of *Neurospora crassa. Journal of Biological Chemistry* **247**, 1536–45.

LAZARUS, C. M. & TURNER, G. (1976). Novel extranuclear chloramphenicol-resistant phenotypes in *Aspergillus nidulans. Heredity* **37**, 150–1 (Abstract).

LÓPEZ-PÉREZ, M. J. & TURNER, G. (1975). Mitochondrial DNA from *Aspergillus nidulans. FEBS Letters* **58**, 159–63.

MARCOU, D. (1961). Notion de longévité et nature cytoplasmique du déterminant de la sénescence chez quelques champignons. *Annales des Sciences Naturelles 12ᵉ Série (Botanique)* **2**, 653–763.

MARCOU, D. & SCHECROUN, J. (1959). La sénescence chez *Podospora* pourrait être due à des particules cytoplasmiques infectantes. *Comptes Rendus de l'Academie des Sciences* **248**, 280–3.

MASON, J. R. & TURNER, G. (1975). Transmission and recombination of extranuclear genes during sexual crosses in *Aspergillus nidulans. Molecular and General Genetics* **143**, 93–9.

MITCHELL, M. B. & MITCHELL, H. K. (1952). A case of 'maternal' inheritance in *Neurospora crassa. Proceedings of the National Academy of Sciences of the United States of America* **38**, 442–9.

MITCHELL, M. B. & MITCHELL, H. K. (1956). A nuclear gene suppressor of a cytoplasmically inherited character in *Neurospora crassa. Journal of General Microbiology* **14**, 84–9.

MITCHELL, M. B., MITCHELL, H. K. & TISSIERES, A. (1953). Mendelian and non-Mendelian factors affecting the cytochrome system in *Neurospora crassa. Proceedings of the National Academy of Sciences of the United States of America* **39**, 606–13.

NAGLEY, P. & LINNANE, A. W. (1972). Biogenesis of mitochondria. XXI. Studies on the nature of the mitochondrial genome in yeast: The degenerative effects of ethidium bromide on mitochondrial genetic information in a respiratory competent strain. *Journal of Molecular Biology* **66**, 181–93.

NASS, M. M. K. & NASS, S. (1963a). Intramitochondrial fibers with DNA characteristics. I. Fixation and electron staining reactions. *Journal of Cell Biology* **19**, 593–611.

NASS, M. M. K., NASS, S. & AFZELIUS, B. A. (1965). The general occurrence of mitochondrial DNA. *Experimental Cell Research* **37**, 516–39.

NASS, S. & NASS, M. M. K. (1963b). Intramitochondrial fibers with DNA characteristics. II. Enzymatic and other hydrolytic treatments. *Journal of Cell Biology* **19**, 613–29.

ORGEL, L. E. (1963). The maintenance of the accuracy of protein synthesis and its relevance to ageing. *Proceedings of the National Academy of Sciences of the United States of America* **49**, 517–21.

PITTENGER, T. H. (1956). Synergism of two cytoplasmically inherited mutants in *Neurospora crassa. Proceedings of the National Academy of Sciences of the United States of America* **42**, 747–52.

REICH, E. & LUCK, D. J. L. (1966). Replication and inheritance of mitochondrial DNA. *Proceedings of the National Academy of Sciences of the United States of America* **55**, 1600–8.

REIJNDERS, L., KLEISEN, C. M., GRIVELL, L. A. & BORST, P. (1972). Hybridization studies with yeast mitochondrial RNAs. *Biochimica et Biophysica Acta*, **272**, 396–407.

RHOADES, M. M. (1946). Plastid mutations. *Cold Spring Harbor Symposia on Quantitative Biology* **11**, 202–7.

RIFKIN, M. R. & LUCK, D. J. L. (1971). Defective production of mitochondrial ribosomes in the *poky* mutant of *Neurospora crassa. Proceedings of the National Academy of Sciences of the United States of America* **68**, 287–90.

RIZET, G. (1952). Les phénomènes de barrage chez *Podospora anserina*. I. Analyse génétique des barrages entre souches S et s. *Revue de Cytologie et Biologie Végétale* **13**, 51–91.

RIZET, G. (1953). Sur l'impossibilité d'obtenir la multiplication végétative ininterrompue et illimitée de l'Ascomycete *Podospora anserina. Comptes Rendus de l'Academie des Sciences* **237**, 838–40.

RIZET, G. (1957). Les modifications qui conduisent à la sénescence chez

Podospora anserina sont-elles de nature cytoplasmique? *Comptes Rendus de l'Academie des Sciences* **244**, 663–5.

RIZET, G. & SCHECROUN, J. (1959). Sur les facteurs cytoplasmiques associés au couple des gènes S-s chez le *Podospora anserina*. *Comptes Rendux de l'Academie des Sciences* **249**, 2392–4.

ROBBERSON, D., ALONI, Y., ATTARDI, G. & DAVIDSON, N. (1972). Expression of the mitochondrial genome in HeLa cells. VIII. The relative position of ribosomal RNA genes in mitochondrial DNA. *Journal of Molecular Biology* **64**, 313–17.

ROWLANDS, R. T. & TURNER, G. (1973). Nuclear and extranuclear inheritance of oligomycin resistance in *Aspergillus nidulans*. *Molecular and General Genetics* **126**, 201–16.

ROWLANDS, R. T. & TURNER, G. (1974a). Physiological and biochemical studies of nuclear and extranuclear oligomycin-resistant mutants of *Aspergillus nidulans*. *Molecular and General Genetics* **132**, 73–88.

ROWLANDS, R. T. & TURNER, G. (1974b). Recombination between the extranuclear genes conferring oligomycin resistance and cold sensitivity in *Aspergillus nidulans*. *Molecular and General Genetics* **133**, 151–61.

ROWLANDS, R. T. & TURNER, G. (1975). Three-marker extranuclear mitochondrial crosses in *Aspergillus nidulans*. *Molecular and General Genetics* **141**, 69–79.

ROWLANDS, R. T. & TURNER, G. (1976). Interactions between nuclear and extranuclear oligomycin-resistant mutants of *Aspergillus nidulans*. *Heredity* **37**, 151–2 (Abstract).

SAGER, R. (1972). *Cytoplasmic genes and organelles.* New York and London: Academic Press.

SANDERS, J. P. M., HEYTING, C. & BORST, P. (1975). The organization of genes in yeast mitochondrial DNA. I. The genes for large and small ribosomal RNA are far apart. *Biochemical and Biophysical Research Communications* **65**, 699–70.

SCHONBAUM, G. R., BONNER, W. D. Jr., STOREY, B. T. & BAHR, J. T. (1971). Specific inhibition of the cyanide-insensitive respiratory pathway in plant mitochondria by hydroxamic acids. *Plant Physiology* **47**, 124–8.

SCOTT, W. A. & MITCHELL, H. K. (1969). Secondary modification of cytochrome c by *Neurospora crassa*. *Biochemistry* **8**, 4282–9.

SMITH, J. R. & RUBENSTEIN, I. (1973a). The development of 'senescence' in *Podospora anserina*. *Journal of General Microbiology* **76**, 283–96.

SMITH, J. R. & RUBENSTEIN, I. (1973b). Cytoplasmic inheritance of the timing of 'senescence' in *Podospora anserina*. *Journal of General Microbiology* **76**, 297–304.

SRB, A. M. (1963). Extrachromosomal factors in the genetic differentiation of *Neurospora*. *Symposia of the Society for Experimental Biology* **17**, 175–87.

THOMAS, D. Y. & WILKIE, D. (1968). Inhibition of mitochondrial synthesis in yeast by erythromycin: cytoplasmic and nuclear factors controlling resistance. *Genetical Research* **11**, 33–41.

TISSIERES, A., MITCHELL, H. K. & HASKINS, F. A.. (1953). Studies on the respiratory system of the poky strain of *Neurospora*. *Journal of Biological Chemistry* **205**, 423–33.

TURNER, G. & ROWLANDS, R. T. (1976). Cytochrome abnormalities and cyanide-resistant respiration in extranuclear mutants of *Aspergillus nidulans*. *Journal of Bacteriology* **125**, 389–97.

TZAGOLOFF, A. & MEAGHER, P. (1972). Assembly of the mitochondrial membrane system. VI. Mitochondrial synthesis of subunit proteins of the rutamycin-sensitive adenosine triphosphatase. *Journal of Biological Chemistry* **247**, 594–603.

UPSHALL, A. (1971). Phenotypic specificity of aneuploid states in *Aspergillus nidulans*. *Genetical Research* **18**, 167–71.

WALDRON, C. & ROBERTS, C. F. (1973). Cytoplasmic inheritance of a cold-sensitive mutant in *Aspergillus nidulans*. *Journal of General Microbiology* **78**, 379–81.

WARING, R. B. & SCAZZOCCHIO, C. (1976). Suppressors of a mitochondrially inherited mutation in *Aspergillus nidulans*. *Heredity* **37**, 153 (Abstract).

CHAPTER 22

Fungal Development and Metabolite Formation
J. F. MARTIN and A. L. DEMAIN

22.1 Introduction

The empirical exploitation of the biochemical activities of fungi by humans is an old phenomenon. Alcoholic fermentations were carried out in biblical times while the use of moulds to saccharify rice in the Koji process dates back at least to 700 AD (Sakaguchi, 1972). However, deliberate culture of fungi for large scale production of industrial products did not begin until the first world war. Unfortunately, large scale production of fungal metabolites has not been accompanied by biosynthetic studies at the enzymological level or by investigations of the relationship between production of metabolites and morphological changes in the producer organism. The cellular sites of synthesis and the mechanisms of secretion are generally unknown. The physiological roles of these metabolites, often excreted in huge amounts, in the producing cells are unknown. A large number of metabolites of industrial importance are produced by filamentous fungi. A selection is shown in Table 22.1.

22.2 Development and Differentiation

Differentiation has been defined by Pasternak (1970) as the stable development of an altered structure and function. This criterion, which emphasizes the irreversibility of the process, excludes transient changes (*e.g.* enzyme regulation) which are referred to as adaptation. Although many differentiation processes are to a certain extent reversible (*e.g.* sporulation up to a

Table 22.1 Metabolites of industrial importance

Metabolites	Producer organism	Industrial use
Ergot alkaloids	*Claviceps purpurea*	Medical/Pharmaceutical
Penicillins	*Penicillium chrysogenum*	Antibiotics
Cephalosporin	*Cephalosporium acremonium*	Antibiotics
Griseofulvin	*P. patulum*	Antibiotic
Gibberellins	*Gibberella fujikuroi*	Plant growth hormone
Kojic acid	*Aspergillus oryzae*	Food flavouring
Muscarine	*Clitocybe rivulosa*	Pharmaceutical
Transformed steroids	*Rhizopus* and other fungi	Pharmaceutical
Riboflavin	*Ashbya gossypii, Eremothecium ashbyii*	Vitamin
Gallic acid	*A. niger, P. glaucum*	Preparation of dyes
Citric acid	*Penicillium* and *Aspergillus*	Chelating agent; acidulant
Gluconic acid	*A. niger* and *Penicillium* sp.	Pharmaceutical
Itaconic acid	*A. itaconicus, A. terreus*	Synthetic polymers
Amylase	*R. delemar* and *Mucor* sp.	Enzyme
Pectinase	*P. glaucum*	Enzyme (juice clarification)
Glucose oxidase, catalase	*A. niger*	Enzymes

certain stage), this definition is essentially correct. Differentiation of fungal structures has been reviewed from several points of view by Bonner (1973) and Smith & Berry (1974).

Morphologically, fungi are a diverse group in nature, including yeasts, filamentous fungi (moulds) and macroscopic mushrooms (higher fungi). Differentiation in fungi is generally associated with the end of active growth (Smith & Galbraith, 1971). Indeed, sporulation in lower fungi, which is a typical differentiation process, remains unexpressed as long as growth is possible (Schaeffer, 1969). This may be interpreted to mean that alternative metabolic pathways are involved in growth and sporulation, making them incompatible. Morphological and physiological phases of development, frequently unique for a particular fungus, are determined by nutritional factors. A body of evidence has accumulated during recent years indicating that production of fungal metabolites of industrial importance is related to the development of the producer fungus.

According to Bartnicki-Garcia (1973), the morphology of fungi is determined by the presence of the rigid outer cell wall. Specific changes in cell wall composition and layer distribution during fungal development are related to cell wall morphogenesis in *Penicillium notatum* (Martin, Nicolas & Villanueva, 1973; Martin, Uruburu & Villanueva, 1973) and in a series of morphological mutants of *Aspergillus nidulans* (Zonneveld, 1974). However, morphological development in fungi cannot be attributed merely to cell wall morphogenesis. Underlying the process of morphogenesis are the mechanisms by which genetic information coded in DNA itself controls the formation of cellular enzymes. Two different mechanisms of gene expression appear to be involved. The first, a short-term mechanism, involves the timing at which differentiation is initiated and determination of particular pathways which are active at a given time. This is dictated by the concentration of nutrients which in turn influence the intracellular concentrations of substrates and low molecular weight regulatory effectors (Hanson, 1975).

Such a control mechanism is only transient in the everchanging environment of batch growth and therefore is not the only one responsible for changes occurring in differentiation. The second mechanism is a long-term effect dependent on changes in the enzyme composition of the cell which are under genetic control.

Solid media

Surface growth of filamentous fungi is heterogeneous and includes aerial, surface and substrate hyphae, each type requiring particular physiological conditions. Aerial hyphae must acquire their nutrients from those mycelia which are in contact with the substrate. A model of colonial growth of *Rhizopus stolonifer*, *Mucor racemosus*, *Penicillium chrysogenum*, *Aspergillus wentii*, *A. nidulans* and *A. niger* was suggested by Trinci (1971). He defined two growth zones in the colony: a central zone and a peripheral zone. Growth in the central zone results in an increase in hyphal density and the formation of fruiting structures but does not contribute to the radial expansion of the colony. Hyphae in the peripheral zone grow exponentially and supply protoplasm for the apical extension of the leading hyphae of the colony. The radial growth rate of these colonies was found to be a function of the length of the leading hyphae spanning the peripheral growth zone and the specific growth rate of these hyphae.

A correlation between fungal growth on solid media, differentiation, and product formation could be established, but because of the heterogeneous nature of this type of mycelial growth it is preferable to develop such a relationship in submerged culture.

Liquid media

In liquid media, the environment is much more homogeneous than on solid media and nutrients are readily available by diffusion. The exception is oxygen which has a very low solubility in water and usually becomes limiting unless high levels of oxygen transfer are maintained. In submerged culture, growth kinetics are dependent on the growth form (Smith & Anderson, 1973). Diffuse filamentous mycelia grow exponentially (Trinci, 1969; Carter & Bull, 1969), but very frequently, pellet formation occurs in static or poorly agitated flasks. The phenomenon of fungal pelleting has been reviewed by Whitaker & Long (1973). Formation of pellets depends on the particular strain, the size of inoculum, the growth medium and the physical environment. Yanagita & Kogane (1963) studied the structure of pellets by autoradiography after incorporation of ^{32}P into pellets of *Aspergillus niger*. Only the actively growing outer cortex of the pellets accumulated ^{32}P. A detailed model of the structure of pellets and kinetics of their growth has been proposed by Trinci (1970). The growth of the pellet is restricted by the diffusion of nutrients. Cube root kinetics of growth (rather than logarithmic growth) were maintained as long as the pellet grew at a constant rate. To avoid pellet formation, the shearing action of the shaking system may be increased by increasing agitation or by using baffled flasks. Pellet formation in liquid culture is an important factor in the production of certain fungal metabolites such as citric acid, penicillin, and itaconic acid (see later).

The biochemical basis of pellet formation is unknown. Pellet formation in *Penicillium chrysogenum* is due to the formation of short distorted hyphae

(Pirt & Callow, 1959). According to Righelato (1975), pellet formation occurs at high growth rates coinciding with a high degree of branching. Pellet formation is initiated by aggregation of conidia during germination. Changes in cell wall composition during spore swelling and germ tube elaboration appear to be responsible for pellet formation (Galbraith & Smith, 1969; Martin et al., 1973a, b). Restricted growth with a high incidence of branching has been described in a large number of colonial mutants. Some of these mutants have a greatly reduced content of wall peptide or carbohydrate, although a causal connection between modified cell wall composition and pellet formation has not been demonstrated (Scott, Mishra & Tatum, 1973; Wrathall & Tatum, 1974; Scott & Tatum, 1970; Brody & Nye, 1970; Scott & Salomon, 1975).

22.3 Primary and Secondary Metabolism

Although some consider the classical concepts of primary and secondary metabolism to be oversimplified, they are nevertheless useful. Primary metabolism involves an interrelated series of enzyme-mediated catabolic, amphibolic, and anabolic reactions which provide biosynthetic intermediates and energy, and convert biosynthetic precursors into essential macromolecules such as DNA, RNA, protein and polysaccharides. Primary metabolism is finely balanced and intermediates are rarely accumulated. On the other hand, secondary metabolites (idiolites) are those metabolites which are often produced in a phase subsequent to growth, have no function in growth (although they may have a survival function), are produced by certain restricted taxonomic groups of organisms, have unusual chemical structures, and are often formed as mixtures of closely-related members of a chemical family. Whereas primary metabolism is basically the same for all living systems, secondary metabolism is restricted to plants and microorganisms and is often species or strain specific. Bu'Lock (1975) interprets secondary metabolism as a manifestation of differentiation which accompanies unbalanced growth. In a few instances, processes have been devised in which primary metabolites such as citric acid accumulate in large amounts. Cultural conditions are often critical for their accumulation and in this sense, their accumulation resembles that of secondary metabolites.

Fungi are prolific producers of idiolites. Turner (1971) lists about a thousand secondary metabolites. Despite the large number, they are synthesized from only a few key precursors in pathways that comprise a relatively small number of reactions and which branch off from primary metabolism at a limited number of points (Bu'Lock, 1961, 1965). Acetyl-CoA is the most important precursor in fungal secondary metabolism, leading to polyketides, terpenes, steroids and metabolites derived from fatty acids. Other secondary metabolites are derived from intermediates of the shikimic acid pathway, the tricarboxylic acid cycle, and from amino acids. The regulation of the biosynthesis of secondary metabolites is similar to that of the primary processes, involving induction, feedback regulation and catabolite repression (Demain, 1968, 1972).

It is difficult to make a comprehensive summary of the production of fungal metabolites in relation to development since little if anything is known in most cases about the relevant biosynthetic processes or morphological differentiation in the producer strains. For this reason, we have restricted ourselves to a few representative primary and secondary metabolites which have industrial importance and/or have been the subject of significant research.

22.4 Phases of fermentations

Growth and metabolism of fungi in submerged batch culture pass through several phases. Borrow *et al.* (1961) established that the type and sequence of change in mycelial composition are determined by the initial composition of the medium. Three different stages were distinguished: balanced phase, storage phase and maintenance phase. During the balanced phase, growth occurs exponentially until an essential nutrient, usually containing nitrogen or phosphorus, is exhausted. Such balanced growth has been termed 'unlimited' by Righelato (1975), *i.e.* the rates of substrate uptake and utilization are independent of the nutrient concentration in the medium and are as high as possible for the system. Exhaustion of any nutrient makes balanced growth impossible and commits the cell to biochemical differentiation. The balanced growth phase has been called the 'trophophase' by Bu'Lock who considers nutrient exhaustion as a gradual process rather than a sudden event (Bu'Lock, 1975). The progressive utilization of substrates carries the culture through a series of stages characterized by increasingly limited growth. The decline in growth rate is therefore controlled by the rate at which the limiting substrate is depleted from the broth. Different degrees of nutrient exhaustion are required to initiate formation of different secondary metabolites. Although secondary metabolism is governed by external factors which bring about the limitation of growth, in those cases where irreversible differentiation ensues (*e.g.* sporulation), subsequent phases of development may be intrinsically rather than externally controlled.

Following balanced growth, a storage phase has been implicated by Borrow *et al.* (1961) during which cellular mass continues to increase due to accumulation of storage materials such as fatty acids and carbohydrates. Production of secondary metabolites starts in this phase. The third stage, according to Borrow *et al.* (1961) is the maintenance phase during which cellular mass remains constant but uptake of glucose and biosynthesis of secondary metabolites continues. Bu'Lock (1967) distinguished only two phases: trophophase which corresponds to the balanced phase of growth and idiophase (or secondary metabolism phase) which includes the storage and maintenance phases of Borrow *et al.* (1961).

Separation of phases according to changes in cellular mass may be of limited value in those cases where cell dry weight increases during idiophase. It must be realized that dry weight is often a poor indicator of true growth. Cellular mass is the summation of replicatory material (required for nucleic acid replication and protein synthesis) and assimilatory material; the latter results in heavier mycelium without true growth. In *Gibberella fujikuroi*, deposition of polysaccharide represents a major part of biomass accumulation during idiophase (Bu'Lock, 1975). The significance of replicatory growth is quite different from assimilatory processes with respect to product formation in idiophase (Bu'Lock, 1975). Growth and secondary metabolism should be understood as processes which compete for key metabolic intermediates rather than as mutually exclusive phenomena.

22.5 Acetate-derived Secondary Metabolites

Secondary metabolites derived from acetyl-CoA via the polyketide (acetate–malonate) pathway form one of the largest group of fungal metabolites (Turner, 1971). They include polyketide antibiotics such as cyanein produced by *Penicillium cyaneum* (Koman, Barath & Betina, 1969), patulin by *P. patulum* and *P. urticae*, citrinin by *P. notatum*, alternariol by *Alternaria utilis*, and orsenillic acid by *P. madriti*. Polyketides are usually formed during idiophase (Bu'Lock, 1967). When nutrient exhaustion occurs, tricarboxylic acid cycle activity is retarded since neither intermediates nor reduced pyridine nucleotides are removed from the cycle. Acetyl-CoA accumulates under these conditions and is converted into citric acid, free fatty acids, or secondary metabolites. High concentrations of fatty acids may in turn regulate the accumulation of other secondary metabolites (Martin, 1976).

Citrinin

Betina, Baunt, Hajnicka & Nadova (1973) reported two distinct maxima in rate of production of the pentaketide citrinin by *Penicillium notatum*. The first productive period occurred in trophophase and the second (which was greater) took place in idiophase following nutrient exhaustion. In media supporting high production, the initial production phase appeared to follow nitrogen depletion in the medium while the second was subsequent to exhaustion of reducing sugars. Mycelial lipids reached their maximum concentration between the two phases of citrinin synthesis (Koman & Betina, 1973). Maximum synthesis of fatty acids was also observed between trophophase and idiophase. It was speculated that the regulation of citrinin biosynthesis was coupled to that of fatty acids, although no direct supporting data were presented.

Patulin

Among the best studied examples of secondary metabolite production in relation to fungal development is that of patulin, a compound which is both an antibiotic and a mycotoxin. *Penicillium urticae* produces a series of polyketide metabolites. The parent metabolite is 6-methylsalicylic acid (6-MSA) from which a range of compounds including patulin are derived. The enzyme system responsible for the synthesis of 6-MSA is a multienzyme complex structurally and functionally similar to, but not identical with, fatty acid synthase (Dimroth, Walter & Lynen, 1970; Dimroth, Gruell, Seyffert & Lynen, 1972). Bu'Lock *et al.* (1965) showed two clear cut physiological phases in cultures of *P. urticae* producing patulin (Fig. 22.1). In the trophophase, mycelial nitrogen, phosphorus, SH-groups and RNA reached maximum values and glucose oxidation, mainly by the hexose monophosphate pathway, was very rapid. In idiophase, assimilation of N and P was reduced, RNA and SH-groups levels dropped, glucose was oxidized more slowly (mainly via the Embden–Meyerhof pathway), and fatty acids and patulin accumulated. The phase transition occurred sharply with a minimum of respiratory activity and acetate utilization.

Stages in the biosynthesis of patulin beginning with the synthesis of 6-MSA make their appearance successively. As shown in Fig. 22.2, labelled

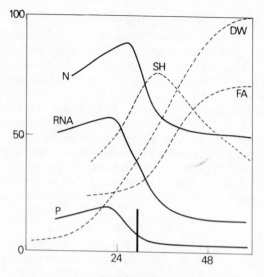

Fig. 22.1 Physiological phases accompanying the production of 6-methyl-salicyclic acid in *Penicillium urticae*. N, mycelial nitrogen; P, mycelial protein; RNA, mycelial ribonucleic acid; FA, mycelial fatty acids; all in μg/mg dry mycelial weight. DW, mycelial dry weight (mg/30 ml); SH, mycelial sulphydryl groups (counts/0.5 second). Vertical bar at first production of 6-methylsalicyclic acid (After Bu'Lock *et al.*, 1965).

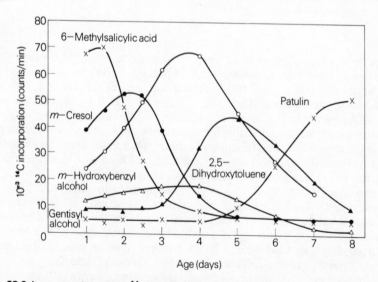

Fig. 22.2 Incorporation of $[1-{}^{14}C]$ acetate into intermediates of the patulin biosynthetic pathway. Mycelia after varying lengths of fermenter growth were incubated in fresh medium containing $[{}^{14}C]$ acetate. ${}^{14}C$-labelled metabolites excreted after 2 h were extracted and analysed by radioactive gas-liquid chromatography (After Murphy *et al.*, 1974).

acetate in incorporated by *Penicillium patulum* initially into 6-MSA, (24–48 h), then into *m*-cresol (48–72 h), *m*-hydroxybenzylalcohol (72–96 h), gentisyl alcohol (96–144 h) and finally into patulin (after 144 h) (Murphy, Vogel, Kripphahl & Lynen, 1974). Bu'Lock *et al.* (1969), with the aid of selective inhibitors of protein synthesis, concluded that 6-MSA synthase is produced during trophophase before 6-MSA synthesis. However, Light (1967) found that single colony isolates of *P. patulum* differ considerably in their timing of 6-MSA synthase formation. After a shift down from rich medium to Czapek–Dox medium, 'early' strains produced high synthase activity soon after transfer while 'late' strains produced low levels of 6-MSA synthase 20 h after the transfer. Once formed, 6-MSA synthase was metabolically stable supporting 6-MSA synthesis when protein synthesis was blocked with inhibitors. 6-MSA synthesis is absent when *P. urticae* grows in a rich medium, indicating regulation by catabolite repression or the need of substrate limitation for initiation of 6-MSA synthesis. The enzymes effecting further metabolism of 6-MSA seem to be formed in idiophase after 6-MSA synthesis has started; they are metabolically labile since continued protein synthesis is required to maintain their activity. Two NADPH-dependent aromatic dehydrogenases catalyzing the reversible conversion of gentisyl alcohol and *m*-hydroxybenzyl alcohol to their corresponding aldehydes have been partially purified (Scott & Beadling, 1974). One of these enzymes, *m*-hydroxybenzylalcohol dehydrogenase, has been characterized by Forrester & Gaucher (1972). The enzyme appears to be inducible but it is not induced by early precursors of the pathway in the presence of fresh medium. The significance of the lack of induction in fresh medium is unclear. It would be desirable to know if early precursors do induce the enzyme in spent medium.

If 6-MSA synthase is formed during growth without immediate formation of 6-MSA, it is interesting to ask what mechanisms control its activity. Bu'Lock *et al.* (1969) suggested either a simple increase in substrate availability during transition from trophophase to idiophase (an hypothesis supported by the results of Murphy *et al.* (1974) on the sequential formation of intermediates) or some type of allosteric activation. Regulation during idiophase of 'late' enzymes in the patulin sequence appears to involve both induction by 6-MSA and a nutritionally-mediated repression by a critical level of some catabolite (Bu'Lock *et al.*, 1969). Thus, phase-linked patulin biosynthesis in *Penicillium urticae* may be controlled by counterbalancing systems of induction and catabolite repression. Of interest here is the observation that dibutyryl cyclic AMP promoted a significant increase in *m*-hydroxybenzylalcohol dehydrogenase activity (Forrester & Gaucher, 1972).

22.6 Gibberellins

The fungal gibberellins form part of a large family of compounds also found in higher plants. The production of gibberellic acid and other gibberellins by *Gibberella fujikuroi* (conidial state of *Fusarium moniliforme*) has practical interest because of the activity of these compounds as plant regulators and their use in the malting process. A comprehensive study of the metabolism of *G. fujikuroi* under glucose, nitrogen, phosphorus and magnesium limitation was published by Borrow *et al.* (1961). The growth pattern and mycelial composition of *G. fujikuroi* in submerged culture were controlled by the rate and sequence in which individual nutrients were exhausted from the culture medium. As mentioned above, Borrow *et al.* (1961) distinguished a balanced phase, storage phase and a maintenance phase. No gibberellic acid was produced during the balanced phase, which was a period of proliferation lasting until nitrogen exhaustion. Production of gibberellic acid started early in the storage phase after growth had ceased. Some residual increase in dry weight could be attributed largely to the accumulation of carbohydrates and triglycerides which reached maximum values of 32 to 45% of the dry weight. Only negligible amounts of gibberellic acid were produced in glucose-limited fermentations. In nitrogen-limited fermentations, production continued linearly throughout the maintenance phase if glucose remained available. When glucose was exhausted, the culture underwent hyphal breakdown, a decrease in dry weight and liberation of mycelial components into the medium (Borrow *et al.*, 1964).

Hyphae were sparsely branched during early growth. Branching increased concomitant with a decrease in length of the hyphae as growth continued. Different morphological changes occurred in *Gibberella* after nutrient exhaustion depending upon the particular nutrient involved. After nitrogen exhaustion induced gibberellic acid production, lipid bodies increased in size and number coalescing later in the fermentation. Glucose exhaustion was followed by rapid vacuolization, giving rise to cells with the cytoplasm restricted to a thin layer inside the cell wall (Borrow *et al.*, 1964).

Bu'Lock, Detroy, Hostalek & Munin-Al-Shakarchi (1974) studied two secondary processes in *Gibberella* which were found to be regulated independently even though both use acetyl-CoA as precursor. One process leads through the diterpene pathway to the gibberellins; the other utilized a multienzyme polyketide synthase to the bikaverin pigments. Both types of product are produced in idiophase but the onset of bikaverin synthesis occurs before that of gibberellin synthesis (Fig. 22.3). Working with batch and chemostat cultures, Bu'Lock and co-workers concluded that each biosynthetic process responds to different levels of a common growth-linked regulator. Full derepression of gibberellic acid synthesis occurs only at the very lowest level of non-carbon nutrient whereas bikaverin synthesis is fully derepressed at levels which are markedly higher, although still well below the optimum for growth. Rather than a clearcut separation of trophophase from idiophase, there is a graded relationship between expression of growth-linked and nongrowth-linked metabolism. Using the same concept, it is possible to explain the early formation of 6-MSA synthase and the late formation of patulin synthases.

22.7 Alkaloids

Ergot alkaloids

The ergot alkaloids are the best known of the plant alkaloids because their formation by fungi makes their biosynthesis more amenable to experimental research (Floss, Robbers & Heinstein, 1974). Until recently, ergot alkaloids were obtained exclusively by extraction from sclerotia formed by parasitic species of the genus *Claviceps*. Today large quantities of ergot alkaloids are produced in submerged culture by these fungi.

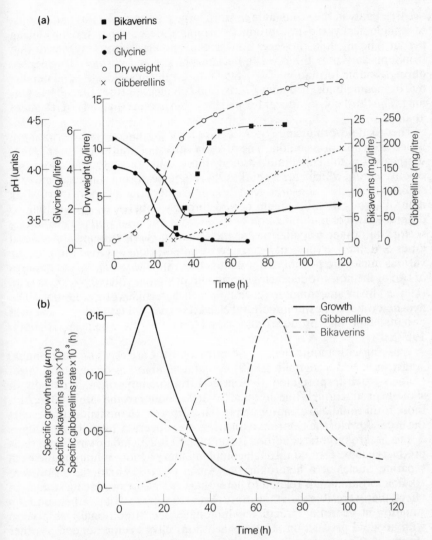

Fig. 22.3 Phase development accompanying bikaverin and gibberellin synthesis by *Gibberella fujikuroi*. (a) Batch culture data on a medium limited by glycine. (b) Data from (a) calculated as specific (gravimetric) rates (After Bu'Lock *et al.*, 1974).

Early studies on the production of alkaloids by various species of *Claviceps* (Taber, 1964; Amici *et al.*, 1967a) indicated that formation of ergot alkaloids occurs in the idiophase following a distinct trophophase. Taber (1964) described a very late onset of alkaloid production in both complex and synthetic media, after cessation of primary metabolism and while dry weight was actually decreasing. Amici *et al.* (1967a) compared the physiological and morphological differentiation of a strain of *Claviceps purpurea* producing a large amount of peptide alkaloids with three non-producer strains derived from it. It was observed that alkaloid production occurred only in the strain able to store large amounts of lipids and sterols. Morphologically the non-producer strains differed from the producing parent. The alkaloid-producer grew as thin and filamentous mycelia during trophophase. Later the mycelia became thicker and partially fragmented during alkaloid formation. Two of the non-producer strains showed numerous branching buds, irregular forms, and fragmented from the very beginning. The third non-producer formed long thin hyphae that did not fragment at all.

The alkaloid-producing strain of *Claviceps purpurea* is a heterokaryon which does not form conidia. The mycelia consist of multinucleate cells that segregate on solid medium forming three different coloured classes of colony sectors which correspond to three non-sectoring single component strains that are non-producers. The disjunction of non-sporulating heterokaryons in solid media occurs through unbalanced growth due to different rates of multiplication of the various nuclei; it is made evident by sectors and other irregularities of the cultures. Homokaryons and several kinds of heterokaryons characterized by different quantitative ratios of the various nuclei are possible after disjunction (Spalla, 1973). Mixed cultures of two of the non-producer single component strains isolated by Amici *et al.* (1967*a*) form anastomoses reaching a neo-heterokaryotic condition. The neo-heterokaryon exhibits cultural characteristics similar to that of the wild parental type and produces alkaloids (Amici, Scotti, Spalla & Tognolli, 1967*b*).

Working with another heterokaryotic strain of *Claviceps purpurea*, Spalla, Amici, Scotti & Tognolli (1969) found that it occasionally sporulates. Isolates from unsporulated mycelia and from conidia differ drastically in alkaloid production. While mycelial isolates produce from 700–2000 μg/ml, those from conidia are unproductive. The appearance of conidia represents the breakdown of the heterokaryon and the loss of ergot-producing ability.

The heterokaryotic condition is common in fungi, and enzymes and their products can be formed through complementary action of mutant alleles in separate nuclei of a heterokaryon. Amici *et al.* (1967*b*) suggested that alkaloid production is related to heterokaryosis of the producing strains. In effect, they were able to demonstrate that heterokaryosis is a rather common situation in sclerotia, structures which have been traditionally associated with alkaloid production. These conclusions have been extended to other strains of *Claviceps purpurea* and *C. paspali* (Spalla *et al.*, 1969; Spalla, 1973).

The production of ergot alkaloids by *Claviceps purpurea* in relation to cellular differentiation has been studied by Rehacek and co-workers (Rahacek *et al.*, 1971; Vorisek, Ludvik & Rehacek, 1974; Rehacek & Kozova, 1975) and Mantle and collaborators (Mantle & Tonolo, 1968; Mantle, Morris & Hall, 1969). Vorisek *et al.* (1974), studying the ultrastructure of an alkaloid-producing culture of *C. purpurea*, found that it was polymorphic. Submerged cultures of *C. purpurea* grown on simple defined medium consisted of hyphae and two types of ovoid cells: conidia and chlamydospores. The conidia had one nucleus and were characterized by granular cytoplasm with lipid bodies and large vacuoles. Following the fifth day of culture, conidia germinated into dividing vegetative cells. The chlamydospores differed from the conidia in that their homogeneous cytoplasm contained numerous lipid inclusions but no vacuoles. Chlamydospores were primarily dumb-bell shaped and usually had two nuclei. They were formed by conversion of terminal hyphal cells. Chlamydospores reached their maximum number at five days when they represented the greater part of the culture and prevailed during alkaloid formation which started on the seventh day of cultivation. Mantle & Tonolo (1968) studying the tissue of cells of natural (parasitic) ergot sclerotia and cells in submerged culture, termed cells similar to that of natural tissue as 'plectenchymatic cells'. This term refers to fungal tissue formed by polygonal cells which are derived from filamentous hyphae but have lost the filamentous arrangement. Mantle & Tonolo (1968) and Mantle et al. (1969) established a correlation between alkaloid production in submerged culture and the formation of plectenchymatic pellets or short thick hyphal filaments.

Study of the ultrastructure of chlamydospores in alkaloid-synthesizing cells revealed an interesting organelle that Vorisek et al. (1974) termed 'lipid-adsorbing bodies'. They are spherical electron-dense bodies, 0.3–1.0 μm in diameter, whose membrane boundaries are frequently penetrated by adhering lipid droplets. Although the direction of movement of the lipid droplets across the membrane of lipid-adsorbing bodies could not be established, it was speculated that these bodies were related to the formation of alkaloids by mobilizing cell lipid reserves. In a later publication, Rehacek & Kozova (1975) compared cytodifferentiation (formation new internal structures) and biochemical differentiation (formation of alkaloid-related enzymes) in Claviceps purpurea during alkaloid production. They concluded that cytodifferentiation preceded biochemical differentiation although their data were not conclusive. The onset of alkaloid synthesis was characterized by predominance of chlamydospores, the presence of vegetative cells with reduced or arrested proliferation, maximum acetyl-CoA carboxylase activity, maximum activity of fatty acid synthesis, and an enlarged pool of tryptophan. Tryptophan is a building block and an inducer of alkaloid production (see review by Floss et al., 1974). Rehacek & Malik (1971) and Rehacek, Kozova, Sajdl & Vorisek (1974) established an inverse relationship between conidia formation and alkaloid synthesis. Retardation of conidia formation and accentuated alkaloid biosynthesis were observed in the presence of increased tryptophan pool levels.

Alkaloids are typical secondary metabolites. What is known about the biochemical differentiation leading to the production of alkaloid synthases and their regulation during the fermentation process has been reviewed by Floss et al. (1974). Alkaloid biosynthetic enzymes appear to be repressed during the rapid growth phase and later induced by tryptophan. Growth-promoting concentrations of phosphate suppresses alkaloid synthesis, a phenomenon common in secondary metabolism. Phosphate inhibition of alkaloid synthesis is overcome by exogenous tryptophan (Robbers et al., 1972). The intracellular pool of tryptophan experiences a temporary two to three-fold increase prior to alkaloid biosynthesis in normal fermentation media (Fig. 22.4). Heinstein, Lee & Floss (1971) found the first enzyme of alkaloid biosynthesis, dimethylallylpyrophosphate: tryptophan dimethylallyl transferase, to appear just before the onset of alkaloid biosynthesis.

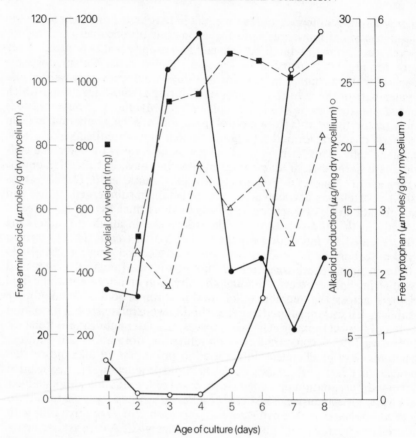

Fig. 22.4 Tryptophan pool in relation to onset of ergot alkaloid production in *Claviceps* sp. SD-58 (After Roberts *et al.*, 1972).

Quinoline alkaloids

The alkaloid-synthesizing system of *Penicillium cyclopium*, which produces the benzodiazepines cyclopenin and cyclopenol and derives from them the quinoline alkaloids viridicatin and viridicatol, has been studied by Luckner and co-workers (Nover & Lucker, 1974; Wilson & Luckner, 1975; Nover & Muller, 1975). Differentiation proceeding during idiophase of *P. cyclopium* involves the transformation of penicilli, the detachment and ripening of conidia, and the formation of enzymes catalyzing alkaloid biosynthesis. While in *Claviceps purpurea*, the formation of conidia usually leads to termination of alkaloid biosynthesis, the formation of alkaloids in *P. cyclopium* seems to coincide with conidial formation. Prominent morphogenic features at the beginning of idiophase development are the formation of penicilli and conidia. The ripening of the conidia coincides with the formation of cyclopenase, an enzyme of alkaloid metabolism which is restricted to conidia (Nover & Luckner, 1974). This enzyme has been identified as a constituent of the inner side of the plasma membrane of conidiospores of *P. cyclopium* (Wilson & Luckner, 1975). The appearance of cyclopenase activity in this system follows 12–24 h after conidial detachment and parallels alkaloid production (Nover & Luckner, 1974) (Fig. 22.5).

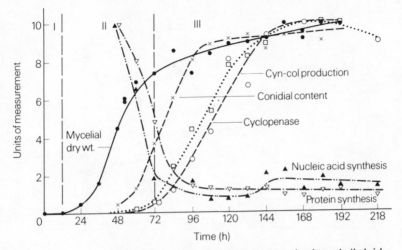

Fig. 22.5 Relationship between conidiation, cyclopenase synthesis and alkaloid formation by *Penicillium cyclopium*. Replacement cultures. I, germination phase; II, trophophase; III, idiophase. Mycelial dry weight (10 = 540 mg/100 cm^2 culture area); conidia (10 = 650 mg conidia/100 cm^2 culture area); cyclopenase (10 = 110/100 cm^2 culture area); rate of cy-col (cyclopenin + cyclopenol) production (10 = 1.2 mg/h/100 cm^2 culture area); rates of protein and nucleic acid synthesis (10 = 100% corresponding to maximal incorporation of ^{3}H-phe or ^{32}Pi in 48 h) (After Nover & Luckner, 1974).

Biosynthesis of alkaloids by partially synchronized emerged (wet surface) cultures of *Penicillium cyclopium* requires protein biosynthesis (Nover & Muller, 1975). Inhibition of protein synthesis by cycloheximide immediately stops alkaloid biosynthesis. However, subsequent removal of the inhibitor results in extremely high alkaloid biosynthetic rates surpassing those in control cultures. To explain the superinduction phenomenon, the existence of a regulatory protein interfering with the formation of the rate-limiting enzyme of alkaloid biosynthesis was postulated.

22.8 Cephalosporins

An interesting case of morphological differentiation in relation to production of secondary metabolites is that of formation of arthrospores and production of cephalosporins by *Cephalosporium acremonium*. The time-course of cephalosporin C fermentation in chemically defined medium shows that the antibiotic is an idiophase product (Demain, 1963) (Fig. 22.6). Glucose is essentially exhausted during the growth phase which takes place during the first 2.5 days. Sucrose is not utilized during this period. After glucose exhaustion, sucrose is utilized for cephalosporin production at a slower rate than that of glucose. This has been interpreted as catabolite repression of cephalosporin biosynthetic enzymes (Demain, 1968). A similar pattern of fermentation was described by Nüesch, Treichler & Liersch

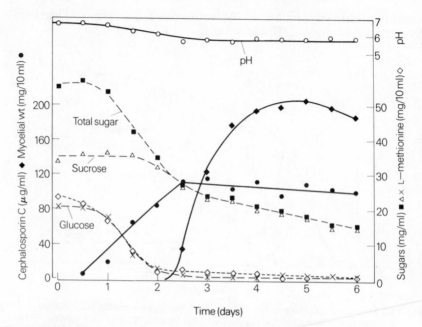

Fig. 22.6 Time course of cephalosporin C production by *Cephalosporium acremonium* mutant 8650 in chemically-defined medium (After Demain, 1963).

(1973). Studies on the biosynthesis of cephalosporin by cell suspensions in phosphate buffer indicate that the capacity of the mycelia to synthesize a certain amount of cephalosporin is strictly dependent on the physiological age of the mycelium. Sequential formation of cephalosporin C and from it, deacetylcephalosporin C, has been shown by Fujisawa *et al.* (1975). While accumulation of cephalosporin started on the second day and reached a maximum on the fourth to fifth day, quantitative conversion into deacetyl-cephalosporin C took place only after the fifth day.

Submerged cultures of *Cephalosporium acremonium* contain various morphological types representing stages in the growth cycle. Observations of morphological differentiation in *Cephalosporium* were initially linked to the onset of antibiotic production by Abraham, Newton & Warren (1964) and Caltrider & Niss (1966). Smith, Warren, Newton & Abraham (1967) showed that the maximum rate of cephalosporin C production coincided with the conversion of slender hyphal filaments to swollen hyphal fragments, *i.e.* arthrospores.

Methionine stimulates biosynthesis of cephalosporin although the stimulatory mechanism is not clear. On the basis of structural analogy, it was first believed that methionine might be a precursor of the α-aminoadipic acid side chain. However, radiotracer studies with labelled methionine indicated that no radioactivity was incorporated into the side chain of cephalosporins (Abraham *et al.*, 1964). Demain & Newkirk (1962) confirmed the stimulation of cephalosporin C biosynthesis by methionine, especially by the D-isomer. In defined medium, norleucine (a non-sulphur structural analogue of methionine) could replace methionine (Demain,

Newkirk & Hendlin, 1963). It was suggested that methionine could have a regulatory role as enzyme repressor or inducer. Caltrider & Niss (1966) studying the incorporation of ^{35}S-methionine into cephalosporin obtained data indicating that methionine plays an essential role as a sulfur donor for the cephalosporin nucleus. Methionine and norleucine were found to act by a common mechanism (Drew & Demain, 1973). Using mutants blocked in the trans-sulphuration process, Drew & Demain (1975) obtained data indicating that although methionine acts as a sulphur donor, this is not the reason for its ability to stimulate biosynthesis of cephalosporin C.

Methionine not only enhances antibiotic synthesis but also increases fragmentation of the hyphae (Caltrider, Huber & Day, 1968). Morphologically the mycelia grown in sulphate medium are generally filamentous, while mycelia grown in methionine-supplemented medium are highly fragmented, swollen, irregular and sporulate to a great degree. Other sulphur amino acid intermediates such as cysteine, cystathionine and homocysteine, which do not stimulate cephalosporin synthesis, yield filamentous growth (Nüesch *et al.*, 1973). Nash & Huber (1971) separated four major morphological types by centrifugation in sucrose gradients: hyphae, arthrospores (fragmented hyphae), conidia, and germinated conidia (germlings) (Fig. 22.7). The distribution of these forms suggested that the growth cycle of *Cephalosporium acremonium* in submerged culture involves a period of initial hyphal development followed after 48 h by extensive fragmentation into arthrospores and formation of conidia. Although total separation of the different forms could not be achieved, the arthrospore-rich fraction was more active in incorporating ^{14}C-valine into β-lactam antibiotics. These findings confirmed earlier observations showing that differentiation of the hyphae into arthrospores coincided with maximum rates of cephalosporin formation. Strains with enhanced cephalosporin synthesis might be expected to have increased differentiation, and environmental conditions and mutational events which enhance β-lactam antibiotic synthesis should also stimulate the formation of arthrospores. This hypothesis was tested by Nash & Huber (1971) and a direct relationship between formation of arthrospores and synthesis of β-lactam antibiotics by mutants of different production ability was found, suggesting that increased differentiation is a determinant in strains with enhanced antibiotic production. The fact that a mutant incapable of producing cephalosporin formed arthrospores does not disturb this relationship since differentiation occurs before antibiotic formation. A nonarthrospore forming mutant capable of producing cephalosporin has not yet been reported. Drew, Winstanley & Demain (1976) showed recently that norleucine mimics the effect of methionine on differentiation. Supplementation of defined medium with either methionine or norleucine resulted in increased fragmentation of the hyphae. Since norleucine contains no sulphur, it suggested that the role of methionine in regulation of differentiation is divorced from its role as sulphur donor for antibiotic production. Therefore, the superior synthetic potential of arthrospores appears to be related to some metabolic characteristic which permits them to be better producers of β-lactam antibiotics than other morphological forms.

Carbohydrate metabolism seems to be important for dimorphism in yeast-like mutants of *Cephalosporium sp.* (Jicinska, 1974). However, no data are available on any possible correlation of morphological differentiation and antibiotic production in these yeast-like mutants.

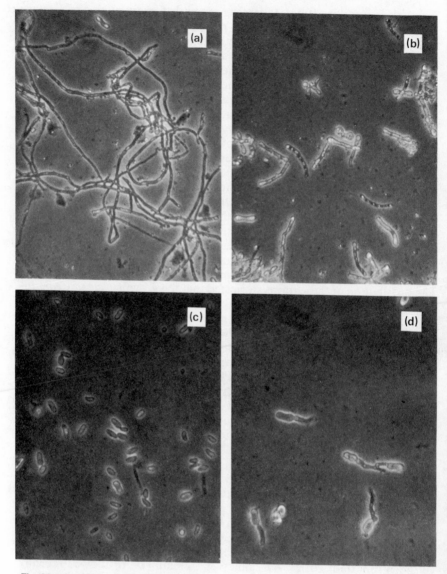

Fig. 22.7 Various morphological forms observed in submerged cultures of *Cephalosporium acremonium* (A) hyphae; (B) arthrospores; (C) conidia; (D) germinating conidia (After Nash & Huber, 1971).

22.9 Penicillins

Biosynthesis of penicillin has received a great deal of attention (see reviews by Lemke & Brannon, 1972; Abraham, 1974; Demain, 1974). Since the early days of the penicillin fermentation, this process has been divided into three phases (Koffler, Emerson, Perlman & Burris, 1945; Hockenhull, 1959): (i) trophophase; (ii) idiophase; and (iii) a phase of senescence with cellular autolysis. An analysis of the dynamics of batch penicillin fermentations showed that the specific rate of penicillin synthesis was greatest when the specific rates of growth, carbohydrate utilization and oxygen uptake declined (Gaden, 1955). Glucose is readily metabolized by *Penicillium chrysogenum* and the level of glucose during the fermentation greatly influences penicillin formation (Stefaniak, Gailey, Jarvis & Johnson, 1946; Johnson, 1952). Rapid utilization of glucose supports a high growth rate but is poor for antibiotic production. Lactose, a slowly-utilized dissacharide, is an excellent carbon source for antibiotic formation. When a mixture of glucose and lactose is used as carbon source, glucose is rapidly used during growth followed by lactose utilization during penicillin production (Demain, 1968). Slow-feeding of glucose has the same beneficial effect as batch-feeding of lactose (Soltero & Johnson, 1953). Maximum penicillin production by early strains of *P. chrysogenum* took place at a rather slow glucose feed rate (0.03% of sugar/h) (Soltero & Johnson, 1954).

Although data from batch cultures indicate that antibiotic is produced in an idiophase in which little or no growth occurs, glucose-limited continuous culture experiments indicate that penicillin can be produced at moderately high growth rates (Pirt & Callow, 1960; Pirt & Righelato, 1967). The restriction of penicillin synthesis to the period after the rapid growth of the organism in batch culture may thus reflect a dependency of penicillin biosynthesis on something other than low growth rates. In continuous culture, the specific production rate of penicillin was constant at growth rates between 0.023 and 0.086 h^{-1}. When growth was eliminated by restricting the glucose supply to a 'maintenance ration' (*i.e.* the glucose requirement at zero growth rate with all the other nutrients in excess), the specific production rate of penicillin fell linearly to zero at a rate inversely related to the previous growth rate. In other words, when the specific growth rate was decreased to below one eighth of the maximum value, the penicillin biosynthetic activity could not be maintained. Glucose supplied in excess inhibited the decay of penicillin production. It appears that in a good batch fermentation, an essential condition is a fast initial growth phase to give a high organism concentration followed by a phase of slow growth to minimize the decay rate of penicillin biosynthesis.

The biosynthesis of penicillin in continuous culture in relation to the metabolic activity and morphology of *Penicillium chrysogenum* has been studied by Righelato, Trinci, Pirt & Peat (1968). Filamentous mycelia predominated at growth rates between 0.023 to 0.075 h^{-1}. The occurrence of pellets and swollen organisms increased as the growth rate was increased. At very low growth rates, conidiation occurred. The most marked morphological change in non-growing cultures supplied with the maintenance ration of glucose was an increase in hyphal vacuolization including the hyphal tip. Another morphological change occurring while glucose was supplied at the maintenance ration was the appearance of phialides and conidia. Only one spore was attached to each phialide; the formation of spore chains characteristic of growth on solid medium was apparently prevented by the vigorous agitation of the culture. Morphological differentiation was accompanied by decreased oxidation of glucose and increased breakdown and resynthesis of nucleic acids. Autolysis of *P. chrysogenum* did not occur while supplying the maintenance ration of glucose but the mould lysed when glucose feed was stopped. Trinci & Righelato (1970) studied the changes in constituents and ultrastructure of hyphal compartments during autolysis of glucose-starved chemostat cultures of *P. chrysogenum*. Protein, RNA, DNA and endogenous respiration decreased rapidly during the first hours of starvation, but carbohydrate decreased only after two days of starvation. Lysis of organelles in individual hyphal compartments occurred synchronously. Ribosomes were rapidly degraded, but membranes were resistant to breakdown. Autolysis of separate compartments in the same hypha was not synchronized. Some cytologically normal hyphal compartments were present at all stages, suggesting that maintenance and/or cryptic growth of some hyphae occurred at the expense of others.

Pellets are sometimes formed in penicillin fermentations in baffled flasks and in large-scale fermenters. The highest penicillin yields are obtained when the mycelia are present as short fragments, but the agitation must not be intense enough to cause excessive mechanical damage. Pellet formation is undesirable in penicillin fermentations (Phillips, 1966). Pirt & Callow (1959) found that pellets in continuous culture were unable to maintain a steady state of growth and their rate of penicillin production was low. The inner core of the pellet is oxygen-limited and probably undergoes lysis.

According to Camici, Sermonti & Chain (1952) the formation of pellets *vs.* filamentous forms in the penicillin fermentation depends upon the number of germinating conidia allowed to grow freely in submerged culture. The higher the number of germinating conidia, the smaller is the final diameter of the pellet. When the number of germinating conidia is very high in relation to the growth supporting capacity of the medium (above 10^6/ml), growth takes place mainly as filaments.

In pellets, autolysis starts at the centre and gradually proceeds toward the periphery leaving an empty space in the centre and a hollow shell of living hyphae oriented outwardly (Camici *et al.*, 1952).

22.10 Citric Acid

Citric acid, a tricarboxylic acid cycle intermediate, is accumulated at very high levels, comparable to that of the secondary metabolites, by *Aspergillus*. Citric acid is widely used in the food and pharmaceutical industries. Because of its industrial importance, the biosynthesis of citric acid by moulds has been the subject of numerous investigations. Citric acid in *Aspergillus niger* is synthesized by condensation of acetyl-CoA and oxalacetate. Sugar in molasses is broken down to pyruvate and acetate which are quantitatively converted into citrate. The accumulation of citric acid appears to be due to the disappearance of aconitase and isocitrate dehydrogenase and the increase in citrate synthase after the phase of rapid growth (Ramakrishnan, Steel & Lentz, 1955; Martin, 1957).

Production of citric acid is related to morphology during growth in liquid medium. Under conditions used in submerged fermentation of ferrocyanide-treated beet molasses, *Aspergillus niger* develops as pellets. According to Clark, Ito & Horitsu (1966), the filamentous growth form has little capacity to produce citric acid and its occurrence always results in poor yields. Steel, Martin & Lentz (1954) varied the size of inoculum, pH and concentration of potassium ferrocyanide in the medium and obtained a complete range of morphological types from filamentous mycelia to pellets. Filamentous growth was obtained with low concentrations of ferrocyanide or an acidic pH. These results were confirmed by Clark (1962). Large soft filamentous clumps were obtained in absence of ferrocyanide and hard pellets in concentrations of ferrocyanide above 10 μg/ml.

Inoculum size also affects the morphology of pellets (Yanagita & Kogane, 1963). Use of large inocula yields white globose pellets with smooth surfaces while small inocula result in large ovoid pellets with rough surfaces. However, it should be realized that changes in morphology and pellet aggregation do occur as growth proceeds during a citric acid fermentation (Martin & Waters, 1952). Galbraith & Smith (1969) confirmed earlier reports on the extreme dependence of pellet formation of *Aspergillus niger* on the pH of the culture. Filamentous growth is obtained at a pH below 2. It was proposed that pH affects cell wall structure and surface charges interacting with pellet formation.

Carilli, Chain, Gualandi & Morisi (1961) observed that the filamentous form of *Aspergillus niger* reduced the oxygen level in the medium to zero after 20 h while growth of pellets resulted in a dissolved oxygen level which was considerably higher.

In recent years, control of pellet formation has been sought by addition of chelating agents (Choudhary & Pirt, 1965). Metal-complexing agents, such as ethylenediaminetetraacetic acid (EDTA) and ferrocyanide, change the morphology to small discrete pellets instead of the gelatinous conglomerates existing in cultures without complexing agents. Although it is generally accepted that pellet formation is a desirable aspect of the citric acid fermentation, Takahashi, Hidaka & Yamada (1965) suggested that the filamentous form of *Aspergillus niger* obtained by addition of surface-active agents to the medium results in higher citric acid production.

Pulpy (filamentous) mycelia of *Aspergillus niger* with increased respiration rates (up to 200% of that in the control) and high cellular growth have been obtained by addition to the culture of the water-soluble surfactant polymer, carboxypolymethylene (carbopol) (Elmayergi, Scharer & Moo-Yung, 1973). Increased amylase production was observed with this type of filamentous growth. Only anionic polymers dissociating into free ions influenced the degree of flocculation of spores prior to germination. It seems that the ionized carboxyl groups of the polymer induced electrostatic repulsion among the spores, initiating pulpy growth (Elmayergi, 1975). Production of α-galactosidase by pellets of *Mortierella vinacea* decreased with the increase of pellet size. Pellets smaller than 250 μm in diameter had almost the same activity as the ground mycelium but for larger units, production decreased with the increase of the pellet size (Kobayashi & Suzuki, 1972).

Formation of clumps by conidia of *Aspergillus niger* is promoted by cAMP and 5-GMP but not by AMP, ADP, dibutyryl cAMP, adenosine, cGMP, guanosine, GDP or GTP (Wold & Suzuki, 1973a). The reason is unclear at this time but it is probably related to cyclic AMP-mediated control of morphology and dimorphism in fungi (Larsen & Sypherd, 1974; Scott & Solomon, 1975). It is also of interest that cAMP stimulates citric acid production in *A. niger* (Wold & Suzuki, 1973b).

The preparation of this chapter was supported by National Science Foundation grant BMS 75-17527.

22.11 References

ABRAHAM, E. P. (1974). *Biosynthesis and enzymatic hydrolysis of penicillins and cephalosporins.* Tokyo: University of Tokyo Press.

ABRAHAM, E. P., NEWTON, G. G. F. & WARREN, S. C. (1964). Studies on the biosynthesis of the cephalosporins. *IAM symposium of Applied Microbiology (Tokyo)* **6**, 79–96.

AMICI, A. M., SCOTTI, T., SPALLA, C. & TOGNOLLI, L. (1967*b*). Heterokaryosis and alkaloid production in *Claviceps purpurea. Applied Microbiology* **15**, 611–15.

AMICI, A. M., MINGHETT, A., SCOTTI, T., SPALLA, C. & TOGNOLLI, L. (1967*a*). Ergotamine production in submerged culture and physiology of *Claviceps purpurea. Applied Microbiology* **15**, 597–602.

BARTNICKI-GARCIA, S. (1973). Fundamental aspects of hyphal morphogenesis. *Symposium of the Society of General Microbiology* **23**, 245–67.

BETINA, V., BAUNT, S., HAJNICKA, V. & NADOVA, A. (1973). Diphasic production of secondary metabolites by *Penicillium notatum* Westling S-52. *Folia Microbiologica* **18**, 40–8.

BONNER, J. T. (1973). Development in lower organisms. *Symposium of the Society of General Microbiology* **23**, 1–7.

BORROW, A., JEFFREYS, E. G., KESSEL, R. H. J., LLOYD, E. C., LLOYD, P. B. & NIXON, I. S. (1961). The metabolism of *Gibberella fujikuroi* in stirred culture. *Canadian Journal of Microbiology* **7**, 227–76.

BORROW, A., BROWN, S., JEFFREYS, E. G., KESSEL, R. H. J., LLOYD, E. C., LLOYD, P. B., ROTHWELL, A., ROTHWELL, B. & SWIFT, J. C. (1964). The kinetics of metabolism of *Gibberella fujikuroi* in stirred culture. *Canadian Journal of Microbiology* **10**, 407–44.

BRODY, S. & NYC, J. F. (1970). Altered fatty acid distribution in mutants of *Neurospora crassa. Journal of Bacteriology* **104**, 780–6.

BU'LOCK, J. D. (1961). Intermediary metabolism and antibiotic synthesis. *Advances in Applied Microbiology* **3**, 293–342.

BU'LOCK, J. D. (1965). *The biosynthesis of natural products.* London: McGraw Hill.

BU'LOCK, J. D. (1967). *Essays in biosynthesis and microbial development.* London: John Wiley.

BU'LOCK, J. D. (1975). Secondary metabolism in fungi and its relationship to growth and development. In *The filamentous fungi.* Vol. 1. *Industrial mycology,* pp. 33–58, edited by J. E. Smith & D. R. Berry. London: Edward Arnold.

BU'LOCK, J. D., SHEPHERD, D. & WINSTANLEY, D. J. (1969). Regulation of 6-methylsalicylate and patulin synthesis in *Penicillium urticae. Canadian Journal of Microbiology* **15**, 279–85.

BU'LOCK, J. D., DETROY, R. W., HOSTALEK, Z. & MUNIN-AL-SHAKARCHI, A. (1974). Regulation of secondary biosynthesis in *Gibberella fujikuroi. Transactions of the British Mycological Society* **62**, 377–89.

BU'LOCK, J. D., HAMILTON, D., HULME, M. A., POWELL, A. J., SHALLEY, H. M., SHEPHERD, D. & SMITH, G. N. (1965). Metabolic development and secondary biosynthesis in *Penicillium urticae. Canadian Journal of Microbiology* **11**, 765–78.

CALTRIDER, P. D. & NISS, H. F. (1966). Role of methionine in cephalosporin synthesis. *Applied Microbiology* **14**, 746–53.

CALTRIDER, P. G., HUBER, F. M. & DAY, L. E. (1968). Effect of methionine and sulfate on the metabolism of *Cephalosporium acremonium. Applied Microbiology* **16**, 1913–18.

CAMICI, L., SERMONTI, G. & CHAIN, E. B. (1952). Observations on *Penicillium chrysogenum* in submerged culture. *The Bulletin of the World Health Organization* **6**, 265–76.

CARRILLI, A., CHAIN, E. B., GUALANDI, G. & MORISI, G. (1961). *Scientific Reports of the Instituto Superior di Sanita* **1**, 177.

CARTER, B. L. A. & BULL, A. T. (1969). Studies of fungal growth and intermediary carbon metabolism under steady and non-steady state conditions. *Biotechnology and Bioengineering* **11**, 785–804.

CHOUDHARY, A. Q. & PIRT, S. J. (1965). Metal complexing agents as metal buffers in medium for the growth of *Aspergillus niger. Journal of General Microbiology* **41**, 99–107.

CLARK, D. S. (1962). Submerged citric acid fermentation of ferrocyanide-treated beet molasses: morphology of pellets of *Aspergillus niger. Canadian Journal of Microbiology* **8**, 133–6.

CLARK, D. S., ITO, K. & HORITSU, H. (1966). Effect of manganese and other heavy metals on submerged citric acid fermentation of molasses. *Biotechnology and Bioengineering* **8**, 465–71.

DEMAIN, A. L. (1963). Synthesis of cephalosporin C by resting cells of *Cephalosporium* sp. *Clinical Medicine* **70**, 2045–51.

DEMAIN, A. L. (1968). Regulatory mechanisms and the industrial production of microbial metabolites. *Lloydia* **31**, 395–418.

DEMAIN, A. L. (1972). Cellular and environmental factors affecting the synthesis and excretion of metabolites. *Journal of Applied Chemistry & Biotechnology* **22**, 345–62.

DEMAIN, A. L. (1974). Biochemistry of penicillin and cephalosporin fermentations. *Lloydia* **37**, 147–67.

DEMAIN, A. L. & NEWKIRK, J. F. (1962). Biosynthesis of cephalosporin C. *Applied Microbiology* **10**, 321–5.

DEMAIN, A. L., NEWKIRK, J. F. & HENDLIN, D. (1963). Effect of methionine, norleucine and lysine derivatives on cephalosporin C formation in chemically defined media. *Journal of Bacteriology* **85**, 339–44.

DIMROTH, P., WALTER, H. & LYNEN, F. (1970). Biosynthese von 6-methylsalicylsaure. *European Journal of Biochemistry* **13**, 98–110.

DIMROTH, P., GREULL, G., SEYFFERT, R. & LYNEN, F. (1972). 6-Methyl-salicylic acid synthetase. *Hoppe Seyler's Zeitschrift fur Physiologische Chemie* **353**, 126.

DREW, S. W. & DEMAIN, A. L. (1973). Methionine control of cephalosporin C formation. *Biotechnology and Bioengineering* **15**, 743–54.

DREW, S. W. & DEMAIN, A. L. (1975). Stimulation of cephalosporin production by methionine peptides in a mutant blocked in reverse transulfuration. *Journal of Antibiotics* **28**, 889–95.

DREW, S. W., WINSTANLEY, D. J. & DEMAIN, A. L. (1976). Effect of norleucine on mycelial fragmentation in *Cephalosporium acremonium*. *Applied and Environmental Microbiology* **31**, 143–5.

ELMAYERGI, H. (1975). Mechanisms of pellet-formation of *Aspergillus niger* with an additive. *Journal of Fermentation Technology* **53**, 722–9.

ELMAYERGI, H., SCHARER, J. M. & MOO-YUNG, M. (1973). Effect of polymer additives on fermentation parameters in a culture of *Aspergillus niger*. *Biotechnology & Bioengineering* **15**, 845–59.

FLOSS, H. G., ROBBERS, J. E. & HEINSTEIN, P. F. (1974). Regulatory control mechanisms in alkaloid biosynthesis. *Recent Advances in Phytochemistry* **8**, 141–78.

FORRESTER, P. I. & GAUCHER, G. M. (1972). *m*-Hydroxybenzyl alcohol dehydrogenase from *Penicillium urticae*. *Biochemistry* **11**, 1108–14.

FUJISAWA, Y., SHIRAFUJI, H., KIDA, M., NARA, K., YONEDA, M. & KANZAKI, T. (1975). New findings in cephalospor in C biosynthesis. *Nature, New Biology* **246**, 154–5.

GADEN, E. L. JR. (1955). Fermentation kinetics and productivity. *Chemistry & Industry*, 154–9.

GALBRAITH, J. C. & SMITH, J. E. (1969). Filamentous growth of *Aspergillus niger* in submerged shake culture. *Transactions of the British Mycological Society* **52**, 237–46.

HANSON, R. S. (1975). Role of small molecules in regulation of gene expression and sporogenesis in bacilli. In *Spores*, Vol. 6, pp. 318–326, edited by P. Gerhardt, R. N. Costilow and H. L. Sadoff. American Society for Microbiology, Washington, D.C.

HEINSTEIN, P. F., LEE, S. L. & FLOSS, H. G. (1971). Isolation of dimethylallyl-pyrophosphate: tryptophan dimethylallyl transferase from the ergot fungus (*Claviceps* sp.) *Biochemical & Biophysical Research Communications* **44**, 1244–51.

HOCKENHULL, D. J. D. (1959). The influence of medium constituents on the biosynthesis of penicillin. *Progress in Industrial Microbiology* **1**, 3–27.

JOHNSON, M. J. (1972). Recent advances in penicillin fermentation. *The Bulletin of the World Health Organization* **6**, 99–121.

JICINSKA, E. (1974). Dimorphic and yeast-like mutants of the genus *Cephalosporium* Cda. *Folia Microbiologica* **19**, 1–4.

KOBAYASHI, H. & SUZUKI, H. (1972). Studies on the decomposition of raffinose by α-galactosidase of mold. *Journal of Fermentation Technology* **50**, 625–32.

KOFFLER, H., EMERSON, R. L., PERLMAN, D. & BURRIS, R. H. (1945). Chemical changes in submerged penicillin fermentations. *Journal of Bacteriology* **50**, 517–48.

KOMAN, V. & BETINA, V. (1973). Diphasic production of secondary metabolites by *Penicillium notatum* Westling S-52. *Folia Microbiologica* **18**, 133–41.

KOMAN, V., BARATH, Z. & BETINA, V. (1969). Fatty acid, lipid and cyanein production by *Penicillium cyaneum*. *Archiv für Mikrobiologie* **65**, 172–80.

LARSEN, A. F. & SYPHERD, P. S. (1974). Cyclic adenosine 3′-5′-monophosphate and morphogenesis in *Mucor racemosus*. *Journal of Bacteriology* **117**, 432–8.

LEMKE, P. A. & BRANNON, D. R. (1972). Microbial synthesis of cephalosporin and penicillin compounds. In *Cephalosporins and penicillins—chemistry and biology*, pp. 370–437. edited by E. H. Flynn, New York: Academic Press.

LIGHT, R. J. (1967). Effects of cycloheximide and amino acid analogues on biosynthesis of 6-methylsalicylic acid in *Penicillium patulum*. *Archives of Biochemistry and Biophysics* **122**, 494–500.

MANTLE, P. G. & TONOLO, A. (1968). Relationship between the morphology of *Claviceps purpurea* and the production of alkaloids. *Transactions of the British Mycological Society* **51**, 499–505.

MANTLE, P. G., MORRIS, L. J. & HALL, S. W. (1969). Fatty acid composition of sphacelial and sclerotial growth forms of *Claviceps purpurea* in relation to the production of ergoline alkaloids in culture. *Transactions of the British Mycological Society* **53**, 441–7.

MARTIN, J. F. (1976). Fatty acid synthesis and polyene macrolide production by *Streptomyces*. *Developments in Industrial Microbiology* **17**, 223–31.

MARTIN, J. F., NICOLAS, G. & VILLANUEVA, J. R. (1973a). Chemical changes in the cell walls of conidia of *Penicillium notatum* during germination. *Canadian Journal of Microbiology* **19**, 789–96.

MARTIN, J. F., URUBURU, F. & VILLANUEVA, J. R. (1973b). Ultrastructural changes in the conidia of *Penicillium notatum* during germination. *Canadian Journal of Microbiology* **19**, 797–801.

MARTIN, S. M. (1957). Citric acid production by submerged fermentation. *Industrial & Engineering Chemistry* **49**, 1231–2.

MARTIN, S. M. & WATERS, W. R. (1952). Production of citric acid by submerged fermentation. *Industrial & Engineering Chemistry* **44**, 2229–33.

MURPHY, G., VOGEL, G., KRIPPAHL, G. & LYNEN, F. (1974). Patulin biosynthesis: the role of mixed function oxidases in the hydroxylation of *m*-cresol. *European Journal of Biochemistry* **49**, 443–55.

NASH, C. H. & HUBER, F. M. (1971). Antibiotic synthesis and morphological differentiation of *Cephalosporium acremonium*. *Applied Microbiology* **22**, 6–10.

NOVER, L. & LUCKNER, M. (1974). Expression of secondary metabolism as part of the differentiation processes during the idiophase development of *Penicillium cyclopium* Westling. *Biochemie und Physiologie der Pflanzen* **166**, 293–305.

NOVER, L. & MULLER, W. (1975). Influence of cycloheximide on the expression of alkaloid metabolism in partially synchronous emerged cultures of *Penicillium cyclopium* Westling. *FEBS Letters* **50**, 17–20.

NÜESCH, J., TREICHLER, H. J. & LIERSCH, M. (1973). The biosynthesis of cephalosporin C. In *Genetics of industrial microorganisms; actinomycetes and fungi*. Edited by Z. Vanek, Z. Hostakek and J. Cudlin, pp. 309–334. Amsterdam: Elsevier.

PASTERNAK, C. A. (1970). *Biochemistry of differentiation*. London: Wiley Interscience.

PIRT, S. J. & CALLOW, D. S. (1959). Continuous flow culture of the filamentous mould *Penicillium chrysogenum* and the control of its morphology. *Nature* **184**, 307–10.

PIRT, S. J. & CALLOW, D. S. (1960). Studies on the growth of *Penicillium chrysogenum* in continuous flow culture with reference to penicillin production. *Journal of Applied Bacteriology* **22**, 87–98.

PIRT, S. J. & RIGHELATO, R. C. (1967). Effect of growth rate on the synthesis of penicillin by *Penicillium chrysogenum* in batch and chemostat cultures. *Applied Microbiology* **15**, 1284–90.

PHILLIPS, D. H. (1966). Oxygen transfer into mycelial pellets. *Biotechnology and Bioengineering* **8**, 456–60.

RAMAKRISHNAN, C. J., STEEL, R. & LENTZ, C. P. (1955). Mechanism of citric acid formation and accumulation in *Aspergillus niger*. *Archives of Biochemistry and Biophysics* **55**, 270–3.

REHACEK, Z. & KOZOVA, J. (1975). Production of alkaloids and differentiation in a submerged culture of *Claviceps purpurea*. *Folia Microbiologica* **20**, 112–17.

REHACEK, Z. & MALIK, J. A. (1971). Cell-pool tryptophan phases in ergot alkaloid formation. *Folia Microbiologica* **16**, 359–63.

REHACEK, Z., KOZOVA, J., SAJDL, P. & VORISEK, J. (1974). The physiology of conidia formation in submerged cultures of *Claviceps purpurea* producing alkaloids. *Canadian Journal of Microbiology* **20**, 1323–9.

REHACEK, Z., SAJDL, P., KOZOVA, J., MALIK, K. A. & RICICOVA, A. (1971). Correlation of certain alterations in metabolic activity with alkaloid production by submerged *Claviceps*. *Applied Microbiology* **22**, 949–56.

RICHELATO, R. C. (1975). Growth kinetics of mycelial fungi. In *The filamentous fungi*, Vol. 1 *Industrial mycology*, pp. 79–103, edited by J. E. Smith and D. R. Berry. London: Edward Arnold.

RIGHELATO, R. C.,TRINCI, A. P. J., PIRT, S. J. & PEAT, A. (1968). The influence of maintenance energy and growth rate on the metabolic activity, morphology and conidiation of *Penicillium chrysogenum*. *Journal of General Microbiology* 50, 399–412.

ROBBERS, J. E., ROBERTSON, L. W., HORNEMANN, K. M. JINDRA, A. & FLOSS, H. G. (1972). Physiological studies on ergot: further studies on the induction of alkaloid synthesis by tryptophan and its inhibition by phosphate. *Journal of Bacteriology* 112, 791–6.

SAKAGUCHI, K. (1972). Historical background of industrial fermentation in Japan. In *Fermentation technology today*, pp. 7–12, edited by G. Terui. Osaka, Society of Fermentation Technology.

SCHAEFFER, P. (1969). Sporulation and the production of antibiotics, exoenzymes and exotoxins. *Bacteriological Reviews* 33, 48–71.

SCOTT, A. I. & BEADLING, L. (1974). Biosynthesis of patulin. Dehydrogenase and dioxygenase enzymes of *Penicillium patulum*. *Bioorganic Chemistry* 3, 281–301.

SCOTT, W. A. & SOLOMON, B. (1975). Adenosine 3'-5'-cycle monophosphate and morphology in *Neurospora crassa*: drug induced alterations. *Journal of Bacteriology* 122, 454–63.

SCOTT, W. A. & TATUM, E. L. (1970). Glucose 6-phosphate dehydrogenase and *Neurospora* morphology. *Proceedings of the National Academy of Science, USA* 66, 515–22.

SCOTT, W. A., MISHRA, N. C. & TATUM, E. L. (1973). Biochemical genetics of morphogenesis in *Neurospora*. *Brookhaven Symposium of Biology* 25, 1–18.

SERMONTI, G. (1969). *Genetics of antibiotic producing micro-organisms*. London: Wiley-Interscience.

SMITH, B., WARREN, S. C., NEWTON, G. G. F. & ABRAHAM, E. P. (1967). Biosynthesis of penicillin N and cephalosporin C: antibiotic production and other features of metabolism of a *Cephalosporium* sp. *Biochemical Journal* 103, 877–90.

SMITH, J. E. & ANDERSON, J. G. (1973). Differentiation in the aspergilli. *Symposium of the Society of General Microbiology* 23, 295–337.

SMITH, J. E. & BERRY, D. R. (1974). *An introduction to biochemistry of fungal development*. London: Academic Press.

SMITH, J. E. & GALBRAITH, J. C. (1971). Biochemical and physiological aspects of differentiation in the fungi. *Advances in Microbial Physiology* 5, 45–134.

SOLTERO, F. V. & JOHNSON, M. J. (1953). Effect of the carbohydrate nutrition on penicillin production by *Penicillium chrysogenum* Q-176. *Applied Microbiology* 1, 52–7.

SOLTERO, F. V. & JOHNSON, M. J. (1954). Continuous addition of glucose for evaluation of penicillin producing cultures. *Applied Microbiology* 2, 41–4.

SPALLA, C. (1973). Genetic problems of production of ergot alkaloids in saprophytic and parasitic conditions. In *Genetics of industrial microorganisms*; *actinomycetes and fungi*. pp. 393–403, edited by Z. Vanek, Z. Hostalek and J. Cudlin. Amsterdam: Elsevier.

SPALLA, C., AMICI, A. M., SCOTTI, F. & TOGNOLI, L. (1969). Heterokaryosis of alkaloid producing strains of *Claviceps purpurea* in saprophytic and parasitic conditions. In *Fermentation advances*, pp. 611–628, edited by D. Perlman. New York: Academic Press.

STEEL, R., MARTIN, S. M. & LENTZ, C. P. (1954). A standard inoculum for citric acid production in submerged culture. *Canadian Journal of Microbiology* 1, 150–7.

STEFANIAK, J. J., GAILEY, F. B., JARVIS, F. G. & JOHNSON, M. J. (1946). The effect of environmental conditions on penicillin fermentation with *Penicillium chrysogenum* X-1612. *Journal of Bacteriology* 52, 119–27.

TABER, W. A. (1964). Sequential formation and accumulation of primary and secondary shunt metabolic products in *Claviceps purpurea*. *Applied Microbiology* 12, 321–6.

TAKAHASHI, J., HIDAKA, H. & YAMADA, K. (1965). Effect of mycelial forms on citric acid fermentation in submerged mold culture. *Agricultural & Biological Chemistry* 29, 331–6.

TRINCI, A. P. J. (1969). A kinetic study of the growth of *Aspergillus nidulans* and other fungi. *Journal of General Microbiology* 57, 11–24.

TRINCI, A. P. J. (1970). Kinetics of the growth of mycelial pellets of *Aspergillus nidulans*. *Archiv für Mikrobiologie* 73, 353–67.

TRINCI, A. P. J. (1971). Influence of the width of the peripheral growth zone on the radial growth rate of fungal colonies on solid media. *Journal of General Microbiology* **67**, 325–44.

TRINCI, A. P. J. & RIGHELATO, R. C. (1970). Changes in constituents and ultrastructure of hyphal compartments during autolysis of glucose-starved *Penicillium chrysogenum*. *Journal of General Microbiology* **60**, 239–49.

TURNER, W. B. (1971). *Fungal metabolites*. London: Academic Press.

VORISEK, J., LUDVIK, J. & REHACEK, Z. (1974). Morphogenesis and ultrastructure of *Claviceps purpurea* during submerged ,alkaloid formation. *Journal of Bacteriology* **120**, 1401–8.

WHITAKER, A. & LONG, P. A. (1973). Fungal pelleting. *Process Biochemistry* **8**, (11), 27–31.

WILSON, S. & LUCKNER, M. (1975). Cyclopenase, ein Lipoproteid der protoplasma Membran von Konidiosporen des Pilzes *P. cyclopium* Westling. *Zeitschrift für Allgemeine Mikrobiologie* **15**, 45–51.

WOLD, W. S. M. & SUZUKI, I. (1973*a*). Promotion of conidia aggregation in *Aspergillus niger* by cyclic AMP and 5′-GMP. *Biochemical and Biophysical Research Communications* **55**, 824–30.

WOLD, W. S. M. & SUZUKI, I. (1973*b*). Cyclic AMP and citric acid accumulation by *Aspergillus niger*. *Biochemical and Biophysical Research Communications* **50**, 237–44.

WRATHALL, C. R. & TATUM, E. L. (1974). Hyphal wall peptides and colonial morphology in *Neurospora crassa*. *Biochemical Genetics* **12**, 59–68.

YANAGITA, T. & KOGANE, F. (1963). Cytochemical and physiological differentiation of mould pellets. *Journal of General and Applied Microbiology* **9**, 179–87.

ZONNEVELD, B. J. M. (1974). α-1,3 Glucan synthesis correlated with α-1,3 glucanase synthesis, conidiation and fructification in morphogenetic mutants of *A. nidulans*. *Journal of General Microbiology* **81**, 445–51.

Species Index

Subject Index